Image-Based Computational Modeling of the Human Circulatory and Pulmonary Systems

Krishnan B. Chandran · H.S. Udaykumar ·
Joseph M. Reinhardt
Editors

Image-Based Computational Modeling of the Human Circulatory and Pulmonary Systems

Methods and Applications

Foreword by Peter Hunter

Springer

Editors
Krishnan B. Chandran
Department of Biomedical Engineering
College of Engineering
1138 Seamans Center
The University of Iowa
Iowa City, IA 52242, USA
chandran@engineering.uiowa.edu

H.S. Udaykumar
Department of Mechanical and Industrial Engineering
College of Engineering
2408 Seamans Center
The University of Iowa
Iowa City, IA 52242, USA
hs-kumar@uiowa.edu

Joseph M. Reinhardt
Department of Biomedical Engineering
College of Engineering
1402A Seamans Center
The University of Iowa
Iowa City, IA 52242, USA
jmr@engineering.uiowa.edu

ISBN 978-1-4419-7349-8 e-ISBN 978-1-4419-7350-4
DOI 10.1007/978-1-4419-7350-4
Springer New York Dordrecht Heidelberg London

Printed on acid-free paper

Springer is part of Springer Science+Business Media (www.springer.com)

To
Vanaja, Aruna and Kelly, Anjana and Jaime

KBC

To
H. N. S. Murthy

HSU

To
Jennifer, Eliza, and William

JMR

Foreword

This book is very timely. The medical imaging community has long used numerical techniques for extracting anatomical *structure* from clinical images, but the use of anatomically and biophysically based computational models to interpret physiological *function* in a clinical setting is relatively new. The book is focused on the cardiovascular and pulmonary systems but the imaging and computational approaches discussed here are equally applicable across many other organ systems.

Many of the authors are experts in clinical image analysis, as well as computational methods, so not surprisingly the starting point for modeling clinical structure–function relations is often an image of some sort—MRI, CT, or ultrasound for anatomical imaging and PET or SPECT for functional images. Part I therefore reviews image acquisition and analysis for all of these techniques. Part II then deals with the physics and computation of soft tissue mechanics, fluid mechanics, and fluid–structure interaction. Multi-scale approaches for understanding blood flow mechanics are also discussed in this section. Part III focuses on the use of image-based computational analysis of cardiopulmonary disease. Applications to diagnostics, therapeutics, surgical planning, and the design of medical devices are considered throughout the book.

The computational methods described and used here include the continuum-based finite element and finite difference techniques, familiar to engineers, and the particle-based lattice Boltzmann methods more familiar to applied mathematicians. The corresponding equations (e.g., Navier–Stokes, Fokker–Planck) are all derived from the same underlying physical laws: the advantages of the former are partly to do with the widespread availability of appropriate constitutive laws and highly developed computer codes; the advantages of the latter are more apparent when dealing with multi-scale physics in which material continuum properties emerge out of the statistical behavior of interacting particles.

Material properties are a recurring theme in this book and, in fact, in any book dealing with an engineering physics analysis of function, because at the level of tissues the properties of biological materials such as soft tissue or blood (and their characterization via constitutive laws) are essential summaries of complex structure–function relations at smaller spatial scales. Several chapters address the important question of how to derive material properties with meso-scale modeling from our knowledge of the structure and function of tissue components. One reason why this is such an important task is because diseases are processes operating mechanistically at the protein level but manifest clinically at the tissue level. Understanding this multi-scale connection and, in particular, the growth and remodeling of tissue under changed loading environments is therefore vital.

This book is an immensely valuable contribution to the computational analysis of structure–function relations in health and disease. It is relevant to current clinical medicine but, perhaps more importantly, provides a guide to computational approaches that will undoubtedly underpin future evidence-based treatment of disease.

Auckland Bioengineering Institute
Auckland, New Zealand

Peter Hunter FRS

Preface

Physiological processes in living systems involve the complex interactions of electrical activity, chemical reactions, and physical phenomena such as mass, momentum, and energy transport. Deviations of any of these processes from their normal states may result in the initiation of diseases. A thorough understanding of physiological processes as they occur in the normal, healthy state as well as under pathological conditions is necessary so that diseases can be detected early enough for interventions to be efficacious. An understanding of complex physiological functions is also vital in the design and development of implants such as vascular stents and heart valve prostheses in human circulation and similar devices in other organ systems. In vivo and in vitro experimental studies, and in recent decades computer simulations, have provided valuable insight into the complex functional physiology and pathophysiology, and our knowledge in these areas continues to grow rapidly. However, in vivo experiments in human subjects and animals require ethical considerations, and the data obtained from in vivo experiments are limited due to practical considerations. In vitro experiments often require expensive equipment such as particle image velocimetry (PIV) systems, and yet there remain limitations in data acquisition due to restrictions of optical access to the areas of interest.

Recent advances in medical imaging instruments, such as magnetic resonance (MR), computed tomography (CT), and ultrasound imaging systems, have improved both the spatial and temporal resolution of the image data that can be acquired. With the appropriate acquisition protocol, these instruments can acquire 3D (volumetric) and even 4D (volume data plus time) data with exquisite anatomic detail. The image data can be visualized using computer graphics techniques to show geometric information and can be processed to provide realistic anatomic models for subsequent computer simulations that explore physiologic function.

With the advent of high-speed computers, computational simulations are increasingly playing a major role in our ability to analyze the physiological processes in the visceral organs and in the human musculoskeletal system. Computational simulations, with appropriate experimental validation, are being increasingly employed for various applications in human health care and have enabled us to reduce the number of animal models required for such studies. It is clear, however, that modeling of biological systems is an extremely challenging enterprise, given the complexity of such systems and the essential roles played by genetic factors and biological variability. Therefore, while a truly "accurate" model of a physiological system or process is very difficult to achieve, there is immense value in developing computational models that can capture essential features of the behavior of a system under well-defined physicochemical conditions. Computer simulations (1) are relatively inexpensive; (2) can cover wide ranges of parameter spaces; (3) can be improved over time with improved inputs and other information from experiments or with advances in modeling techniques, numerical methods, and computer hardware; and (4) can provide information on flow and stress fields that are difficult to measure or visualize.

The development of computational techniques and advances in hardware in terms of speed and memory have therefore established computer simulations as a strong source of knowledge regarding the behavior of biological systems. In fact, the current phase of computational developments is directed toward enabling increasingly sophisticated representations of biological systems. A particular case is that of multi-scale modeling of such systems. Physiological processes in living systems vary over a wide range of temporal and spatial scales. For example, chemical reactions that take place at a subcellular level require analysis at a timescale on the order of nanoseconds and at spatial dimensions on the order of nanometers. On the other hand, functional physiology of visceral organs such as the human heart involves a timescale on the order of seconds and at dimensions on the order of centimeters. Disease processes such as atherosclerosis, a common arterial disease in humans, develop during a time span of several years. Computational simulations on spatial and temporal scales ranging from nanometers to meters and nanoseconds to years are continuing to be developed, and strategies for integrating both spatial and temporal scales are being explored. In the last five decades, the explosion of new imaging modalities for structural and functional imaging of organs in the human body has also provided additional information for simulations attempting to model complex anatomy and physiology. It can be anticipated that computational simulations will increasingly play a vital role in the area of human health care.

In this book, we address the current status and possible future directions of simulations that have been employed and are continuing to be developed for applications in the human cardiovascular and pulmonary systems. In these two systems, simulations involve the description of the complex fluid flow (blood flow in the cardiovascular system and air flow in the pulmonary system), the mechanics of the soft tissue (vessel and airway walls, cardiac structures, and lung tissue), and the constant interaction between fluids and soft tissue. Typical disease processes, such as atherosclerosis in the human arteries and emphysema in the human lungs, result from alterations at the microstructural level with alterations in viscous properties and mass transport within local regions. Realistic simulation of the physiology and alterations resulting in the initiation and development of disease processes requires the following:

a. Acquisition of images of the organs of interest employing appropriate imaging modality, employment of state-of-the-art image processing and segmentation, and reconstruction of morphologically realistic three-dimensional (3D) geometry of the region of interest as a function of time.

b. Appropriate boundary conditions (pressures, flow rates, etc.) obtained from physiological measurements.
c. Development of computational techniques for the fluid flow (e.g., to represent blood rheology in the human circulation and turbulent compressible flow analysis for transport of air in the lung airways), the soft tissue (nonlinear anisotropic material description for the cardiac and blood vessel structures and the pulmonary airways from the trachea to the alveolar sacs), and the fluid–structural interaction analyses.
d. Validation of the computational techniques with appropriate experimental or computational simulations, before the application of the simulations, to describe the various physiological and pathophysiological processes.

The chapters to follow in this work are divided into three sections:

Part I deals with image data acquisition and geometric reconstruction commonly employed in the diagnosis and treatment of cardiovascular and pulmonary diseases. Chapter 1 discusses commonly employed imaging modalities used for anatomical and functional imaging of these two-organ systems, as well as trade-offs between spatial and temporal resolution, invasiveness of the imaging technique, and the use of ionizing vs. non-ionizing radiation. Chapter 2 focuses on contemporary image analysis and data processing techniques in order to identify anatomic structures in the images, delineate region boundaries, and construct three-dimensional geometric representations of regions of interest to be employed in the simulations.

Part II consists of discussions of state-of-the-art computational techniques for biological soft tissue, biological fluid, and the analysis of interaction between the fluid and the surrounding tissue. Chapter 3 presents the numerical approaches for solving the Navier–Stokes equations at two distinct scales, viz., the large-scale system that applies at the level of large blood vessels and prosthetic devices, and the small-scale systems that apply to the microvasculature. Chapter 4 details the modeling and solution of the equations governing the dynamics of soft tissue in the cardiovascular system. Chapter 5 focuses on the issue of fluid–structure interactions and distinguishes three types of techniques used to simulate the presence of structures immersed in blood flow. Issues pertaining to the behavior of the fluid–structure coupled solutions as they are influenced by the properties of the immersed solid are discussed.

The majority of the simulations published to date are focused mainly at the organ level where the biological soft tissue as well as the fluid can be treated as a continuum. There are limitations imposed on such simulations due to various practical constraints, including computer memory, processing speed, modeling uncertainties and complexity, biological variability. Even with the increasing speeds and memory densities of state-of-the-art computers, with the finest possible mesh density in the computational simulations, organ systems can at best be resolved down to dimensions in the order of millimeters—i.e., cellular and subcellular phenomena need to be modeled. However, in the last several decades, our knowledge of the physiological functions and pathological processes at the cellular and subcellular levels has also increased significantly. On the horizon of the computational landscape lies the possibility of linking computational analyses from the organ level (i.e., at the length scale of meters) all the way through to the cellular and subcellular levels (at the length scales of microns) and in time from nanoseconds to disease evolution scales. For example, numerous studies have focused on the relationship between the shear stress induced by the blood flow on the endothelial cells and the shear stresses computed by the simulations at the various arterial segments, as these are related to morphologically observed sites of atherosclerotic plaque development. Numerous experimental studies and simulations have also been employed at the level of endothelial cells in order to understand the response of the cells to external stimuli in the form of structural changes as well as to understand chemical alterations and the release of various growth factors and other enzymes. Recognizing that it is beyond the capabilities of even state-of-the-art high-performance computers to incorporate events at the subcellular level to those at the organ level through direct numerical computations, multi-scale simulation techniques are being investigated. Chapter 6 attempts to sketch the outlines of such a multi-scale modeling effort as it applies to the transport of blood at the micro- and mesoscales. The challenge of connecting these efforts to the large-scale blood flow simulations detailed in Chapters 3 and 5 lies at the frontier of multi-scale modeling.

The focus of Part III is on the application of computational simulations to a range of problems often encountered in the human circulatory and pulmonary systems. Chapter 7 addresses the current status of the simulations on our understanding of the arterial blood flow and the relationship between fluid-induced stresses and atherosclerotic plaque development. Topics include three-dimensional reconstruction of coronary arterial segments and simulation of coronary flow dynamics, flow simulations in the aorta and arterial bifurcations, and image-based simulation of abdominal aortic aneurysms (AAA). Models to analyze the endovascular implants for treating AAA and bypass grafting for the treatment of arterial occlusions are also discussed in this chapter. Detailed treatment of the biomechanics of both AAA and cerebral aneurysms is the topic of Chapter 8. The biomechanical modeling of aneurysm segments includes the effect of the material property of diseased arterial segments and prediction of rupture of aneurysms. The effect of alterations in the fluid flow on the biomechanics of the aneurysms is also discussed in detail in this chapter. Chapter 9 deals with the application of computational simulations for interventional treatments. Topics addressed in this chapter include application of modeling and simulation to assess atheromatous plaque vulnerability to rupture, mechanical effects of balloon angioplasty, and the design of endovascular stents that are implanted after angioplasty to open occluded arterial segments. In the second part of the chapter, the use of simulations for the surgical planning of single ventricle heart defect (SHVD) is described. As discussed with specific examples, rapid development of other patient-specific applications on interventional techniques and surgical planning is anticipated in the near future. The focus of Chapter 10 is on the application of modeling and simulation toward an understanding of the biomechanics of the respiratory system and the complex relationships between pulmonary anatomy, tissue dynamics, and respiratory function, as well as on how these relationships can change in the presence of pathological processes. Chapters 11 and 12 deal with the biomechanics of the heart valve function. The native heart valves have a complicated three-dimensional geometry. Since diseases of the valves are predominant in the left heart, the functional biomechanics of the aortic and mitral valves are of interest in increasing our understanding of the function of the healthy valve, the mechanical factors that contribute to the valvular diseases—such as calcification of the leaflets—and valvular regurgitation. These dynamic simulations have potential applications in the planning of patient-specific valvular repair strategies as well as in the development of tissue-engineered valve replacements. The dynamic simulations of the heart valves are challenging, requiring the inclusion of the entire valvular apparatus including the annulus, leaflets, and the ascending aorta for the aortic valves, and the leaflets, annulus, chordae and the papillary muscles for the mitral valves, as well as the development of accurate fluid–structure interaction analysis. These topics are covered in Chapter 11 along with the potential applications on the improved designs for biological valve prostheses. Simulations to understand the cause of thrombus deposition, a continuing and significant problem associated with mechanical valve prostheses, and the simulations toward our understanding of the fluid mechanical factors responsible for the same is the topic of Chapter 12.

Iowa City, Iowa

Krishnan B. Chandran
H.S. Udaykumar
Joseph M. Reinhardt

Contents

Contributors

Cyrus Aidun George W. Woodruff School of Mechanical Engineering, Georgia Institute of Technology, Atlanta, GA 30332, USA, cyrus.aidun@me.gatech.edu

Iman Borazjani Mechanical and Aerospace Engineering Department, SUNY University at Buffalo, Buffalo, NY 14260, USA, iman@buffalo.edu

Krishnan B. Chandran Department of Biomedical Engineering, College of Engineering, 1138 Seamans Center, The University of Iowa, Iowa City, IA 52242, USA, chandran@engineering.uiowa.edu

Jennifer K.W. Chesnutt Department of Mechanical Engineering, The University of Texas at San Antonio, San Antonio, TX, USA, jennifer.chesnutt@utsa.edu

Tom Claessens BioMech Research Group, Faculty of Applied Engineering, Department of Mechanics, University College Ghent, B-9000 Ghent, Belgium, tom.claessens@hogent.be

Ashley A. Colletti University of Toledo School of Medicine, Toledo, OH 43614, USA, ashley.colletti@rockets.utoledo.edu

Joris Degroote Faculty of Engineering, Department of Flow, Heat and Combustion Mechanics, Ghent University, B-9000 Ghent, Belgium, joris.degroote@ugent.be

Richard W. Downe Department of Electrical and Computer Engineering, Iowa Institute for Biomedical Imaging, The University of Iowa, Iowa City, IA 52242, USA, richard-downe@uiowa.edu

David W. Kaczka Harvard Medical School, Beth Israel Deaconess Medical Center, Boston, MA 02215, USA, dkaczka@bidmc.harvard.edu

Hyunggun Kim Division of Cardiology, Department of Internal Medicine, The University of Texas Health Science Center at Houston, Houston, TX 77030, USA, hyunggun.kim@uth.tmc.edu

Kyungmoo Lee Department of Electrical and Computer Engineering, Department of Biomedical Engineering, Iowa Institute for Biomedical Imaging, The University of Iowa, Iowa City, IA, kyungmoo-lee@uiowa.edu

Jia Lu Department of Mechanical and Industrial Engineering, The University of Iowa, 2137 Seamans Center, Iowa City, IA 52242, jia-lu@uiowa.edu

Robert MacMeccan George W. Woodruff School of Mechanical Engineering, Georgia Institute of Technology, Atlanta, GA 30332, USA, robert.macmeccan@milliken.com

Jeffrey S. Marshall School of Engineering, University of Vermont, Burlington, VT 05405, USA, jeffm@cems.uvm.edu

Tim McGloughlin Department of Mechanical and Aeronautical Engineering, Materials and Surface Science Institute, Centre for Applied Biomedical Engineering Research, University of Limerick, National Technological Park, Castletroy, Limerick, Ireland, tim.mcgloughlin@ul.ie

Mark E. Olszewski Philips Healthcare, CT Clinical Science, Highland Heights, OH, USA, mark.olszewski@philips.com

Madhavan L. Raghavan Department of Biomedical Engineering, 1136 Seamans Center, College of Engineering, The University of Iowa, Iowa City, IA 52242-1527, USA, ml-raghavan@uiowa.edu

Michael S. Sacks Department of Bioengineering, Swanson School of Engineering, University of Pittsburgh, Pittsburgh, PA 15219, USA; School of Medicine, The McGowan Institute, University of Pittsburgh, Pittsburgh, PA 15219, USA, msacks@pitt.edu

Patrick Segers bioMMeda Research Group, Faculty of Engineering, Ghent University, B-9000 Ghent, Belgium, patrick.segers@ugent.be

Hidenori Shikata Ziosoft Inc., Redwood City, CA, USA, hidenori@shikatas.net

Brett A. Simon Harvard Medical School, Beth Israel Deaconess Medical Center, Boston, MA 02215, USA, bsimon@bidmc.harvard.edu

Milan Sonka Department of Electrical and Computer Engineering, Iowa Institute for Biomedical Engineering, The University of Iowa, Iowa City, IA 52242, USA

Fotis Sotiropoulos St. Anthony Falls Laboratory, Department of Civil Engineering, University of Minnesota, Minneapolis, MN 55414, USA, fotis@umn.edu

Brooke N. Steele 2148 Burlington Nuclear Engineering Laboratories, NC State University, Raleigh, NC 27695-7115, USA, bnsteel@gmail.com

Merryn H. Tawhai Auckland Bioengineering Institute, The University of Auckland, Auckland, New Zealand, m.tawhai@auckland.ac.nz

Daniel R. Thedens Department of Radiology, 0446 John W. Colloton Pavilion, The University of Iowa, Iowa City, IA 52242, USA, dan-thedens@uiowa.edu

Juerg Tschirren Department of Electrical and Computer Engineering, Iowa Institute for Biomedical Imaging, The University of Iowa, Iowa City, IA 52242, USA, juerg@vidadiagnostics.com

H.S. Udaykumar Department of Mechanical and Industrial Engineering, 2408 Seamans Center, College of Engineering, The University of Iowa, Iowa City, IA 52242, USA, hs-kumar@uiowa.edu

Soumik Ukil Imaging and Video Services, Nokia India Pvt. Limited, Bangalore, India, soumik.ukil@nokia.com

Peter Van Ransbeeck BioMech Research Group, Faculty of Applied Engineering, Department of Mechanics, University College Ghent, Ghent, B-9000, Belgium, peter.vanransbeeck@ugent.be

Pascal Verdonck bioMMeda Research Group, Faculty of Engineering, Ghent University, B-9000, Ghent, Belgium, pascal.verdonck@ugent.be

Jan Vierendeels Faculty of Engineering, Department of Flow, Heat and Combustion Mechanics, Ghent University, Ghent, B-9000, Belgium, jan.vierendeels@ugent.be

S.C. Vigmostad Department of Biomedical Engineering, 1420 Seamans Center, College of Engineering, The University of Iowa, Iowa City, IA 52242, USA, svigmost@engineering.uiowa.edu

David A. Vorp Departments of Surgery and Bioengineering, University of Pittsburgh, Pittsburgh, PA 15219, USA, vorpda@upmc.edu

Andreas Wahle Department of Electrical and Computer Engineering, Iowa Institute for Biomedical Imaging, The University of Iowa, Iowa City, IA 52242, USA, andreas-wahle@uiowa.edu

Michael T. Walsh Department of Mechanical and Aeronautical Engineering, Materials and Surface Science Institute, Centre for Applied Biomedical Engineering Research, University of Limerick, National Technological Park, Castletroy, Limerick, Ireland, michael.walsh@ul.ie

Ajith P. Yoganathan Wallace H. Coulter School of Biomedical Engineering, Georgia Institute of Technology and Emory University, 313 Ferst Drive, Atlanta, GA 30332-0535, USA, ajit.yoganathan@bme.gatech.edu

Diane A. de Zélicourt Petit Institute of Bioengineering, Georgia Institute of Technology and Emory University, Atlanta, GA 30332, USA, diane.dezelicourt@gmail.com

Honghai Zhang Department of Electrical and Computer Engineering, Iowa Institute for Biomedical Imaging, The University of Iowa, Iowa City, IA 52242, USA, honghai-zhang@uiowa.edu

Fei Zhao Department of Electrical and Computer Engineering, Iowa Institute for Biomedical Imaging, The University of Iowa, Iowa City, IA 52242, USA, zhaof@ge.com

Part I
Cardiac and Pulmonary Imaging, Image Processing, and Three-Dimensional Reconstruction in Cardiovascular and Pulmonary Systems

Chapter 1
Image Acquisition for Cardiovascular and Pulmonary Applications

Daniel R. Thedens

Abstract Medical imaging hardware can provide detailed images of the cardiac and pulmonary anatomy. High-speed imaging can be used to acquire time sequences showing tissue dynamics or can capture a snapshot of the changing anatomy at an instant in time. Some imaging modalities can also provide functional information, such as perfusion, ventilation, and metabolic activity, or mechanical information, such as tissue deformation and strain. With the appropriate acquisition protocol, some of these imaging devices can acquire 3D (volumetric) and even 4D (volume data plus time) data with excellent anatomic detail. This image data can be visualized using computer graphics techniques to show geometric information, and the data can be processed to provide realistic anatomic models for subsequent computer simulations that explore physiologic function. This chapter describes the most commonly used imaging modalities for cardiovascular and pulmonary applications and describes some of the advantages and disadvantages of the different modalities. New, emerging modalities that may be important imaging tools in the future are introduced.

1.1 Introduction to Imaging

The field of medical imaging has its origin in the discovery of x-rays by Wilhelm Roentgen in 1895, a feat which earned him the first Nobel Prize in Physics in 1901. The very first x-ray image was of the hand of his wife. By the first decade of the twentieth century, x-rays were being used for medical diagnosis, and the specialty of radiology was established. X-ray imaging remained essentially the only diagnostic imaging technique available until the 1950s.

D.R. Thedens (✉)
Department of Radiology, 0446 John W. Colloton Pavilion, The University of Iowa, Iowa City, IA 52242, USA
e-mail: dan-thedens@uiowa.edu

K.B. Chandran et al. (eds.), *Image-Based Computational Modeling of the Human Circulatory and Pulmonary Systems*, DOI 10.1007/978-1-4419-7350-4_1,

The postwar produced a proliferation of new imaging techniques as assistive technologies in electronics and computers were applied. Nuclear medicine, which introduces radioactive elements into the body and uses a gamma camera to detect their distribution in the organs, appeared in the 1950s. The principles of sonar, an important technology developed during World War II, were applied to diagnostic imaging in the form of ultrasound imaging in the 1960s and have been a staple of diagnostic imaging ever since.

Advances in computer technology throughout the 1970s and 1980s brought an explosion of tomographic imaging techniques. Computed tomography (CT) used x-ray imaging as a basis to generate two-dimensional (2D) tomographic "slices" of the body, which eliminated many of the limitations of x-ray projection imaging. Magnetic resonance imaging (MRI) similarly took the underlying technique of nuclear magnetic resonance (NMR) to generate true 2D and three-dimensional (3D) images of the body. MRI is distinguished by the wide variety of contrast mechanisms that can be generated in an exam with a single scanner. The inventors of each of these modalities were honored with a Nobel Prize (Godfrey Hounsfield and Allan Cormack for CT in 1979, and Paul Lauterbur and Peter Mansfield for MRI in 2003).

All of these imaging modalities have continued to benefit from advances in computer technology over the past 25 years. The speed and quality of the images produced by ultrasound, CT, and MRI have improved by orders of magnitude since their initial development, and this trend is likely to continue into the foreseeable future. As a result, the value and importance of diagnostic imaging will continue to grow and expand into new areas of application.

Among the fundamental differences between the various imaging modalities considered in this chapter is the type of energy used. Imaging methods based on x-rays, such as CT, utilize electromagnetic waves of very high energy. The energy level is sufficiently high to cause changes and damage to living tissues. X-rays are thus an example of *ionizing radiation*, and there is a risk of long-term damage to organs if the exposure to this form of radiation is too great. MRI also uses electromagnetic energy in the formation of images, but the energy level used is orders of magnitude lower, corresponding to the radiofrequency range of the spectrum (roughly the same range as the FM radio band). As a result, the energy in MRI is *non-ionizing* and cannot inflict damage on tissues. Ultrasound uses mechanical energy in the form of high-frequency sound waves, which is also considered non-ionizing as no tissue damage is done. Both MRI and ultrasound are safe for repeated use, whereas exposure to x-rays should be limited due to the potentially harmful effects.

A wide variety of diagnostic information can be generated from all of these imaging modalities. Broadly, the types of information acquired can be categorized as anatomic imaging or functional imaging. Anatomic imaging is concerned with the depiction and distinguishing of the anatomical structures in the body. The information may be qualitative in terms of the appearance of normal or abnormal tissues or quantitative by measuring the size, shape, and density of the body tissues and organs. Functional imaging is concerned with measuring physiologic activity such as metabolism, blood flow, or chemical changes. Some functional measures may be derived from anatomic data, such as calculation of cardiac function metrics like ejection fraction from a time series of anatomic images.

1.1.1 Invasive Techniques

While noninvasive imaging techniques such as CT and MRI progress to produce ever more detailed views of anatomy and function, the "gold standard" imaging techniques for some clinical questions remain more invasive methods. For cardiovascular assessment, x-ray angiography is still the preferred technique for diagnosing disease of the vasculature.

The procedure involves introduction of a catheter into a major vessel, which is then guided toward the vessels of interest. A radiodense contrast dye is injected into the bloodstream and continuously imaged with x-ray imaging. The flow of the contrast agent can be followed to locate areas of vessel blockage or narrowing. The direct targeting of the vessels of interest with the contrast injection and x-ray imaging means that the vessel can be seen with exquisite detail and the flow (or lack of it) is viewable and reviewable in real time. Frequently the diagnosis and treatment of such blockages can be performed at the same time. However, it must be recognized that a significant number of these procedures will result in a negative diagnosis, exposing the patient to unnecessary risks from the invasive nature of the procedure.

1.1.2 Role of Noninvasive Imaging

Because of the risks of invasive procedures, noninvasive imaging techniques based on ultrasound, CT, or MRI, which do not involve any sort of surgical procedure, have been widely developed to replace more and more of these previously invasive techniques. Each of these modalities operates on different physical principles. Ultrasound uses acoustic waves to probe tissue characteristics and movement. CT forms 2D and 3D images from x-ray projections, thus depicting density. MRI images the distribution and characteristics of hydrogen protons throughout the body, which is essentially a map of the water content within the body. The three modalities represent different trade-offs in acquisition time and complexity, resolution, image quality, and versatility. All three remain important tools for diagnosing and assessing treatment of cardiopulmonary disease, in many cases providing complementary information.

While the ultimate dream of a single "one-stop shop" protocol for cardiac and pulmonary imaging for complete assessment of the cardiopulmonary system has not yet materialized, all three of the primary imaging techniques continue to thrive and advance. The result has been more detailed and higher quality diagnostic information available with a drastic reduction in the need for surgical or invasive procedures. In the sections that follow, a description of the principles of each of these modalities as well as their roles in cardiovascular and pulmonary assessment will be presented.

1.2 Ultrasound/Echocardiography

Ultrasound imaging has its origins in the research and development that produced advances in underwater sonar (an acronym for SOund Navigation And Ranging) for detection of submarines. Ultrasound itself refers to acoustic or sound waves at frequencies that are above the range of human hearing. Like all of the imaging modalities described here, ultrasound has advanced tremendously since its original inception. Early diagnostic techniques provided only one-dimensional (1D) profiles of body tissues. But with the development of more sophisticated equipment and data processing, ultrasound has become capable of 2D real-time imaging of anatomy as well as flow and velocity, and even of 3D imaging. The portability, safety, and low cost of ultrasound has created an ever-widening set of applications for its use both as an initial screening tool and for quantitative assessment of morphology, function, and flow throughout the cardiovascular system.

1.2.1 Principles of Ultrasound

This section will describe the basic principles and techniques commonly used in ultrasound imaging. A thorough treatment of the physics of ultrasound imaging is given in [1, 2]. The propagation of acoustic energy in body tissues depends on its material properties, and ultrasound imaging uses the propagation and reflection of this energy to build tomographic images of these acoustic waves and their interactions with body tissues.

Fundamentally, an acoustic wave is created by mechanical compression and rarefaction ("stretching") of an elastic medium. An acoustic wave is created by a continuous cyclic "back and forth" motion of a transducer that begins these alternating compressions and rarefactions. This pattern of displacement will then propagate through the medium with a characteristic velocity for that material. The wave propagates by transferring the mechanical energy of the compressions to adjacent material in the medium, so that the actual movement of material as the wave moves is very small.

In ultrasound imaging, a *transducer* is used to produce the high-frequency acoustic waves as well as to detect the signal that returns to produce the images. The transducer consists of a piezoelectric material that can convert electrical energy into mechanical vibrations that create the acoustic wave, with frequencies typically in the range of 3–15 MHz. The transducer also functions as a detector, converting the returning acoustic waves back into electrical energy for image formation, either directly or via digitization of the signal and subsequent processing. For imaging, the ultrasound probe typically contains an array of up to 512 such transducers in a variety of configurations.

The velocity c at which sound propagates through a material is determined by its bulk modulus B (related to its stiffness and compressibility) and density ρ (mass per unit volume) as

$$c = \sqrt{\frac{B}{\rho}} \tag{1.1}$$

The speed of sound in air (which is highly compressible) is about 330 m/s, whereas the velocities in soft tissues are on the order of 1800 m/s, and in a stiff material such as bone they rise to 4000 m/s. Another important tissue characteristic is the *acoustic impedance* Z, measured in $\mathrm{kg/m^2\ s}$ and defined as

$$Z = \rho c \tag{1.2}$$

with ρ and c the material density and speed of propagation, respectively. Again, the acoustic impedance varies widely for differing materials, from $0.4\times10^3\ \mathrm{kg/m^2\ s}$ for air to approximately $1.5\times10^6\ \mathrm{kg/m^2\ s}$ for many soft tissues and $7.8\times10^6\ \mathrm{kg/m^2\ s}$ for bone.

The formation of ultrasound images relies on the interactions of acoustic waves occurring at the *boundaries* of different tissues (Fig. 1.1). When an acoustic wave reaches a boundary between tissues with differing acoustic impedances, some of the energy of the wave will be reflected, and this reflected energy returns back to the source of the acoustic wave, where it can be measured and a projection or image can be reconstructed. For two tissues of impedance Z_1 and Z_2, the fraction of reflected intensity at a particular interface is given as

$$R_\mathrm{I} = \left(\frac{Z_2 - Z_1}{Z_2 + Z_1}\right)^2 \tag{1.3}$$

R_I will thus range from zero (no reflection) to one (complete reflection and no transmission).

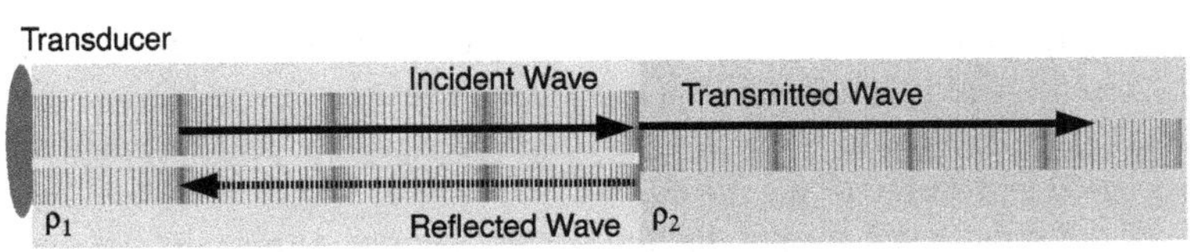

Fig. 1.1 An ultrasound transducer transmits a mechanical wave into the tissue. When the wave is incident on a boundary between tissues of different density ρ_1 and ρ_2, part of the wave is transmitted into the second tissue and part of the wave is reflected back toward the transducer. The reflected energy is detected and used to form a projection or an image

This reflected energy from the tissue boundaries is necessary to detect changes in the tissues. Between different types of soft tissues, such as fat and muscle, the amount of reflected energy is small (on the order of 1%). This is sufficient to detect and display, while the majority of the energy in the beam continues to propagate forward. Boundaries where there is a large difference in impedances (such as between air or lung and soft tissues) will result in nearly all of the energy being reflected and none of it propagating further. As a result, material beyond air pockets and solid materials such as bone are unobservable with ultrasound. For observing body tissues, a path to the anatomy of interest must be found that does not travel through an air-filled space, and a conducting gel is used on the face of the ultrasound transducer to eliminate air pockets between the transducer and the skin for just this reason.

In addition to reflection, several other types of interactions between acoustic waves and tissues may occur. Refraction occurs when the incident acoustic wave is not perpendicular to the boundary between tissues. The result is a change in the direction of the beam and a violation of the assumption that the wave reflection is in a straight line, which can cause artifacts in the resulting image. In body tissues containing very small structural elements (on the order of the wavelength of the acoustic wave), the reflections become diffuse and *scatter.* This yields a "rough" appearance to the tissue boundaries. However, since most organs have a very characteristic structure, the pattern of scatter will be distinctive for specific organs, and this can yield diagnostically important information about the tissue. Further, since the scatter depends on the wavelength of the acoustic wave, adjusting the frequency of the ultrasound beam can provide additional tissue characterization based on a characteristic "texture" pattern in the tissue appearance. Finally, not all of the energy of the ultrasound beam will be transmitted and reflected through the tissue. Some of the energy is lost as heat in the tissues, and the beam is said to be *attenuated.* The degree of attenuation is roughly proportional to the frequency of the beam and also varies with the type of tissue, meaning that this effect can also be of diagnostic value.

1.2.1.1 M-Mode

The fundamental method of image formation in ultrasound imaging is the *pulse echo* method. A pulse of ultrasonic energy (with duration on the order of 1 μs) is rapidly generated and transmitted into the body tissues. As described above, the interactions of the acoustic wave with tissues of differing acoustic properties produce reflections and scatter; these are then returned to and detected by the transducer after the pulse transmission has been turned off. This process of transmission and signal recording is repeated anywhere from 500 to 15,000 times per second. Each detected signal is then processed and stored or displayed.

One of the most basic display techniques is *M-mode* ultrasound, where the M refers to motion. This in turn is based on *B-mode* ultrasound, where the B stands for brightness. In B-mode, the signal returned from a single directional ultrasound beam is displayed as a projection whose brightness is proportional to the returned signal. Because the timing of the returned signal relates to the depth from which it originates, the B-mode display represents a 1D projection of the tissue interfaces along the beam.

In M-mode ultrasound, the same beam direction is repeatedly displayed from the B-mode information. The result is a time-resolved display of the motion occurring along this projection. This display method has commonly been used to show the motion of heart valves, as the repeated sampling of the projection can occur very rapidly, yielding excellent temporal resolution. However, as 2D echocardiographic techniques have advanced, M-mode display has declined in importance, overtaken by truly 2D and 3D display methods for most of the same information. Figure 1.2 shows a sample M-mode acquisition used to derive functional parameters over the heart cycle.

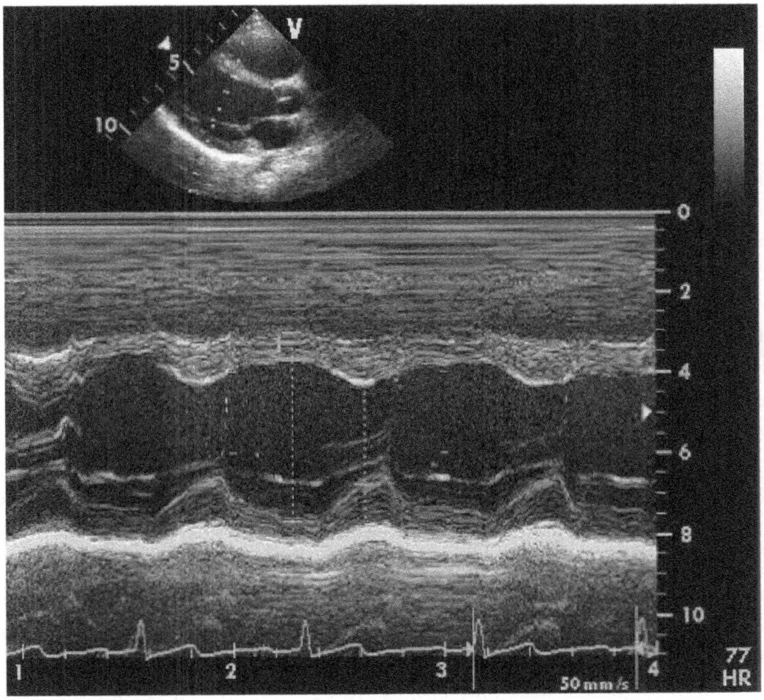

Fig. 1.2 M-mode echocardiogram acquired over multiple cardiac cycles. The *dashed lines* estimate systolic and diastolic parameters to derive functional indices such as stroke volume and ejection fraction

1.2.1.2 2D Ultrasound

With the advent of transducer arrays and rapid signal processing, B-mode and M-mode ultrasound have been almost completely supplanted by 2D imaging acquisitions. The fundamental principle of 2D ultrasound is precisely the same as that of B-mode. A pulse of acoustic energy is transmitted into the tissues, and the returned reflections or "echoes" are recorded and displayed, providing a display of the boundaries between tissues and characteristic scatter patterns of organs.

In 2D ultrasound, a complete image is formed by sweeping over all of the transducer elements, each generating a single line of image data. The lines are assembled into a complete 2D image for display, and the process is repeated. The transducer may be linear, where the elements are lined up in a straight line along the transducer to produce parallel beams and a rectangular image field of view. Another configuration is a curvilinear array, which uses a smaller transducer head with a convex shape. The individual elements are fanned out over the transducer and the resulting images have a trapezoidal field of view.

1.2.2 Echocardiography

On modern ultrasound equipment, the rate at which images are formed is fast enough to generate real-time imaging of the beating heart. This has made ultrasound an extremely effective imaging modality for observing cardiac morphology and function. When applied to cardiac imaging, ultrasound is usually referred to as *echocardiography*.

For cardiac imaging, the transducer array used is convex, generating the trapezoidal field of view described above. This allows the transducer to be small enough to be positioned in the limited set of locations that permit an unobstructed window for acquiring images of the heart (in particular, without crossing air spaces in the lungs), such as between the ribs.

Since echocardiography remains predominantly a 2D imaging technique, a complete echocardiographic exam utilizes multiple imaging planes to acquire a complete description of the anatomy and function of the heart. The *long-axis* view of the heart runs parallel to the long axis of the heart (as the name suggests) and depicts the left atrium, left ventricle (LV), septum, and posterior wall of the LV. It can also be oriented to show right ventricular (RV) inflow and outflow tracts. The *short-axis* plane runs perpendicular to the long axis, showing the LV in cross section. By varying the positioning of the imaging plane, the morphology and function of the heart can be assessed from base to apex. The *four-chamber* view cuts through the heart from apex to base to show all four chambers, as well as the mitral and tricuspid valves in a single view.

1.2.2.1 Morphologic Imaging

The capabilities of echocardiography to generate high-resolution and real-time depictions of the cardiac anatomy from multiple vantage points are also useful for generating quantitative assessments of many indices of cardiovascular health. The most basic measurements relate to the size of the various chambers and outflow tracts of the heart. In the LV, dimensions of the posterior wall and septum can be taken either from the 2D images or directly from an M-mode projection. There are multiple methods for estimating LV mass from echocardiograms. Short-axis dimensions can be used with a simple geometric formula to produce a reasonably accurate mass measurement, though this entails some assumptions about the geometry of the chamber. Alternatively, two long-axis views can be used with a Simpson's rule derivation for a more accurate assessment of LV mass and volumes. Calculating dimensions of the left-and right ventricular outflow tracts is also possible from the appropriate long-and short-axis views, respectively.

With similar methods, RV volumes and thickness can be measured as well as those of the left atrium. In general, an appropriate orientation needs to be captured over the heart cycle, followed by identification of the standard locations for measurement of wall or chamber dimensions.

1.2.2.2 Function

The dynamic nature of echocardiographic image acquisition makes it well-suited for studying the functional dynamics of the heart. Several functional indices can be derived directly from the appropriate morphological measurements. Since echocardiography produces images of high spatial and temporal resolution, such indices are of primary importance.

LV ejection fraction (LVEF) is the single most important index of cardiac function and can be derived directly from measurements of end-systolic and end-diastolic volumes measured as outlined above. Stroke volume is another important parameter of diagnostic interest and is measured from the same parameters.

1.2.2.3 Flow (Doppler)

Doppler echocardiography utilizes the *Doppler effect* to ascertain the rate and direction that material is moving. Primarily, this is used to generate quantification of the rate of blood flow, but in principle it can be applied to any moving tissue.

The Doppler effect arises when the acoustic waves of the ultrasound beam are reflected by moving red blood cells. Stationary tissues may reflect ultrasonic waves, but their frequency will not be affected. When a wave is reflected by a moving material, the frequency of the wave will be increased if the material is moving toward the transducer and decreased if it is moving away. The frequency shift depends on the speed of sound in the moving material c, the frequency of the transmitted wave f_0, the velocity of the material v, and the angle between the beam and the motion θ. The frequency shift Δf is then

$$\Delta f = \frac{2 f_0 v \cos\theta}{c} \tag{1.4}$$

Thus, once the measurement of the frequency change is taken, the velocity is computed as

$$v = \frac{\Delta f c}{2 f_0} \tag{1.5}$$

assuming the beam is parallel to the flow direction (if not, the flow rate will be underestimated).

Color Doppler imaging is the most widely used form of Doppler flow imaging at present. Flow measurements are continuously taken by alternately pulsing and recording the frequency shifts returned. The frequency information is mapped to a predefined color map to display the flow information atop the anatomic data. By convention, red is used for flow toward the transducer and blue for flow away from the transducer, with lighter shades indicating greater velocity. In areas of turbulent flow where the flow directions and velocities are highly variable, green is displayed.

The acquisition and display of flow information make it easy to identify cardiac abnormalities, many of which are characterized by disturbances in flow patterns. These include valve disease where backflow may be seen and pathologies of diastolic function where flow patterns into the ventricles may be readily observed. An example of color Doppler imaging in a subject with an atrial septal defect is shown in Fig. 1.3.

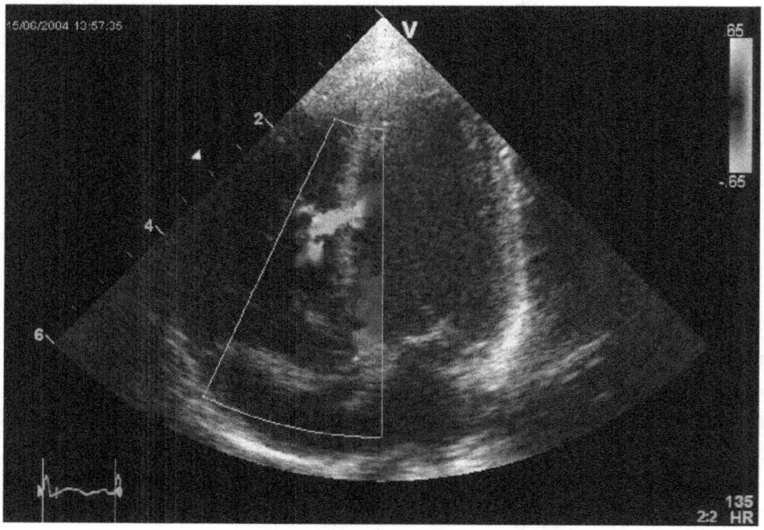

Fig. 1.3 Doppler mode echocardiogram of the heart with color-coded overlay indicating flow velocity and direction in this subject with an atrial septal defect

Doppler imaging is not limited to observation of blood flow. Measurements of the velocities of myocardial tissues can also be used to estimate velocities and myocardial strain rates. The velocities under consideration are much lower than those from flowing blood, and the translational motion of the heart may cause errors in the velocity measurements, so care must be taken in interpreting the derived values.

1.2.2.4 TTE Versus TEE

As noted previously, because of the position of the lungs relative to the heart and the need for an air-free path to visualize the heart with ultrasound, the placement and orientation of the ultrasound probe is limited to a few external positions on the chest (where it is called *transthoracic echocardiography* or *TTE*). An alternative approach is to use a special ultrasound probe that can be passed into the esophagus, which places the probe in much closer proximity to the heart (which sits within millimeters of the esophagus). This is known as *transesophageal echocardiography* (TEE). The result is much increased reflection energy and reduced attenuation and a corresponding improvement in image quality. Several conditions involving the left atrium, mitral valve, pulmonary artery, and thoracic aorta are best seen on TEE.

Obviously, TEE is a much more invasive procedure than TTE, as it requires a fasting patient and conscious sedation to position the probe in the esophagus. TEE is therefore not a routine initial screening tool, but does provide valuable anatomic and flow information that cannot be acquired with regular TTE.

1.2.3 Vascular/Peripheral

Because of the quantitative information about flow velocities and abnormal flow patterns that can be generated with Doppler ultrasound, its use has expanded beyond the heart to nearby great vessels and peripheral vessels as well, in order to assess vascular function throughout the body.

Numerous pathologies of the aorta are routinely imaged and diagnosed by ultrasound, both standard TTE and TEE. Aortic aneurysms and aortic dissection are often assessed with both TTE and TEE, and atherosclerotic plaque is also a common finding from TEE. Doppler ultrasound is becoming increasingly popular for initial assessments of carotid artery disease. The Doppler measurements of blood flow velocity can help identify the significance of plaques and lesions in the artery. Doppler imaging has found uses in peripheral vessels as well, where it serves as an inexpensive screening tool.

Vascular assessment can also be performed using *intravascular ultrasound* (IVUS). IVUS utilizes a miniature ultrasound probe attached to the end of a catheter which can then be inserted inside the lumen of a vessel. Images can then be acquired from the inside of a blood vessel to depict its lumen and wall. In particular, this permits direct discrimination of atherosclerotic plaque contained within the vessel wall and quantification of both the degree of narrowing and the total plaque volume contained therein. IVUS can also determine plaque tissue characterization, as the calcified, fibrous, and lipid components of a lesion can be distinguished based on their appearance on ultrasound. IVUS is most commonly applied to the coronary vasculature, where it can be used to measure plaque burden and plan treatment prior to angioplasty or to assess stent placement or restenosis. Because of its ability to quantify plaque burden, IVUS is also useful for assessing efficacy of treatments for coronary atherosclerosis.

While IVUS can provide unique information on the state of blood vessel lumen and walls, it is an invasive technique compared to the other tomographic imaging methods discussed here, though it does not require a contrast agent as conventional angiography does. Imaging is also limited to vessels large enough to accommodate the probe, and positioning within large vessels may result in oblique cross sections due to angulation of the probe. Nevertheless, IVUS has proven highly valuable in understanding the characteristics and development of atherosclerotic lesions.

1.3 Computed Tomography (CT)

CT was the first of the tomographic imaging technologies to permit generation of images representing cross-sectional "slices" of the internal anatomy. The first CT scanner was installed in 1972 and the initial application was in brain imaging. The first images required a 4–5 min scan time and produced images with an 80×80 pixel matrix. Subsequent advances in x-ray tubes, detectors, and computer hardware have improved on these characteristics by many orders of magnitude. CT scanners are now capable of acquiring and reconstructing large 3D data sets in a few seconds, making it possible to visualize minute structures and dynamic processes with exquisite clarity.

1.3.1 Principles of CT

This section provides an overview of the principles of CT imaging. For a more comprehensive treatment, see the relevant chapters of Ref. [3]. CT is based on the principles of x-ray imaging. X-rays are generated in a vacuum tube by firing electrons at a target (the *anode*) which produces a beam of electromagnetic radiation in the x-ray spectrum. The beam is directed toward the body, and a detector on the opposite side (which may be film or some type of solid-state or digital device) records the amount of x-ray energy that passed through the body. In essence, the detector serves to measure the attenuation experienced by the x-ray beam as it passes through the body and creates a 2D projection image of the 3D anatomy. Contrast between body tissues is developed because high-density tissues such as bone will absorb greater amounts of energy than low-density soft tissues.

A fundamental limitation of x-ray imaging is the projective nature of the resulting image, meaning that structures in the third "depth" dimension are overlaid on each other, requiring multiple views to elucidate the arrangement of structures. To overcome this limitation, tomography was developed in the early 1900s, exploiting principles of projective geometry. The x-ray source and detector are simultaneously rotated around a central focus point as the x-rays are generated. Structures at the focal point remain in focus throughout this motion, while structures away from the focus will be blurred out and appear as noise. The result is an image showing only the internal structures at this focal point. This form of tomography can be considered to be a precursor of modern CT imaging, which has almost entirely supplanted it.

Similarly, the mathematical underpinnings needed for CT image formation have a long history, originating in the work of Joseph Radon in a paper published in 1917. The *Radon transform* and its inverse describe the relationship between an unknown object and a set of line integrals or projections through the object. The remaining development necessary for modern CT imaging to become feasible was the application of the digital computer to perform the numerical computations required to generate a tomographic image from a set of angular projections. CT imaging is thus one of the first imaging modalities made possible by advances in computer technology.

1.3.1.1 Basic CT

The basic process of image formation in CT imaging is the collection of a set of projections taken at multiple angles which can be subsequently reconstructed into a 2D imaging slice through the same region of the body. In conventional CT imaging, a *fan-beam* geometry is commonly used to generate the measurements. The x-ray generator can be considered to be a point source, and a set of diverging *rays* are emitted and pass through the body. On the opposite side of the scanner bore, detectors are arranged to measure the x-ray attenuation over the entire beam.

The attenuation of the beam follows the relationship

$$I_{\text{detected}} = I_{\text{transmitted}} e^{-\mu x} \tag{1.6}$$

where I represents an x-ray intensity and x is the thickness. The transmitted intensity is also measured at the detector as a reference value. The resulting attenuation μ measured will be an average over the path of the x-ray beam at each location. The attenuation coefficient can then be computed as

$$\mu t = log_e(I_{\text{transmitted}}/I_{\text{detected}}) \tag{1.7}$$

Since the transmitted intensity is available at the detector, this relationship is inherently normalized for the intensity of the beam, leaving only dependence on the attenuation characteristics of the body tissues. As a result of this transformation, the intensity values displayed on a CT image have a physical meaning in terms of the attenuation coefficients at each location in the generated slice. For computation and display, the attenuation coefficient is further normalized to the *Hounsfield scale* (named for one of the Nobel-winning inventors of CT), measured in *Hounsfield Units* (HU), which relates the attenuation to that of distilled water as

$$\text{HU} = \frac{\mu_{\text{tissue}} - \mu_{\text{water}}}{\mu_{\text{water}}} \times 1000 \tag{1.8}$$

On this scale, pure water has an attenuation of 0 HU, while air has an attenuation of -1000 HU. The use of this scale permits direct identification of tissue types in images based on their measured value of HU and known attenuation characteristics and is useful for diagnosis and subsequent image processing and visualization.

Formation of a complete image requires recording attenuation measurements over a full 180° range of angles. Numerous techniques have been developed over the years as the sophistication of detectors and control of the hardware have improved. Early "first-generation" scanners utilized a single detector and required a sequence of translate–rotate motions. As the number of detectors that could be incorporated into the scanner increased, this gave way to systems with a few hundred detectors that required only rotational motion, with source and detector array rotated on opposite sides of the patient. Subsequent generations of scanners utilize even more detectors, completely encircling the bore of the scanner (a few thousand in total) and requiring motion of the x-ray source only, which allows for faster scan times. The most recent advances in imaging include helical CT, whereby data can be acquired continuously while the table moves through the bore, rather than needing to stop and start for each set of slices. With this combination of technologies, imaging in the span of a single breath hold became possible.

1.3.1.2 Multidetector CT

The present state of the art in CT scanning is focused on the use of multiple detector arrays to further increase the speed and efficiency of the acquisition. This arrangement is known as *multidetector CT* (MDCT). MDCT retains the ring of detectors surrounding the bore of the scanner, but instead of single detectors at each location, an array of densely packed detectors is assembled (Fig. 1.4). Thus, for a stationary location within the scanner bore, a number of images equal to the number of elements in the array can be recorded and reconstructed. The dense packing of the detectors also means that the slice thickness achievable is now dependent on the size of the detector that can be constructed, rather than on the width of the x-ray beam produced. Alternatively, the data from multiple elements of the array may be combined together to generate thicker slices of higher quality than would be generated by the single elements of the detector array.

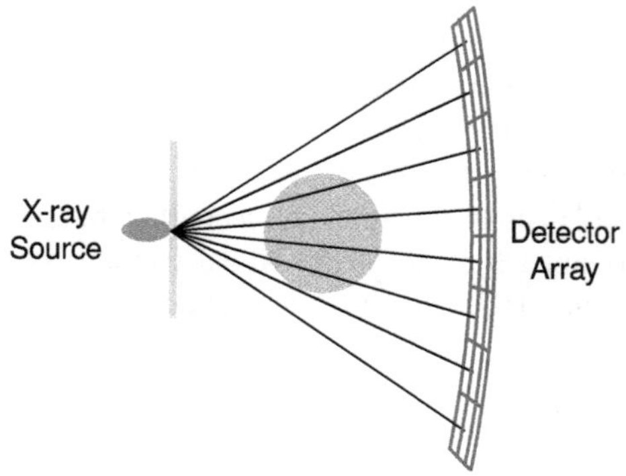

Fig. 1.4 In a multidetector CT system, the collimated beam from an x-ray source passes through the material being imaged and an array of detectors records the transmitted energy at the different projection angles in parallel

As of this writing, MDCT scanners with arrays capable of generating 64 slices are becoming commonplace, and even larger arrays are appearing on the market, with 256-slice arrays now available from multiple manufacturers. Because of their highly parallel nature, 64-slice and higher scanners can take a complete 4D data set (3D spatial information with 0.5 mm acquired slice thickness and on the order of 150 ms reconstructed temporal resolution) over the heart in a short breath hold.

1.3.2 Cardiac CT

The capability to rapidly generate such comprehensive visualizations of the beating heart has vaulted MDCT to a premier position for assessment of cardiac and coronary anatomy.

1.3.2.1 Coronary Arteries

A primary use of cardiac CT is to assess the coronary vasculature for stenoses and calcifications associated with heart disease [4]. The current generation of 64-slice MDCT scanners now has sufficient spatial and temporal resolution to permit accurate assessment of coronary artery stenosis. High spatial resolution is needed to identify coronary artery disease in at least the major coronary vessels. High temporal resolution is needed to be able to acquire this level of resolution in a short breath hold as well as to accommodate high heart rates and arrhythmias.

The resolution capabilities of 64-slice and higher MDCT scanners are approaching the resolution of conventional (invasive) angiography. MDCT can now realize resolutions on the order of 0.4 mm slices, compared to a nominal resolution of about 0.2 mm for conventional angiography. As a result, MDCT is increasingly used to assess the severity of disease and to reduce the need for conventional angiography in patients who do not show severe stenosis and can thus rule out coronary artery disease. Figure 1.5 shows an example of MDCT used for detecting coronary artery stenosis. MDCT also has the advantage of the possibility of determining tissue characteristics of stenoses, such as calcification. Three-dimensional reconstructions can aid in localization of lesions and planning of interventions.

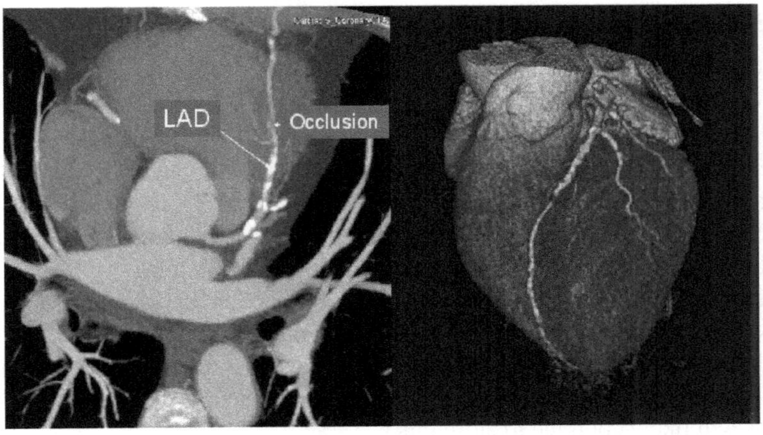

Fig. 1.5 MDCT of the left coronary artery in a patient with an occlusion in this vessel. The *left panel* displays the artery in a single plane view, while the *right panel* shows a 3D reconstruction of the heart and coronary vessels

Nevertheless, conventional angiography remains the gold standard measure of coronary status, particularly for collateral vessels. This may continue to be revisited as MDCT continues to improve in resolution and scan time.

A secondary use of MDCT in coronary arteries is *coronary artery calcification* scoring. Calcifications in the coronary arteries are readily visualized on MDCT because of their high density, and many studies have shown a high degree of correlation between calcium scoring and overall plaque burden, which in turn may predict the risk of future cardiac events. Again, the use of 64-slice scanners makes the acquired resolution detailed enough to eliminate many of the partial voluming effects and other limitations of previous generations of scanners and yield greater accuracy in this assessment, though the risk of false-positive results from these measures has not been eliminated.

1.3.2.2 Aorta

MDCT with intravenous iodinated contrast is also widely used for detecting and assessing problems in the thoracic and abdominal aorta [5]. The high resolution and volume coverage of MDCT can serve as the basis for 3D visualization of the lumen of the aorta.

Aortic dissection is a tear in the wall of the aorta, which permits blood to flow between the layers of the wall and further forces them apart, with the risk that the aorta itself will rupture with fatal consequences. MDCT (along with MRI) detects dissection with a high degree of sensitivity and specificity. Though MRI remains the gold standard for this condition, the more rapid scan time and higher resolution of MDCT may be preferable in many instances. Similarly, aortic aneurysms are well visualized and followed on MDCT [6]. Detection of aneurysms again relies on contrast-based examination of the vessel.

1.3.2.3 Cardiac Function

The rapid scan times, with temporal resolution approaching 60 ms, along with the resolution of MDCT in the heart, has generated interest in its use for cardiac function assessment [7]. Presently, MRI is the primary standard for measuring such indices as left ventricular (LV) ejection fraction, end-diastolic and end-systolic volumes, and LV mass. The submillimeter slice thickness possible with 64-slice CT has brought its accuracy for these measures to a sufficient level such that MDCT is making inroads for LV function assessment because of its rapid scan times.

The present limitations of MDCT for cardiac function are the temporal resolution (which is limited by the rotation speed of the scanner gantry) and the short scan times, where there may be variability in these parameters from heartbeat to heartbeat. Further advances in CT such as larger numbers of slices and dual source systems will likely continue to close this gap.

A dual-source CT scanner is equipped with two x-ray sources and two corresponding detectors, oriented at 90° to each other. The two sources and detectors can operate simultaneously, acquiring twice as much data at a time compared to a single-source system. This permits complete image acquisitions in half the time, a particular advantage for cardiac imaging as it improves the temporal resolution and reduces the required time for a breath hold.

1.3.3 Pulmonary CT

The high resolution and rapid scan times now available with MDCT have opened up its use for a variety of indications in the lungs. In order to prevent artifacts from motion in the lungs, CT imaging of the lungs requires a breath-holding protocol. Previously, the length of the scan required to generate the needed resolution would require breath-holding durations that were prohibitively long for many patients with lung disease. With MDCT now permitting high-resolution imaging within seconds, such acquisitions now become feasible, and MDCT has become the modality of choice for numerous clinical and research questions. The resolution of MDCT imaging is the highest among tomographic imaging methods suitable for lung imaging.

1.3.3.1 Parenchyma

CT is among the only imaging modalities capable of generating high-resolution tomographic imaging of the lung parenchyma, as the tissue densities are unsuitable for imaging with ultrasound or MRI. As a result, CT is the preferred method of imaging for assessment of lung nodules and staging of treatment. Additionally, CT can be used to quantify the extent and severity of chronic obstructive pulmonary disease (COPD) and emphysema based on areas of reduced attenuation that result from these conditions [8].

The continuous coverage generated by CT imaging coupled with the potential for very thin slice acquisitions permit CT acquisitions to be arbitrarily reconstructed into 3D volumes to generate 3D models of the airways. These features are used in applications such as CT bronchography and virtual bronchoscopy. These methods are enhanced by the use of 3D rendering techniques and visualizations that allow both qualitative and quantitative assessments of airways. As the resolution of MDCT continues to improve, smaller and smaller airways can be visualized and measured for ever-growing understanding of the normal and diseased lung.

The use of inhaled xenon gas as a CT contrast agent has yielded improvements in measurements of regional pulmonary ventilation [10]. Xenon has a high atomic number and is thus much more radiodense in proportion to its concentration compared to air or soft tissues; it therefore yields high contrast against such tissues. Xenon-enhanced CT (Xe-CT) involves inhaling and exhaling the gas during a time series of imaging acquisitions. Local and regional ventilation time constants can then be derived from the rate of the gas movement. While presently Xe-CT is primarily a research tool, it provides unique and valuable information on lung structure and function.

1.3.3.2 Pulmonary Angiography

Assessment of the pulmonary vasculature has also become a routine use of MDCT in the lungs. As with other modalities, pulmonary angiography generally involves the introduction of a contrast agent (an iodinated agent in the case of CT). The primary use of pulmonary angiography is in the identification of pulmonary embolism [9]. Figure 1.6 shows a sample MDCT acquisition depicting a "saddle" embolism in the left and right pulmonary arteries.

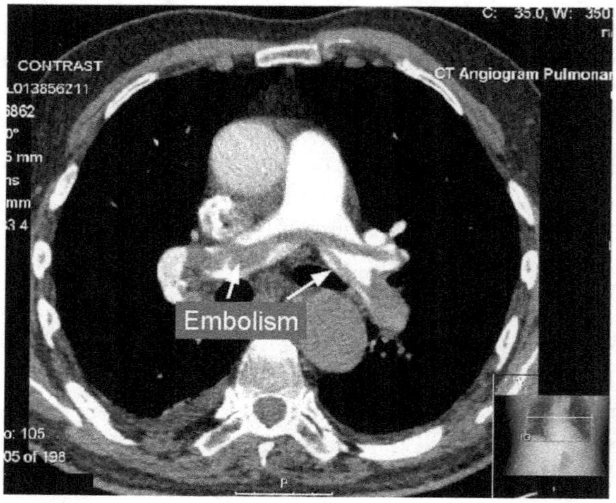

Fig. 1.6 MDCT image through the pulmonary artery demonstrating a pulmonary embolism

1.4 Magnetic Resonance Imaging (MRI)

1.4.1 Principles of MRI

This section describes the basic principles behind the formation of images with MRI. Detailed descriptions of the physics and instrumentation are provided in Refs. [11, 12].

Magnetic resonance imaging (MRI) relies on the phenomenon of nuclear magnetic resonance to generate image contrast. The hydrogen atom (along with other species having an odd number of protons or neutrons, such as sodium and phosphorous) possesses a spin angular momentum. The single proton of the hydrogen atom (often referred to in this context as a spin) is by far the most abundant and thus is considered in the vast majority of imaging applications. Most importantly, for the purposes of imaging, the spins will give rise to a magnetic moment and will act like microscopic bar magnets. As a result, when the protons are placed in a strong static magnetic field, at equilibrium they tend to line up in the same direction as the external field. The net effect of all the spins lined up in this way generates a small but measurable magnetization along the longitudinal direction of the large external field. The magnitude of this magnetization increases as the strength of the external field is increased.

1.4.1.1 Signal Generation

By itself, the magnetization does not give much useful information about the distribution of the protons within the object. The application of a second small (relative to the primary strong) magnetic field oscillating in the radiofrequency (RF) range sets up a resonance condition and will perturb the spins away from their equilibrium state, "tilting" them away from their alignment with the main field into the transverse plane. Much like a gyroscope, this will excite the spins, causing them (and their magnetic fields) to precess about the direction of the main field, and the rate at which the spins precess is directly proportional to the strength of the main magnetic field. Figure 1.7 shows the relationship between the two magnetic fields and the resulting perturbation of the magnetization vector. A fundamental principle of electromagnetics is that a time-varying magnetic field can induce an electric current in an appropriately placed coil of wire, generating a signal that can measure the distribution of the spins within the object. Since the rate of precession depends on the magnetic field strength, slightly varying the field strength across the bore of the magnetic with gradient fields yields a spatially varying rate of precession. When the RF field is removed, the spins begin to return toward their equilibrium state, aligned with the strong static magnetic field.

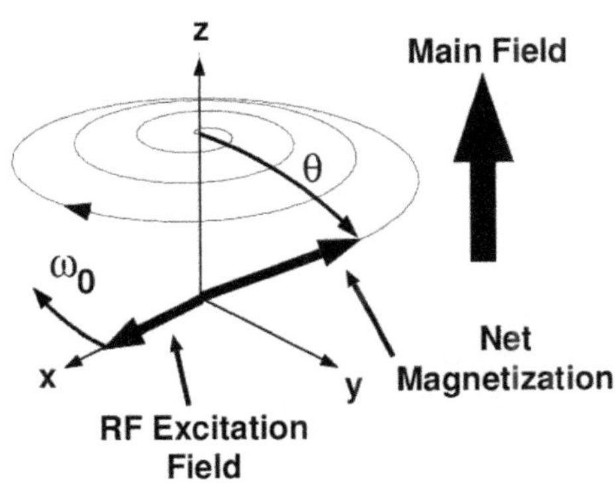

Fig. 1.7 A rotating RF magnetic field with frequency ω_0 is applied perpendicular to the direction of the main magnetic field. This causes the aligned spins to tip away from the main field direction and precess at the same frequency ω_0, producing a detectable signal from the spins

The rate of return of spins to their equilibrium state is governed by two time constants intrinsic to different tissue types, T1 and T2. T2 determines how long it will take for the signal generated by the "tipped" spins to decay away. T1 measures the amount of time it takes for the spins to completely return to their equilibrium alignment with the main magnetic field. Because of this signal decay, an MR imaging experiment generally must consist of several cycles of signal generation followed by signal measurement or acquisition.

1.4.1.2 General Techniques and Contrast Mechanisms

The signal measured from a tissue will thus depend on its density of protons as well as its T1 and T2 relaxation parameters. Motion and flow also contribute to the final signal generated. The remarkable ability of MRI to generate a wide variety of tissue contrast arises from the fact that the imaging experiment can be designed to vary the relative weight of each of these parameters in the measured signal. For example, muscle and fat have very different T1 and T2 parameters, and by varying the timing of the applied RF excitation pulses, maximum contrast between the two can be achieved. Other strategies may enhance or suppress flowing blood compared to stationary tissues.

1.4.1.3 Morphology

The most basic use of cardiac MRI is to depict the structure or morphology of the heart. Two general classes of imaging techniques are widely used for cardiac imaging, commonly referred to as black-blood and bright-blood techniques.

Black-Blood Imaging

Black-blood images are produced by T2-weighted spin-echo (SE) imaging sequences [13], in which two RF excitations (an excitation pulse and an inversion pulse) are applied to the imaged volume. After the excitation pulse, the excited spins begin to lose coherence due to slight variations in their resonant frequencies, resulting in a rapid loss of overall signal.

The inversion pulse "flips" the magnetization about one of the axes, permitting these spins to regain their coherence and generate an echo when the signal has been restored. When the two pulses are separated by a sufficient interval, flowing blood experiences only one of these pulses and thus does not produce a restored signal echo, leaving a flow void in the chambers of the heart. The timing of the two RF pulses sets the echo time (TE) at which the signal refocuses (and data are acquired) and determines the precise signal and contrast features of the image. For black-blood imaging, a TE of at least 20 ms is usually used. A longer TE yields greater contrast based on T2 characteristics of the tissues, which may be useful to identify such lesions as acute myocardial infarction or myocardial scar. This comes at the expense of reduced overall signal due to signal decay. Standard SE sequences show excellent contrast among myocardium (medium intensity), epicardial fat (high intensity), and flowing blood (low intensity). The signal void created by SE sequences generates images with especially good contrast in endocardial regions, valves, and vessel walls.

The main limitation of standard SE sequences is the acquisition time required in a cardiac-triggered exam, which results in poor temporal resolution and the prospect of significant respiratory motion artifact. Fast SE (FSE) sequences overcome this limitation by applying multiple inversion pulses and additional signal readouts during a single cardiac cycle. Speedups of an order of magnitude are possible in this way. However, the longer readout times degrade the image contrast due to the more complex dependence on relaxation times.

The currently preferred black-blood technique for imaging cardiac morphology is a T2-weighted inversion recovery (IR) pulse sequence. This sequence applies additional RF excitation pulses to effectively null the signal from blood (and possibly fat as well) based on its T1 relaxation parameters. This is usually followed by a FSE sequence that can be acquired in 15–20 heartbeats, suitable for a breath-held acquisition and yielding a robust black-blood sequence with T2 contrast.

Bright-Blood Imaging

Bright-blood images originate from gradient echo (GRE) imaging sequences which only use a single RF excitation, relying on the gradient hardware instead of an inversion pulse to refocus the signal for data acquisition. Much shorter TE times (1–10 ms) are used, and the excitation and data readouts can be repeated more frequently (every 10–20 ms). Because the blood need only experience the single RF pulse to generate a signal, it appears brighter than myocardium on GRE acquisitions. The short TE between excitation and data readout enhances this effect since there is less time for signal decay due to relaxation. Additional flow-compensation pulses can also be applied to further enhance blood signal and improve contrast with nearby myocardium. As with FSE imaging, the fastest imaging sequences utilize multiple excitations and data readouts over an extended interval (80 ms is a typical duration) synchronized to the cardiac cycle to generate images that can be acquired within a breath-holding interval. Contrast between blood and myocardium is generally not as good as with SE imaging, as varying flow profiles may result in heterogeneous blood pool.

The availability of faster gradient hardware has seen a resurgence of techniques based on steady-state free precession (SSFP) [14]. SSFP maximizes the use of signal from blood by applying rapid excitations repeated at very short intervals. The resulting contrast is a function of relaxation parameters as T1/T2. The short repetition times greatly reduce flow effects and show a more homogeneous depiction of myocardial blood pool, which in turn improves contrast with myocardium and visualization of papillary muscles. Rapid excitations also permit better temporal resolution, or the time savings can be traded off for higher resolution at the same time resolution. As state-of-the-art MR gradient hardware proliferates, SSFP will likely become even more common.

The rapid repetition of readouts in both GRE and SSFP means that several images at the same location can be taken at different time points within the heart cycle. Alternatively, the imaging time can be used to acquire multiple slices at a reduced temporal resolution. Using segmented acquisitions, a multi-slice multi-phase view of the cardiac morphology can be acquired within a single breath hold of 15–20 heartbeats.

1.4.1.4 Function

Many of the techniques mentioned above for imaging of cardiac morphology, including both black-blood and bright-blood imaging, are also suitable for measuring cardiac function indices. Compared to other modalities, MRI has the advantage that completely arbitrary image orientations can be chosen, guaranteeing that true long-axis or short-axis views serve as the basis for quantitative measurements.

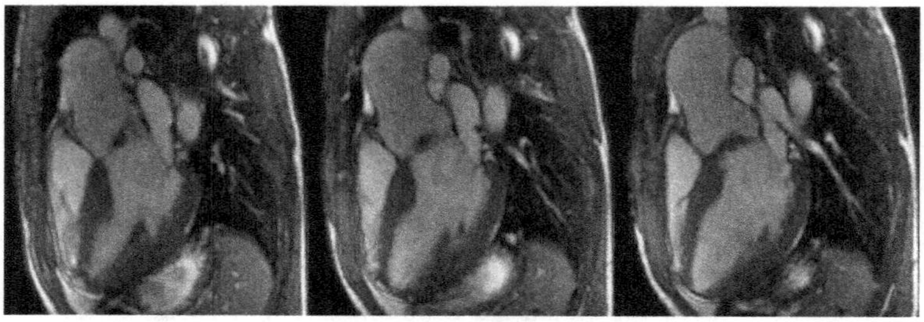

Fig. 1.8 Cardiac MRI two-chamber views of the left atrium and left ventricle

Figure 1.8 shows an example of a two-chamber view of the left ventricle. The availability of 3D information in the form of multiple parallel slices eliminates the need for any geometric assumptions about ventricular anatomy when estimating masses and volumes, a significant advantage over x-ray and ultrasound.

Bright-blood GRE imaging is more commonly used for evaluation of ventricular function. The shorter acquisition time permits a greater number of slices to be acquired during the cardiac cycle, which can be used for higher temporal resolution (more frames per cycle) or for a greater volume coverage (more slice locations). The acquisition of images at multiple phases of the cardiac cycle is known as cine MRI (example shown in Fig. 1.9). With present system hardware, a complete multi-slice multi-phase cine data set suitable for quantitative analysis can be acquired in a single breath-hold interval. The limiting factor with standard GRE imaging is the contrast between medium-intensity myocardium and the bright-blood pool. Areas of slower flowing blood will demonstrate reduced intensity, making delineation of the endocardial contours difficult.

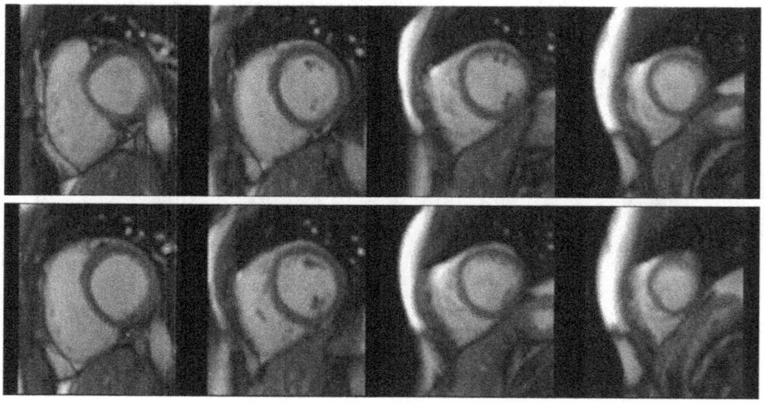

Fig. 1.9 Sequence of cardiac MRI short-axis images of the left ventricle. Slices are shown from base to apex (*left* to *right panels*) in the diastolic phase of the cardiac cycle

The recent advances in SSFP imaging cited above may solve this problem to some degree with more robust contrast. The faster repetition time used in SSFP also increases the frame rates possible in a cine study. With state-of-the-art gradient hardware, truly 3D cine MRI with no gaps between slices is now possible within a single breath-hold interval.

Improving gradient and computing hardware has now made real-time imaging feasible for functional imaging. Rates of 16 frames per second or more can be continuously obtained, much like x-ray fluoroscopy. The scan plane can be modified directly on the real-time images, dramatically reducing the time required for "scout" scans to find the proper short-axis orientation. At such rates, cardiac gating and breath holding are unnecessary, which permits imaging of patients with arrhythmias. Presently, spatial resolution of real-time studies remains comparatively limited, but a number of ongoing developments in image reconstruction techniques are improving this. Two general strategies exploit the widespread use of multiple receiver coils. Simultaneous acquisition of spatial harmonics (SMASH)and sensitivity encoding (SENSE) use the spatially varying response of a group of coils as an additional means of spatial encoding to reduce the time needed to acquire a given resolution image. Other techniques analyze the temporal dimension of the acquisition to reduce the acquisition of redundant information and enhance either temporal or spatial resolution.

Each of these forms of cine and real-time MRI data is useful for computing several global measures of cardiac function. Accurate and reproducible quantitative measurements of ventricular volumes at both systole and diastole, masses, and ejection fraction (difference between the diastolic and systolic ventricular volumes) are all computable with multi-slice or volume data sets. In each case, myocardial border identification is necessary to extract quantitative results. Compared to x-ray and ultrasound, MRI also accurately depicts epicardial borders, again eliminating the geometric assumptions that often must be made in competing modalities. As a result, regional myocardial function assessments can also be made with cine techniques. This may be done subjectively, viewing cine or real-time "loops," or through quantitative measurements of regional wall thickness and strain.

Regional measurements of 3D strain are possible using myocardial tagging. This imaging method excites myocardium with a pattern of lines or grids whose motion can then be tracked over the heart cycle, providing a precise depiction of the deformations occurring within the myocardial tissues. Analysis of these deformations in short- and long-axis views gives 3D strain measurements useful for determining local myocardial function. A promising rapid technique is harmonic phase (HARP) imaging, which has potential as a real-time technique.

1.4.1.5 Perfusion/Ischemia

Another important indicator that can be assessed by MRI is regional blood flow (or perfusion) in the myocardium [15]. This may indicate areas of damage to myocardium from a cardiac event or insufficient blood flow resulting from a significant arterial stenosis. Determination of blood flow within the myocardium depends on the use of contrast agents (usually gadolinium-based) that change the relaxation characteristics of blood, particularly the T1 relaxation time. Gadolinium causes a considerable shortening of the T1 relaxation time, meaning that magnetization returns to equilibrium much more rapidly. When RF excitation pulses are applied in rapid succession, tissues with short T1 relaxation will still have time to recover and generate greater signal for subsequent excitations. Longer T1 relaxation times means that little magnetization has returned to the equilibrium state, so later excitations result in much less signal. Appropriate timing of a pair of RF pulses can maximize the signal difference between two tissues with known T1 relaxation times.

Perfusion is mostly measured during the "first pass" into the myocardium after injection of the contrast agent. Areas of myocardium with adequate blood flow will have enhanced intensity from the shortened T1 of the inflowing blood. Perfusion deficits will not receive this material and remain at lower intensity. The time of the imaging window is limited as contrast material may soon begin to diffuse from normal to deficit regions, and the contrast agent will recirculate with the blood within 15 s. Hence, rapid GRE sequences are used to image quickly and permit multiple slices to be obtained over a volume. T1 contrast is maximized by applying an RF "preparation" pulse that initially excites or saturates all of the blood and tissues. After a delay time that causes contrast-enhanced material to return toward equilibrium while the longer T1 tissues recover much less magnetization to yield strong T1 contrast, a standard fast GRE imaging sequence is applied. The result is bright signal in normal tissue and low-intensity regions of perfusion deficit. Acquisition of several time frames during this process permits quantitative measurements of the severity of these perfusion abnormalities. Further myocardial tissue characterization is possible using gadolinium contrast agents by waiting an extended duration (20 min or more) before imaging. Gadolinium contrast will eventually move to the extracellular space and accumulate more in areas of non-viable myocardium, resulting in enhanced signal in these areas on T1-weighted images compared to normal tissue.

1.4.2 MR Angiography

In addition to imaging of the heart, MRI has also been widely applied to imaging vessels throughout the body [16]. Its advantages over conventional x-ray angiography go beyond the fact that it is much less invasive. MRI can also collect true 3D data, permitting arbitrary selection of views and slices in post-processing to optimize the visualization of vessels. This is especially helpful in complex vascular trees where tracing the vessel of interest may be difficult. Contrast for MR angiography can be developed in two ways. Pulse sequences may exploit the different signal properties of flowing and stationary tissues to produce images. Other sequences rely on the relaxation characteristics of arterial and venous blood, usually enhanced by T1-shortening contrast agents as described for myocardial perfusion. In both cases, the goal is to generate images of the vessel lumen suitable to detect and evaluate stenoses.

Two flow-based imaging techniques are in common use for MR angiography and both effectively produce "bright-blood" images of the vessel lumen. Phase-contrast (PC) imaging takes advantage of the fact that flowing blood will move during the data acquisition readout. Since spatial information is encoded by a spatially varying magnetic field gradient, flowing spins experience a changing magnetic field as they move, resulting in a phase change in their signal compared to stationary tissues. By applying an appropriate encoding gradient pattern prior to imaging, flowing blood can be selectively viewed. PC imaging can also quantitatively measure flow velocities. Time-of-flight (TOF) imaging uses the continuous replacement of flowing blood in the imaged slice to differentiate it from static tissue. Rapid repetition of excitation pulses covering the imaged slice saturates and eventually eliminates signal from stationary material because there is not enough time to regain any equilibrium magnetization. Flowing blood retains signal since fresh unsaturated blood is constantly flowing into the slice to be excited and flows away again before saturation can be complete. The result produces high signal from flowing blood against the low intensity of background structures.

Reliance on flow for image contrast may introduce artifacts where flow patterns are not ideal. Such anomalies will affect both PC and TOF sequences. Areas of slow flow may have reduced signal, either due to reduced phase changes for PC or saturation in TOF. Complex flow patterns and turbulence can also cause reduced intensities within the vessel lumen in both cases. The consequences could include stenoses that are overestimated or a false appearance of an occlusion of the vessel.

The limitations of flow-based angiography have made flow-independent techniques more prevalent. It is possible to create high-contrast angiographic images using only the intrinsic T1 and T2 relaxation characteristics of blood through a variety of "prepared contrast" pulses that saturate or suppress one or more background tissues. However, injectable contrast agents such as those based on gadolinium compounds have proven to be safe and well tolerated and are widely available. These contrast agents dramatically reduce the T1 relaxation time of blood and greatly enhance its signal on TOF images. Much of MR angiography is now dominated by contrast-agent-based protocols.

Once again, the main limiting factor in contrast studies is the time before the contrast agent leaks outside the blood vessels and begins to enhance the signal in tissues other than blood. Successful contrast angiography therefore requires careful timing of contrast injection and image acquisition and a rapid acquisition technique to minimize artifacts due to contrast dispersion and respiratory motion. Fast 3D GRE imaging is most commonly used to acquire T1-based contrast to yield bright contrast-enhanced blood pool. Subtraction of a non-contrast-enhanced volume may also be used to further suppress background structures. A variety of strategies have been employed to reduce the imaging time to acquire a 3D data set even further and assure accurate timing of the acquisition. Partial acquisition methods which acquire 60–75% of a full data set and synthesize the rest based on mathematical assumptions can help reduce imaging times. More extreme versions of this have been applied to radial sampling patterns to reduce acquisition time even further, trading the shortened time for some increased and coherent background noise.

The timing of the acquisition relative to the injection of contrast agent is also crucial. If the data acquisition occurs too early, signal will not yet be enhanced, while a late acquisition will show poor contrast because of heightened signal from other tissues or veins. For many applications, a fixed time delay based on previous experience may be sufficient, although increased doses of contrast often accompany this technique to increase the window of enhancement. A much smaller dose of contrast may be given and tracked with a sequence of rapid 2D images that may be used to pinpoint the transit time prior to a full 3D acquisition. Automatic monitoring of the signal at a predefined location upstream from the desired location has also been implemented. The use of real-time imaging to monitor contrast passage is another possibility.

The limited volume imaging time available because of the dispersion of contrast agent into other tissues is currently being addressed. New intravascular contrast agents that do not leak into tissues during the course of a typical MR exam are being perfected by a number of researchers. As a result, their T1 shortening properties can be utilized for longer or multiple exams without the enhancement of background tissues. MR angiograms in higher resolution 3D or over the whole body then become possible. The longer persistence in the blood pool does mean that both arteries and veins will be displayed for longer 3D scan durations. Some means of separating the two may be needed for diagnostic use of such images.

Coronary artery imaging may be a particular beneficiary of such contrast agents, as the necessity of high resolution, 3D coverage, and motion correction requires longer scan times than are feasible with standard contrast material. The flow and saturation effects that often compromise 3D techniques are also improved with such contrast agents. Perfection of a minimally invasive coronary MR imaging exam is of particular interest because of the number of highly invasive x-ray angiography procedures that are performed that show no clinically significant disease.

1.4.3 Pulmonary MRI: Emerging Techniques

As noted previously, MRI of the lungs has generally been limited to contrast-enhanced angiography of the pulmonary vasculature, as the lung parenchyma has low proton density and significant susceptibility effects from the air–tissue interfaces.

Hyperpolarized gas MRI has emerged in recent years as a promising new approach to imaging of lung structure and function [17]. Rather than using the inherently low proton density of lung tissue as a basis for imaging, an inhaled MR-sensitive gas (such as ^{3}He) is used as a contrast agent. To overcome the extremely low density of spins, the gas first undergoes the process of *hyperpolarization*, which aligns the spins and produces the magnetization that would normally be accomplished with the main magnetic field of the scanner. The normal level of polarization (the number of excess spins preferentially aligned with the magnetic field and thus capable of generating a signal) for a clinical strength magnet is on the order of 5–10 parts per million (0.0005%); the processed gas typically reaches polarization levels of 25%, an increase of a factor of 50,000. This more than compensates for the density differences and permits generation of high-quality (high SNR) images.

An important limitation of hyperpolarized gas imaging is that (unlike standard proton imaging), the polarization is not renewed during the scan. That is, after an excitation, the magnetization generated returns to its thermal equilibrium state rather than the hyperpolarized state (which is effectively zero). As a result, imaging pulse sequences for hyperpolarized media must carefully manage the polarization to generate images of sufficient quality.

Hyperpolarized ^{3}He imaging can provide detailed maps of lung ventilation, as shown in Fig. 1.10. Because only the helium gas produces signal in such images, the signal intensity in the acquired images is related directly to the distribution of the gas within the lungs. In addition to ventilation information, additional imaging techniques such as diffusion imaging can yield more structural and functional information. Diffusion imaging produces a map of the apparent diffusion coefficient (ADC). This parameter will be affected by the size of the lung structures where the gas is located, with small structures restricting the range over which the gas may diffuse. Increases in ADC can indicate a local loss of structural integrity and are useful for assessing conditions such as emphysema. Figure 1.10 also shows an example of an ADC map from a volunteer corresponding to the ventilation scan in the middle panel.

As with Xe-CT, hyperpolarized ^{3}He imaging remains a research tool that does provide unique information that cannot otherwise be obtained noninvasively. Hyperpolarized xenon gas can also be used for MRI to generate similar information. Future work may see an emerging role for these gas-based contrast agents for noninvasive diagnosis and treatment assessment.

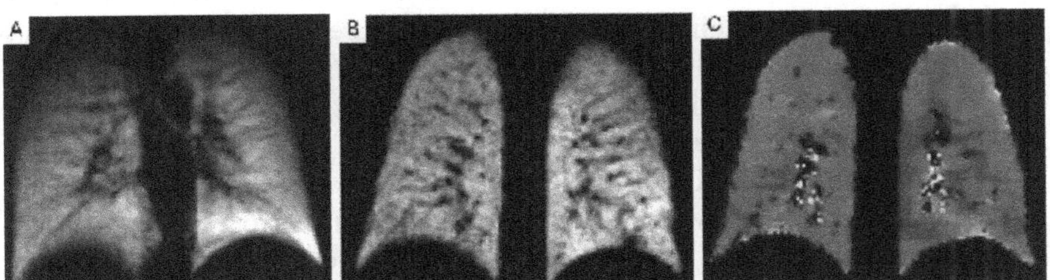

Fig. 1.10 Depiction of lung ventilation using hyperpolarized ^{3}He MRI. Ventilation images are shown for two slice thicknesses (*left* and *center panels*), along with a corresponding map of apparent diffusion coefficient (ADC), which relates to dimensions of the lung microstructure and is sensitive to disease states

1.5 Other Techniques

Ultrasound (echocardiography), MRI, and MDCT represent the primary tomographic imaging methods used for assessment of cardiac morphology and function, as they provide the most detailed imagery among widely available techniques. Nevertheless, there are additional imaging techniques that fill important niches in the realm of cardiac imaging. Single photon emission computed tomography (SPECT) utilizes a gamma camera to generate 3D views of the gamma rays emitted by radioactive tracer elements. Positron emission tomography (PET) operates on similar principles, but relies on the generation of positrons from radioactive decay and their subsequent annihilation, which produces gamma rays.

1.5.1 SPECT

SPECT relies on the introduction of radioactive tracer elements into the body and measures their distribution based on the x-rays or gamma rays emitted as they decay. In cardiac imaging, the most common utilization is in evaluating myocardial perfusion to localize and quantify areas of perfusion defects. The elements used in these exams are usually thallium-201 or a technetium-based tracer. They may be injected while the subject is at rest or with exercise or stress-inducing agents.

The principles of image formation in SPECT are very similar to those used in CT. Detection of the emitted gamma rays is accomplished with one or more scintillation cameras. The cameras rotate around the patient, acquiring a sequence of equally spaced projection images of the gamma ray distribution emitted from the patient. A tomographic slice can then be reconstructed from these projections using the same image formation methods as in CT. Typically, multiple SPECT slices are acquired over a 3D volume, permitting the resulting slices to be reformatted into arbitrarily oriented views. In cardiac imaging, these would typically be short- and long-axis views to account for anatomical variations in the position of the heart within the chest. The resulting cardiac SPECT images are used to identify and quantify metabolic function and myocardial perfusion defects at rest and under stress, yielding a useful tool for assessing the extent and severity of coronary artery disease and the functional consequences.

The resolution of SPECT images is low compared to other tomographic imaging modalities such as CT and MRI. Images are typically formed from 60 to 120 projections, resulting in an image matrix size of 128×128 pixels and a resolution on the order of 1 cm per pixel. SPECT is therefore not widely used for anatomic descriptors of cardiac health or functional measures derived from anatomical images such as LV mass or ejection fraction. Recently, combined SPECT-CT scanners have appeared on the market, permitting direct registration of high-resolution anatomic information from CT and metabolic function and perfusion information from SPECT in a single exam.

1.5.2 PET

Like SPECT, PET generates images of the distribution of radioactive tracer elements. While SPECT utilizes a variety of radionuclides that primarily emit gamma rays, in PET imaging positron emitters are introduced as tracer elements and *annihilation photons* arising from the interaction of an emitted positron and an electron are detected and utilized in the formation of an image.

Positrons may be emitted during the radioactive decay of certain isotopes. When these positrons lose energy, they may eventually interact with an electron. The resulting interaction (called *annihilation*) converts the entire mass of both particles into photon energy. This energy is emitted as a pair of photons of energy 511 keV (the equivalent energy as described in the mass–energy equivalence relationship $E = mc^2$) going in opposite directions (separated by 180°). The two photons generated by an annihilation event are thus characterized by the facts that they have a very specific energy (511 keV), they are generated at the same time, and their paths form a single line from the point at which the annihilation occurred. These traits are used as the basis for image formation in PET imaging.

In a PET scanner, several rings of detectors surround the subject. The detectors are designed to be maximally sensitive to the 511 keV photons generated. When photons are detected, only those at the appropriate energy are recorded, along with the time at which they were detected. For the photons with the proper energy to have arisen from an annihilation, *coincidence detection* is performed. Photons that are detected at the same time at two different detectors are presumed to come from the same annihilation event. This event must also have occurred along a line connecting the two detector locations. Thus, each coincidence that is detected generates a *line of response*. During the scan, all of the lines of response are recorded, and once the acquisition is complete, these projection data are used to generate a tomographic slice of the distribution of the decaying radionuclide. Once again, the reconstruction process is very similar to that used in CT and SPECT. While imaging resolution in PET is better than in SPECT, it is still considerably lower compared to CT and MRI. Typical resolutions are on the order of 5 mm per pixel.

By far the most common radionuclide used in PET imaging is 18-fludeoxyglucose (FDG), which is a glucose analog. The distribution of FDG can thus be used for assessment of glucose metabolism in tissues. Its principal use is in diagnosis and monitoring of cancer, as most malignant tumors have a high rate of metabolic activity. PET imaging with FDG is suitable for determining myocardial viability, that is, areas of reversible ventricular dysfunction. This can help assess if surgical interventions such as coronary artery bypass are warranted. PET is also useful for cardiac perfusion studies. In this use case, the common tracer elements are ^{13}N-labeled NH_3 or ^{15}O-labeled H_2O. ^{82}Rb has also been used for perfusion, with the added advantage that it does not require an on-site cyclotron to generate the needed tracer. Nevertheless, because of the complexity and expense of PET imaging, perfusion studies are usually indicated only when SPECT results are of poor quality or inconclusive.

As with SPECT, new scanners integrating PET detectors into a high-resolution imaging scanner such as MRI or CT are appearing to yield hybrid scanners that can noninvasively capture both anatomic information with high detail and accurately registered functional information in a single study.

1.6 Summary

With the rapid development of many tomographic imaging techniques, the fields of cardiac and pulmonary imaging have advanced tremendously in recent years. Each of the primary imaging modalities described above has its own unique strengths and limitations for answering research and clinical questions related to cardiopulmonary anatomy and function. In many cases, there is overlap in the information that can be acquired from each technique. This leaves the selection of the appropriate trade-offs in such aspects as cost, resolution, scan time, and radiation exposure at the discretion of the investigator or clinician.

Each of these imaging modalities continues to advance toward faster and more accurate image generation. High-resolution and 3D ultrasound probes are proliferating for echocardiology, MDCT scanners are being built with an ever-increasing number of slices as well as multiple energy levels, and higher field MRI systems with greater degrees of parallel coils are appearing. The fusion of high-resolution anatomic imaging with MDCT and MRI with functional and metabolic imaging from SPECT and PET in a single device also has the potential to change the volume and types of information that can feasibly be acquired in a single study. It can be expected that the quantity and quality of cardiopulmonary imaging data will continue to increase, permitting the development of a greater understanding of cardiac and pulmonary disease and treatment.

References

1. Hedrick WR, Hykes DL, Starchman DE (2004) Ultrasound physics and instrumentation, 4th edn. Mosby, St. Louis, MO
2. Matthew JP, Ayoub CM (2005) Clinical manual and review of transesophageal echocardiography, 1st edn. McGraw-Hill, New York
3. Bushberg JT, et al (2002) The essential physics of medical imaging, 2nd edn. Lippencott Williams & Wilkins, Philadelphia, PA
4. Bastarrika G, Lee YS, Huda W, Ruzsics B, Costello P, Schoepf UJ (2009) CT of coronary artery disease. Radiology 253:317–338
5. Litmanovich D, Bankier AA, Cantin L, Raptopoulos V, Boiselle PM (2009) CT and MRI in diseases of the aorta. AJR Am J Roentgenol 193:928–940
6. Agarwal PP, Chughtai A, Matzinger FR, Kazerooni EA (2009) Multidetector CT of thoracic aortic aneurysms. Radiographics 29:537–552
7. Orakzai SH, Orakzai RH, Nasir K, Budoff MJ (2006) Assessment of cardiac function using multidetector row computed tomography. J Comput Assist Tomogr 30:555–563
8. Ley-Zaporozhan J, Ley S, Kauczor HU (2008) Morphological and functional imaging in COPD with CT and MRI: present and future. Eur Radiol 18:510–521

9. Kuriakose J, Patel S (2010) Acute pulmonary embolism. Radiol Clin North Am 48:31–50
10. Simon BA (2005) Regional ventilation and lung mechanics using x-Ray CT. Acad Radiol 12:1414–1422
11. Bernstein MA, King KF, Zhou XJ (2004) Handbook of MRI pulse sequences. Academic, Boston, MA
12. Haacke EM, Brown RF, Thompson M, Venkatesan R (1999) Magnetic resonance imaging: physical principles and sequence design. Wiley, New York NY,
13. Abdel-Aty H, Simonetti O, Friedrich MG (2007) T2-weighted cardiovascular magnetic resonance imaging. J Magn Reson Imaging 26:452–459
14. Chavhan GB, Babyn PS, Jankharia BG, Cheng HL, Shroff MM (2008) Steady-state MR imaging sequences: physics, classification, and clinical applications. Radiographics 28:1147–1160
15. Bandettini WP, Arai AE (2008) Advances in clinical applications of cardiovascular magnetic resonance imaging. Heart 94:1485–1495
16. Ersoy H, Rybicki FJ (2008) MR angiography of the lower extremities. AJR Am J Roentgenol 190:1675–1684
17. Fain SB, Korosec FR, Holmes JH, O'Halloran R, Sorkness RL, Grist TM (2007) Functional lung imaging using hyperpolarized gas MRI. J Magn Reson Imaging 25:910–923

Chapter 2
Three-dimensional and Four-dimensional Cardiopulmonary Image Analysis

Andreas Wahle, Honghai Zhang, Fei Zhao, Kyungmoo Lee, Richard W. Downe, Mark E. Olszewski, Soumik Ukil, Juerg Tschirren, Hidenori Shikata, and Milan Sonka

Abstract Modern medical imaging equipment can provide data that describe the anatomy and function of structures in the body. Image segmentation techniques are needed to take this raw data and identify and delineate the relevant cardiovascular and pulmonary anatomy to put it into a form suitable for 3D and 4D modeling and simulation. These methods must be able to handle large multi-dimensional data sets, possibly limited in resolution, corrupted by noise and motion blur, and sometimes depicting unusual anatomy due to natural shape variation across the population or due to disease processes. This chapter describes modern techniques for robust, automatic image segmentation. Several applications in cardiovascular and pulmonary imaging are presented.

2.1 Introduction

The heart and lungs are vital organs that ensure sufficient oxygenation of the body. Deoxygenated blood arrives through the vena cava in the right atrium of the heart and is pumped through the right ventricle into the lungs, where carbon dioxide is exchanged for oxygen. The now oxygen-rich blood flows from the lungs back to the left atrium of the heart, from which it is pumped through the left ventricle into the aorta and distributed throughout the body. There are many modern imaging modalities available (x-ray computed tomography, magnetic resonance, ultrasound, x-ray angiography) to image the cardiovascular and pulmonary systems.

This chapter presents essential cardiopulmonary image processing and modeling methods in Section 2.2, based on statistical shape modeling, region growing, fuzzy connectivity, and optimum graph search. Section 2.3 presents in detail the application of statistical models in cardiac segmentation and disease classification,

A. Wahle (✉)
Department of Electrical and Computer Engineering, Iowa Institute for Biomedical Imaging, The University of Iowa, Iowa City, IA, USA
e-mail: andreas-wahle@uiowa.edu

K.B. Chandran et al. (eds.), *Image-Based Computational Modeling of the Human Circulatory and Pulmonary Systems*, DOI 10.1007/978-1-4419-7350-4_2,

while Section 2.4 summarizes several cardiovascular studies of connective tissue disorder in the aorta, the analysis of aortic thrombi, and the development of plaque in coronary arteries. Finally, Section 2.5 provides details on the segmentation and quantitative analysis of airway and vascular trees as well as the lung lobes.

2.2 Segmentation and Modeling Methodology

2.2.1 Active Shape and Appearance Models

Active shape and active appearance models (ASM/AAM) are statistical models that capture average shape and appearance as well as modes of variation from a set of training data. In general, the complex data are reduced to a set of shape and appearance modes, ideally allowing the model to represent any sample of the training set as the combination of its modal parameter values *(modal indices)*. This model can then be applied to previously unseen data and iteratively converges to a best match by identifying the modal indices for which the model has the overall smallest distance to the given shape or appearance. ASM and AAM have been used for various applications such as 2D+time LV segmentation of ultrasound data [1], 3D LV or LV+RV segmentation of ultrasound and MR data [2], bi-temporal LV segmentation of MR data [3], 3D LV segmentation of MR and CT data [4], 4D LV shape fitting of MR and CT data [5], and modeling four heart chambers of 4D CT data [6]. Andreopoulos and Tsotsos [7] achieved LV segmentation of 4D MR images using 3D AAM followed by 2D+time ASM. Zhu et al. [8] performed LV segmentation of 4D MR data in a phase-by-phase manner where the statistical and dynamic models were combined in 3D segmentation. In the following, the 3D case is discussed, where in general the method is not restricted to any given number of dimensions.

The basis of an ASM or AAM is *Principal Component Analysis* (PCA) [9], which in brief determines the principal axes from a set of samples. This is illustrated in Fig. 2.1. The example shows a scatter cloud which is clearly clustered in one direction. Thus, the most significant changes can be expected along the principal axis U of the first mode. Orthogonal to U, the second-mode axis V is defined. Rather than by a coordinate (x, y), each sample can now be represented in this alternate coordinate system as (u, v). The main information is still retained even if only the u modal index is considered, but the v index is neglected. Thus, retaining only the most significant modes reduces the amount of data while nevertheless capturing the major information contained in the datasets.

Developed by Cootes and Taylor [10], the statistical shape model contains 3D shapes and their variations, where a 3D sample shape is represented by a fixed number of surface points called *landmarks*. The landmarks are required to be evenly distributed on the sample surface and dense enough so that their triangular mesh is close to the original sample surface. Most importantly, landmarks placed on different samples must correspond to each other.

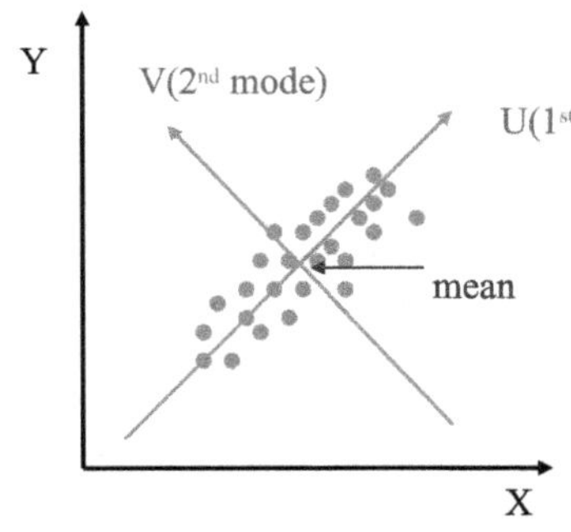

Fig. 2.1 Example for determination of the modes of variation from a 2D scatter plot of samples: The first mode follows the overall direction of the scatter cloud, whereas the second mode provides the ability to distinguish samples in the direction orthogonal to the main direction

2.2.1.1 Building a 3D Statistical Shape Model

A 3D sample shape with N landmarks can be represented by a shape vector with dimensionality of $3N$, $\mathbf{s} = [x_1, y_1, z_1, \ldots, x_N, y_N, z_N]^T$, where (x_i, y_i, z_i) are the coordinates of the ith landmark. To align all the sample shape vectors to a common frame of reference, *Procrustes Analysis* is applied to an M-sample population, $\{\mathbf{s}_1, \mathbf{s}_2, \ldots, \mathbf{s}_M\}$, also allowing the determination of the mean shape vector $\bar{\mathbf{s}}$ [11]. The PCA is then applied to generate the 3D statistical shape model and produce a new representation for any shape vector as $\mathbf{s} = \bar{\mathbf{s}} + \mathbf{\Phi b}$, where $\mathbf{\Phi}$ contains all the eigenvectors, $\boldsymbol{\phi}_i$, of the covariance matrix $\mathbf{C}$ defined as

$$\mathbf{C} = \frac{1}{M}\sum_{i=1}^{M}(\mathbf{s}_i - \bar{\mathbf{s}})(\mathbf{s}_i - \bar{\mathbf{s}})^T \tag{2.1}$$

The term $\mathbf{\Phi}$ $\mathbf{b}$ represents the shape variation of the sample shape from the mean shape. The eigenvectors $\boldsymbol{\phi}_i$ are sorted with respect to the magnitude of their associated eigenvalues λ_i such that $\lambda_i \geq \lambda_{i+1}$. Since the smaller the eigenvalue, the less its influence on the sample shape, the lesser significant eigenvalues do not contribute much to a shape and may actually represent noise rather than information. Consequently, the sizes of $\mathbf{\Phi}$ and $\mathbf{b}$ are frequently reduced to the z most significant modes of variation. The cutoff z is determined such that

$$\sum_{i=1}^{z}\lambda_i \geq \alpha\sum_{i=1}^{3N}\lambda_i, \qquad 0 \leq \alpha \leq 1 \tag{2.2}$$

with a typical value of $\alpha = 97\%$. After the reduction of $\mathbf{\Phi}$ and $\mathbf{b}$, any sample shape can be approximated by

$$\mathbf{s} \approx \bar{\mathbf{s}} + \Phi\mathbf{b} \tag{2.3}$$

Since $\bar{\mathbf{s}}$ and $\mathbf{\Phi}$ are common for all samples in the population, the only unique vector for any sample is its modal parameter vector $\mathbf{b}$, whose dimensionality is much smaller than that of $\mathbf{s}$. This also implies that the *exact* shape of a given training sample may no longer be recoverable from the modal indices due to lack of the least-significant modes. The ith element of $\mathbf{b}$, b_i, is referred to as the ith mode or the ith modal parameter and its standard deviation is $\sigma_i = \sqrt{\lambda_i}$. It represents the distance of the given shape from the mean shape in the direction of that mode, where it has to be kept in mind that this does not relate to any specific axis in 3D space but is the combination of variations determined to be ith significant over the training set.

2.2.1.2 Extension to Higher Dimensions

This could be extended to an n-D problem. For example, a 3D + time model (thus, 4D) would consist of the concatenated 3D coordinates for each time instance. The 4D shape can be represented by a fixed number of 3D shapes. For a P-phase 4D shape that contains P 3D shapes, its 4D shape vector can be written as $\mathbf{s} = [\hat{\mathbf{s}}_1{}^T, \hat{\mathbf{s}}_2{}^T, \cdots, \hat{\mathbf{s}}_P{}^T]^T$, where $\hat{\mathbf{s}}_i$ is a 3D shape vector as defined above. It is not necessary for $\hat{\mathbf{s}}_i$ and $\hat{\mathbf{s}}_j$ $(i \neq j)$ to have the same number of landmarks. However, it is still required that all samples at each phase have the same number of landmarks. The 4D shape model can be generated using the same procedure described above.

2.2.1.3 Combining Shape and Appearance

Thus far, only the coordinates of the landmarks were taken into account to retrieve the shape information from the given sample. Similarly, the intensity pattern—the texture vector $\vec{t}$—of the object can also be modeled by a statistical texture model once the correspondence of all interior voxels is established, thus introducing an active *appearance* model. Combining modes of variation from shape ($\vec{b}_s$) and texture ($\vec{b}_t$) with appropriate weighting $\mathbf{W}_\mathrm{s}$ results in an appearance vector $\vec{a}$ that can be statistically modeled.

$$\vec{a} = \begin{bmatrix} \mathbf{W}_\mathrm{s}\vec{b}_\mathrm{s} \\ \vec{b}_\mathrm{t} \end{bmatrix} = \begin{bmatrix} \mathbf{W}_\mathrm{s}\Phi_\mathrm{s}^T(\vec{s} - \bar{\vec{s}}) \\ \Phi_\mathrm{t}^T(\vec{t} - \bar{\vec{t}}) \end{bmatrix}, \vec{a} \approx \Phi_\mathrm{a}\vec{b}_\mathrm{a} \qquad (2.4)$$

Both AAM and ASM use an iterative process for segmentation where the model is fit to the target image and refined based on the model and image features. Figure 2.2 shows an example for the matching of a 2D + time AAM simultaneously in multiple phases, yielding good correspondence with the manual tracing [1]. The AAM and ASM were also combined in a multi-stage hybrid approach [12] and its usage in MR ventricular segmentation yielded better results than AAM or ASM alone.

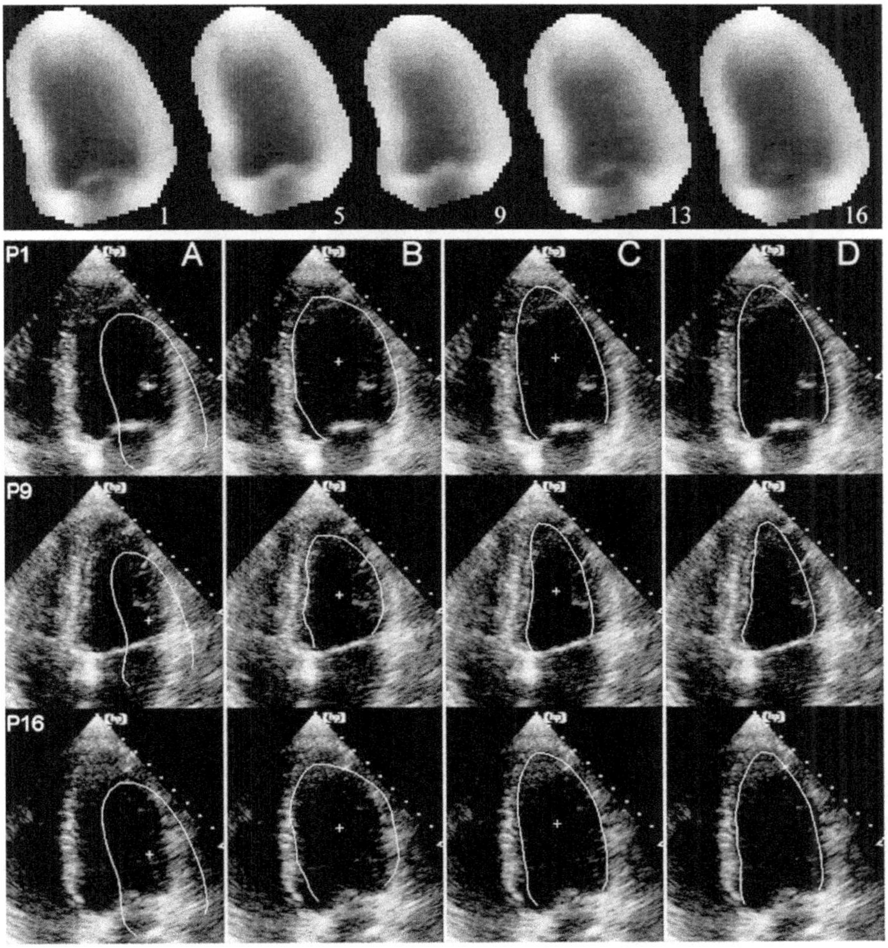

Fig. 2.2 Mean shape of a 2D + time (*phase*) active appearance motion model (*top row*) and convergence process in three of the phases (**a–c**, *bottom row*); (**d**) shows the ground truth based on manual tracing [1, ©2002 IEEE]

2.2.1.4 Robust ASM and AAM Implementations

When segmenting image data, the problem frequently occurs that parts of the structure to be segmented are missing or distorted. This introduces ambiguities into the analysis and may lead a conventional AAM to fail. A *robust* AAM (RAAM) analyzes the residual errors of a match and identifies clusters of residuals which are associated with the lack of information. It then removes those clusters from the analysis to find the best match for the remaining residuals [13, 14]. Figure 2.3 shows an example comparing AAM results with those of an RAAM, which demonstrate that in cases where the image data are distorted by spot-like occlusions the RAAM nevertheless can perform well.

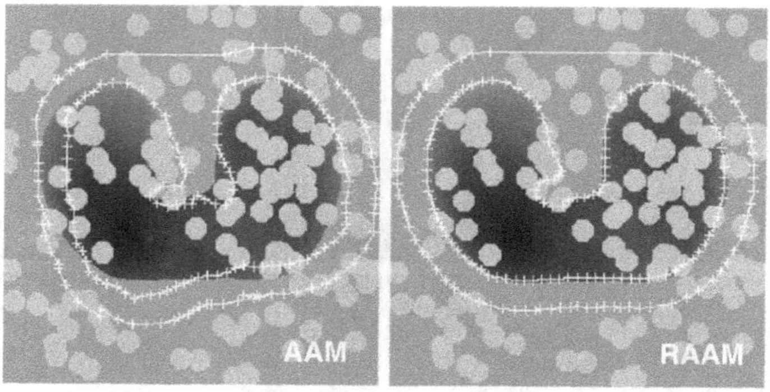

Fig. 2.3 Regular AAM vs. robust RAAM on synthetic diaphragm images with spot-like occlusions [12, ©2005 IEEE]; the RAAM succeeds with a match where the regular AAM clearly fails

Such approaches still assume, though, that the *training* of the model was performed on complete datasets. Robust PCA (RPCA) methodology provides the ability to also obtain a feasible statistical model with incomplete training data, though the regional quality of the model may vary depending on the amount of samples available with sufficient confidence for that specific region. Skočay et al. [15] have developed an RPCA model by combining an expectation–maximization (EM) method to estimate the principal axes with a probabilistic PCA [16] algorithm. First, the principal axes and coefficients are determined using a standard PCA on the training samples. Local outliers are determined based on an objective function and labeled as missing data. The EM algorithm is applied to replace those missing data and the process iteratively repeated until the set of outliers is minimized:

E-Step: Estimate modal indices **b** to represent the given samples based on the currently assumed principal axes represented by $\mathbf{\Phi}$;

M-Step: Compute new principal axes $\mathbf{\Phi}$ to maximize the combined likelihood that the modal indices **b** represent the given samples using $\bar{\mathbf{s}} + \Phi\mathbf{b}$ (repeat until the error is minimized).

While the RAAM considers both shape and texture information, the computationally more efficient robust ASM (RASM) utilizes the shape information only and may use classifiers to include local or regional image information [17–19].

2.2.2 Region Growing and Fuzzy Connectivity Segmentation

2.2.2.1 Region Growing

Region growing-based segmentation can bring desirable results in noisy images, where borders may be extremely difficult to detect. Homogeneity is an important property of regions and is used as the main segmentation criterion in region growing, whose basic idea is to divide an image into zones of maximum homogeneity. The criteria for homogeneity can be based on gray level, color, texture, shape, model (using semantic information), etc. Properties chosen to describe regions influence the form, complexity, and amount of prior information in the specific region growing segmentation method.

In region-based segmentation, the regions must satisfy the following conditions:

$$H(R_i) = \text{TRUE}\,, \quad i = 1, 2, \ldots, S \tag{2.5}$$

$$H(R_i \cup R_j) = \text{FALSE}\,, \quad i \neq j\,, R_i \text{ adjacent to } R_j \tag{2.6}$$

where S is the total number of regions in an image and $H(R_i)$ is a binary homogeneity evaluation of the region R_i. Resulting regions of the segmented image must be both homogeneous and maximal, where by "maximal" we mean that the homogeneity criterion would not be true after merging a region with any adjacent region.

There are three primary versions of region growing—region merging, region splitting, and split-and-merge approaches. Of special interest are the homogeneity criteria, whose choice is the most important factor affecting the methods mentioned; general and specific heuristics may also be incorporated. The simplest homogeneity criterion uses an average gray level of the region, its color properties, simple texture properties, or an m-dimensional vector of average gray values for multi-spectral images. Importantly, 3D implementations are easily possible and thus these approaches are directly applicable to volumetric medical images. Three-dimensional filling represents its simplest form and can be described as a 3D connectivity-preserving variant of thresholding.

The most natural method of region growing is to begin the growth in the raw image data, each pixel representing a single region. These regions almost certainly do not satisfy the condition of Eq. (2.6), and so regions will be merged as long as Eq. (2.5) remains satisfied. As a dual approach, region splitting starts with the entire image representing a single region, which is sequentially split into smaller child subregions as long as the parent (sub)regions do not satisfy the criterion of homogeneity. Region split-and-merge approaches then combine these two basic principles. A somewhat separate approach to region growing is the watershed image segmentation approach that represents image intensity as a topographic surface that is flooded by a raising water level [20, 21]. Many individual variants of region growing and/or watershed segmentation methods exist, and an introductory overview can be found in Ref. [22].

2.2.2.2 Fuzzy Connectivity-Based Segmentation

The above-referenced region growing segmentation methods are based on crisp (or *hard-coded*) relationships between or within the individual regions to be segmented. In many cases, however, these relationships may vary across the image due to noise, intensity variations across images, limited spatial resolution, etc. The *fuzzy connectivity* segmentation approach takes these uncertainties into consideration. Rather than defining crisp relations, it attempts to describe the segmentation task with fuzzy rules such as *if two regions have about the same gray value and if they are relatively close to each other in space, then they likely belong to the same object.* A framework for such a reasoning approach is called *fuzzy logic.*

Fuzzy connectivity segmentation attempts to mimic the analysis strategy of a trained human observer who is typically able to perform a segmentation task by hand, frequently considering the likelihood of whether nearby image elements belong together. If they seem to belong to the same object based on their image and spatial properties, they are assigned to the same region. In other words, the image pixels seem to *hang together* when forming an object. The *hanging togetherness* property is then described using fuzzy logic. Early work in the fuzzy connectivity field was reported in Refs. [23, 24], followed by Refs. [25, 26]. Udupa et al. stated an important concept that voxels belonging to the same objects tend to *hang together*, thus defining objects by a combination of the spatial relationship of its elements (pixels, voxels), while at the same time considering local image properties [27–30]. The spatial relationships should be determined for *each* pair of image elements in the entire image. To accomplish this, local and global image properties are considered. An overview of the fuzzy connectivity segmentation principles with accompanying algorithm outlines and step-by-step examples can be found in Ref. [22].

2.2.3 Graph-Based Segmentation

Graph-based segmentation methods have a long history of being applied in medical image processing to extract anatomical features from various modalities. In the cardiovascular domain, an early problem was the automated detection of vascular borders from x-ray angiograms, e.g., to determine the degree of stenosis in coronary arteries [31–33]. In general, the problem splits into three tasks: (a) select an algorithm to determine a "minimum cost" path through the graph, (b) define the structure of the graph, and (c) define the "cost" associated with each node or edge of the graph which represents a feature of the image.

2.2.3.1 Approaches Based on Rectangular Graph Structures

For the problem of identifying arteries in 2D images, *dynamic programming* [22] is an example of a widely used graph-based method. The structure of the graph is implied by this choice to a certain extent as an array of scan lines orthogonal to the vessel, representing its cross sections over a segment, and searching for its boundaries at the left- and right-hand side of the vessel [31]. For example, looking at a border pixel candidate at location j in line i, the "cheapest" continuation from either $\{j-1, j, j+1\}$ in line $i-1$ is determined which results in the lowest overall cost (Fig. 2.4). This is done for all scan lines and all candidates, and the path with

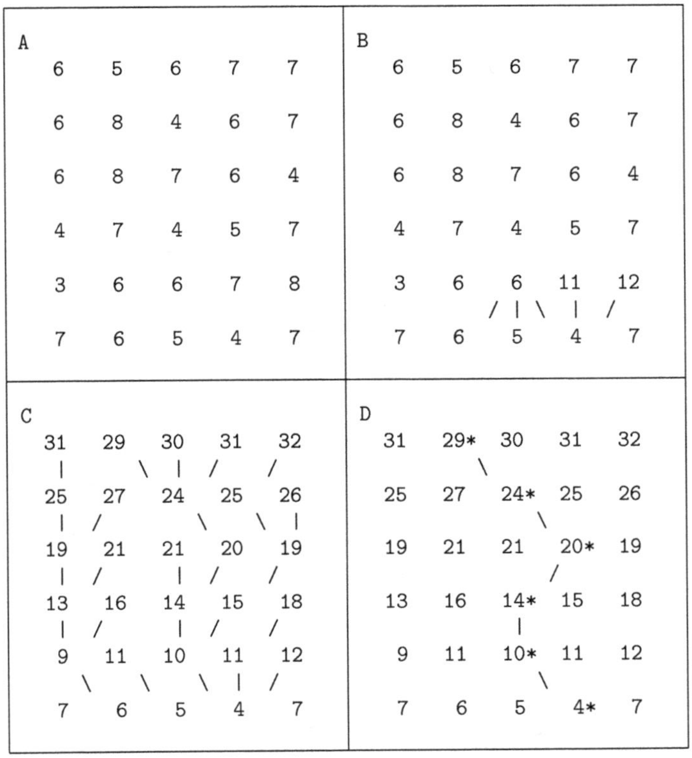

Fig. 2.4 Graph with node costs, dynamic programming search for a path from *bottom* to *top*: (a) initial cost assignments, (b) the cost of the "cheapest" neighbor is added to the local cost, (c) final cost assignment, (d) lowest cost in top row traced back results in the desired path

the lowest sum of costs is the result. Since the vessel appearance is brighter or darker than the background, a gradient-based cost function is used [32], i.e., the cost is lowest at locations of the highest gradient or where the highest probability is seen that a pixel represents a border point. The cost function can have multiple components, e.g., a mix of first and second derivatives is usually employed to avoid bias for over- or underestimation of the vessel diameter [33] (Fig. 2.5).

The example of artery detection also shows that multiple related structures (left and right borders in this case) may be aimed for. Rather than detecting those in two separate steps (e.g., by employing two different cost functions based on the gradient direction), a simultaneous approach can be used by extending the graph into a third dimension [35, 36]. In this way, a single optimum path seeking minimum cost represents the location for both the left and right borders along the vessel segment. What remains a problem in either single- or multiple-contour approach is the need to derive the graph from the image data, where the artery may be arbitrarily oriented and has to be resampled first into a suitable geometry.

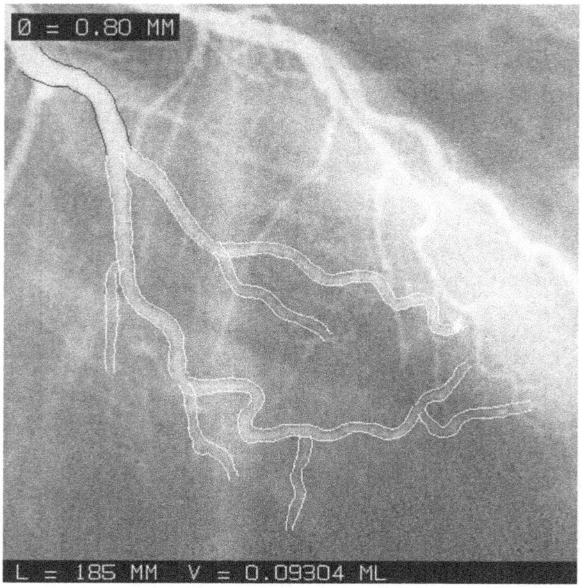

Fig. 2.5 Dynamic programming result applied to coronary x-ray angiography to detect the vessel borders based on first and second derivatives as cost function [34, © 1995 IEEE]

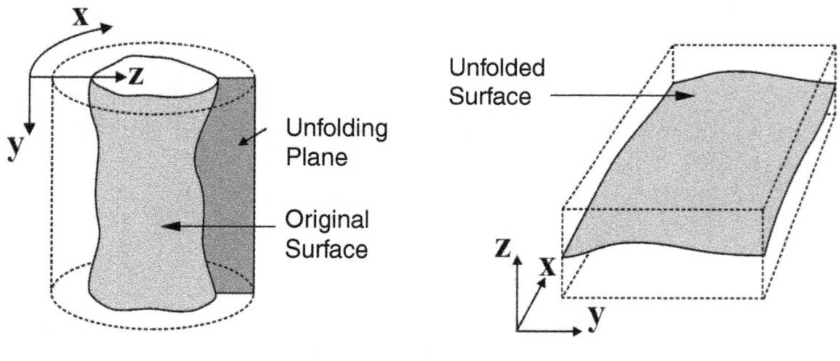

Fig. 2.6 Example for unwrapping a tubular structure to a rectangular grid for segmentation [37, 39, ©2005 SPIE, ©2006Elsevier]

In the case of the artery, a preliminary vessel centerline is usually determined and used to define the scan lines for building the graph rows to ensure that those are in the right order for the detection. In the easiest case, this preliminary centerline can be just a straight line from the first to the last point, with perpendicular sampling of the image data to apply graph search. More difficult resampling may be necessary, e.g., in the case of tubular structures such as in intravascular ultrasound [37] or aortic 3D MRI images [38]. Starting from the centerline, radial scan lines are used to "unwrap" the structure to a rectangular coordinate system (Fig. 2.6) with a scan *plane* in 3D, rather than just lines in the 2D case. Where closed contours are required, the two ends of the scan plane have to wrap around.

2.2.3.2 Minimum-Cut Approaches

A more recent approach to graph-based segmentation is optimum graph search based on $s-t$ cuts [40–42]. This approach is in general n-dimensional and can be directly used for multiple-surface and closed-surface problems. Starting from an $(n-1)$-dimensional base surface, columns are defined perpendicular to that surface, and the cost for each node on the column derived from the respectively resampled image data.

As illustrated in Fig. 2.7a, each node of the graph has a cost associated with it as before (the surface to be segmented runs horizontally in this example). To formalize, a weighted graph $G = (V, E)$ is composed of a node set V and an arc set E. The nodes $v \in V$ correspond to image pixels (or voxels) and arcs $< v_i, v_j > \in E$ connect the nodes v_i, v_j. Every arc $< v_i, v_j > \in E$ has a cost (or weight) which represents some measure of preference that the corresponding pixels belong to the object of interest. The arc costs are derived from the node costs. There are three types of arcs to connect the nodes (Fig. 2.7b): *intra*-column arcs, *inter*-column arcs, and for the multi-surface problem, also inter-surface arcs. Let us assume, for the 3D case, that $I(\mathbf{x}, \mathbf{y}, \mathbf{z})$ is a volumetric image and $\mathrm{Col}(x, y)$ is a node subset $\{V(x, y, z)|z \in \mathbf{z}\}$. For intra-column arcs (E^{a}), along each column $\mathrm{Col}(x, y)$, every node $V(x, y, z)(z > 0)$ has a directed arc to the node $V(x, y, z-1)$. For inter-column arcs (E^{r}), along the x-direction and for any $x \in \mathbf{x}$, a directed arc is constructed from each node $V(x, y, z) \in \mathrm{Col}(x, y)$ to node $V(x+1, y, \max(0, z-\Delta_x)) \in \mathrm{Col}(x+1, y)$. The connections can be used to define *smoothness constraints*, i.e., the maximum difference in terms of nodes from one column to a neighboring column or between two surfaces. In this example, Δ_x is the smoothness constraint in the x-axis. For inter-surface arcs (E^{s}), for any node $V_1(x, y, z)$ in $\mathrm{Col}_1(x, y)$ with $z \geq \delta^u$, a directed arc connecting $V_1(x, y, z)$ to $V_2(x, y, z-\delta^u)$ is constructed. Also, for each node $V_2(x, y, z)$ in $\mathrm{Col}_2(x, y)$ with $z < Z - \delta^l$, a directed arc connecting $V_2(x, y, z)$ to $V_1(x, y, z+\delta^l)$ is created. δ^u and δ^l are the maximum and minimum surface separation constraints, respectively [41]. For the following $s-t$ cut algorithm, the node costs are transformed by subtracting for each node $V(x, y, z)(z > 0)$ the cost of the node $V(x, y, z-1)$ underneath. A special case is handled for $z = 0$, where one node is selected (here d) and the sum of all node costs in $z = 0$ plus 1 subtracted from it (i.e., $3+8 = 11 \Rightarrow 3-12 = -9$).

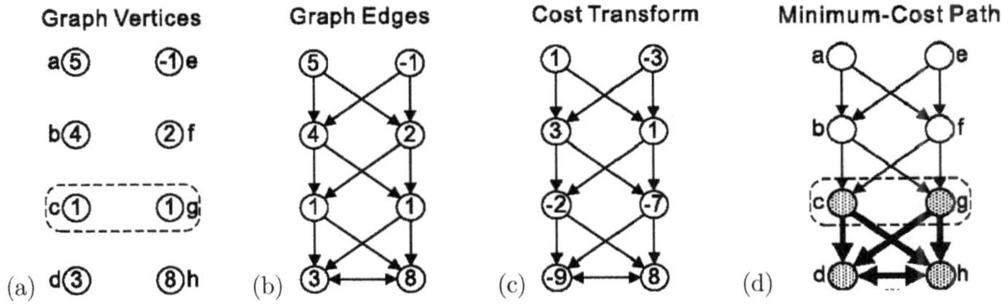

Fig. 2.7 Minimum $s-t$ cut: (**a**) nodes with costs, (**b**) connecting arcs, (**c**) cost subtraction, (**d**) resulting minimum closed set

Various methods to solve $s-t$ cut problems exist [40, 43], which in brief connect all nodes with a *negative* cost to a "source" (s) and all nodes with a *positive* cost to a "target" or "sink" (t), where the initial edge costs are derived from the node costs. In an iterative process, the edges are saturated and removed until no path from s to t is left. The remaining nodes accessible from s build the *minimum closed set* of nodes below the sought surface, which can then be mapped back into the original image data based on the known transformation used to define the columns (Fig. 2.7d). It is also possible to detect *multiple* surfaces at once and to define smoothness and distance constraints between them. This is done by adding more dimensions to the graph. An example segmenting six surfaces in groups from retinal OCT scans is presented in Ref. [44].

The description above uses x,y,z coordinates for illustration. In general, the topology of the graph is not restricted to rectangular structures as they were achieved by resampling in the examples for dynamic programming in 2D image data or tubular structures. Any surface which can be triangulated can be used to define the columns and their respective connections (Fig. 2.8). For each triangle, the inter-column nodes can be directly derived from the triangle itself, thus the graph topology is straightforward. This is a great advantage in modeling arbitrary surfaces, where the graph does not need any resampling but can be directly based on the surface mesh and also includes closed surfaces and branching structures [45, 46]. Care has to be taken though that the columns do not intersect in space, which may lead to invalid surfaces. A possible solution to this problem is the use of medial sheets to prevent columns from crossing into the region of another column [47, 48].

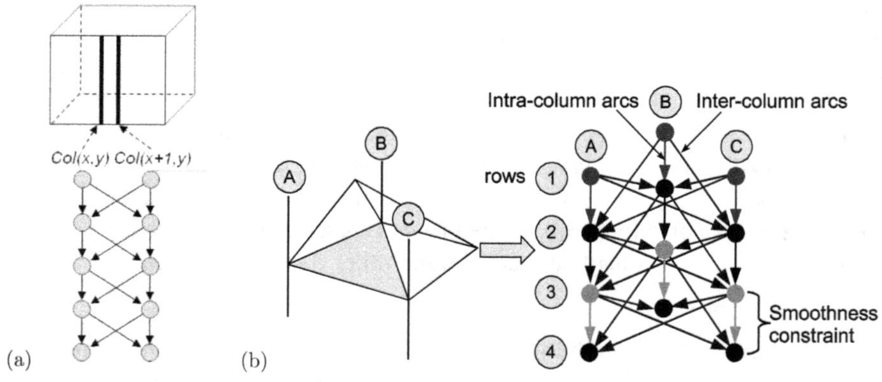

Fig. 2.8 Graph design: (**a**) two columns with connecting arcs based on a rectangular plane, (**b**) triangulated preliminary surface defining columns and their arcs within a given triangle

2.2.3.3 Cost Functions

Aside from the graph algorithm employed for segmentation, the definition of the cost or objective function is of utmost importance for identifying the correct border. While traditionally edge-based functions like gradients are used to determine changes in intensity, the underlying image data may be complex and require an approach more suited for the specific problem. For example, ultrasound images exhibit a very specific noise pattern (speckle) that needs to be accounted for. Frequently, that "noise" is removed by filtering [49]. Another way is to utilize the distribution of that speckle noise to determine boundaries (e.g., Rayleigh distribution [37, 50, 51]). A class of region-based functions are the Chan–Vese terms [52], thus the edge to detect for segmentation becomes in fact the difference between the gray-level distributions of two adjacent regions (Fig. 2.9).

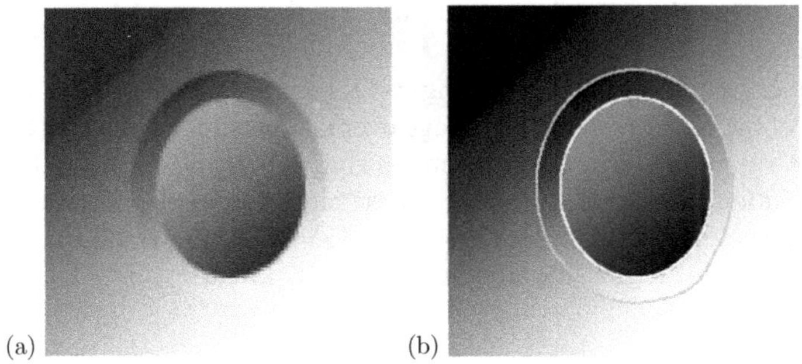

Fig. 2.9 Region-based segmentation without distinct borders [41, ©2006 IEEE]: (**a**) original image, (**b**) segmented inner and outer contours

Rather than determining the cost in an analytical way by defining a function and modifying its parameters to segment a given structure, it is also possible to *learn* where the boundary of an object is located from expert-traced examples [53]. This is not unsimilar to the active shape/active appearance models discussed earlier; however, it is not necessarily based on principal component analysis. Instead, a given pattern can be compared against a set of patterns derived from training to determine the probability that it is a border pattern [37].

It is frequently desired to put certain constraints on a segmentation. The smoothness constraint has already been mentioned earlier and can be implemented in a way that simply restricts the number of nodes that can be skipped between columns. Another option is to add penalties to the costs, which would not require hard constraints such as the graph topology, but which makes certain alternatives less attractive. To introduce a bias toward a straight line, for example, edges connecting a node j in column i with the same node $(i+1, j)$ may have a lesser cost than edges from (i, j) to $(i, j \pm 1)$.

2.3 Cardiac Applications

While the previous sections described the background for various segmentation methods commonly used in medical imaging, the following will focus on the actual application for clinically relevant questions in cardiovascular and pulmonary imaging. As the first example, the ventricles of the heart have been chosen to demonstrate how active shape and active appearance models, which were introduced in Section 2.2.1, can be employed. Most importantly, they can be used beyond segmentation of the anatomical structure. Statistical models can also be utilized to support the diagnosis of a specific disease (here Tetralogy of Fallot, which is a congenital disease requiring monitoring for complications after surgery [54]). While segmentation methods based on analytically defined cost functions do not necessarily require a training phase, a statistical model by definition requires a set of samples from which to derive the point distribution and subsequent modal indices relative to a mean shape. Therefore, the first section will focus on the modeling step based on expert tracings.

Magnetic resonance (MR) imaging is a common modality to analyze the ventricles of the heart, with the advantage over computed tomography (CT) or conventional x-ray ventriculography that no ionizing radiation is involved. It has a high image resolution and favorable signal-to-noise ratio. Ultrasound is another modality of choice, however, with higher noise levels. Also, the access window for ultrasound is rather limited due to ribs, the sternum, and other structures.

2.3.1 Modeling and Quantitative Analysis of the Ventricles

Conventional analysis of cardiac ventricular magnetic resonance images is performed using short-axis images and does not guarantee completeness and consistency of the ventricle coverage. The commonly used ventricle function index measured in MR image analysis is the ejection fraction (EF) calculated from end-diastole (ED) volume and end-systole (ES) volume measured from manual segmentations performed on the short-axis images. The ejection fraction is simply a volumetric measurement without ventricle shape information. It has long been noticed that the inclusion or exclusion of the short-axis images at the mitral valve and tricuspid valve planes is the main source of error in the ventricle volumetric measurements [55, 56]. Thus, the long-axis images should also be used to provide clear location information of the valve planes and apexes of the ventricles [57]. While clinical images usually involve acquisition of both short-axis and long-axis datasets, obtaining a complete 3D or 4D representation of the left and right ventricles from the combined image data involves several steps [58].

2.3.1.1 Manual Ventricle Segmentation

The manual ventricle segmentation was performed using Catmull–Rom spline [59] contours as shown in Fig. 2.10. Two closed contours, the left ventricle (LV) endocardial and epicardial borders, and one open contour for the right ventricle (RV) were manually traced on the long-axis and short-axis images and the control points of the contours were recorded.

Conventional ventricular MR images are acquired from several cardiac cycles. The commonly existing breathing artifact in MR images is visible as misalignments

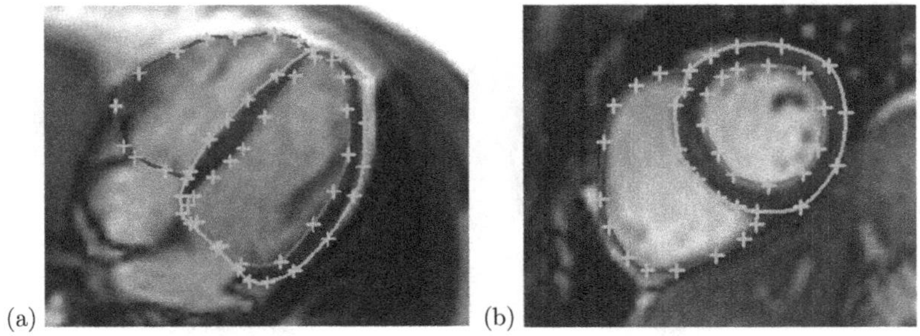

Fig. 2.10 Manual ventricle segmentation examples of (a) a long-axis image, and (b) a short-axis image. Points marked "+" are the control points of the 2D contour splines [58]

of the ventricles and needs to be corrected prior to image analysis. The images from both long axis and short axis were used in an *iterative mutual registration* algorithm. First, two isotropic 3D images from the original long-axis and short-axis 2D images are generated in a common coordinate system using cubic B-spline interpolation [60]. For each original short-axis 2D image $\mathbf{I}_s$ that is acquired from image plane $\mathbf{P}_s$,

- Sample the interpolated long-axis 3D image at the image plane $\mathbf{P}_s$. The sampled image is $\mathbf{I}_t$.
- Register image $\mathbf{I}_s$ to the sampled image $\mathbf{I}_t$. The transform of $\mathbf{I}_s$ is restricted to translation.
- Move the control points of manual segmentation contours on $\mathbf{I}_s$ to new locations specified by registration.

The interpolated short-axis 3D images are then regenerated from the new translated short-axis 2D images, the original long-axis 2D images are registered to the images sampled from the interpolated short-axis 3D image at corresponding image planes, and the contour control points are moved as described before. Then, the interpolated long-axis 3D images are regenerated from the new translated long-axis 2D images. This process is continued until convergence is reached.

Thus far, the ventricle is represented by basically independent 2D manual segmentation contours from the long-axis and short-axis images. To obtain the complete shape, those are first translated into two isotropic 3D labeled volumes, using three labels, LV+RV, LV, and LV blood pool. All pixels enclosed by LV endocardial border are labeled as LV blood pool, and so on. A 2D distance map is calculated for each label and 2D distance transforms applied to the boundary of each labeled region. Isotropic 3D distance maps are obtained for each label by interpolating the 2D distance maps using cubic B-spline interpolation. Finally, the 3D distance maps are translated into isotropic 3D labeled volumes by labeling voxels with negative distances. The resulting labeled volume has three types of voxel labels, LV blood pool, LV myocardium, and RV.

The ventricle surfaces of the two 3D labeled volumes generated from the long-axis and short-axis segmentations are shown in Fig. 2.11a where neither labeled

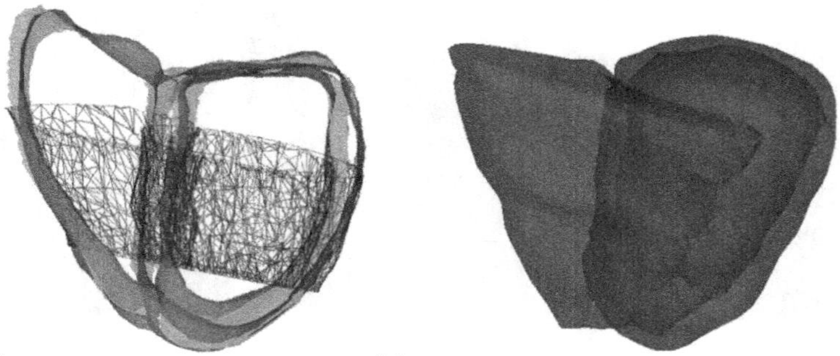

(a) (b)

Fig. 2.11 Complete ventricle shape generation. (**a**) Three-dimensional ventricle shape created from the long-axis (*solid surface*) and short-axis (*mesh*) manual segmentations. (**b**) The complete ventricle shape after approximating the base and apex [58]

volume covers the ventricles completely. Consequently, a combined labeled volume has to be generated from the long-axis and short-axis volumes with approximated base and apex. The short-axis volume is used as the middle portion of ventricles of the combined volume. The short-axis left ventricle volume is extended to base and apex specified by the long-axis volume, where base or apex are approximated by scaling and translating the closest 2D slice of the short-axis volume. The scaling factor and translation parameters are estimated from the long-axis volume. Similarly, the short-axis right ventricle volume is extended by cropping the right ventricle base at the location where the left and right ventricles begin to separate on the long-axis volume.

2.3.1.2 3D Shape Generation

The resulting ventricle surface of the combined volume is shown in Fig. 2.11b. An example of a 3D image for which a breathing artifact was corrected and interpolated is shown in Fig. 2.12a on two 2D slices. The same 3D image overlaid with combined labeled volume (overlaid shadows in images) is shown in Fig. 2.12b, where the

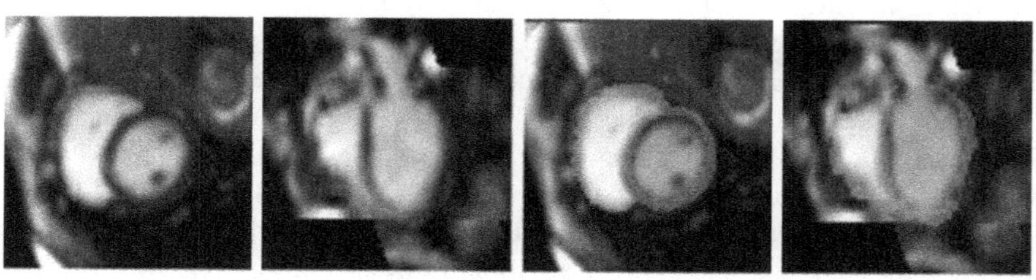

(a) (b)

Fig. 2.12 An example of breathing artifact correction and complete ventricle shape generation shown on two 2D slices of (**a**) the isotropic 3D image generated by breathing artifact correction and interpolation and (**b**) the isotropic 3D image overlaid with combined 3D labeled volume [58]

coverage of ventricles is complete. With labeled volumes of samples available, the landmarks were automatically generated using a template-based algorithm proposed by Frangi et al. [61] consisting of three steps: generation of a mean labeled volume as a template, generation of landmarks on the mean labeled volume, and propagation of landmarks back to the sample labeled volumes. The flowchart of this algorithm is shown in Fig. 2.13.

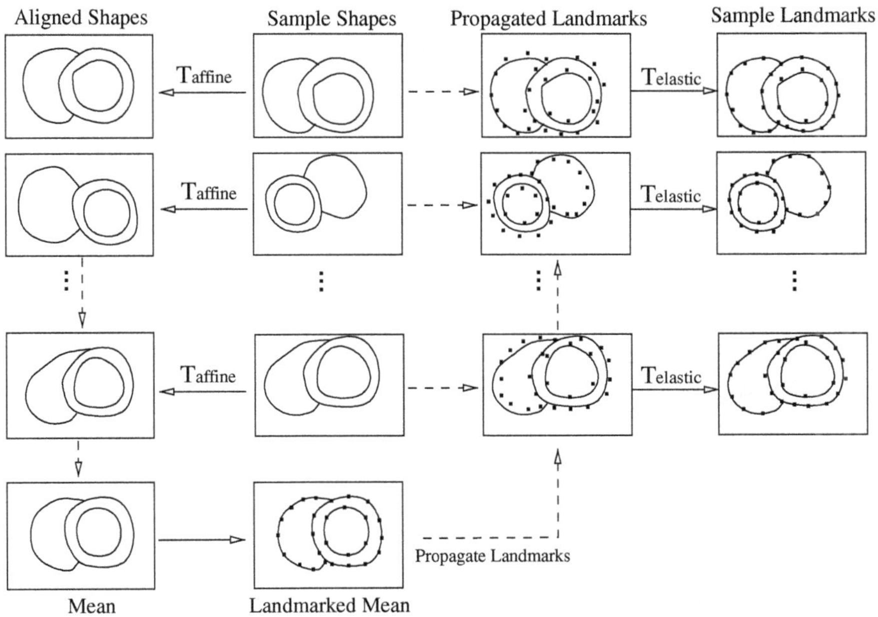

Fig. 2.13 Flowchart of the automatic landmark generation [58] (2D slices shown for simplicity)

From an arbitrarily selected sample volume as the initial mean volume, each sample volume is registered to the mean volume using an affine transform, $\mathcal{T}_{\text{affine}}$, which is restricted to translation, rotation, and scaling. The *incorrect voxel ratio*, defined as the ratio of incorrectly matched non-background voxels to the total number of non-background voxels, is minimized. Since the surface of the aligned sample volume shall be as close to that of the mean volume as possible, the scaling of $\mathcal{T}_{\text{affine}}$ is an *anisotropic* scaling. Distance maps of all aligned sample volumes are created [62] and the mean volume generated using distance-based shape merging. After calculating the average 3D distance map over all samples and translating it back to the labeled volume, the process is repeated with the newly generated mean volume until convergence. To obtain a triangulated mesh with landmarks, a *marching tetrahedrons* [63] algorithm customized for the labeled volume data representation can be utilized. After smoothing, the number of triangles (and thus final landmarks) is reduced by *quadric triangle decimation* [64]. As the final step, the generated landmarks on the mean volume are propagated back onto the samples by applying their individual inverse affine transforms along with a series of elastic *B*-spline

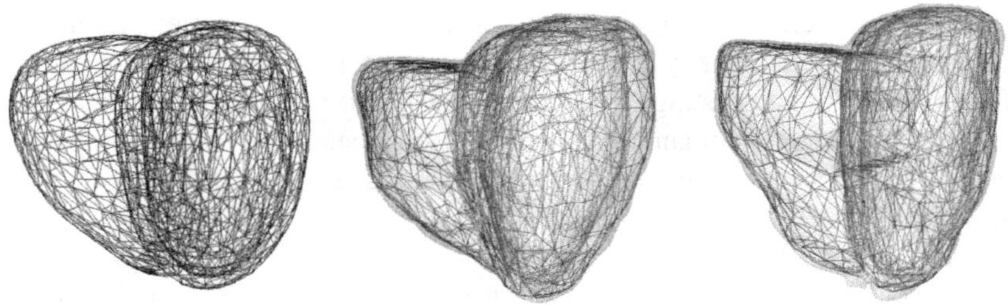

(a) (b) (c)

Fig. 2.14 Examples of automatic landmark generation. (**a**) The triangular mesh of the landmarks generated on the mean labeled volume. (**b, c**) Two examples of sample ventricles (*solid surface*) with propagated and transformed landmarks (*vertices* of *triangular mesh*); some small landmark propagation errors can be seen at ventricle base and apex[58]

transforms. Examples for propagated and transformed sample landmarks are shown in Fig. 2.14 in one phase.

After PCA was performed over the landmarks of all samples as described in Section 2.2.1.1, the modes of variation can be visualized. Figure 2.15 shows the 3D shape variations in one phase introduced by varying the values of the first two strongest modes of the 16-phase 4D model constructed from 10 Tetralogy of Fallot patients and 25 control subjects. The most important shape feature of patients' RV—its enlargement—was captured by these two strongest modes of the model.

2.3.2 Tetralogy of Fallot Classification

Tetralogy of Fallot (TOF) is a common congenital heart disease characterized by four typical features: right ventricular hypertrophy, right ventricular outflow tract (RVOT) obstruction, ventricular septal defect (VSD), and overriding aorta. TOF infants are often treated by initial palliative surgery to augment the RVOT and close the VSD, but patients often suffer from pulmonary valve insufficiency after surgery. Over time, this can lead to right ventricular dilation and dysfunction, eventually requiring pulmonary valve replacement (PVR) in the adult years [54, 65, 66]. Although TOF patients often have similar life expectancy as healthy people with an estimated prevalence of 80,000 in the USA [67], they need to be monitored throughout their lives (most frequently with MR imaging) to identify whether further surgery is required. The traditional functional indices measured in clinical analysis such as ventricular volumes and ejection fraction may not be sufficient to allow timely patient management and prevent long-term sequelae [68, 69] because they do not fully utilize the rich information provided by the MR data to describe the complex ventricular remodeling process. The modal indices obtained from modeling with an ASM or AAM can directly be used to *classify* the status of a disease, assuming that the modes of variation capture distinct features of — in this case — Tetralogy of Fallot.

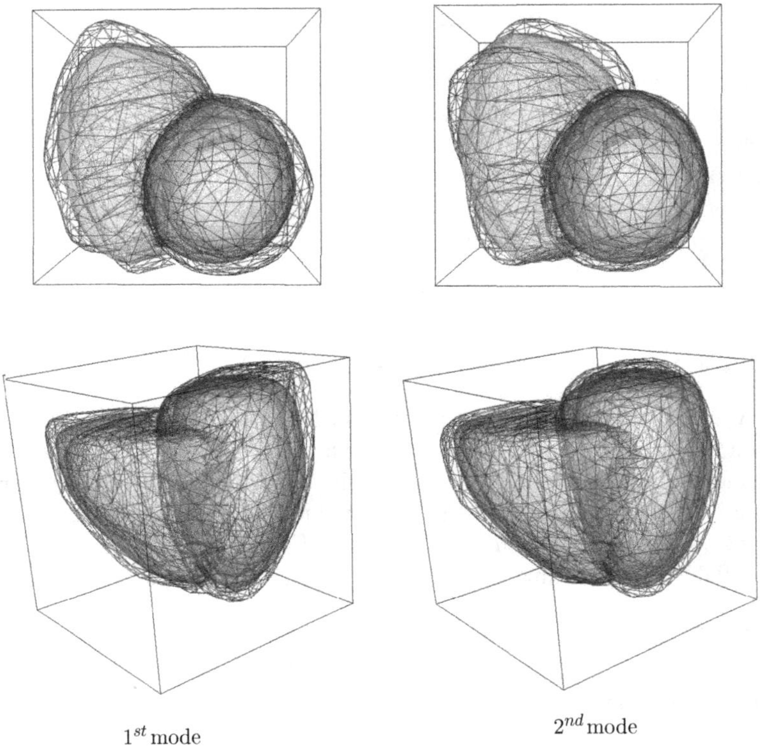

Fig. 2.15 The 3D shape variations corresponding to varied values of the two strongest shape modes within $[-2\sigma, 2\sigma]$. Solid surface represents the mean shape and the triangular meshes represent shape variations introduced by Mean-2σ and Mean$+2\sigma$ [58]

2.3.2.1 Study Population and Experimental Methods

Steady-state free precession MR imaging of the heart was performed on 25 patients with repaired TOF and resultant pulmonic regurgitation. Our study population also included 25 normal subjects. The original 2D long- and short-axis images were manually traced by an expert cardiologist with three ventricular borders: the LV endocardial and epicardial borders, and the RV border. The normal and TOF groups were modeled separately by two sets of 4D AAMs containing 16 phases per cardiac cycle. For each subject group, the LV and RV were modeled and segmented together as well as separately.

The segmentation performance was evaluated using *hold out* cross validation. For each subject group, an AAM was created and trained using a training set of 20 hearts. Then the models were used to segment the testing set of the remaining hearts. When segmenting each heart or ventricle, the AAMs for normal and TOF hearts were both used and the segmentation with better model fitting (smaller RMS of intensity difference) was chosen as the segmentation result. This procedure was repeated with different training and testing sets until all the hearts were segmented. The segmentation results were represented as the segmented landmarks and the segmentation errors were determined as the distances between the segmented landmarks and the ventricular surface created from the manual segmentation. By convention, a segmented landmark outside the manually segmented ventricle was considered having a positive error. Our method achieved mostly subvoxel signed surface positioning errors of 0.2±1.1 voxels for normal left ventricle, 0.6±1.5 voxels for normal right ventricle, 0.5±2.1 voxels for TOF left ventricle, and 1.3±2.6 voxels for TOF right ventricle [70].

2.3.2.2 Novel Ventricular Function Indices

The AAM segmentation results were translated back to ventricular volume representation for functional analysis. All the normal and TOF hearts (combined LV and RV segmentations) were analyzed by a single 4D statistical shape model and the resulting PCA shape modes were analyzed by linear discriminant classification and *leave-one-out* testing. The body surface areas of 22 normal and 24 TOF subjects were available. Volume–time curves (VTC) were created for each subject by plotting the ventricular volumes, normalized by individual body surface area, for each of the 16 phases. The VTC of LV and RV were combined such that the VTC for each subject is a 32-element vector. PCA was applied to the VTC and the resulting PCA modes were analyzed by linear discriminant classification and *leave-one-out* testing.

The distribution of the two strongest PCA shape modes is shown in Fig. 2.16a. All hearts were classified as normal/TOF with 100% correctness. The distribution of the two strongest PCA modes of the LV+RV VTC is shown in Fig. 2.16b. All hearts with available body surface area information were also identified as normal/TOF with 100% correctness when using more than two PCA modes in classification.

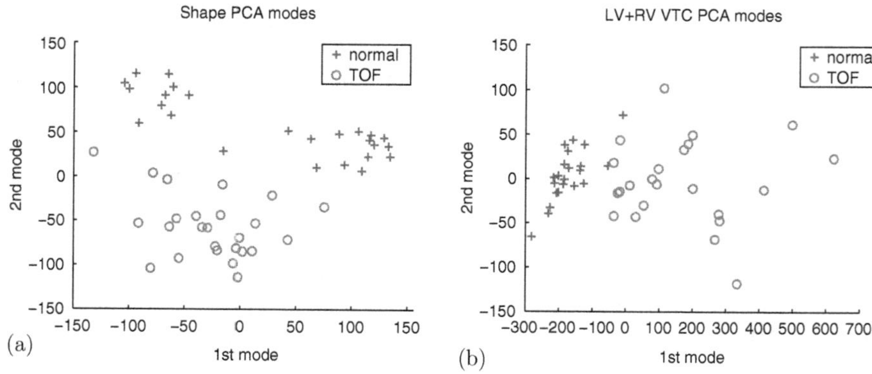

Fig. 2.16 The two strongest PCA modes of (**a**) shape model of LV+RV, and (**b**) volume–time curves of LV+RV [70]

Since the TOF hearts have higher variability of shape and motion compared to normal hearts, as shown in Fig. 2.16, using these novel indices to identify TOF disease progression is indeed feasible.

The novel ventricular function indices of PCA shape modes and VTC contain additional temporal information about the cardiac cycle and better describe the ventricular motion than the conventional indices derived only from end-diastole and end-systole such as the ejection fraction. Their strength was shown by achieving 100% normal/TOF identification. These new indices have the potential to be used for staging of TOF disease progression by their association with the higher variability of ventricular shape and motion in TOF subjects compared to the normal hearts.

2.4 Vascular Applications

2.4.1 Connective Tissue Disorder in the Aorta

Aortic aneurysms and dissections are the 18th leading cause of death in the USA, representing 0.6% of all deaths in 2006 [71]. Persons with certain congenital connective tissue disorders, such as Marfan syndrome and Familial Thoracic Aortic Aneurysm syndrome, are at increased risk of developing aortic aneurysm and/or dissection. Therefore, early diagnosis of connective tissue disorders is of great importance.

Many aortic segmentation techniques were developed in 3D using computed tomography (CT) or MR images. Rueckert [72] used geometric deformable models (GDM) to track the ascending and descending aorta. Behrens [73] obtained a coarse segmentation using randomized Hough transform (RHT). Bruijne [74] introduced an adapting active shape model (AASM) for tubular structure segmentation. Subasic [75] utilized a level-set algorithm for segmentation of abdominal aortic aneurysms. Considerable work has been done to establish the abnormalities in vessel mechanics that exist in patients with connective tissue disorders [76–78]. Included among the parameters that have proven useful in assessing the risk of aortic disease progression in Marfan patients is the determination of aortic pulse wave velocity [79]. In a closely related research direction, several authors have proposed techniques for tracking the cardiac movement in 4D cardiac images. Bardinet [80] presented an algorithm for tracking surfaces in 4D cardiac images based on a parametric model. Chandrashekara [81] built a statistical model derived from the motion fields in the hearts of several healthy volunteers to track the movement of the myocardium.

During the last decade, cardiovascular magnetic resonance (MR) imaging contributed substantially to cardiovascular disease diagnosis by providing gated 4D images. Usually two complementary sets of images are acquired (Fig. 2.17): the candy cane (CC) view optimizes the acquisition slices to obtain a good overall view of the entire aorta, whereas the left-ventricular outflow tract (LVOT) view provides the details at the aortic root, relevant for the assessment of connective tissue

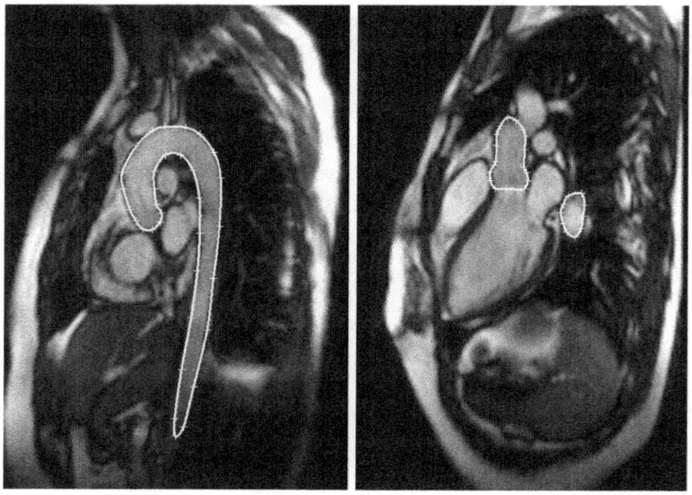

Fig. 2.17 MR image data with manual tracing of the aorta [38, ©2009 Elsevier]

disorders like Marfan syndrome. The method discussed here [38] consists of two main stages: aortic segmentation and connective tissue disorder assessment. Surface segmentation of the aortic lumen from the MR images is obtained with an automatic 4D segmentation method. Next, a quantitative method to detect the differences in the aortic 4D function between normal and connective tissue disorder patients is employed to provide quantitative descriptors used in a disease classification step.

2.4.1.1 4D Segmentation of Aortic MR Image Data

For the presegmentation, a 4D fast marching level-set method simultaneously yields approximate 4D aortic surfaces. From the resulting volumetric representation, the aortic centerlines are determined from each approximate surface by skeletonization. The centerline is used to resample the presegmented lumen into slices perpendicular to the aorta. Finally, the accurate aortic surface is obtained using 4D optimal graph search as discussed in Section 2.2.3

The 4D fast marching method of Malladi and Sethian was used to obtain the preliminary surface [82]. Starting with an interactively identified seed point within the 4D aortic image, the initial surface propagates in an outward direction. To minimize the tendency of the level-set segmentation to leak into the extra-aortic space in the presence of imperfect border properties, a 3D tubular structure enhancement filter is applied to the original image [83]. The filtering step enhances the appearance of tubular structures and suppresses the appearance of non-tubular structures. The fast marching algorithm stops the surface in the vicinity of object boundaries, thus yielding the approximate object surface. The tubular structure enhancement filter limits the level-set segmentation leaks to local areas and the resulting inaccuracy of the preliminary segmentation is resolved in the subsequent surface detection step. A 3D thinning algorithm [84] is applied to the result of presegmentation to extract an approximate aortic centerline for each phase. During each iteration, this method first marks all surface border voxels that are simple points and not line-end points as potential deletable points, then rechecks those points and deletes the point which is still simple and not line-end point after the deletion of some previously visited marked points. This procedure is performed recursively until a one voxel-wide centerline is produced. A minimum length threshold is also set to prune the unwanted small branches.

In order to construct the 4D aortic surface detection graph, a coordinate transformation is applied to each single-phase image and by resampling the image using cubic B-spline interpolation in the plane perpendicular to the centerline [85]. The aorta is straightened by stacking the resampled cross sections to form a new volume. Each cross section in the resampled volume is unfolded into polar coordinates to transfer the cylindrical surface into a terrain-like surface (Fig. 2.18). After coordinate transformation, the desired 3D surfaces in an n-phase 4D aortic image are determined using our multi-surface segmentation algorithm [41]. Instead of resorting to sequential segmentations of time series of 3D images, our segmentation method obtains a globally optimal 4D surface, where all 3D surfaces are segmented simultaneously and the temporal interrelationship of the 3D surfaces is considered in the global optimization.

The cost function designed for this application takes advantage of the edge information distinguishing the aortic surface from the aortic lumen and background [22]. Additionally, aortic wall is visible in portions of the aorta. To take advantage of this double-surface appearance, a combination of the edges of opposite orientations is used in parts of the aortic surface [86, 87]. More specifically, the aortic wall appearance is different for the ascending and the descending aorta portions on MR images. Therefore, the aorta was divided into two parts—the first 20% of the length between the aortic valve and the diaphragm was labeled as the ascending aorta and the remaining part included the aortic arch and the descending aorta. To deal with the different image appearance of the aortic wall in these two aortic parts, two different cost functions were developed as outlined below. First, a 3D edge image of the aorta is computed from the straightened image data. To detect the accurate surface,

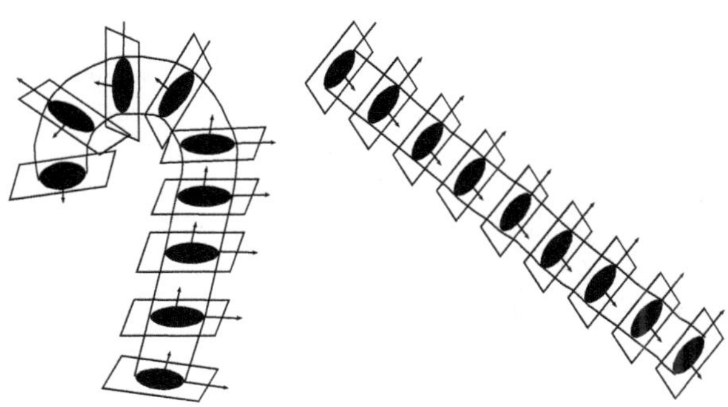

Fig. 2.18 The unfolding process, transforming aorta into a straight cylinder [38, ©2009 Elsevier]

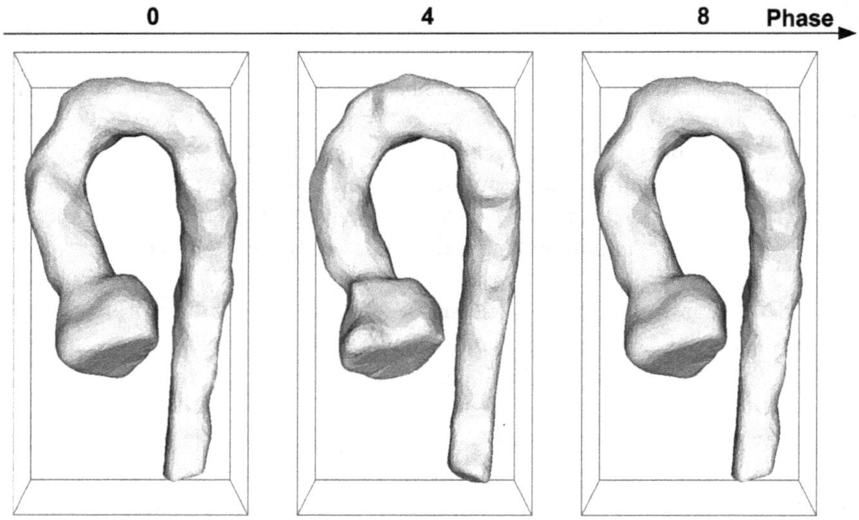

Fig. 2.19 Volumetric representation of computer-based segmentation of a diseased aorta [38, ©2009 Elsevier]

a combination of first and second derivatives (3×3 Sobel edge detector and 5×5 Marr–Hildreth edge detector) was employed in the individual cross sections [88].

2.4.1.2 Disease Detection

The disease assessment method is directly based on the analysis of the 4D segmentation result (Fig. 2.19). First, a point distribution model (PDM) is built representing the aortic shape and its motion during the cardiac cycle. Then, the modal indices of the PDM are used as input to a support vector machine (SVM) classifier. This is similar to the process described in Section 2.3 on assessment of Tetralogy of Fallot. The landmarks on the template shape were generated phase by phase, then mapped back onto the original volumes by using the inverse affine transform T^{-1}_{affine} of the initial mapping, followed by a B-spline elastic transform [89] to propagate the landmarks onto the individual shapes. Since the point correspondence of landmarks generated on different sample instances is guaranteed, each resulting shape sample is represented by a shape vector $s_t = (x_1, y_1, z_1, ..., x_m, y_m, z_m)$, consisting of m pairs of (x, y, z) coordinates of the landmark points. With the shape vectors $\{s_t\}_{t \in [0,n-1]}$ for every single-phase 3D image, the final shape vector for the 4D image can be obtained by concatenating all shape vectors $\{s_t\}_{t \in [0,n-1]}$ into a single vector $Y = (s_0, s_1, ..., s_{n-1})$. These shape vectors are suitable for principal component analysis, yielding a combination of modal indices for each shape as described in Section 2.2.1.

With each 4D aortic instance represented by the modal indices **b** describing the observed shape and motion variations, an efficient classification algorithm was developed for pattern recognition using a support vector machine classifier (SVM)

[90–93]. The classifier is used to classify 4D aortic instances into classes of normal and connective tissue disorder subjects. Given M input training samples $d \in \Re^n$ with class labels $g \in \{-1, 1\}$, the SVM maps a sample d into a high-dimensional space using a kernel function $K(d, d_i)$ and constructs an optimal hyperplane separating the two classes in this space. The optimal hyperplane is identified so that it maximizes its distance from the training samples (maximal margin in the high-dimensional space).

2.4.1.3 Accuracy of Segmentation and Classification

From 104 image datasets available, a subset of 21 datasets was randomly selected and used for validating the segmentation performance. Limiting the validation set to 21 datasets was motivated by the extraordinary labor intensity of manual tracing in 4D. Once validated, the aortic segmentation was performed on the entire set of 104 subjects for functional studies as described below. Therefore, 21 subjects (7 patients and 14 normal subjects) were used to assess the performance of the automated 4D segmentation. Aortic luminal surfaces were compared to the expert-traced independent standard. The independent standard was defined by manual tracing of 4 randomly selected slices in each of 5 randomly selected phases in the 21 subjects (total of 420 manually traced slices). Surface positioning errors were defined as the shortest distances between the manually traced contours and computer-determined surfaces in the 4D aortic images. The diagnostic performance of the modal indices of aortic shape and motion was assessed for (a) a single end-diastolic 3D segmentation, (b) 2-phase end-diastolic and end-systolic 3D segmentation, and (c) 16-phase 4D segmentation.

All 4D aortic MR images used for the segmentation performance assessment were successfully segmented by our 4D segmentation algorithm. Comparison of the computer-determined and expert-traced surfaces showed a good agreement. The segmentation produced aortic surfaces with subvoxel accuracy and a very minimal bias as judged by the signed surface positioning errors of -0.07 ± 1.16 voxel (-0.10 ± 2.05 mm) and unsigned positioning errors of 0.88 ± 0.81 voxel (1.55 ± 1.44 mm). The classification performance was assessed in 52 datasets after the method's parameters were independently optimized in a separate design dataset (Table 2.1). For the single-phase case, 300 landmarks were automatically generated on each aortic luminal surface. The classifier based on eight most significant principal components ($\kappa = 4$, $T = 4$) exhibited an overall classification correctness of 82.7%. For the two-phase case (end-systole and end-diastole), the two cardiac phases were considered a single shape/motion instance and 600 landmarks were generated. The classifier based on eight most significant principal components ($\kappa = 1$, $T = 4$) exhibited correctness of 88.5%. For the 16-phase 4D case, the 16 cardiac phases were considered a single shape/motion instance and 4800 landmarks were generated. The first 10 principal components were selected as SVM input features. The classifier ($\kappa = 0.25$, $T = 32$) exhibited correctness of 90.4%.

Table 2.1 Classification correctness of the 1/2/16-phase models [38, ©2009 Elsevier]

	Predicted, 1 phase		Predicted, 2 phases		Predicted, 16 phases	
Disease status	Diseased	Normal	Diseased	Normal	Diseased	Normal
Diseased	25	1	23	3	22	4
Normal	7	19	3	23	1	25

The achieved results suggest that the functional (shape and motion) information is an important contributor to the ability to distinguish between the normal and connective tissue disorder subjects. Clearly, the functional information (shape changes, motion) missing in the 1 phase 3D model is partly present in the 2-phase model and more completely in the 16-phase model. The additional information is reflected in the corresponding improvement of classification performance.

After analyzing the data, it is clear that the 16-phase method is very specific with respect to correctly classifying normal controls. This prompts the question of why 3 of 26 suspected CTD subjects were not correctly classified by the 2-phase approach and why 4 of 26 were not correctly identified using the 16-phase approach. The most likely explanation of this behavior comes from the subject selection. All subjects included in the patient group are established or suspected future CTD patients. Some of our study subjects were referred for an MRI because of a family history of aortic dissection. Due to the believed genetic basis of the connective tissue diseases that cause aortic dissection, first-degree relatives of individuals who have had an aortic dissection underwent MRI screening of their aortic dimensions. The "patients" who were inaccurately classified fall into this category due to family history, where follow-up clinical evaluations have (so far) not demonstrated any evidence of connective tissue disease in either of these cases.

2.4.2 Aortic Thrombus and Aneurysm Analysis

An abdominal aortic aneurysm (AAA) is an abnormal expansion of the abdominal aorta resulting from a weakening of the vessel wall. AAA is frequently observed in males over 60 and develops very slowly over years with no symptoms. Surgery is required if an AAA is bigger than 5.5 cm in diameter, which is considered susceptible to rupture. Of a variety of medical modalities, computed tomography angiography (CTA) has been widely used to diagnose AAA since it provides detailed anatomical information about the aortic lumen, thrombus, and calcifications [94] (Fig. 2.20).

AAA thrombus segmentation in CTA images and quantitative analysis are of paramount importance for diagnosis, risk assessment, and determination of treatment options. The lumen segmentation is simplified in CTA images since the contrast media enhances the difference in CT intensity between the lumen and other

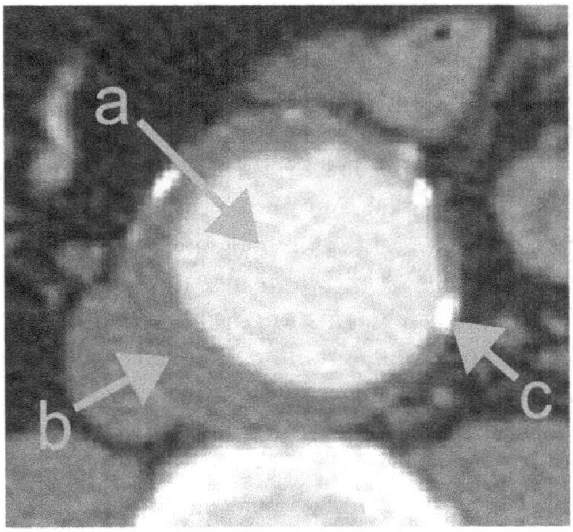

Fig. 2.20 Example of an abdominal CTA image, cropped to show a section of the aorta, (**a**) lumen, (**b**) thrombus, and (**c**) calcification [46]

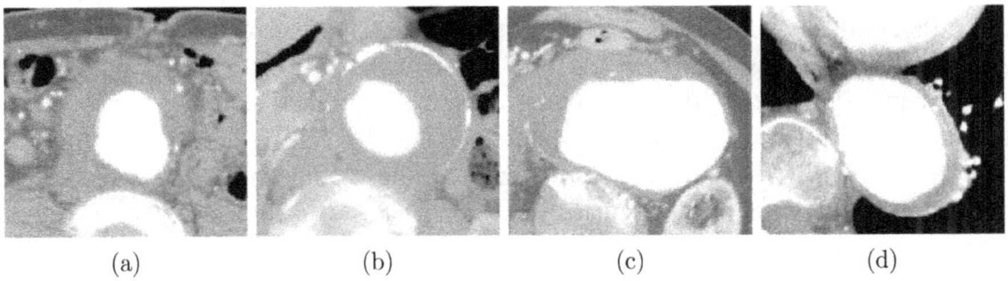

Fig. 2.21 CTA image variability makes thrombus segmentation challenging. (**a**) Outer thrombus surface not visible. (**b**) Similar CT intensities between the thrombus and other tissues. (**c**) Irregular thrombus geometry. (**d**) Strong intensity gradients near the spine, the heart, and the lungs [46]

structures. On the other hand, thrombus segmentation in CTA images is challenging due to the thrombus outer surface being locally obscured in some cases, the CT intensity of the thrombus being similar to that of other tissues surrounding it, an irregular geometric structure of the thrombus, or the proximity of other structures with pronounced surface intensity gradients (Fig. 2.21).

Various approaches have been proposed to automatically segment an AAA thrombus. Loncaric et al. [95] introduced a 3D deformable model (DM)-based approach utilizing a 3D level-set algorithm to detect the aortic lumen, but they did not describe a method for thrombus segmentation, which is more challenging. Tek et al. [96] proposed a mean shift-based ray propagation method combined with a smoothness constraint. The primary advantage of this method is its fast processing speed. The ray propagation is much faster than active contour update or level-set evolution. However, their study focused on the segmentation of the 2D luminal boundary and did not deal with a double-surface structure. Subasic et al. [97] proposed another 3D deformable model-based technique using a level-set algorithm.

While minor user intervention was required, such as providing the sphere center and radius to define an initial surface for the 3D level-set algorithm, they did not mention the quantitative thrombus segmentation performance, and the reported thrombus segmentation results, which are overlapped on the original data, were not accurate locally. De Bruijne et al. [98] introduced an active shape model (ASM) segmentation scheme, in which a statistical shape model composed of landmark points was fitted to the image iteratively. Although the reported thrombus segmentation results were promising, this approach required substantial user interaction. The manual segmentation of the first slice was required for the initialization of the ASM, and if contours in subsequent slices were not reasonable, they needed to be modified by the user. Olabarriaga et al. [94] proposed a 3D discrete deformable model-based method for thrombus segmentation and adopted a nonparametric statistical gray level appearance model built from training data with a supervised pattern classification technique. In order to train a k-NN classifier, however, samples inside, outside, and at the given boundary positions needed to be collected manually.

The method by Lee et al. [46] discussed here is based on the 3D/4D optimal graph search surface detection method of Li et al. [41, 45], which is capable of detecting multiple interacting surfaces simultaneously. The original approach is appropriate for the segmentation of a tubular object once the data are unfolded based on the centerline to construct a graph. However, it is limited to being employed to the structure which has multiple branches, such as iliac bifurcation, since thrombi sometimes exist in the iliac arteries. Therefore, the method has been extended to construct a graph directly from a triangulated surface, as discussed in Section 2.2.3. The preliminary surface is determined by a region growing algorithm as introduced in Section 2.2.2.

2.4.2.1 Initial Luminal Surface Segmentation

In order to construct a triangle mesh-based graph, we need a smooth luminal surface, which is used to construct normal graph columns. The method for initial luminal surface segmentation is composed of five steps: anisotropic diffusion, region growing, region smoothing, marching cube surface definition, and vertex smoothing. Anisotropic diffusion [49, 99] is applied to original image data to reduce the noise in the lumen and thrombus regions while preserving the boundary. In this step, we need two mouse clicks to identify the lumen and thrombus regions. The conductance parameter (k) of the anisotropic diffusion is calculated as the difference between I_{lumen} and I_{thrombus}, which are the average CT intensities at the lumen and thrombus regions, respectively. The number of iterations is 5, and the time step is 1/7. Then, a rough lumen region is identified by employing region growing with the following stopping criteria:

$$\text{Condition 1:} \qquad max|\nabla I| < (I_{\text{lumen}} - I_{\text{thrombus}}) \times 80\% \tag{2.7}$$

$$\text{Condition 2:} \qquad I > I_{\text{thrombus}} + (I_{\text{lumen}} - I_{\text{thrombus}}) \times 60\% \tag{2.8}$$

where I is the original volumetric data, and $\max|\nabla I|$ is the maximum gradient magnitude in 27 directions at each voxel of I. Condition 1 accounts for 20% uncertainty of $(I_{\text{lumen}} - I_{\text{thrombus}})$, and the CT intensities on real lumen regions should be larger than $I_{\text{thrombus}} + (I_{\text{lumen}} - I_{\text{thrombus}}) \times 60\%$ to prevent leakage in the regions having low gradient responses (Condition 2). In order to get a smooth luminal surface, the regions obtained from the region growing are smoothed by applying a 5×5×5 averaging filter mask three times. The initial luminal surface is built by using the marching cube algorithm. Finally, the luminal surface is smoothed by averaging adjacent vertices since graph columns approximately perpendicular to the rough luminal surface may cause the failure of the graph construction due to column intersections. In this study, 30 iterations of vertex smoothing are performed. Figure 2.22 shows the initial luminal surface segmentation results on two datasets.

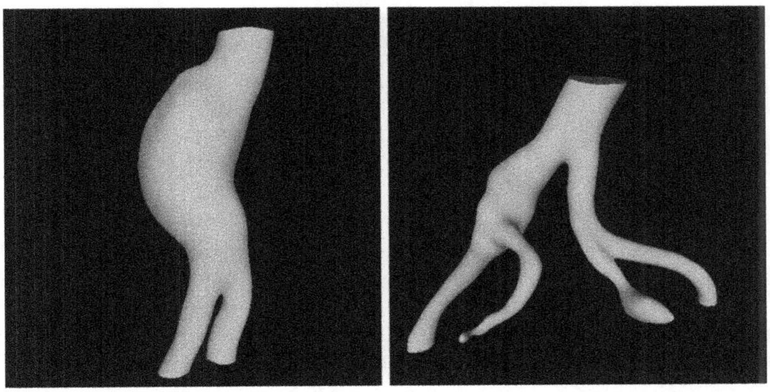

Fig. 2.22 Segmented initial luminal surface for graph search [46]

2.4.2.2 Graph Search and Cost Function Design

In order to create the graph, orthogonal columns have to be created with respect to the preliminary surface determined by the previous steps. Thus, columns are resampled using tri-linear interpolation along the averaged normal direction of surrounding triangles on the initial luminal surface obtained by the marching cube algorithm. The adjacencies of columns are already known from triangle meshes of the initial luminal surface. The segmentation of two coupled surfaces is formulated as computing a minimum closed set in a 3D geometric graph constructed from two cost functions. In this study, the smoothness constraint is 1, the minimum surface separation constraint is 2, and the maximum surface separation constraint is large enough to cover the thrombus outer surface.

As an input to the graph search, an edge-based cost function for both surfaces of the thrombus is used. For the *inner* surface, the original volumetric data are smoothed by averaging 3×3×3 neighbor voxels to reduce noise, and the maximum gradient magnitude in 27 directions at each voxel is calculated. Then, these gradient magnitudes are smoothed again to improve the connectivity and smoothness of the thrombus inner surface edge representation. The gradient magnitudes are inverted to determine the minimum costs. At this stage, it can be assumed that the real luminal surface is very close to the initial luminal surface, so we need a mask to ignore the regions located more than 3 voxels from the initial luminal surface. Anisotropic diffusion, region growing, and region smoothing come from the initial lumen segmentation in Section 2.4.2.1. To apply binary dilation, smoothed regions are thresholded by the maximum gradient magnitude divided by 5, which was obtained experimentally. The mask was created by applying two iterations of binary dilation to the thresholded initial lumen region. Finally, we can get the cost function for the real thrombus inner surface by combining the inverted mask with the inverted gradient magnitudes (Fig. 2.23).

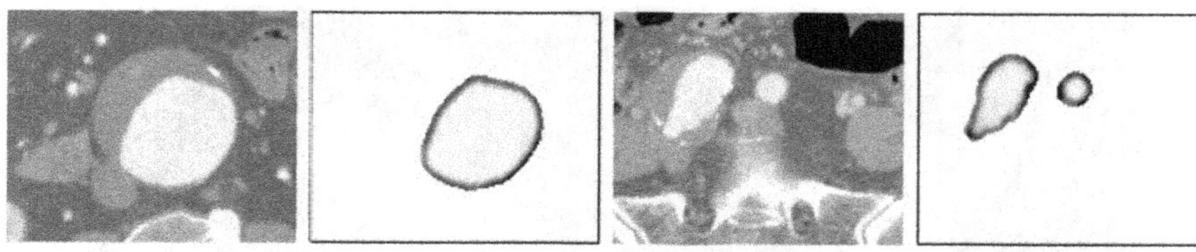

Fig. 2.23 Original image and cost function for the thrombus inner surface [46]

For the thrombus *outer surface*, a more complex cost function was designed. It is difficult to detect the accurate thrombus outer surface by using only gradient magnitudes since locations around the thrombus, spine, heart, and lungs have very strong edge responses. In order to solve this problem, we used local histogram equalization, which can reduce such strong edge responses, as well as locally enhance the weak edge responses of the thrombus outer surface. For the local histogram equalization, a $5\times5\times5$ window is used. Another important component utilizes a distance transform. Since the thrombus outer surface tends to exist relatively close to the thrombus inner surface, more weight is put on the regions close to the thrombus inner surface. Figure 2.24 shows the two original datasets, distance maps, and cost functions for thrombus outer surfaces. Because of the local lack of the thrombus outer surface evidence, sometimes a small number of control points are required to modify the cost function, thus guiding the outer wall surface detection.

2.4.2.3 Data and Results

The reported thrombus segmentation method was tested in nine MDCT image datasets from nine patients imaged using both 16- and 64-channel systems (Brillance CT, Philips Healthcare, Cleveland, OH, USA). Out of these, eight datasets exhibited AAA, and one dataset depicted an aneurysm in the iliac artery. The size of each dataset was between $512\times512\times90$ and $512\times512\times135$ voxels. The voxel

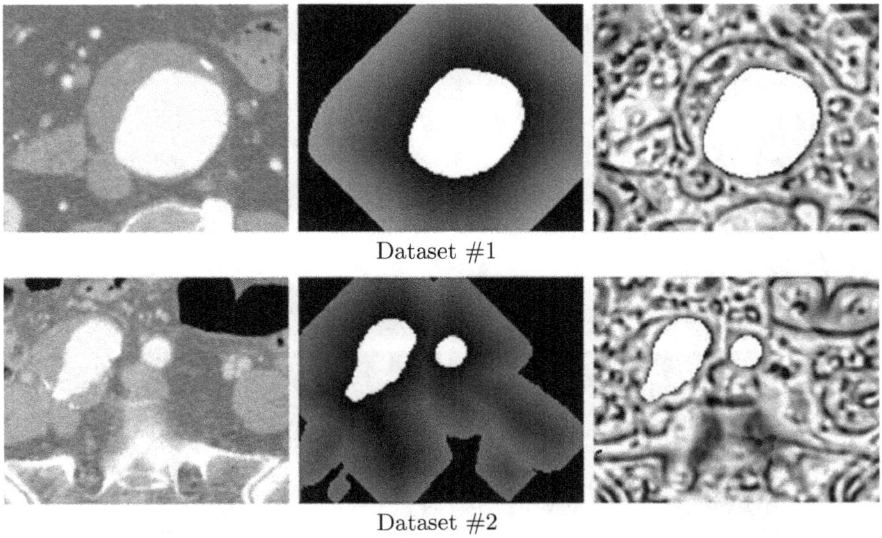

Fig. 2.24 Original image, distance map, and cost function for the thrombus outer surface [46]

spacing ranged from 0.52×0.52×1.0 to 0.78×0.78×1.0 mm. The total number of image slices to be segmented was 951 across the nine subjects.

Figure 2.25 shows the results of thrombus segmentation in two example datasets. The red line represents the segmented thrombus inner surface, and the green line shows the segmented thrombus outer surface. The 3D thrombus model is reconstructed using the complete volumetric segmentation. Local thrombus thickness is color-coded on the inner thrombus surface. The quantitative assessment of our method revealed visually acceptable aortic lumen and thrombus segmentation in eight out of nine datasets. The ninth dataset, in which thrombus segmentation failed, is atypical in the sense that the thrombus is highly eccentric. Our segmentation method does not cover this case since the graph columns cannot include a thrombus region that is located so far from the initial luminal surface. In three out of eight datasets, our method performed fully automatically with no user interaction. In the remaining five datasets, in which the automated method locally failed due to image ambiguity, 7.80 ± 2.71 mouse clicks per case (0.083 ± 0.035 mouse clicks per image slice) were needed to obtain complete 3D segmentations with no local surface failures.

In summary, this approach utilizes the power and flexibility of the optimal triangle mesh-based 3D graph search method and provides a highly automated tool for the segmentation of the aortic inner and outer surfaces, including the iliac bifurcation. For dealing with local failures caused by image ambiguity, a simple interaction was implemented to guide the computer segmentation without the need to trace borders manually. The detected thrombus characteristics are quantitatively described by reporting local thrombus thickness and its overall volume.

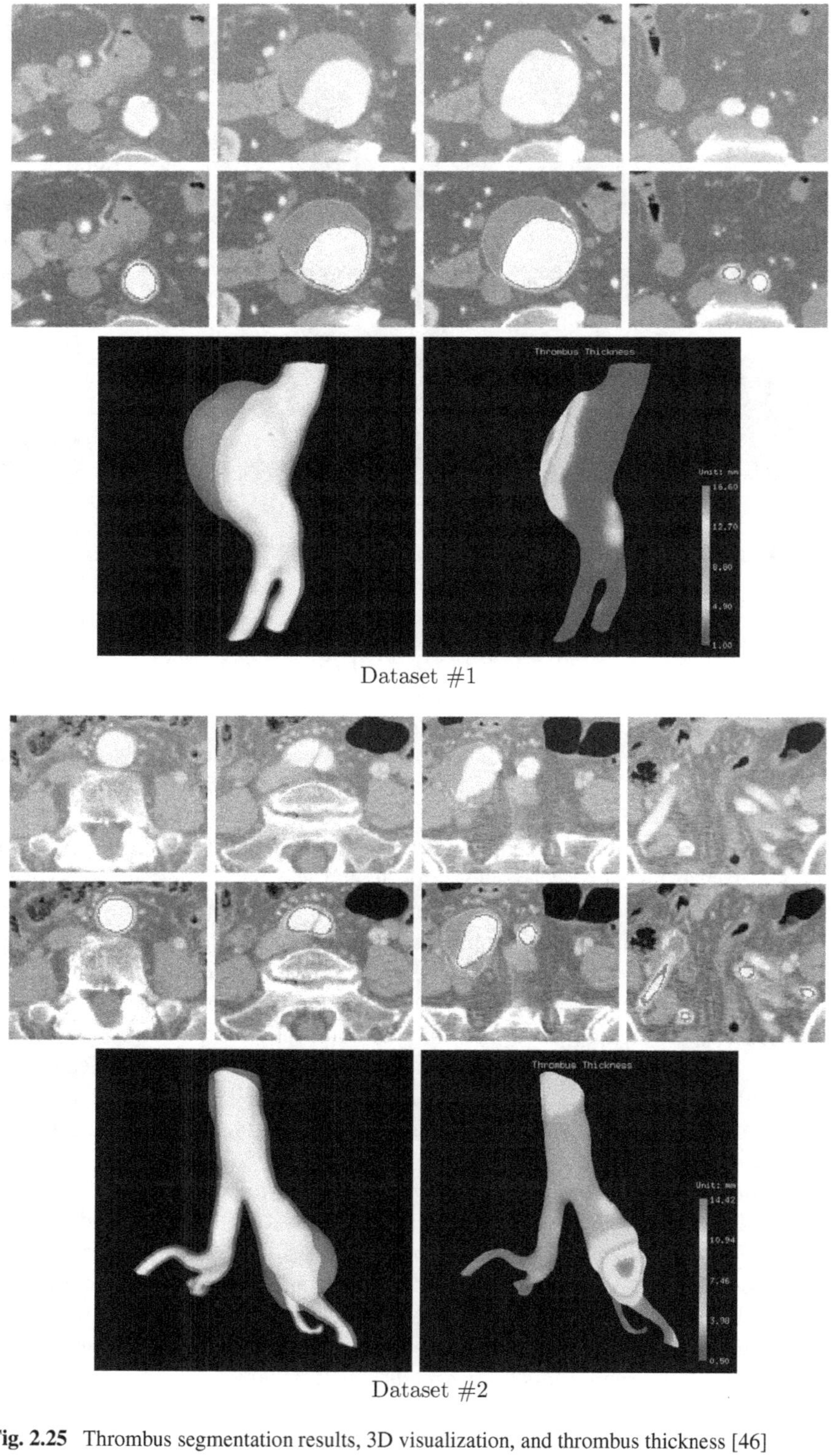

Fig. 2.25 Thrombus segmentation results, 3D visualization, and thrombus thickness [46]

2.4.3 Plaque Distribution in Coronary Arteries

The coronary arteries are a system of vessels supplying the heart itself with oxygenated blood, and they originate from the aorta. The left coronary artery splits into the *left anterior descending* (LAD) and *left circumflex* (LCX) arteries. On the opposite side of the aorta, the *right coronary artery* (RCA) supplies the right side of the myocardium [100]. Coronary atherosclerosis starts early in life and is a major cause of death in industrialized countries. Therefore, the study of plaque development and progression is of high interest. The relationships among vascular geometry, hemodynamics, and plaque development in the coronary arteries are complex and not yet entirely understood.

An early stage of plaque development is intimal thickening [101]. However, the lumen size is preserved during the early atherosclerotic process due to *compensatory enlargement*. Figure 2.26a–c illustrates this process. Glagov et al. determined that luminal narrowing (stenosis) generally occurs after the plaque area exceeds about 40% of the cross-sectional vessel area [102]. Obstructive stenoses are most frequently treated by percutaneous transluminal coronary angioplasty (PTCA) and subsequent stenting to restore the lumen. Previous studies have linked plaque development with low wall shear stress [103], which in turn depends on the vessel geometry. Friedman et al. [104] have reported that the intima was generally thicker at sites exposed to lower wall shear stresses in a coronary artery branch. Ideally, the wall shear stress distribution at the onset of atherosclerosis should be analyzed and then related with subsequent plaque progression. In vessel segments with initially low wall shear stress one would expect larger plaque accumulation; however, there are only a limited number of longitudinal studies that have compared plaque distribution in relation to local wall shear stress distribution at multiple points in

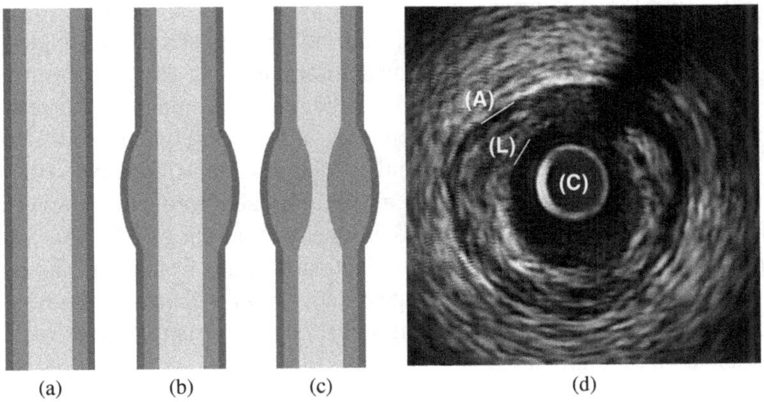

Fig. 2.26 Development of atherosclerotic plaque: lumen—*light gray*, plaque/intima—*mid gray*, media—*dark gray*: (**a**) vessel without any stenosis; (**b**) compensatory enlargement; (**c**) luminal narrowing; (**d**) IVUS image with C = catheter, L = lumen/plaque, and A = media/adventitia borders

time. While Wentzel et al. [106] used a relatively short interval of 6 weeks to follow-up, Stone et al. [107] compared distributions in wall shear stress and plaque development over a period of 6 months.

Selective x-ray angiography has been the method of choice for diagnostic and interventional cardiology for decades, but using angiography alone allows only a limited analysis of the vessel lumen by quantitative coronary angiography [108, 109]. Similarly, 3D analyses from biplane angiography can only offer elliptical approximations of the lumen shape [34]. Substantially more detailed cross-sectional information can be obtained by intravascular ultrasound (IVUS), an established complement to x-ray angiography [110]. An example of an IVUS frame is shown in Fig. 2.26d. Most importantly, IVUS allows the accurate assessment of plaque distribution and volume, as well as the effect of interventional treatment of coronary stenosis on plaque and vessel morphology [109]. In contrast to x-ray angiography, compensatory enlargement is directly visible in IVUS. The most challenging task is the determination of the lumen/plaque and media/adventitia borders in IVUS while limiting the need for user interaction at the same time. Previous methods [86, 111–114] use a variety of segmentation algorithms. The limitations of these approaches are mainly due to the lack of higher-dimensional context and the lack of knowledge-based segmentation criteria. The approach detailed here therefore utilizes 3D optimum graph search with IVUS-specific cost functions [37]. Recent developments in IVUS technology also allow plaque classification [115].

2.4.3.1 Segmentation and 3D Fusion

We have developed a comprehensive system that generates geometrically correct 3D and/or 4D (i.e., 3D plus time) reconstructions of coronary arteries and computes quantitative indices of coronary lumen and wall morphology. The reconstructions serve as input for hemodynamic analyses and allow for interactive visualization [106, 109, 116–118]. A flowchart outlining the system is given in Fig. 2.27. In general, vessel curvature and torsion are derived from biplane (or a pair of single-plane) x-ray angiograms, and the cross-sectional information is obtained from IVUS. Thus, the resulting model accurately reflects the spatial geometry of the vessel and includes any accumulated plaque. Fusion leads to the 3D/4D *plain model*, consisting of the lumen/plaque and media/adventitia contours oriented relative to the IVUS catheter. After tetrahedral meshing, this model is suitable for hemodynamic analyses. Morphologic analyses are performed following resampling of the cross sections orthogonal to the vessel centerline, to eliminate distortions from the position of the IVUS catheter within the vessel. The quantitative results *annotate* the resampled contour model, which is then used for visualization and further analyses.

Segmentation of angiographic images has a long history and is well established in quantitative coronary analysis. Examples of such software include Refs. [31–33, 119]. A methodological overview can be found in Ref. [109]. All of these methods aim for identifying the vessel lumen borders as visible in the angiograms. For the purpose of fusion between angiography and IVUS, the angiographic lumen borders are only utilized to establish a reference for the absolute orientation of the IVUS

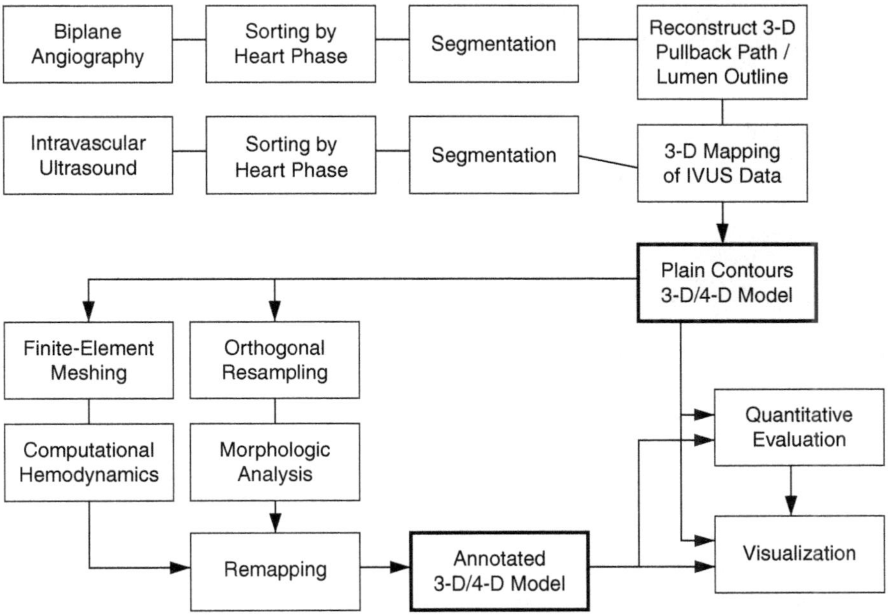

Fig. 2.27 Processing of the data from acquisition over segmentation to the generation of the *plain* model, followed by quantitative analyses used to *annotate* the 3D/4D model [37, 105 ©2005 Springer, ©2006 Elsevier]

data in 3D space [116]. A dynamic programming approach with edge-based cost functions was employed, with a third cost function modeling a gray-value ridge within the lumen to segment the IVUS catheter [37]. The user has to identify the location of the IVUS transducer, shown in its distal position since the angiograms were taken before the pullback started; the proximal endpoint, usually the ostium; and a few intermediate points to create a spline defining the region of interest.

A comprehensive review article on IVUS segmentation and quantitative evaluation has been presented by Klingensmith [120]. Approaches to identify the lumen/plaque and media/adventitia borders include those based on graph search [86] and simulated annealing [111]. Another frequently met approach is the use of active contours *(snakes)*, which iteratively determine the borders based on energy functions [113, 121]. Also, texture information can be taken into account. In this case, interfaces between regions are searched for, rather than explicit edges [122, 123]. While many approaches, e.g., the segmentation method of Li et al. [112], perform an initial longitudinal segmentation over the entire pullback to establish the 3D context, followed by cross-sectional contour detection, the algorithm discussed here operates in a true 3D manner. The IVUS data are acquired in a cylindrical volume, thus the determination of an explicit centerline as with the aortic data in Section 2.4.1 is *not* required. Instead, the images can directly be unfolded to a terrain-like structure suitable for the graph (Fig. 2.28). The graph search is employed in a multiresolution approach for maximum efficiency and robustness to noise or image artifacts. At

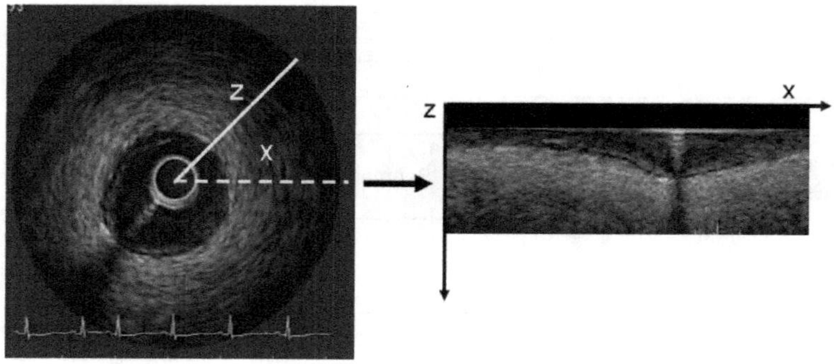

Fig. 2.28 Unfolding an IVUS image frame using a polar transform so that the detection of the cylindrical surfaces of varying radius is transformed to the detection of an elevation map of varying height [37, 39, ©2005 SPIE, ©2006 Elsevier]

each resolution stage within the three-tiered sequence, starting at a low resolution and continuing to a high resolution, the searched area of the graph is reduced so that more efficient use of computation time is made.

The cost function contains three classes of information at both global and local levels: expected border properties, information theoretic criteria based on the statistical properties of ultrasound, and regional homogeneity properties. It has been observed that although noise, speckle, and artifacts tended to confuse previous automated methods, human tracers are still able to correctly segment the vast majority of IVUS images. Thus, learning from experience based on expert-traced samples is a frequently used component in cost functions. In order to perform the learning step, a window is passed over the training images. Within the window, the pattern of pixels is examined. For each pattern that contains a border pixel, the system increments an accumulator entry corresponding to that border pattern, as shown in Fig. 2.29 for a 2D example. An accumulator entry exists for every possible pattern. After this training stage is complete, the learned information can be used to score an image that is to be analyzed. In the same way that the accumulator was

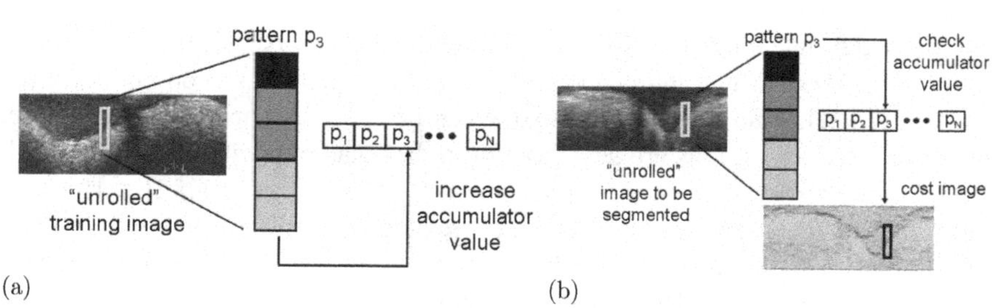

Fig. 2.29 During the learning stage (**a**), accumulator entries corresponding to border patterns are increased with every incidence of that border pattern in the training set; during segmentation (**b**), the accumulator values are used to assign the likelihood of each pattern in the new image being a border pattern [37, 39, ©2005 SPIE, ©2006 Elsevier]

created in the training step, patterns are examined in the image. Then, a cost image is generated by assigning a likelihood value to each pixel based on the value found in the accumulator for the same pattern. It has been known for some time that the developed speckle in ultrasonic images is characterized by a Rayleigh distribution [50, 124]. Recently, this knowledge has been used in the accurate segmentation of the luminal boundary in IVUS images by deformable contours that separate two regions (lumen and vessel wall) based on their global statistical properties [114]. This concept is similarly adapted to our optimal 3D graph approach; the objective function for the contour estimation was approximated to produce a voxel cost at each polar-transformed image location [37]. Degradations in image quality due to severe vascular disease or artifacts can alter the development of a true Rayleigh distribution in the image speckle. Therefore, we chose to add a version of the Chan–Vese minimum variance criterion [52] as a third component of our cost function. This term allows segmentation without the presence of gradients and without the assumption of a particular statistical model. All three components are weighted independently for the lumen/plaque and media/adventitia borders to obtain maximum segmentation efficacy.

After the angiographic and IVUS data are segmented, *fusion* can be performed to yield a 3D model (or a set of 3D models in the 4D case) for the artery. Some fusion systems require constant angiographic supervision of the catheter during pullback to reconstruct the respective position of the IVUS frame from the projections of the transducer [125–127]. Instead, just a single pair of angiograms, depicting the IVUS catheter in its most distal location, suffices to reconstruct the pullback path in 3D and then follow it during the pullback [128, 129]. The catheter path is extracted from the biplane angiograms, reconstructed into 3D space, and then used to map the IVUS images to their locations [116–118]. The fusion problem is twofold: first, the *localization* of the individual IVUS frames in 3D; second, the estimation of the spatial *orientation* of each frame. Figure 2.30 shows a 3D rendering of the fused model in the *Virtual Reality Markup Language* (VRML).

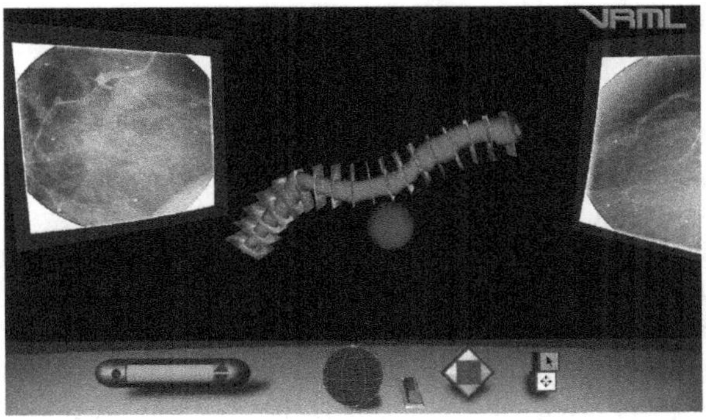

Fig. 2.30 Result of the fusion process as 3D VRML rendering [116, ©1999 IEEE]: Angiograms and some IVUS frames at their geometrically correct locations, lumen surface visualized in red

2.4.3.2 Hemodynamic and Morphologic Analysis

The blood flow through the coronary arteries is simulated and the wall shear stress distribution determined using *computational fluid dynamics* (CFD) methodology. Commercial software is available for meshing the enclosed lumen surface, which we have used to obtain tetrahedral elements (Fig. 2.31). Such unstructured grids provide better flexibility and accuracy than structured grids. The methodology employed in our research is detailed in Refs. [130, 131]. Simulations range from steady-flow analysis in the end-diastolic phase (performed for this specific study due to its limited computational effort) to unsteady flow in a moving grid (at the cost of increased computational expense). Blood is usually treated as an incompressible, homogeneous, and Newtonian fluid. The 3D model of the vessel is extended by 10× the inlet diameter at the ostium and 5× the outlet diameter at the distal end to avoid artifacts at the ends of the modeled segment. Since the flow rate could not be measured in each of the coronary arteries during data acquisition for each patient due to procedural limitations, we assumed a flow rate of 100 ml/min for all the coronary arterial segments employed in this analysis based on previously reported averages [132, 133].

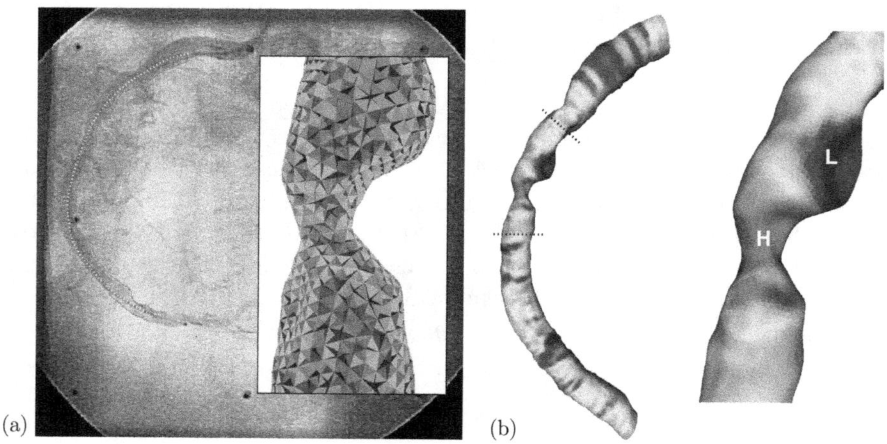

Fig. 2.31 Computational hemodynamics: (**a**) meshing of the 3D surface to obtain a tetrahedral finite-element model [37, ©2006 Elsevier]; (**b**) color-coded wall shear stress, with blue indicating low and red indicating high local wall shear stress

To verify the observation that plaque tends to accumulate at the inner bend of the curvature rather than at the outer bend of the curvature, the relative portion r_{PC} of regions where inner curvature coincides with above-average plaque accumulation, or outer curvature coincides with below-average plaque accumulation, is determined. The ratio r_{PC} represents a "plaque/curvature index" with a value $r_{PC}>0.5$ indicating that more plaque has accumulated circumferentially along the inner curvature as compared to the outer curvature, thus supporting the hypothesis. Four regions are defined, as depicted in Fig. 2.32e: R_{ai} (red), R_{ao} (magenta), R_{bi} (yellow), and R_{bo} (blue); the "neutral" region R_n of local below-threshold

(1)
(2)
(3)
(4)
(5)

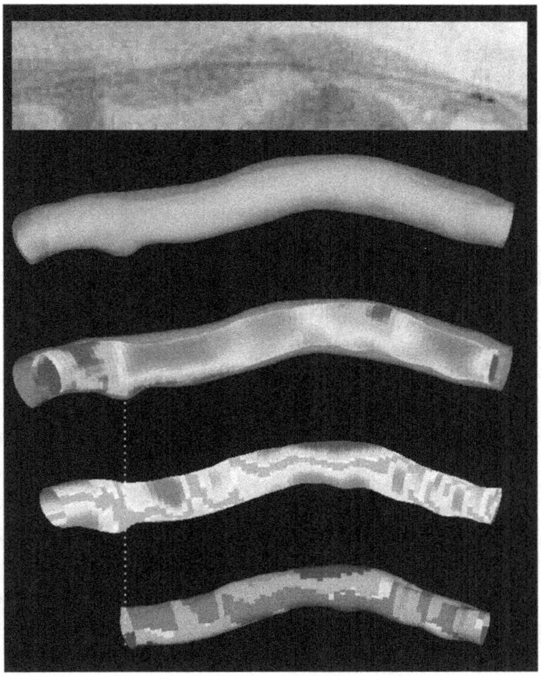

Fig. 2.32 Plaque thickness vs. curvature: (1) angiogram of a left anterior descending artery with the IVUS catheter; (2) 3D model with lumen and adventitia borders from fusion, where the volume between the red and green surfaces represents the vessel wall; (3) plaque-thickness annotation derived from the model shown in (2), where blue color indicates low and red color high wall thickness; (4) curvature-index annotation derived from (2), blue color marks "outer" and red "inner" curvature; (5) after classification into regions correlating the data from (3) and (4), with the branch segment removed from analysis as indicated by the red dotted line—see Section 2.4.3.2 for the definition of the regions

curvature is colored in green. These regions represent pairs distinguishing circumferentially "above-average" plaque thickness (*a*) from "below-average" plaque thickness (*b*), coinciding with either "inner curvature" (*i*) or "outer curvature" (*o*) of the vessel wall. Thus, the plaque/curvature index is defined as $r_{PC} = \|R_{ai} + R_{bo}\| \,/\, \|R_{ai} + R_{bo} + R_{ao} + R_{bi}\|$. A similar measure r_{PW} of elements for which circumferentially above-average plaque thickness coincides with below-average wall shear stress (and vice versa) is determined for each vessel segment. Replacing "inner curvature" (*i*) with "lower-than-average wall shear stress" (*l*) and "outer curvature" (*o*) with "higher-than-average wall shear stress" (*h*) yields the definition for the plaque/wall shear stress index.

2.4.3.3 Studies and Results

The segmentation system was tested on a set of 21 in vivo IVUS pullbacks containing a mixture of each of the three major coronary arteries. In a leave-one-out cross-correlation scheme, segmented borders were compared with independent, expert-identified borders that had been manually traced as the ground truth. The differences between the detected borders and the manually traced borders were computed for 720 corresponding points (every half degree angularly) on each border, showing no bias in segmentation and a maximum positional error of 0.236 mm for the lumen/plaque and 0.300 mm for the media/adventitia borders [37].

The index r_{PC} indicating the relative portion of regions where circumferentially the inner curvature coincided with above-average plaque accumulation, or the outer curvature coincided with below-average plaque accumulation, was determined in a set of 60 vessels. Twelve different threshold values were empirically selected ranging from 2.31 to 22.94°/cm, resulting in 10.1–77.8% of circumferential locations being assigned to the neutral region R_n, as colored green in the example in Fig. 2.32e. While the average r_{PC} over all vessels exceeded the target value of 0.5 for all thresholds, it increases steadily with the increase of the curvature threshold. Thus, the more the regions of low curvature that were included in R_n and, therefore, increased the proportion of higher curvature regions that were included in the calculation of r_{PC}, the more the hypothesis was supported. However, individual vessels varied in their r_{PC} compliance, and the standard deviation increased with the threshold. As shown in Table 2.2, the hypothesis was more strongly supported in LAD arteries, whereas the results in the RCA and LCX arteries were less supportive.

Table 2.2 Results categorized by curvature threshold and vessel, with $r_{PC}>0.5$ was true for all (12), at least half (≥6), at least one (≥1), or none (=0) of the curvature thresholds [37, ©2006 Elsevier]

	Number of curvature thresholds for which $r_{PC} > 0.5$ is satisfied				
Vessel	12	≥6	≥1	=0	
LAD	17	3	1	2	*23*
RCA	10	4	1	8	*23*
LCX	9	1	0	4	*14*
	36	*8*	*2*	*14*	*60*

While disease progression and intervention impact the curvature/plaque relationship to some extent, but appears to retain the basic cause/effect relationship, a substantial impact has to be expected for the wall shear stress distribution. Those patterns are altered when the limits of positive remodeling are reached [134]. Thus, the vessel subsegments for which the area stenosis is between 10 and 40% are of particular interest (the compensatory-enlargement range identified by Glagov [102]). Consequently, we concentrated on whether and how significantly the correlation improves once vessel segments of certain properties are excluded from the analysis. In this way, indirect evidence of which local conditions favor the underlying hypothesis of below-average wall shear stress inducing above-average plaque thickness was sought. We grouped each vessel segment by disease severity, first excluding regions of branching, stenting, or strong calcifications. The remaining subsegments were categorized by their cross-sectional luminal narrowing being below or above the Glagov threshold of 40%. A comparison of the hypothesis compliance was made between all those subsegments regardless of their degree of narrowing and only those segments of the same vessel which were below the Glagov threshold [37].

A ratio g_{PW} determines the increase of r_{PW} for Glagov-range subsegments over the r_{PW} of all subsegments of that vessel. A $g_{PW}>1$ indicates better compliance of segments within the remodeling range with the hypothesis that low wall shear stress (cause) correlates with larger circumferential plaque accumulation (effect) compared to the overall vessel, which also includes subsegments of later disease stages. In other words, the inverse relationship between local wall shear stress and plaque thickness was significantly more pronounced ($p < 0.025$) in vessel cross sections exhibiting compensatory enlargement (positive remodeling) without luminal narrowing than when the full spectrum of disease severity was considered. There is a limit, though, for that observation: when more than 35% of subsegments in a given vessel were beyond the Glagov-remodeling threshold, a significant increase with $g_{PW}>1$ was no longer obtained ($p>0.75$). The findings of this study confirm (in vivo) the hypothesis that relatively lower wall shear stress is associated with early plaque development, where more research is necessary to obtain more direct evidence that this is the case.

2.5 Pulmonary Applications

Multidetector CT (MDCT) scanning is rapidly becoming an important means of assessing patients with respiratory symptoms. Recent advances in MDCT scanner technology enable the scanning of the entire lung with nearly isotropic sub-millimeter voxel dimensions (on the order of 0.4 mm). Common symptoms of breathlessness, coughing up blood, and even dry cough demand that a CT scan of the chest be performed, especially if other clinical features suggest a parenchymal lung disorder. The chest MDCT scan is assessed in conjunction with the patient history and examination findings, with global tests of pulmonary function to confirm a clinical diagnosis and to follow the patient over time.

Lung segmentation is typically performed as the first step of any subsequent quantitative analysis of pulmonary parenchyma, lung lobes, intrathoracic airway, or vascular trees. The goal of lung segmentation is to separate the voxels corresponding to lung tissue from the voxels corresponding to the surrounding anatomy. Many approaches exist, most of them based on gray-level properties of the parenchyma, which is generally darker than surrounding tissues on CT scans [135, 136]. A combination of optimal thresholding [137] and region connectivity and topology analysis is frequently used to identify the lungs. Optimal thresholding [137] is an adaptive method for threshold selection that is data driven and thus adjusts the threshold to the underlying data. As described in Ref. [135], the optimal thresholding step automatically selects the appropriate threshold to separate the lungs from the body and chest wall. The approach reported in Ref. [135] was extensively validated and successfully

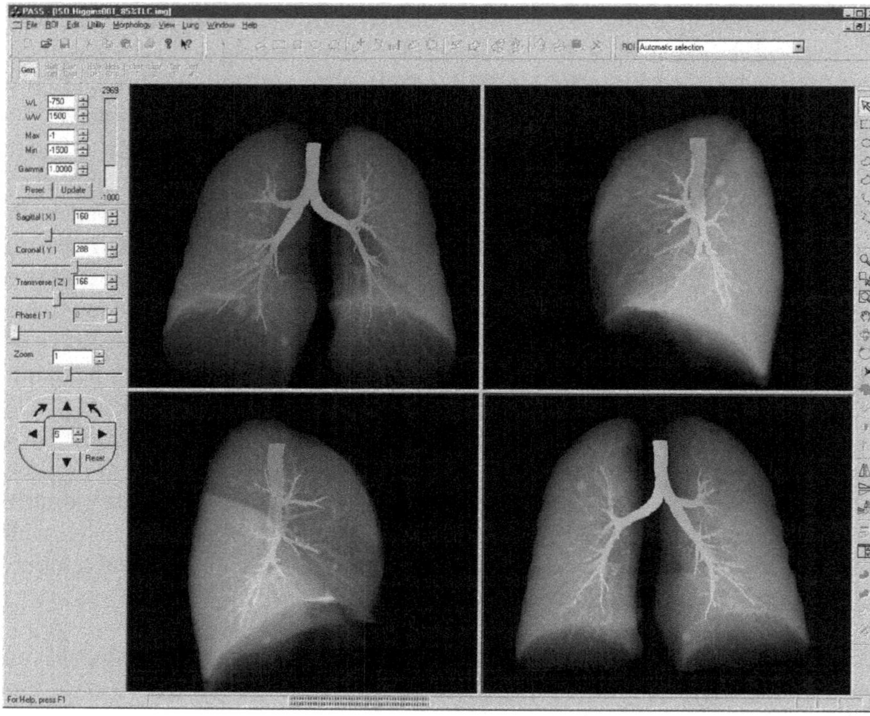

Fig. 2.33 Segmentations of the lungs, lobes, and airway tree extracted from a 3D MDCT image of a normal subject

applied to over 3000 CT datasets acquired from nearly 1000 subjects enrolled in the National Emphysema Treatment Trial (NETT) (Fig. 2.33). After segmentation, the left and right lungs may be separated using dynamic programming.

2.5.1 Segmentation and Quantitative Analysis of Airway Trees

Knowledge about the location of the pulmonary airway tree as well as its geometrical properties is important for the diagnosis, study, and treatment of lung disorders as well as for physiological studies of pulmonary function. Existing airway segmentation approaches mainly include region growing-based methods [138–141], morphology-based methods [142–144], and combinations of the two [145–147]. Other methods proposed in the past include rule-based methods [148, 149], energy function minimization [144], and ROI modification-based techniques [150]. Schlathölter et al. [151] use a front-propagation algorithm for segmenting airway trees. Branchpoints are detected when the front splits up. Diameters are measured during the segmentation and a leak is identified if the diameter suddenly increases.

There are many reasons why airway tree segmentation is difficult to achieve from volumetric CT images in a robust fashion from clinical-quality or low-dose

CT images. Some of the reasons are anatomy related, e.g., airway obstructions; others are caused by heart beat-induced movement artifacts and image reconstruction artifacts. Consequently, one of the biggest problems when segmenting airway trees with automated methods is leakage into the extra-luminal regions (Fig. 2.34). Leaks occur because the airway wall that separates the airway lumen from the surrounding lung parenchyma is relatively thin. The lung parenchyma, on the other hand, exhibits a texture similar to that of small airways. Partial volume effects and image noise greatly decrease the CT visibility of the airway wall. At some places this can cause the segmentation algorithm to "grow" through the airway wall. Once this happens, potentially large regions of lung parenchyma are wrongly identified as airways.

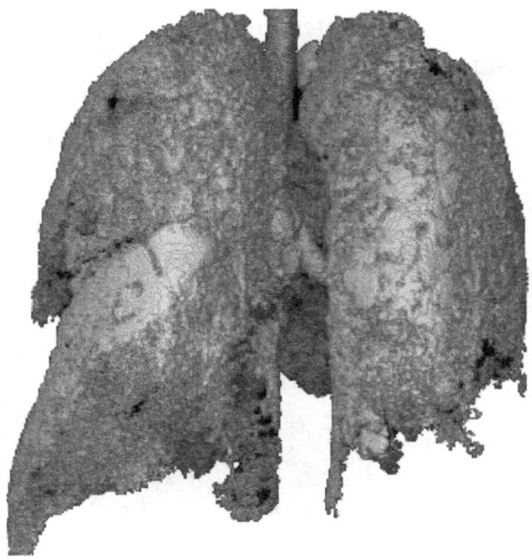

Fig. 2.34 Example of a severe segmentation leak [87, ©2005 IEEE] (Emphysema patient, airway tree segmentation attempted with standard 3D region-grow algorithm—leak was unavoidable)

The quantitative measurement step follows the segmentation, and its purpose is to accurately find the location of the inner airway wall in order to conduct geometrical measurements [152]. In the past, measurements have often been conducted manually [153, 154]. Automated methods [143, 155–159] have been presented in previous work. Reinhardt et al. [155, 156] showed that the accuracy of measurement depends on the size of the airway and proposed a model-based system using a 2D Gaussian function to model the point-spread function of the scanner. Prêteux et al. [143, 157] approximated the inner airway wall with a Butterworth polynomial, but concluded that their method produces a sharply increased error for small airways. Saba et al. [158] used the full-width/half-max method to conduct measurements in 2D and then model the 3D airways based on these measurements.

2.5.1.1 Airway Tree Segmentation

An example [87] of an advanced airway segmentation method based on *fuzzy connectivity* [27, 160] and subsequent quantitative analysis of the airway wall is provided in the following paragraphs. During the execution of this algorithm, two regions—foreground and background—are competing against each other. This method has the great advantage that it can overcome image gradients and noise and is thus well-suited for low-dose MDCT image data. The disadvantage is its relatively high computational complexity. Computing time can be reduced by splitting the segmentation space into a number of smaller subspaces and by keeping the search space as tight as possible around the airway segments. The desire to keep the segmentation within a small area, together with the need to detect possible leaks at their root, leads to the idea of using a relatively small adaptive region of interest (ROI) that follows the airway tree branches as they are segmented. The ROI has a cylindrical shape and adapts its geometrical dimensions, orientation, and position to the predicted size, orientation, and position of the airway branch to be segmented. Fig. 2.35 illustrates this concept. Using a cylindrically shaped ROI (vs. the more common cubical ROI used in other 3D image segmentation tasks) has the advantage that the ROI better adapts to the target shape, which is close to cylindrical. This means less "useless" background voxels have to be analyzed and the computing time can be shortened. A similar approach was independently used by Mori et al. [150].

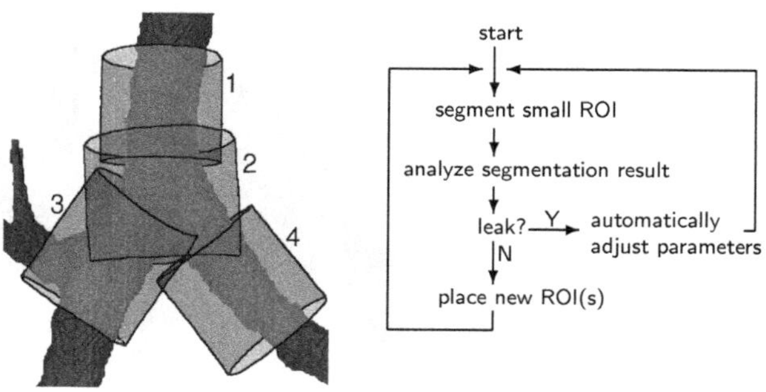

Fig. 2.35 Basic concept of airway tree segmentation. Adaptive cylindrical regions of interest (*light gray*) follow airway tree branches as the segmentation proceeds [87, ©2005 IEEE]. Segmentation is performed in a small area only, which keeps the computing time down. Possible problems (leaks) can be detected early and addressed. The simplified flow diagram to the right does not show all details (for example, the termination criteria)

The basic idea of segmentation with fuzzy connectivity [27, 160] is that the voxels of an input image are compared with a seed-voxel and the similarity/dissimilarity is expressed as a fuzzy membership value. The similarity of two voxels c and d is expressed by the *affinity value* $\psi(c,d) =$ [0, < *CitationRefCitationID* = "*CR*1" > 1 < /*CitationRef* >], which is normally computed based on gray values and defined for adjacent voxels only. In this application we consider 18-connected voxels as adjacent. If c and d are not directly adjacent, then their similarity is compared by looking at all possible "chains" of adjacent voxels that connect c and d. The strongest chain is chosen to represent the similarity. The strength of a chain is defined by the lowest ψ value along its length (weakest link). Voxels are assigned to the foreground region if their ψ value exceeds a predefined value.

The multi-seeded fuzzy connectivity (MFC) method takes this idea one step further by growing two regions (foreground and background) simultaneously and letting them compete for voxels. The method guarantees that both resulting regions, foreground and background, are connected in themselves, i.e., no isolated "islands" may occur.

Consequently, the method depends on the ability to detect leaks. If leaks are not present or resolved, then sequentially identify the seed points for the next ROI to perform the segmentation (Fig. 2.35). Details of the relatively complex process and description of the multi-seeded fuzzy affinity functions, directional affinity, seedpoint selection, positioning of the new cylindrical ROIs, as well as the method of local leak detection can be found in Ref. [87]. This method has shown to significantly outperform the conventional region growing-based approaches and demonstrated applicability to low-dose MDCT image scans [87] (Fig. 2.36 and 2.37).

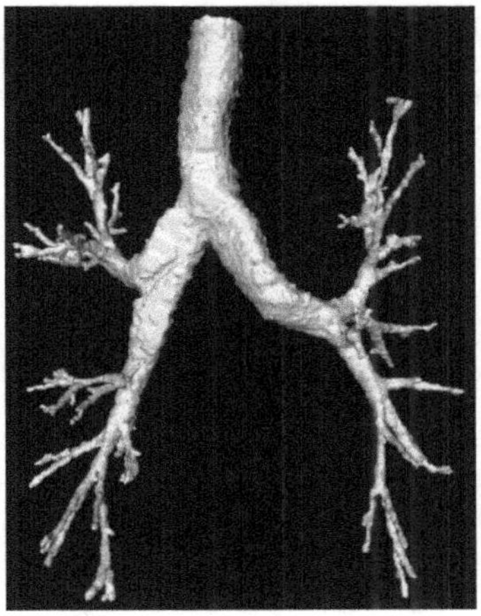

Fig. 2.36 Segmentation result using the new method [87, ©2005 IEEE]. Tree from the same CT scan used in Fig. 2.34

2.5.1.2 Quantitative Analysis of Airway Tree Morphology

Once the airway trees are segmented as described above, the segmentation of the airway lumen borders is not sufficiently accurate for quantitative analysis. The approximate surfaces of the airway tree segments together with the airway tree skeleton centerlines are used to guide the accurate detection of the airway walls. The process is thus divided into three steps (Fig. 2.38):

- Resample 2D slices perpendicular to airway segment centerlines.
- For each 2D slice separately, segment airway wall using 2D dynamic programming—*or*—for the 3D segment simultaneously, segment the inner and outer airway wall surfaces using 3D graph search (Section 2.2.3).
- Conduct measurements on segmentation result.

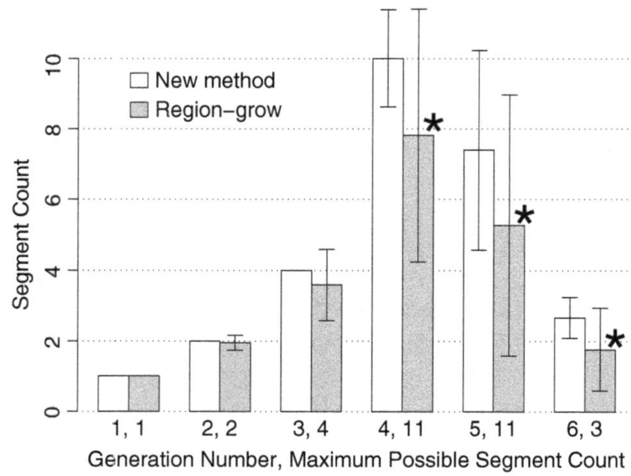

Fig. 2.37 Count of correctly segmented airway tree branches (according to tree branch nomenclature), sorted by generation number [87, ©2005 IEEE]. Mean ± standard deviation is reported. For the generations marked with *, the MFC method retrieves a statistically significant higher count of branches ($p = 0.013$, 0.037, and 0.004 for generations 4, 5, and 6, respectively)

Fig. 2.38 Quantitative analysis: 2D slices are resampled perpendicular to centerline, inner border is detected, cross-sectional area and minor and major diameters are computed [87, ©2005 IEEE]

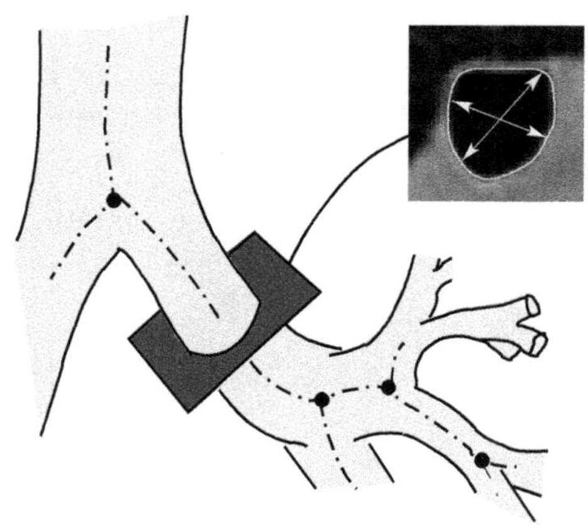

A 2D slice $\mathcal{I}_{\text{gray}}$ is resampled from the original gray-level volume using the centerline information, and a 2D slice $\mathcal{I}_{\text{seg}}$ is resampled from the earlier airway segmentation result. The slices are oriented perpendicular to the centerline of the respective airway segment and one pair of slices is resampled for every centerline voxel position. Resampling at every centerline voxel position assures that every possible position along the airway tree segments is covered by measurements. The perpendicular orientation is determined by computing the tangent to the smoothed centerline (the centerline is obtained from the skeletonization process). Resampling is performed using tri-linear interpolation.

For every voxel along the centerline, the wall segmentation takes $\mathcal{I}_{\text{gray}}$ and $\mathcal{I}_{\text{seg}}$ as input and outputs $\vec{P} = [\vec{p}_0, \vec{p}_1, \ldots, \vec{p}_{N-1}]$ with $\vec{p}_i = (x_i, y_i)$ as output, where $\vec{p}_{0\ldots N-1}$ represent the segmented points along the inner airway border, defined in the Cartesian coordinate system of $\mathcal{I}_{\text{gray}}$. *Dynamic programming* [22] is used for the wall segmentation. The input images are *radially resampled* in order to "stretch" the target border. Starting from the point defined by the centroid of the segmentation result in $\mathcal{I}_{\text{seg}}$, N evenly spaced rays are cast, and along each ray M points are sampled from $\mathcal{I}_{\text{gray}}$ and $\mathcal{I}_{\text{seg}}$.

The *cost function* $\mathcal{I}_{\text{cost}}$ used for the dynamic programming uses the first and second derivatives of the gray-level image and is based on a cost function proposed in Ref. [88]. Strictly speaking the cost function used here should be called the "reward" function since higher values are preferred over lower ones. But to be consistent with the literature the more common term *cost function* is used here—and the dynamic programming is maximizing the cost

$$\mathcal{I}_{\text{cost}} = \omega \cdot \underbrace{\mathcal{I}_{\text{smooth}} \star \mathcal{M}_{\text{Sobel}}}_{\text{first derivative}} \dot{+} (1-\omega) \cdot \underbrace{\mathcal{I}_{\text{smooth}} \star \mathcal{M}_{\text{Marr}}}_{\text{second derivative}} \tag{2.9}$$

where $\star$ symbolizes a 2D convolution, the $\dot{+}$ symbol stands for point-wise summation, $\mathcal{I}_{\text{smooth}}$ is $\mathcal{I}_{\text{gray}}^{\text{rad}}$ smoothed with a 5×5 unit matrix. $\mathcal{M}_{\text{Sobel}}$ is a 5×5 Sobel mask, and $\mathcal{M}_{\text{Marr}}$ is a 5×5 Marr mask. By changing the value of constant ω, the position of the resulting border can be pushed or pulled radially as illustrated in Fig. 2.39. This is based on a well-known behavior of first- and second-derivative edge detectors that consistently position their maximum edge responses on one or the other side of the true edge, respectively. Consequently, weighting their responses can be used for accurate positioning of the edge detection response with respect to the correct edge location. The purpose of this is to adapt the cost function to the estimated size of the current airway segment and thus to maintain a high degree of measurement accuracy across all sizes of airways. The size of the airway is estimated based on a

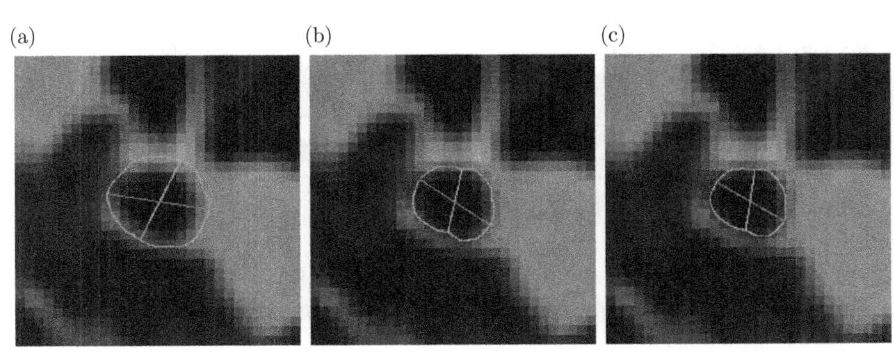

Fig. 2.39 Border position can be modified by adjusting the value of ω in Eq. (2.9) [87, ©2005 IEEE]. Note that the airway size is overestimated in (**a**) and underestimated in (**c**). (a) $\omega = 0.00$, (b) $\omega = 0.25$, (c) $\omega = 1.00$

pixel count in the resampled segmentation result $\mathcal{I}_{\text{seg}}$. The value of ω is determined via the empirically determined piece-wise linear function. Due to the impossibility of determining correct airway wall measurements in vivo, the values of ω must be found by an optimization process during which phantom CT images of tubes with known sizes are used. The phantom sizes must cover the full range of airway sizes and corresponding CT imaging parameters must be used for the phantom scans.

Validation of this approach demonstrated subvoxel accuracy ranging from −0.26 to −0.05 mm in MDCT images with voxel sizes of 0.391 × 0.391 × 0.6 mm. Using this approach for airway lumen analysis, quantitative assessment of luminal diameter as a function of the airway tree generation can be obtained as shown in Fig. 2.40. More detailed results can be seen in Ref. [87].

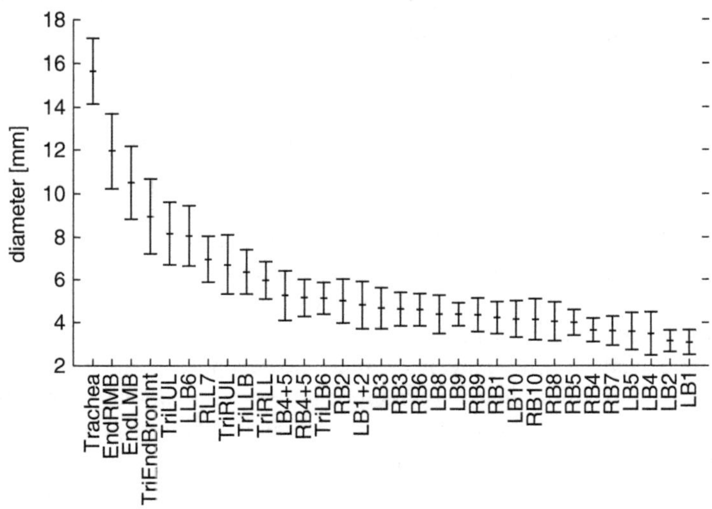

Fig. 2.40 Diameters ± standard deviation of named segments. Based on 22 normal subjects [87, ©2005 IEEE]

When working with the entire segment, the full set of 2D-resampled cross sections can be combined in a single 3D volume and a graph formed as shown in Fig. 2.6. The method described in Ref. [41] can be used to detect both inner and outer surfaces of the airway segments (Fig. 2.9 and 2.41). Until very recently, the airway wall was typically only quantified outside of bifurcations (in the "straight" segments between branches), although methods for cross-bifurcation airway wall analysis are beginning to appear [47, 48], Fig. 2.42.

2.5.2 Quantitative Analysis of Pulmonary Vascular Trees

The pulmonary arterial and venous structures deliver deoxygenated blood to the lung periphery and return oxygenated blood to the systemic circulation. These highly complex branching structures support the primary function of the lung that

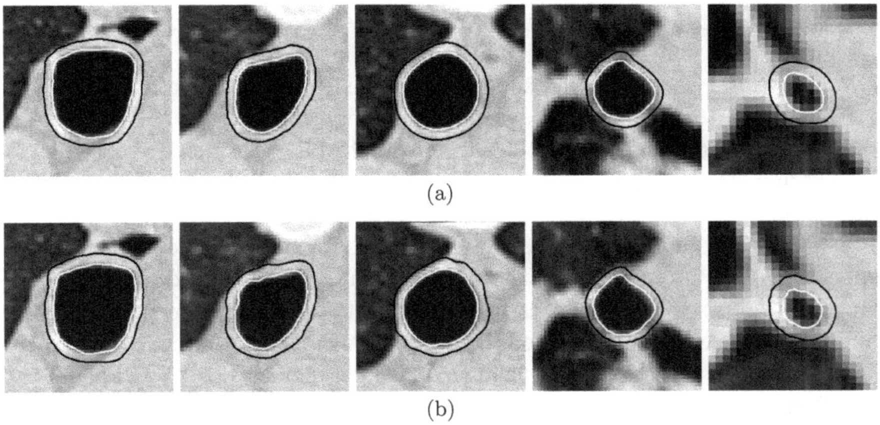

Fig. 2.41 Comparison of observer-defined and computer-segmented inner and outer airway wall borders [41, ©2006 IEEE]. (**a**) Expert-traced borders. (**b**) 3D surface obtained using our double-surface segmentation method

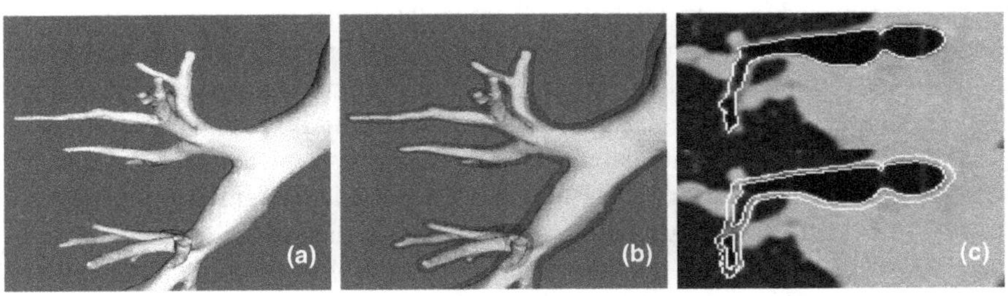

Fig. 2.42 Airway segmentation: (**a**) the inner wall (result of presegmentation). (**b**) Double surfaces after graph search. (**c**) The comparison of preliminary result and graph search result in 2D slices. Note that the inner and outer surfaces are smooth and three-dimensionally accurate across the airway branching

is to bring blood into close proximity with incoming fresh gas delivered to the terminal air sacs (alveoli) through the process of respiration. In clinical practice, it is of great importance, for instance, to be able to characterize the vascular trees for the detection of pulmonary emboli (localized blockages), detection of signs of pulmonary hypertension, and for the differentiation between vasculature and focal opacities (for detection of lung cancer and other localized pathologies). The vascular trees can also serve as a roadmap for the tracking of lung tissues across lung volume changes or across time as the lung is serially monitored. Vessel segments with radii of 2 mm or less are readily detectable in high-resolution MDCT images. While detectable, manual segmentation of these complex tree structures, even if an individual were willing to take the time, has been found to present a near impossible task due to the following reasons:

- It is difficult to determine boundaries of a vessel consistently, especially thin segments, due to the partial volume effects and image noise.
- Volumetric lung scans of the adult human consist of more than 500 slices and the vascular tree, in a bipodial fashion, rapidly branches as one tracks the vessels from their central to peripheral locations, with the full tree structure consisting of more than 23 generations.

In addition, manual measurements of the vessels for assessment of diameters and branching angles are unreliable. The measurement of diameter requires the determination of cross-sectional planes perpendicular to the local segment centerlines. Similarly, a plane that includes both parent and child segments around the branchpoint needs to be localized for branching angle measurements. Both measurements are difficult to perform manually via 2D images because of problems of foreshortening in projected view. Therefore, highly automated segmentation of the pulmonary vascular tree based on 3D image analysis plays an important role in detecting and characterizing the vessel structure.

Several 3D vessel segmentation algorithms have been presented to date. Tube enhancement filters based on a combination of the eigenvalues of a Hessian matrix have been reported in Refs. [161, 162]. Segmentation can be performed simply by thresholding of the filter output. The filters have an ability to handle a range of radii by multi-scale implementation. Lorigo et al. [163] reported a vessel segmentation algorithm based on a "co-dimension two" level-set method. Vasilevskiy and Siddiqi [164] used gradient flows implemented using a level-set method for 2D and 3D vessel segmentation. Aylward and Bullitt [165] reported an intensity ridge traversal method to extract vessels. A tracking direction was estimated by an eigenvector of the Hessian matrix at each tracking front position. Boldak et al. reported model-based vessel tracking [166]. Mayer et al. presented pulmonary vessel segmentation in contrast-enhanced CT data [167]. Fridman et al. [168] used cores [169, 170] to track the vascular tree from a seed point.

The eigenvalues and eigenvectors of the Hessian matrix are implicitly or explicitly used in some of the above algorithms and the algorithms have worked well for extracting vessels in several organ systems imaged by CT or MR. The tube enhancement filters can extract both thin segments and thick segments without using seed points. However, such filters produce disconnections around the junctions since they are based on a cylindrical vessel segment model. Segmentation results obtained from vessel traversal algorithms generally have better connectivity between segments but often miss peripheral thin segments.

An algorithm that extracts peripheral thin segments as well as thick segments from thoracic CT images with better connectivity by integration of the tube enhancement filter and vessel traversal approaches was reported in Ref. [171]. This integrated algorithm consists of three major steps: (1) Tube enhancement based on the cylindrical shape model using an eigenvalue of the Hessian matrix serves as a filter to extract vessels and to produce information that is used to determine a set of seed points in the following vessel traversal step. (2) The traversal step starts from each seed point until one of the eigenvalues of the Hessian matrix changes its sign twice, signifying that the front point of a trajectory has reached a junction. (3) Branchpoint analysis is accomplished by applying a thinning method which then allows for the selection of objects with many branchpoints, serving as a means of distinguishing between vascular trees and noise components.

The vessel enhancement is performed in a standard manner using a Hessian cylindrical model. Based on the cylindrical vessel model, the eigenvalues of the Hessian matrix are commonly employed as efficient criteria to differentiate tube structures from other image components. Let three eigenvalues of the Hessian matrix at a point be λ_1, λ_2, and λ_3 and their corresponding eigenvectors be e_1, e_2, and e_3, respectively. Suppose that the eigenvalues meet the condition $\lambda_1 \leq \lambda_2 \leq \lambda_3$. Based on the model, when a point is close to the center of a segment, λ_1 and λ_2 take on large negative values whereas λ_3 takes on a small value. Figure 2.43 illustrates a cylindrical vessel model, with the eigenvalues/eigenvectors at the center of the model. According to the model, the following criteria are typically used to construct a filter output function:

1. Locality: λ_1 and λ_2 are large negative values.
2. Elongation: λ_3 is much less than λ_1 and λ_2 ($\lambda_3/\lambda_1 \approx 0$ or $\lambda_3/\lambda_2 \approx 0$).
3. Symmetry: λ_1 and λ_2 are of almost the same values when a point is close enough to the center of a segment ($\lambda_2/\lambda_1 \approx 1$).

These criteria capture the characteristics of the model so as to differentiate tube structures from sheet- and blob-shaped structures when a point is within a vessel segment and close to its center. However, pulmonary vascular trees include many junctions, and criteria 2 and 3 are not always satisfied around the branchpoints. Similarly, λ_1 and λ_2 do not always satisfy criterion 3, especially in the thin segments and around the junctions. The filter output function $F(\mathrm{x})$ is defined as follows:

$$F(\mathrm{x}) = \arg\max_{\sigma_f \in \mathbf{S}} -\sigma_f^2 \lambda_2 / I(\mathrm{x}) \tag{2.10}$$

where σ_f is the standard deviation of a Gaussian function convoluted with an image so as to take second derivatives in the volume coordinate and $I(\mathrm{x})$ is the intensity value at the point. $\mathbf{S}$ is a discrete set of σ_f for multi-scale integration. This equation only takes into account criterion 1. Thus, it may also enhance blob structures and not only the desirable cylindrical structures, and a post-processing is needed to overcome this limitation as described below.

Initial segmentation is typically obtained by thresholding the filter output. Note that different values are needed for scans of different inspiration levels due to the varying air content in the lungs. Since the segmentation works on a voxel by voxel basis, the result also contains local disconnections and small holes. These are caused mainly by image noise and the difference of the intensity distribution on a cross-sectional plane between straight segments and junctions. The cross-sectional shape of a vessel contour at a junction becomes an ellipse. It causes lower filter output and

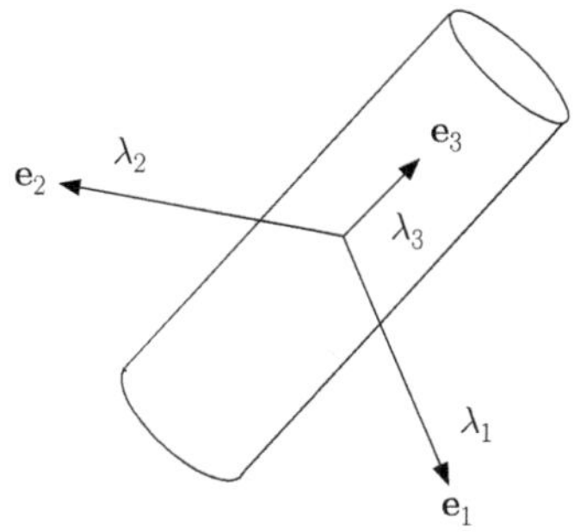

Fig. 2.43 An illustration of a vessel model. It is assumed that the model has a cylindrical shape with a 2D Gaussian-like intensity distribution at each cross section. The *arrows* show the eigenvectors of a Hessian matrix at the center. The length of the *arrows* represents the absolute value of corresponding eigenvalues

local disconnections around junctions, especially where very thin segments bifurcate from a thick segment. The vessel traversal approach is suitable for improving connectivity and for filling small holes in a segment. The connectivity improvements include the following processes: (1) initialization, including automated localization of seed points; (2) tracking terminations at junctions; and (3) radius estimation for boundary recovery. In the first step, seed points are preferentially identified at segment centers. According to the vessel model, the intensity function takes a local maximum at the center of a segment. The local maximum position $\mathbf{x}_{max}$ in the volume coordinate can then be calculated by the following equation [172]:

$$\mathrm{x}_{max} = \mathrm{x} + \mathrm{p} = \mathrm{x} - \frac{(\nabla I, \mathrm{e}_1)}{\lambda_1}\mathrm{e}_1 - \frac{(\nabla I, \mathrm{e}_2)}{\lambda_2}\mathrm{e}_2 \tag{2.11}$$

where ∇I is a gradient vector at $\boldsymbol{x}$, e_1 and e_2 are eigenvectors corresponding to the eigenvalues λ_1 and λ_2. The operator $(,)$ takes the inner product of the two vectors in the equation. ∇I is estimated by using σ_f, which maximizes Eq. (2.10) in S. Since estimation of x_{max} is sensitive to image noise, $\boldsymbol{x}$ is considered to be located close enough to the center when the following conditions are met:

$$p_x \leq 0.5, \quad p_y \leq 0.5, \quad p_z \leq 0.5 \tag{2.12}$$

where p_x, p_y, and p_z are the components of the vector $\boldsymbol{p}$. Figure 2.44 outlines the principles of the tracking process. After tracking, spheres with estimated radii are drawn at each tracking front position to fill possible holes and gaps between two segments.

The vascular tree can be characterized as an object with many branchpoints. An object that contains small numbers of branchpoints is likely part of other structures. Therefore, a thinning algorithm [173] is applied to each object in the result to obtain the number of branchpoints in the object from its graph representation. Any

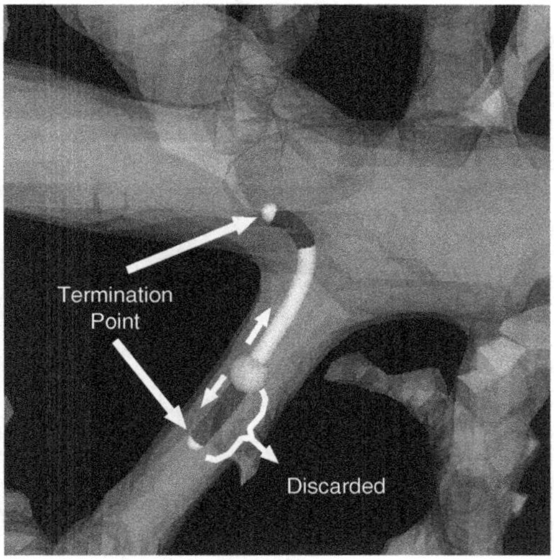

Fig. 2.44 An illustration of the tracking process from a thin segment. The seed point is shown by a *gray* sphere. At the seed point, the tracking front can take both e_3 and $-e_3$ as the initial direction. Either way, tracking will be terminated after λ_3 changes its sign twice and becomes a positive value. Then the intensity value at the termination points is examined and either trajectory whose intensity value at the termination point is less than the other will be discarded. The other will be used for boundary recovery with the radius estimation

object with less than 100 branchpoints is thus discarded and not considered part of a vascular tree.

A quantitative validation was performed with more than 1000 manually identified points selected from inside the vessel segments in 44 scans from 22 human subjects imaged via MDCT during breath holds in total lung capacity (TLC, inspiration) and functional residual capacity (FRC, expiration) scans to assess true positives (TPs). Similarly, 1000 points were randomly placed outside of the vessels to evaluate false positives (FPs) in each case. On average, for both the high and low volume lung images, 98% of the points were properly marked as vessel and 2% of the points were assessed as FPs.

Figure 2.45 shows volume-rendered images of the segmentation results from both TLC and FRC scans of one subject. Both results include peripheral thin segments close to the pleural surface. Since the scale of these images is the same, they also show how the lung geometry changes between the two volumes scanned.

An interesting byproduct of the vascular tree segmentation is that it depicts the pulmonary fissures as regions void of the vessel segments. Pulmonary fissures are visible structures in the lung which separate each lung into lobes. As shown in Fig. 2.49 and discussed in Section 2.5.3, this information can be used for lobar segmentation. It is important to realize that the described approach does not distinguish between arteries and veins. While it is of great interest to separate the vascular trees in disjoint arterial and venous sub-trees, this is a very difficult task due to many areas of artery–vein merges in the image data, caused by finite voxel size and partial

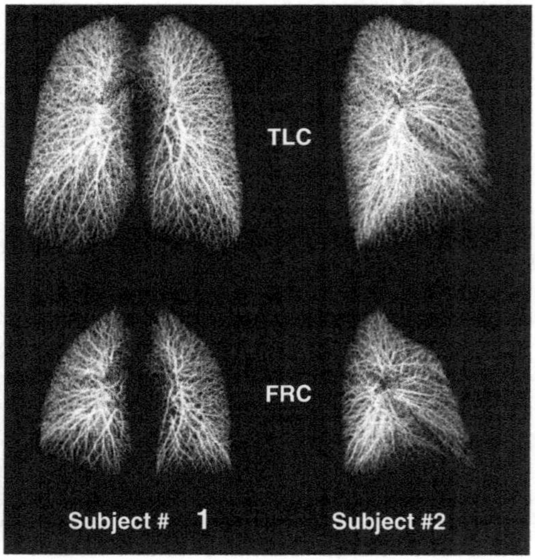

Fig. 2.45 Two examples of segmentation results with TLC and FRC scans of two subjects. All images are shown at the same scale. Note the breathing-related changes of lung vasculature caused by regional lung parenchymal expansion between the TLC and FRC lung volumes

volume effect of MDCT acquisition. A promising attempt to separate arterial and venous trees in pulmonary MDCT images was recently reported in Ref. [174].

2.5.3 Segmentation of Lung Lobes

The human lungs are divided into five distinct anatomic compartments called *lobes*. The separating junctions between the lobes are called the *lobar fissures*. The left lung consists of the upper and lower lobes, which are separated by the left *oblique* or *major* fissure. The right lung consists of the upper, middle, and lower lobes: the upper and middle lobes are separated by the *horizontal* or *minor* fissure; the middle and upper lobes are separated from the lower lobe by the right oblique (major) fissure [175, 176]. The branching patterns of the airway and vascular trees are closely related to the lobar anatomy. Although there are some exceptions, mainly in the case of *incomplete fissures*, each lobe is served by separate airway and vascular networks.

CT imaging can be used to study the lobar anatomy. A major challenge to the automatic detection of the fissures is the fact that the fissures have low contrast and variable shape and appearance in CT imagery, which sometimes makes it difficult even for manual analysts to mark their exact location. Usually, the fissures appear as a thin bright sheet, dividing the lung parenchyma into two parts (see Fig. 2.46).

Table 2.3 summarizes recent CT-based lobe segmentation studies. Previous approaches to lobe segmentation can be roughly divided into two classes: direct and indirect. The former approaches consist of methods that search for the fissures based on gray-level information present in the data, while the latter approaches consist of methods that use information from other anatomical structures to approximate

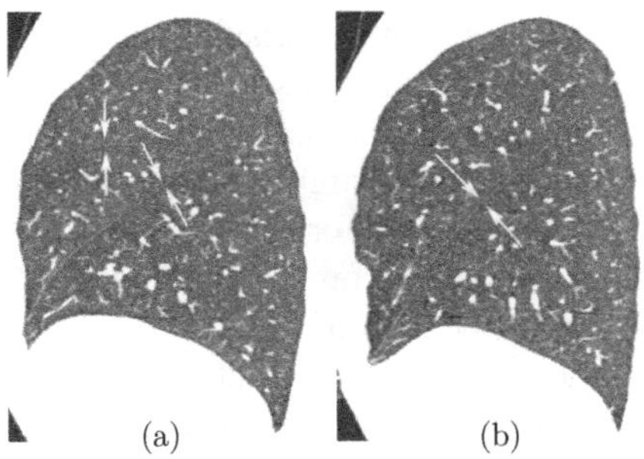

Fig. 2.46 Lung lobes on MDCT [177, ©2009 IEEE]. (**a**) Sagittal slice from the right lung showing the right oblique and horizontal fissures—the horizontal fissure is oriented horizontally, while the oblique fissure is tilted away from the vertical axis. (**b**) Sagittal slice from the left lung showing the left oblique fissure

Table 2.3 High-resolution CT-based studies on automatic lobar fissure segmentation

Authors	Reported Results	Comments
van Rikxoort et al. [178]	95% classifier accuracy	Direct, voxel classifier, needs post-processing steps to define lobar surfaces
Zhang et al. [179]	1.96 mm mean RMS error	Direct, atlas based, computationally intensive, horizontal fissure semi-automatic
Wang et al. [180]	1.01 mm average distance error	Direct, shape based, semi-automatic, oblique fissures only
Wiemker et al. [181]	No validation	Direct, 3D sheet filtering
Kuhnigk et al. [182]	No validation	Indirect, watershed transform based, semi-automatic
Zhou et al. [183]	No validation	Indirect, Voronoi division based
Ukil et al. [177]	1–2 mm mean RMS error	Direct, multi-stage, uses anatomical information from airway and vascular trees followed by optimal graph search for fissure surfaces

the location of the fissures. Of the direct methods, Zhang et al. [179] presented a method for automatic segmentation of the oblique fissures using an atlas-based initialization procedure, followed by a two-step graph searching procedure to delineate the fissures. One drawback of the atlas-based initialization is the time-intensive atlas construction, which involves manually delineating the fissures on a number of subjects, and the computationally expensive deformation of the atlas fissures onto a template image. Wang et al. [180] proposed a method for segmentation of the fissures using a 2D shape-based curve growing model. Their method involves a semi-automatic initialization, after which the oblique fissures are segmented in 2D slices automatically using a Bayesian formulation influenced by both image data and similarity to a shape prior. Wiemker et al. [181] proposed an automatic segmentation approach based on 3D filtering of the image data. The Hessian matrix and structure tensor-based filters are used to enhance the fissures. More recently, van Rikxoort et al. [178] described a nearest-neighbor classifier approach to identify voxels on the fissures. Since their method identifies fissure voxels, a post-processing step is needed to define a fissure surface separating the lobes.

Among the indirect approaches, Kuhnigk et al. [182] proposed a method for the indirect estimation of the lobes, using information from the segmented vasculature. They presented a method for semi-automatic segmentation of the lobes using a watershed transform on a distance map of the vasculature. Their method required that the user manually place markers on vessels to guide the segmentation. The method is fast and interactive, with the ability to edit the results in real time. In a similar approach, Zhou et al. [183] reported a method based on the Voronoi division of the lungs using the lobar bronchi.

The lobar segmentation approach of Ukil et al. deserves special attention [177]. The basic idea of this method is to use the anatomical information provided by the segmentation and analysis of the airway and vascular trees to guide the segmentation of the fissures. A two-step approach is followed: in the first step an approximate fissure region of interest (ROI) is generated using the airway and vascular trees' anatomical information; in the second refinement step, the fissures are accurately

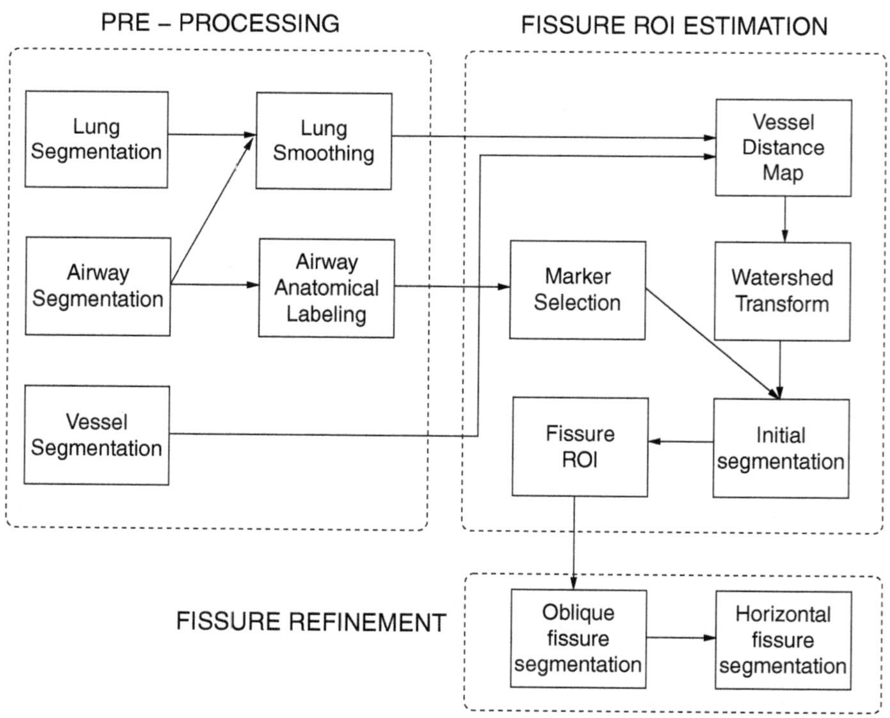

Fig. 2.47 Flow diagram of the fissure segmentation process reported in Ref. [177, ©2009 IEEE]. First, the lungs, airways, and vessels are detected. Using this anatomic information, an approximate fissure ROI is estimated. This ROI is refined using the optimal 3D graph search approach based on grayscale information

located using the available contrast information in the ROI via optimal graph searching (Section 2.2.3, [41]). The algorithm (Fig. 2.47) is robust with respect to common variations in scan parameters such as x-ray dosage, lung volume, and CT reconstruction kernels and has been applied to images of both normal and diseased subjects. As an add-on, a fully automatic method was also developed to detect incomplete fissures, using a fast marching-based segmentation of a projection of the optimal fissure surface computed by the graph search. Once detected, the incomplete fissure can be used to extrapolate and smoothly complete the fissure surface. Figures 2.48, 2.49, and 2.50 show the intermediate steps of the approach and the resulting lobar segmentation. See also Fig. 2.33 for a comprehensive visualization of the segmented lungs, lobes, and airway tree.

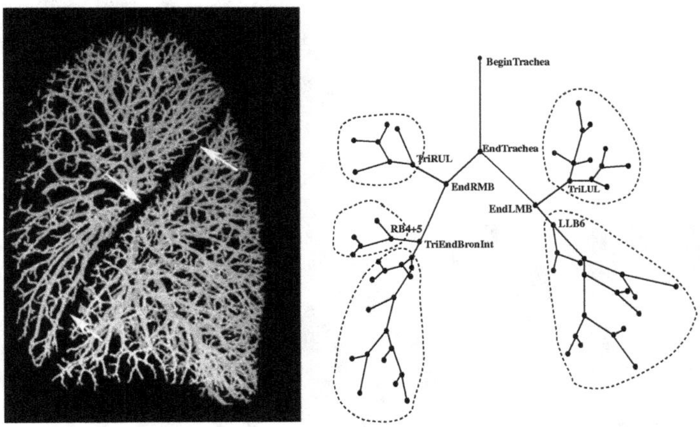

Fig. 2.48 Anatomical information used for lobar segmentation [177, ©2009 IEEE]. (**a**) Volume-rendered vascular tree showing the relationship between the vascular tree and the lobar boundaries; (**b**) anatomically labeled airway tree (from [184, ©2005 IEEE]) showing lobar sub-trees. The airway branchpoints *TriLUL*, *LLB6*, *TriRUL*, *RB*4+5, and *TriEndBronInt* correspond to the *left upper* lobe, *left lower* lobe, *right upper* lobe, *right middle* lobe, and *right lower* lobe, respectively

2.6 Discussions and Conclusions

In this chapter, we have presented various methods and applications for the assessment of cardiovascular and pulmonary structure and function, eventually leading to disease assessment. The vast information provided by modern imaging modalities can be utilized efficiently by 3D and 4D methodology for both segmentation of the structures of interest and quantification of the disease severity and its progression or regression. Applications are diverse, ranging from basic studies on cause and effect of a disease to aiding the diagnosis of a physician for a specific patient. Conventional analysis of the image data by visual inspection or simple 2D measurements, as is frequent practice in clinical routine, may result in incomplete assessment of a patient's condition. The advanced methodology to classify the status of disease and to allow

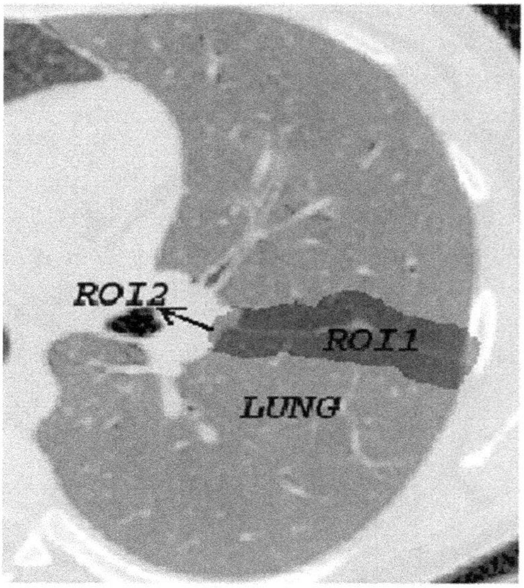

Fig. 2.49 Fissure ROI for left oblique fissure [177, ©2009 IEEE]. This ROI is derived from the airway and vascular tree segmentation via watershed analysis. The final fissure surface is detected in this ROI by employing the optimal graph searching approach

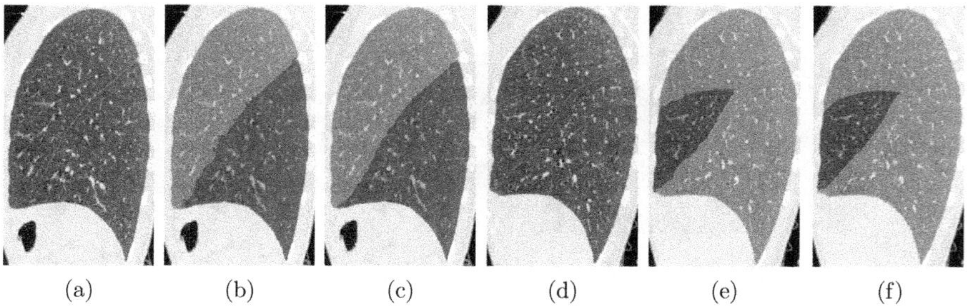

Fig. 2.50 Lobar segmentation results performed in 3D [177, ©2009 IEEE]. (**a, d**) CT data, (**b, e**) watershed segmentation, (**c, f**) optimal surface detection

the calculation of complex compound indices well beyond currently applied metrics promises to improve both patient care and our understanding of the underlying mechanisms of a disease.

Acknowledgments The work presented in this chapter was supported in part by the National Institutes of Health, Bethesda, MD (grants R01 EB004640, R01 HL 063373, R01 HL 064368, R01 HL 071809), and by Philips Medical Systems, Cleveland, OH (Section 2.4.2).

References

1. Bosch JG, Mitchell SC, Lelieveldt BPF, Nijland F, Kamp O, Sonka M, Reiber JHC (2002) Automatic segmentation of echocardiographic sequences by active appearance motion models. IEEE Trans Med Imaging. 21(11):1374–1383
2. Mitchell SC, Bosch JG, Lelieveldt BPF, van der Geest RJ, Reiber JHC, Sonka M (2002) 3-D active appearance models: segmentation of cardiac MR and ultrasound images. IEEE Trans Med Imaging 21(9):1167–1178

3. Stegmann MB, Pedersen D (2005) Bi-temporal 3D active appearance models with applications to unsupervised ejection fraction estimation. In: Michael Fitzpatrick J, Reinhardt JM (eds) Medical imaging 2005: image processing, vol 5747 of SPIE Proceedings, Sau Diego, CA; Bellingham, WA, pp 336–350
4. van Assen HC, Danilouchkine MG, Dirksen MS, Reiber JHC, Lelieveldt BPF (2008) A 3-D active shape model driven by fuzzy inference: application to cardiac CT and MR. IEEE Trans Inf Technol Biomed 12(5):595–605
5. Perperidis D, Mohiaddin RH, Edwards PJ, Rueckert D (2007) Segmentation of cardiac MR and CT image sequences using model-based registration of a 4D statistical model. In: Josien PWP, Reinhardt JM (eds) Medical imaging 2007: image processing, Paper 65121D, vol 6512 of SPIE Proceedings, Sau Diego, CA; Bellingham, WA
6. Ordas S, Oubel E, Leta R, Carreras F, Frangi AF (2007) A statistical shape model of the heart and its application to model-based segmentation. In: Josien PWP, Reinhardt JM (eds) Medical imaging 2007: image processing, vol 6511 of SPIE Proceedings, Sau Diego, CA; Bellingham, WA, p 65111K
7. Andreopoulos A, Tsotsos JK (2008) Efficient and generalizable statistical models of shape and appearance for analysis of cardiac MRI. Med Image Anal 12(3):335–357
8. Zhu Y, Papademetris X, Sinusas AJ, Duncan JS (2008) Bidirectional segmentation of three-dimensional cardiac MR images using a subject-specific dynamical model. In: MICCAI 2008, Springer, Berlin pp 450–457
9. Jolliffe IT (1986) Principal component analysis. Springer, New York, NY
10. Cootes TF, Taylor CJ, Cooper DH, Graham J, (1992) Training models of shape from sets of examples. In: Hogg DC, Boyle RD (eds). British machine vision conference 1992. Springer, London pp 9–18
11. Besl PJ, McKay ND (1992) A method for registration of 3-D shapes. IEEE Trans Pattern Anal Mach Intell 14(2):239–256
12. Mitchell SC, Lelieveldt BPF, van der Geest RJ, Bosch JG, Reiber JHC, Sonka M (2001) Multistage hybrid active appearance model matching: segmentation of left and right ventricles in cardiac MR images. IEEE Trans Med Imaging 20(5):415–423
13. Beichel R, Bischof H, Leberl F, Sonka M (2005) Robust active appearance models and their application to medical image analysis. IEEE Trans Med Imaging 24(9):1151–1169
14. De la Torre F, Black MJ, (2001) Robust principal component analysis for computer vision. In: Proceedings of computer vision, ICCV 2001, vol. 1, pp 362–369, IEEE-CS Press, Los Alamitos, CA
15. Skočaj D, Bischof H, Leonardis A (2002) A robust PCA algorithm for building representations from panoramic images. In: Heyden A, Sparr G, Nielsen M, Johansen P (eds) Computer vision — ECCV 2002, Lecture notes in computer science, vol 2353 Springer, Berlin pp 761–775
16. Tipping ME, Bishop CM, (1999) Mixtures of probabilistic principal component analyzers. MIT Neural Comput 11(2):443–482
17. Abi-Nahed J, Jolly MP, Yang GZ, (2006) Robust active shape models: a robust, generic and simple automatic segmentation tool. In: Larsen R, Nielsen M, Sporring J (eds), Medical image computing and computer-assisted intervention (MICCAI 2006). Lecture notes in computer science, vol 4191 Springer, Berlin, pp 1–8
18. Lekadir K, Merrifield R, Yang GZ (2007) Outlier detection and handling for robust 3-D active shape models search. IEEE Trans Med Imaging. 26(2):212–222
19. Zheng Z, Barbu A, Georgescu B, Scheuering M, Comaniciu D, (2008) Four-chamber heart modeling and automatic segmentation for 3-D cardiac CT volumes using marginal space learning and steerable features. IEEE Trans Med Imaging 27(11) 1668–1681
20. Collins SH (1975) Terrain parameters directly from a digital terrain model. Can Survey 29(5):507–518
21. Soille P, Ansoult M (1990) Automated basin delineation from DEMs using mathematical morphology. Signal Process 20: 171–182

22. Sonka M, Hlavac V, Boyle R, (2008) Image processing, analysis, and machine vision, 3rd edn. Thompson Learning, Toronto, ON
23. Rosenfeld A (1979) Fuzzy digital-topology. Inf Control 40(1):76–87
24. Rosenfeld A (1984) The fuzzy geometry of image subsets. Pattern Recogn Lett 2(9): 311–317
25. Bloch I (1993) Fuzzy connectivity and mathematical morphology. Pattern Recogn Lett 14(6):483–488
26. Dellepiane S, Fontana F (1995) Extraction of intensity connectedness for image processing. Pattern Recogn Lett 16(3):313–324
27. Udupa JK, Samarasekera S (1996) Fuzzy connectedness and object definition: theory, algorithms, and applications in image segmentation. Graphs Models Image Process 58: 246–261
28. Udupa JK, Wei L, Samarasekera S, Miki Y, van Buchem MA, Grossman RI (1997) Multiple sclerosis lesion quantification using fuzzy-connectedness principles. IEEE Trans Med Imaging 16: 598–609
29. Rice BL, Udupa JK (2000) Clutter-free volume rendering for magnetic resonance angiography using fuzzy connectedness. Int J Imaging Systs Technol, 11: 62–70
30. Saha PK, Udupa JK, Odhner D (2000) Scale-based fuzzy connected image segmentation: theory, algorithms, and validation. Comput Vis Image Underst 77: 145–174
31. Kirkeeide RL, Fung P, Smalling RW, Gould KL (1982) Automated evaluation of vessel diameter from arteriograms. In: Proceedings of computers in cardiology 1982, Seattle, WA, pp 215–218, IEEE-CS Press, Los Alamitos, CA
32. Reiber JHC, Kooijman CJ, Slager CJ, Gerbrands JJ, Schuurbiers JCH, den Boer A, Wijns W, Serruys PW, Hugenholtz PG (1984) Coronary artery dimensions from cineangiograms — methodology and validation of a computer-Assisted analysis procedure. IEEE Trans Med Imaging MI-3(3):131–141
33. Beier J, Oswald H, Sauer HU, Fleck E (1991) Accuracy of measurement in quantitative coronary angiography (QCA), In: Lemke HU, Rhodes ML, Jaffe CC, Felix R (eds) Computer assisted radiology (CAR '91), Springer, Berlin pp 721–726
34. Wahle A, Wellnhofer E, Mugaragu I, Sauer HU, Oswald H, Fleck E (1995) Assessment of diffuse coronary artery disease by quantitative analysis of coronary morphology based upon 3-D reconstruction from biplane angiograms. IEEE Trans Med Imaging 14(2):230–241
35. Sonka M, Winniford MD, Collins SM (1992) Reduction of failure rates in automated analysis of difficult images: Improved simultaneous detection of left and right coronary borders. In: Proceedings of computers in cardiology 1992, Durham, NC, IEEE-CS Press, Los Alamitos, CA pp 111–114
36. Sonka M, Winniford MD, Collins SM, (1995) Robust simultaneous detection of coronary borders in complex images. IEEE Trans Med Imaging 14(1):151–161
37. Wahle A, Lopez JJ, Olszewski ME, Vigmostad SC, Chandran KB, Rossen JD, Sonka M (2006) Plaque development, vessel curvature, and wall shear stress in coronary arteries assessed by X-ray angiography and intravascular ultrasound, Med Image Anal—Funct Imaging Model Heart 10(4):615–631
38. Zhao F, Zhang H, Wahle A, Thomas MT, Stolpen AH, Scholz TD, Sonka M (2009) Congenital aortic disease: 4D magnetic resonance segmentation and quantitative analysis. Med Image Anal 13(3):483–493
39. Olszewski ME, Wahle A, Vigmostad SC, Sonka M (2005) Multidimensional segmentation of coronary intravascular ultrasound images using knowledge-based methods. In: Fitzpatrick JM, Reinhardt JM (eds) Medical imaging 2005: image processing, vol 5747, SPIE Proceedings, Bellingham WA, pp 496–504
40. Wu X, Chen DZ (2002) Optimal net surface problems with applications. In: Widmayer P, Ruiz FT, Morales R, Hennessy M, Eidenbenz S, Conejo R, (eds) 29th international colloquium on automata, languages and programming (ICALP'02), Lecture notes in computer science, vol 2380, Springer, Berlin pp 1029–1042

41. Li K, Wu X, Chen DZ, Sonka M (2006) Optimal surface segmentation in volumetric images — a graph-theoretic approach. IEEE Trans Pattern Anal Mach Intell 28(1):119–134
42. Wu X, Chen DZ, Li K, Sonka M (2007) The layered net surface problems in discrete geometry and medical image segmentation. Int J Comput Geom Appl (Selected Papers from the 16th ISAAC Conference) 17(3):261–296
43. Boykov Y, Veksler O, Zabih R (2001) Fast approximate energy minimization via graph cuts, IEEE Trans Pattern Anal Mach Intell 23(11):1222–1239
44. Garvin MK, Abràmoff MD, Kardon R, Russell SR, Wu X, Sonka M (2008) Intraretinal layer segmentation of macular optical coherence tomography images using optimal 3-D graph search, IEEE Trans Med Imaging 27(10):1495–1505
45. Li K, Millington S, Wu X, Chen DZ, Sonka M (2005) Simultaneous segmentation of multiple closed surfaces using optimal graph searching. In: Christensen GE, Sonka M (eds) Information processing in medical imaging (IPMI 2005), Lecture notes in computer science, vol 3565 Springer, Berlin, pp 406–417
46. Lee K, Yin Y, Wahle A, Olszewski ME, Sonka M (2008) 3-D segmentation and quantitative analysis of inner and outer walls of thrombotic abdominal aortic aneurysms. In: Hu XP, Clough AV (eds) Medical imaging 2008: physiology, function, and structure from medical images, vol 6916, SPIE Proceedings, Bellingham, WA pp 691626.1–691626.9
47. Liu X, Chen DZ, Wu X, Sonka M (2008) Optimal graph-based segmentation of 3D pulmonary airway and vascular trees across bifurcations. In: Brown M, de Bruijne M, van Ginneken B, Kiraly A, Kuhnigk JM, Lorenz C, Mori K, Reinhardt JM (eds) The first international workshop on pulmonary image analysis (MICCAI 2008), pp 103–111, Lulu Enterprises, Morrisville, PA
48. Liu X, Chen DZ, Wu X, Sonka M (2009) Optimal graph search based image segmentation for objects with complex topologies. In: Pluim JPW, Dawant BM (eds) Medical imaging 2009: image processing, vol 7259 SPIE Proceedings, Bellingham, WA pp 725915.1–725915.10
49. Perona P, Malik J (1990) Scale-space and edge detection using anisotropic diffusion. IEEE Transs Pattern Anal Mach Intell 12(7):629–639
50. Wagner RF, Smith SW, Sandrick JM, Lopez H (1983) Statistics of speckle in ultrasound b-Scans. IEEE Trans Sonics Ultrason 30(3):156–163
51. Roy Cardinal, MH, Meunier J, Soulez G, Maurice RL, Therasse E, Cloutier G (2006) Intravascular ultrasound image segmentation: a three-dimensional fast-marching method based on gray level distributions. IEEE Transs Med Imaging 25(5):590–601
52. Chan TF, Vese LA (2001) Active contours without edges. IEEE Trans Image Process 10(2):266–277
53. Brejl M, Sonka M (2000) Object localization and border detection criteria design in edge-based image segmentation: automated learning from examples. IEEE Trans Med Imaging 19(10):973–985
54. Anderson RH, Tynan M (1988) Tetralogy of Fallot—a centennial review. Int J Cardiol 21(3):219–232
55. Debatin JF, Nadel SS, Sostman HD, Spritzer CE, Evans AJ, Grist TM (1992) Magnetic resonance imaging—cardiac ejection fraction measurements: phantom study comparing four different methods. Invest radiol 27(3):198–204
56. Pattynama PM, Doornbos J, Hermans J, van der Wall EE, de Roos A (1992) Magnetic resonance evaluation of regional left ventricular function. Effect of through-plane motion. Invest Radiol 27(9):681–685
57. Bloomer TN, Plein S, Radjenovic A, Higgins DM, Jones TR, Ridgway JP, Sivananthan MU (2001) Cine MRI using steady state free precession in the radial long axis orientation is a fast accurate method for obtaining volumetric data of the left ventricle. J Magn Reson Imaging 14: 685–692
58. Zhang H, Walker NE, Mitchell SC, Thomas MT, Wahle A, Scholz TD, Sonka M (2006) Analysis of four-dimensional cardiac ventricular magnetic resonance images using statistical

models of ventricular shape and cardiac motion. In: Manduca A, Amini AA (eds) Medical imaging 2006: physiology, function, and structure from medical images. vol 6143, SPIE Proceedings, Bellingham, WA pp 47–57

59. Catmull E, Rom R (1974) A class of local interpolating splines. In: Barnhill RE, Riesenfeld RF (eds) Computer aided geometric design Academic. New York, NY pp 317–326
60. Unser M (1999) Splines: a perfect fit for signal and image processing. IEEE Signal Process Mag 16(6):22–38
61. Frangi A, Rueckert D, Schnabel J, Niessen W (2002) Automatic construction of multiple-object three-dimensional statistical shape models: application to cardiac modeling. IEEE Trans Med Imaging 21: 1151–1166
62. Maurer CR, Qi R, Raghavan V (2003) A linear time algorithm for computing exact euclidean distance transforms of binary images in arbitrary dimensions. IEEE Trans Pattern Anal Mach Intell 25: 265–270
63. Lorensen WE, Cline HE (1987) Marching cubes: a high resolution 3D surface construction algorithm. Comput Graph 21 163–169
64. Hoppe H (1999) New quadric metric for simplifying meshes with appearance attributes. In: Proceedings visualization ′99. IEEE Press, Piscataway, NJ, pp 59–66
65. Gatzoulis MA, Webb GD, Daubeney PEF (2003) Diagnosis and management of adult congenital heart disease. Churchill Livingstone UK
66. Anderson RH, Weinberg PM (2005) The clinical anatomy of Tetralogy of Fallot. Cardiol Young 15(1):38–47
67. Hoffman JIE, Kaplan S, Liberthson RR (2004) Prevalence of congenital heart disease. Am Heart J 147(3):425–439
68. Therrien J, Siu SC, McLaughlin PR, Liu PP, Williams WG, Webb GD, (2000) Pulmonary valve replacement in adults late after repair of tetralogy of Fallot: are we operating too late? J Am Coll Cardiol 36(5):1670–1675
69. Therrien J, Siu SC, Harris L, Dore A, Niwa K, Janousek J, Williams WG, Webb G, Gatzoulis MA (2001) Impact of pulmonary valve replacement on arrhythmia propensity late after repair of Tetralogy of Fallot. Circulation 103(20):2489–2494
70. Zhang H, Thomas MT, Walker NE, Stolpen AH, Wahle A, Scholz TD, Sonka M (2007) Four-dimensional functional analysis of left and right ventricles using MR images and active appearance models. In: Proceedings of Medical imaging 2007: physiology, function, and structure from medical images, vol. 6511, SPIE Proceedings, Bellingham, WA pp 65111M.1–65111M.10
71. The Centers for Disease Control and Prevention (CDC) (2010) http://www.cdc.gov/ Accessed Oct 2010
72. Rueckert D, Burger P, Forbat S, Mohiaddin R, Yang G (1997) Automatic tracking of the aorta in cardiovascular MR images using deformable models. IEEE Trans Med Imaging, 16(5):581–590
73. Behrens T, Rohr K, Stiehl H (2003) Robust segmentation of tubular structures in 3D medical images by parametric object detection and tracking. IEEE Trans Syst Man Cybern 33(4):554–561
74. de Bruijne M, van Ginneken B, Viergever MA, Niessen WJ (2003) Adapting active shape models for 3D segmentation of tubular structures in medical images. In: Proceedings of IPMI, LNCS, vol 2732, Springer Heidelberg pp 136–147
75. Subasic M, Loncaric S, Sorantin E (2002) 3D image analysis of abdominal aortic aneurysm. In: Sonka M, Fitzpatrick JM (ed) Medical imaging 2002: image processing, vol 4684. SPIE Press, Bellingham, WA, pp 1681–1689
76. Cheng TO (2006) Decreased aortic root distensibility rather than increased aortic root diameter as an important cardiovascular risk factor in the Marfan syndrome. Am J Cardiol, 97:1422
77. Vitarelli A, Conde Y, Cimino E, D'Angeli I, D'Orazio S, Stellato S, Padella V, Caranci F (2006) Aortic wall mechanics in the Marfan syndrome assessed by transesophageal tissue Doppler echocardiography. Am J Cardiol 97:571–577

78. Kardesoglu E, Ozmen N, Aparci M, Cebeci BS, Uz O, Dincturk M (2007) Abnormal elastic properties of the aorta in the mitral valve prolapse syndrome. Acta Cardiol 62:151–155
79. Sandor GG, Hishitani T, Petty RE, Potts MT, Desouza A, Desouza E, Potts JE, (2003) A novel Doppler echocardiographic method of measuring the biophysical properties of the aorta in pediatric patients. J Am Soc Echocardiogr 16:745–750
80. Bardinet E, Cohen L, Ayache N (1996) Tracking and motion analysis of the left ventricle with deformable superquadrics. Med Image Anal 1(2):129–49
81. Chandrashekara R, Rao A, Sanchez-Ortiz GI, Mohiaddin RH, Rueckert D (2003) Construction of a statistical model for cardiac motion analysis using nonrigid image registration. In: Proceedings of IPMI, LNCS, vol 2878, Springer, Heidelberg pp 599–610
82. Malladi R, Sethian JA (1998) A real-time algorithm for medical shape recovery. In: Proceedings of ICCV. Narosa Publishing House, New Delhi, India, pp 304–310
83. Frangi A, Niessen W, Vincken K, Viergever M (1998) Multiscale vessel enhancement filtering, In: Proceedings of MICCAI, LNCS, vol 1496, pp 130–137, Springer Germany
84. Palagyi K, Tschirren J, Sonka M (2003) Quantitative Analysis of Intrathoracic Airway Trees: Methods and Validation, In: Proceedings of IPMI, LNCS vol 2732, Springer, Heidelberg pp 222–233
85. Unser M, Aldroubi A, Eden M (1993) B-Spline signal processing: part II - efficient design and applications. IEEE Trans Signal Process 41:834–848
86. Sonka M, Zhang X, Siebes M, Bissing MS, DeJong SC, Collins SM, McKay CR, (1995) Segmentation of intravascular ultrasound images: a knowledge-based approach. IEEE Trans Med Imaging 14(4):719–732
87. Tschirren J, Hoffman EA, McLennan G, Sonka M, (2005) Intrathoracic airway trees: segmentation and airway morphology analysis from low-dose CT scans. IEEE Trans Med Imaging 24(12):1529–1539
88. Sonka M, Reddy GK, Winniford MD, Collins SM, (1997) Adaptive approach to accurate analysis of small-diameter vessels in cineangiograms. IEEE Trans Med Imaging 16(2): 87–95
89. Lee S, Wolberg G, Shin SY (1997) Scattered data interpolation with multilevel B-splines. IEEE Trans Visual Comput Graph, 3:228–244
90. Boser BE, Guyon I, Vapnik V (1992) A training algorithm for optimal margin classifiers. In: Proceedings of the fifth annual workshop on computational learning theory. ACM, New York, NY, pp 144–152
91. Cortes C, Vapnik V, (1995) Support-vector networks. Mach Learn, 20(3):273–297
92. Cristianini N, Taylor JS (2000) An introduction to support vector machines and other kernel-based learning methods. Cambridge University Press, Cambridge
93. Chang C-C, Lin C-J, (2001) LIBSVM: A library for support vector machines, http://www.csie.ntu.edu.tw/~cjlin/libsvm Accessed Oct 2010
94. Olabarriaga S, Rouet J, Fradkin M, Breeuwer M, Niessen W (2005) Segmentation of thrombus in abdominal aortic aneurysms from CTA with non-parametric statistical grey level appearance modelling. IEEE Trans Med Imaging 24(4):477–486
95. Loncaric S, Subasic M, Sorantin E (2000) 3-D deformable model for abdominal aortic aneurysm segmentation from CT images. In: Proceedings of first int'l workshop on image and signal processing and analysis, pp 139–144
96. Tek H, Comaniciu D, Williams J (2001) Vessel detection by mean-shift based ray propagation. In: Proceedings IEEE workshop on mathematical methods in biomedical image analysis (MMBIA 2001). IEEE Press, Piscataway, NJ, pp 228–235
97. Subasic M, Loncaric S, Sorantin E (2001) 3D image analysis of abdominal aortic aneurysm. In: Sonka M, Hanson K (eds) Medical imaging, medical imaging 2001: image processing, vol 4322, SPIE Press, Bellingham, pp 388–394
98. Bruijne M, Ginneken B, Viergever M, Niessen W (2004) Interactive segmentation of abdominal aortic aneurysms in CTA data. Med Image Anal 8(2):127–138
99. Yu Y, Acton ST (2002) Speckle reducing anisotropic diffusion. IEEE Trans Image Process, 11(11):1260–1270

100. Dodge JT, Brown BG, Bolson EL, Dodge HT (1988) Intrathoracic spatial location of specified coronary segments on the normal human heart. Circulation 78(5):1167–1180
101. Stary HC, Chandler AB, Dinsmore RE, Fuster V, Glagov S, Insull W, Rosenfeld ME, Schwartz CJ, Wagner WD, Wissler RW (1995) A definition of advanced types of atherosclerotic lesions and a histological classification of atherosclerosis: a report from the committee on vascular lesions of the council on arteriosclerosis, american heart association. Arteriosc Thromb Vasc Biol, 15(9):1512–1531
102. Glagov S, Weisenberg E, Zarins CK, Stankunavicius R, Kolettis GJ, (1987) Compensatory enlargement of human atherosclerotic coronary arteries. N Engl J Med 316(22):1371–1375
103. Gibson CM, Diaz L, Kandarpa K, Sacks FM, Pasternak RC, Sandor T, Feldman CL, Stone PH (1993) Relation of vessel wall shear stress to atherosclerosis progression in human coronary arteries. Arteriosc Thromb, 13(2):310–315
104. Friedman MH, Bargeron CB, Deters OJ, Hutchins GM, Mark FF (1987) Correlation between wall shear and intimal thickness at a coronary artery branch. Atherosclerosis 68(1/2):27–33
105. Wahle A, Lopez JJ, Olszewski ME, Vigmostad SC, Chandran KB, Rossen JD, Sonka M, (2005) Analysis of the interdependencies among plaque development, vessel curvature, and wall shear stress in coronary arteries. In: Frangi AF, Radeva PI, Santos A, Hernandez M (eds) Functional imaging and modeling of the Heart (FIMH '05), Lecture Notes in computer science. vol. 3504. pp 12–22, Springer, Berlin
106. Wentzel JJ, Gijsen FJH, Stergiopulos N, Serruys PW, Slager CJ, Krams R (2003) Shear stress, vascular remodeling and neointimal formation. J Biomech, 36(5):681–688
107. Stone PH, Coşkun AÜ, Kinlay S, Clark ME, Sonka M, Wahle A, Ilegbusi OJ, Yeghiazarians Y, Popma JJ, Orav J, Kuntz RE, Feldman CL (2003) Effect of endothelial shear stress on the progression of coronary artery disease, vascular remodeling, and in-stent restenosis in man; In-vivo 6-month followup study. Circulation 108(4):438–444
108. Brown BG, Simpson P, Dodge JT, Bolson EL, Dodge HT, (1991) Quantitative and qualitative coronary arteriography. In: Reiber JHC, Serruys PW, (eds) Quantitative coronary arteriography, Developments in Cardiovascular Medicine. vol. 117. Kluwer, Dordrecht pp 3–21
109. Reiber JHC, Koning G, Dijkstra J, Wahle A, Goedhart B, Sheehan FH, Sonka M (2000) Angiography and intravascular ultrasound. In: Sonka M, Fitzpatrick JM (eds) Handbook of medical imaging — volume 2: medical image processing and analysis, SPIE Press, Bellingham, WA pp 711–808
110. von Birgelen C, de Vrey EA, Mintz GS, Nicosia A, Bruining N, Li W, Slager CJ, Roelandt JRTC, Serruys PW, de Feyter PJ, (1997) ECG-gated three-dimensional intravascular ultrasound: feasibility and reproducibility of the automated analysis of coronary lumen and atherosclerotic plaque dimensions in humans. Circulation 96(9):2944–2952
111. Herrington DM, Johnson T, Santago P, Snyder WE (1992) Semi-automated boundary detection for intravascular ultrasound. In: Proceedings of Computers in cardiology 1992, Durham, NC, IEEE-CS Press, Los Alamitos, CA, pp 103–106
112. Li W, von Birgelen C, Di Mario C, Boersma E, Gussenhoven EJ, van der Putten NHJJ, Bom N (1994/1995) Semi-automatic contour detection for volumetric quantification of intravascular ultrasound. In: Proceedings of computers in cardiology 1994, Bethesda MD, IEEE-CS Press, Los Alamitos, CA, pp 277–280
113. Klingensmith JD, Shekhar R, Vince DG (2000) Evaluation of three-dimensional segmentation algorithms for the identification of luminal and medial-adventitial borders in intravascular ultrasound Images. IEEE Trans Med Imaging 19(10):996–1011
114. Brusseau E, de Korte CL, Mastik F, Schaar J, van der Steen AFW (2004) Fully automatic luminal contour segmentation in intracoronary ultrasound imaging — a statistical approach. IEEE Trans Med Imaging 23(5):554–566
115. Nair A, Kuban BD, Tuzcu EM, Schoenhagen P, Nissen SE, Vince DG (2002) Coronary plaque classification with intravascular ultrasound radiofrequency analysis. Circulation 106(17):2200–2206

116. Wahle A, Prause GPM, von Birgelen C, Erbel R, Sonka M (1999) Fusion of angiography and intravascular ultrasound in-vivo: establishing the absolute 3-D frame orientation, IEEE Trans Biomed Eng—Biomed Data Fusion, 46(10):1176–1180
117. Wahle A, Prause GPM, DeJong SC, Sonka M (1999) Geometrically correct 3-D reconstruction of intravascular ultrasound images by fusion with biplane angiography—methods and validation. IEEE Trans Med Imaging 18(8):686–699
118. Wahle A, Sonka M (2005) Coronary plaque analysis by multimodality fusion. In: Suri JS, Yuan C, Wilson DL, Laxminarayan S (eds) plaque imaging: pixel to molecular level, Studies in Health, Technology and Informatics vol 113, IOS Press, Amsterdam pp 321–359
119. Brown BG, Bolson EL, Frimer M, Dodge HT (1977) Quantitative coronary arteriography; estimation of dimensions, hemodynamic resistance, and atheroma mass of coronary artery lesions using the arteriogram and digital computation. Circulation 55(2):329–337
120. Klingensmith JD, Schoenhagen P, Tajaddini A, Halliburton SS, Tuzcu EM, Nissen SE, Vince DG (2003) Automated three-dimensional assessment of coronary artery anatomy with intravascular ultrasound scanning. Am Heart J 145(5):795–805
121. Kass M, Witkin A, Terzopulous D (1988) Snakes: active contour models. Int J Comput Vis 1(4):321–331
122. Mojsilović A, Popović M, Amodaj M, Babić R, Ostojić M (1997) Automatic segmentation of Intravascular ultrasound images; a texture-based approach. Ann Biomed Eng 25(6): 1059–1071
123. Zhang X, McKay CR, Sonka M (1998) Tissue characterization in intravascular ultrasound images. IEEE Trans Med Imaging 17(6):889–899
124. Burckhardt CB (1978) Speckle in ultrasound B-mode scans. IEEE Trans Son Ultrason SU-25(1):1–6
125. Evans JL, Ng KH, Wiet SG, Vonesh MJ, Burns WB, Radvany MG, Kane BJ, Davidson CJ, Roth SI, Kramer BL, Meyers SN, McPherson DD (1996) Accurate three-dimensional reconstruction of intravascular ultrasound data; spatially correct three-dimensional reconstructions. Circulation 93(3):567–576
126. Pellot C, Bloch I, Herment A, Sureda F (1996) An attempt to 3-D reconstruct vessel morphology from X-Ray projections and intravascular ultrasounds modeling and fusion, Comput Medi Imaging Graph 20(3):141–151
127. Shekhar R, Cothren RM, Vince DG, Cornhill JF (1996) Fusion of intravascular ultrasound and biplane angiography for three-dimensional reconstruction of coronary arteries. In: Proceedings of Computers in Cardiology 1996, Indianapolis, IN, IEEE Press, Piscataway, NJ, pp 5–8
128. Laban M, Oomen JA, Slager CJ, Wentzel JJ, Krams R, Schuurbiers JCH, den Boer A, von Birgelen C, Serruys PW, de Feyter PJ, (1995) ANGUS: a new approach to three-dimensional reconstruction of coronary vessels by combined use of angiography and intravascular ultrasound. In: Proceedings of Computers in cardiology 1995, Vienna, AT, IEEE Press, Piscataway, NJ, pp 325–328
129. Prause GPM, DeJong SC, McKay CR, Sonka M (1996) Semi-automated segmentation and 3-D reconstruction of coronary trees: biplane angiography and intravascular ultrasound data fusion, In: Hoffman EA (eds) Medical imaging 1996: physiology and function from multidimensional images, vol. 2709, SPIE Proceedings, Bellingham, WA, pp 82–92
130. Lai YG, Przekwas AJ (1994) A finite-volume method for fluid flow simulations with moving boundaries. comput Fluid Dyn, 2:19–40
131. Ramaswamy SD, Vigmostad SC, Wahle A, Lai YG, Olszewski ME, Braddy KC, Brennan TMH, Rossen JD, Sonka M, Chandran KB (2004) Fluid dynamic analysis in a human left anterior descending coronary artery with arterial motion. Ann Biomed Eng, 32(12): 1628–1641
132. Sabbah HN, Walburn FJ, Stein PD (1984) Patterns of flow in the left coronary artery. J Biomech Eng 106(3):272–279

133. Perktold K, Hofer M, Rappitsch G, Loew M, Kuban BD, Friedman MH (1998) Validated computation of physiologic flow in a realistic coronary artery branch. J Biomech 31(3): 217–228
134. Wentzel JJ, Janssen E, Vos J, Schuurbiers JCH, Krams R, Serruys PW, de Feyter PJ, Slager CJ (2003) Extension of increased atherosclerotic wall thickness into high shear stress regions is associated with loss of compensatory remodeling. Circulation 108(1): 17–23
135. Hu S, Reinhardt JM, Hoffman EA (2001) Automatic lung segmentation for accurate quantitation of volumetric X-ray CT images. IEEE Trans Med Imaging 20(6):490–498
136. Guo J, Reinhardt JM, Kitaoka H, Zhang L, McLennan G, Hoffman EA (2002) Integrated system for CT-based assessment of parenchymal lung disease. In: 2002 international symposium on biomedical imaging, Washington, DC, pp 871–874, 7–10 July 2002
137. Otsu N (1979) A Threshold selection method from gray–level histograms. IEEE Trans Syst Man Cybern 9(1):62–66
138. Chiplunkar R, Reinhardt JM, Hoffman EA (1997) Segmentation and quantitation of the primary human airway tree. In: SPIE Medical Imaging. San Diego, CA
139. Tozaki T, Kawata Y, Niki N, Ohmatsu H, Kakinuma R, Eguchi K, Kaneko M, Moriyama N (1998) Pulmonary organs analysis for differential diagnosis based on thoracic thin-section CT images. IEEE Trans Nucl Sci 45(12)3075–3082
140. Mori K, Suenaga Y, Toriwaki J (2000) Automated anatomical labeling of the bronchial branch and its application to the virtual bronchoscopy. IEEE Trans Medi Imaging 19(2): 103–114
141. Law TY, Heng PA (2000) Automated extraction of bronchus from 3D CT images of lung based on genetic algorithm and 3D region growing. In: SPIE Proceedings on Medical Imaging. vol 3979. San Diego, CA. pp 906–916
142. Pisupati C, Wolf L, Mitzner W, Zerhouni E (1996) Mathematical morphology and its applications to image and signal processing, Chapter. Segmentation of 3D pulmonary trees using mathematical morphology. Kluwer Dordrecht pp 409–416
143. Prêteux F, Fetita CI, Grenier P, Capderou A (1999) Modeling, segmentation, and caliber estimation of bronchi in high-resolution computerized tomography. J Electron Imaging 8(1):36–45
144. Fetita CI, Prêteux F (2002) Quantitative 3D CT bronchography. In: Proceedings IEEE international symposium on biomedical imaging (ISBI'02). Washington, DC, July, 2002
145. Bilgen D (2000) Segmentation and analysis of the human airway tree from 3D X-ray CT images. Master's thesis, The University of Iowa, IA, USA, December 2000
146. Kiraly AP (2003) 3D Image analysis and visualization of tubular structures. Ph.D. thesis, The Pennsylvania State University, Department of Computer Science and Engineering, May, 2003
147. Aykac D, Hoffman EA, McLennan G, Reinhardt JM (Aug 2003) Segmentation and analysis ofthe human airway tree from 3D X-Ray CT images. IEEE Trans med imaging 22(8): 940–950
148. Sonka M, Sundaramoorthy G, Hoffman EA (1994) Knowledge-based segmentation of intrathoracic airways from multidimensional high resolution CT images. In: Eric AH, Acharya RS (eds) Physiology and function from multidimensional images, medical imaging 1994: physiology and function from multidimensional images. SPIE Press, Bellingham, vol 2168. pp 73–85
149. Park W, Hoffman EA, Sonka M (1998) Segmentation of intrathoracic airway trees: a fuzzy logic approach. IEEE Trans, Med Imaging 17(8):489–497
150. Kitasaka T, Mori K, Hasegawa H-i, Suenaga Y, Toriwaki J-i (2003) Extraction of bronchus regions from 3D chest X-ray CT images by using structural features of bronchus. In: Computer assisted radiology and surgery (CARS) 2003, International Congress Series 1256, Elsevier, 2003, pp 240–245

151. Schlathölter T, Lorenz C, Carlsen IC, Renisch S, Deschamps T (2002) Simultaneous segmentation and tree reconstruction of the airways for virtual bronchoscopy. In: SPIE Medical Imaging 2002. Image processing. San Diego, CA. pp 103–113, February 2002
152. King GG, Müller NL, Parè PD (1999) Evaluation of airways in obstructive pulmonary disease using high-resolution computed tomography. Am J Respir Crit Care Med 159(3):992–1004
153. King GG, Müller NL, Whittall KP, Xiang Q-S, Paré PD (2000) An analysis algorithm for measuring airway lumen and wall areas from high-resolution computed tomographic data. Am J Respir Crit Care Med 161(2):574–580
154. Wood SA, Zerhouni EA, Hoford JD, Hoffman EA, Mitzner W (1993) Quantitative 3-D reconstruction of airway and pulmonary vascular trees using HRCT. In: SPIE proceedings biomedical image processing and biomedical visualization. vol 1905. San Jose, CA, pp 316–323
155. Reinhardt JM, D'Souza ND, Hoffman EA (1997) Accurate measurement of intra-thoracic airways. IEEE Trans Med Imaging 16(12):820–827
156. Reinhardt JM, Park W, Hoffman EA, Sonka M (1997) Intrathoracic airway wall detection using graph search with CT scanner PSF information. In: Proceedings of SPIE conference medical imaging, vol 3033. Newport Beach, CA, 23–28 Feb, pp 93–101
157. Prêteux F, Fetita CI, Grenier P (1997) Modeling, segmentation, and caliber estimation of bronchi in high-resolution computerized tomography. In: Statistical and stochastic methods in image processing II. SPIE proceedings, vol 3167. San Diego, CA, pp 58–69, July 1997
158. Saba OI, Hoffman EA, Reinhardt JM (2000) Computed tomographic-based estimation of airway size with correction for scanned plane tilt angle. In: Chen C-T, Anne VC (eds) Medical Imaging 2000: physiology and function from multidimensional images, vol 3978, pp 58–66
159. Wiemker R, Blaffert T, Bülow T, Renisch S, Lorenz C (2004) Automated assessment of bronchial lumen, wall thickness and bronchoarterial diameter ratio of the tracheobronchial tree using high-resolution CT. In: Computer assisted radiology and surgery (CARS 2004) excerpta medica international congress series, vol 1268. Elsevier, Amsterdam, NL, pp 967–972
160. Herman GT, Carvalho BM (2001) Multiseeded segmentation using fuzzy connectedness, IEEE Trans Pattern Anal Mach Intell. 23(5):460–474
161. Sato Y, Nakajima S, Shiraga N, Atsumi H, Yoshida S, Koller T, Gerig G, Kikins R (1998) Three-dimensional multi-scale line filter for segmentation and visualization of curvilinear structures in medical images. Med Image Anal 2(2):143–168
162. Krissian K, Malandain G, Ayache N, Vaillant R, Trousset Y (2000) Model based detection of tubular structures in 3D images. Comput Vis Image Underst 80(2):130–171
163. Lorigo LM, Faugeras OD, Grimson WEL, Keriven R, Kikinis R, Nabavi A, Westin CF (2001) CURVES: curve evolution for vessel segmentation. Med Image Anal 5(3):195–206
164. Vasilevskiy A, Siddiqi K (2002) Flux maximizing geometric flows. IEEE Trans Pattern Anal Mach Intell 24(12):1565–1578
165. Aylward SR, Bullitt E (2002) Initialization, noise, singularities, and scale in height ridge traversal for tublar object centerline extraction. IEEE Trans Med Imaging 21(2):61–75
166. Boldak C, Rolland Y, Toumoulin C (2003) An improved model-based vessel tracking algorithm with application to computed tomography angiography. J Biocybern Biomed Eng, 23(1):41–63
167. Mayer D, Bartz D, Fischer J, Ley S, del Ro A, Thust S, Kauczor HU, Strasser W, Heussel CP (2004) Hybrid segmentation and virtual bronchoscopy based on CT Images. Acad Radiol 11(5)
168. Fridman Y, Pizer SM, Aylward SR, Bullitt E (2003) Segmenting 3D branching tubular structures using cores. In: Goos G, Hartmanis J, Leeuwen JV (eds) Medical image computing and computer-assisted intervention (MICCAI 2003) (Lecture notes in computer science), vol 2879. Springer, Berlin, pp 570–577

169. Pizer SM, Eberly D, Fritsch DS (January 1998) Zoom-invariant vision of figural shape: the mathematics of cores. Comput Vis image underst 69(1):55–71
170. Morse BS, Pizer SM, Puff DT, Gu C (1998) Zoom-invariant vision of figural shape: effects on cores of image disturbances. Comput Vis Image Underst 69(1):72–86
171. Shikata H, Hoffman E, Sonka M (2004) Automated segmentation of pulmonary vascular tree from 3D CT images. In: Amini AA, Manduca A (eds) Proceedings medical imaging 2004: physiology, function, and structure from medical images. SPIE Press, Bellingham, vol 5369, pp 107–116
172. Sato Y, Tamura S (2000) Detection and quantification of line and sheet structures in 3-D images. In: Delp SL, DiGoia AM, Jaramaz B (eds) Lecture notes in medical image computing and computer-assisted intervention - MICCAI 2000 LNCS 1935, vol 1935. Springer, Berlin, pp 154–165
173. Zhou Y, Toga W (1999) Efficient skeletonization of volumetric objects. IEEE Trans Visual Comput Graph 5(3):196–209
174. Saha PK, Gao Z, Alford S, Sonka M, Hoffman E (2009) A novel multiscale topomorphometric approach for separating arteries and veins via pulmonary CT imaging. In: Josien PWP, Benoit M, Dawant (eds) Medical imaging 2009: image processing, vol 7259. Bellingham, WA, pp 725910–725910
175. Hoffman EA, Ritman EL (1985) Effect of body orientation on regional lung expansion in dog and sloth. J Appl Physiol 59(2):481–491
176. Hubmayr RD, Rodarte JR, Walters BJ, Tonelli FM (1987) Regional ventilation during spontaneous breathing and mechanical ventilation in dogs. J Appl Physiol, 63(6):2467–2475
177. Ukil S, Reinhardt JM (2009) Anatomy-guided lung lobe segmentation in X-Ray CT images. IEEE Trans Med Imaging 28(2):202–214
178. van Rikxoort EM, van Ginneken B, Klik M, Prokop M (2008) Supervised enhancement filters: application to fissure detection in chest CT scans. IEEE Trans Med Imaging 27(1):1–10, doi:10.1109/TMI.2007.900447
179. Zhang L, Hoffman EA, Reinhardt JM (2006) Atlas-driven lung lobe segmentation in volumetric X-ray CT images. IEEE Trans Med Imaging 25(1):1–16
180. Wang J, Betke M, Ko JP (2006) Pulmonary fissure segmentation on CT. Medi Image Analy 10(4):530–547
181. Wiemker R, Bülow T, Blaffert T (2005) Unsupervised extraction of the pulmonary interlobar fissures from high resolution thoracic CT data. In: Proceedings of the 19th international congress and exhibition – computer assisted radiology and surgery (CARS). Berlin, Germany, pp 1121–1126
182. Kuhnigk JM, Hahn HK, Hindennach M, Dicken V, Krass S, Peitgen HO (2003) Lung lobe segmentation by anatomy-guided 3-D watershed transform. In: Milan SJ, Fitzpatrick M (eds) Medical imaging 2003: image processing, vol 5032. SPIE Press, Bellingham, WA, pp 1482–1490
183. Zhou X, Hayashi T, Hara T, Fujita H (2004) Automatic recognition of lung lobes and fissures from multislice CT images. In: Proceedings of SPIE conference medical imaging, vol 5370, San Diego, CA, pp 1629–1633
184. Tschirren J, McLennan G, Palágyi K, Hoffman EA, Sonka M (2005) Matching and anatomical labeling of human airway tree. IEEE Trans Med Imaging 24(12):1540–1547

Part II
Computational Techniques for Fluid and Soft Tissue Mechanics, Fluid–Structure Interaction, and Development of Multi-scale Simulations

Chapter 3
Computational Techniques for Biological Fluids: From Blood Vessel Scale to Blood Cells

Fotis Sotiropoulos, Cyrus Aidun, Iman Borazjani, and Robert MacMeccan

Abstract Simulation of flows in the cardiovascular system faces many challenges. Chief among these is the issue of treatment of blood flow at disparate scales. For blood flows through large vessels a Newtonian homogeneous fluid model can be adequate, while in the capillaries and in orifices and constrictions individual blood cells and interactions among blood cells assume importance. Another important feature of flows in the cardiovascular system or in the presence of cardiovascular prostheses is the interaction of blood with moving boundaries (e.g. arterial walls, heart, heart valves, and ventricular assist devices). Computational fluid dynamics has made significant progress in tackling these challenges to the extent that it is now feasible to calculate flows through parts of the cardiovascular system with a great degree of fidelity and physiological realism. This chapter presents fundamental aspects of the demands on and capabilities of numerical solution techniques for solving a variety of blood flow phenomena. Large scale flows with significant fluid inertia and small scale flows with individual blood cells are covered. Applications of the methods and sample results are shown to illustrate the state-of-the-art of computations in cardiovascular biofluid dynamics.

3.1 Introduction

Hemodynamic stress plays an important biological role in many aspects of the cardiovascular system at the macroscopic as well as the cellular level. The stresses experienced by blood elements and endothelial cells arise due to the combined action of hemodynamic shear stress and structural stresses developing in the vasculature, as well as the local cellular stress on platelets and endothelial cells. The magnitude of these stresses at the cell level is ultimately determined by complex

F. Sotiropoulos (✉)
St. Anthony Falls Laboratory, Department of Civil Engineering, University of Minnesota, Minneapolis, MN 55414, USA
e-mail: fotis@umn.edu

K.B. Chandran et al. (eds.), *Image-Based Computational Modeling of the Human Circulatory and Pulmonary Systems*, DOI 10.1007/978-1-4419-7350-4_3,

fluid–structure interactions (FSIs) that occur across a wide range of scales. At the blood vessel scale (approximately millimeter), FSI between the flowing blood and the compliant vascular walls in conjunction with the complex anatomic geometry of typical blood vessels and the pulsatile nature of blood flow gives rise to a highly three-dimensional and unsteady flow environment. The term "disturbed flow" is often used in the literature to characterize this environment, which is dominated, among others, by aperiodic vortex shedding, regions of recirculation, vortex formation with rich vorticity dynamics, and even regions where transition to turbulence occurs. These complex macro-scale phenomena collectively comprise the ambient flow environment within which blood elements flow and interact with each other and the endothelium. The mechanical stresses that are ultimately experienced by cells, however, are determined by phenomena and interactions that occur at the cellular scale (approximately micrometer) (see Ge et al. [1] for a detailed related discussion in the context of heart valve flows). Therefore, a critical prerequisite for understanding and quantifying the links between mechanical stresses and disease pathways is being able to quantify hemodynamic stresses at physiologic conditions across the entire range of hemodynamically relevant scales: from the macro-scale of the blood vessel diameter to the micro-scale of a blood element or a cell. The only viable alternative for accomplishing this challenging undertaking is to employ computational modeling.

In this chapter we review the state of the art in computational methods for simulating cardiovascular flow phenomena at both the macro- and micro-scales. In Section 3.2 we describe the governing equations of blood flow at the macro-scale and present an overview of numerical techniques for simulating fluid–structure interaction problems involving arbitrarily complex domains with compliant walls and moving immersed boundaries. We also review recent applications of such methods to a few representative cardiovascular flow problems with emphasis on high-resolution numerical simulations that resolve flow phenomena at hemodynamically relevant scales. In Section 3.3 we review numerical methods for carrying out FSI simulations at cell-resolving scales and report results demonstrating the ability of such methods to perform direct numerical simulations of whole blood. In Section 3.4 we discuss future directions, emphasizing the need for developing true multi-scale computational algorithms that couple together methods such as those reviewed in Sections 3.2 and 3.3.

3.2 Computational Methods for Macro-scale Hemodynamics

3.2.1 Governing Equations

3.2.1.1 The Fluid Flow Equations

The flow of blood in the cardiovascular system is governed by the three-dimensional, unsteady, incompressible continuity and Navier–Stokes equations. Using tensor notation (repeated indices imply summation) the governing equations non-dimensionalized by a characteristic velocity U and a characteristic length L read as follows (i=1,2,3):

$$\begin{aligned} \frac{\partial u_i}{\partial x_i} &= 0 \\ \frac{Du_i}{Dt} &= -\frac{1}{\rho}\frac{\partial p}{\partial x_i} + \frac{1}{\mathrm{Re}}\frac{\partial \sigma_{ij}}{\partial x_j} \end{aligned} \tag{3.1}$$

where u_i is the ith component of the velocity vector, ρ is the density of blood, p is the pressure, Re is the Reynolds number, defined as $\mathrm{Re} = UL/\nu$, ν is the kinematic viscosity, D/Dt is the material derivative, defined as $D/Dt(\cdot) = \partial/\partial t(\cdot) + u_j\partial/\partial x_j(\cdot)$, and σ_{ij} is the stress tensor. A constitutive equation for the stress tensor is required to close the governing equations. In sufficiently large vessels, blood behaves as a Newtonian fluid [2], i.e., the stress tensor is a linear function of the strain rate and can be expressed as follows:

$$\sigma_{ij} = 2\mu\gamma_{ij} \tag{3.2}$$

where γ_{ij} is the strain rate tensor:

$$\gamma_{ij} = \frac{1}{2}\left(\frac{\partial u_i}{\partial x_j} + \frac{\partial u_j}{\partial x_i}\right) \tag{3.3}$$

In the above equations, μ is the dynamic viscosity of blood, which is considered constant and equal to 4 cP.

For most macro-scale simulations the Newtonian fluid assumption is satisfactory. Nevertheless, in small arteries and veins or in small-scale design features of various medical devices (e.g., the hinges of a mechanical heart valve), the size of blood cells (~10 μm) may not be negligible relative to the size of the flow domain. In such cases, non-Newtonian effects become important and evidence suggests the emergence of shear-thinning behavior [3]. The non-Newtonian behavior of blood is well studied in the field of hemorheologysee reviews [3–5]—and only a brief overview is presented here. Blood is a two-phase suspension of blood cells (red/white blood cells and platelets) in an aqueous solution called plasma [4]. The apparent viscosity of blood depends on the red blood cells (red blood cell volume fraction, red blood cell aggregation, and mechanical properties of red blood cells), as they constitute the greatest proportion (about 98%) of all cells [4, 6] and microscopic models are required to capture the blood behavior under all conditions. Microscopic models of whole blood that can resolve individual blood cell interactions are discussed extensively in the second part of this chapter; so here we focus only on continuum, macroscopic models of blood rheology that are suitable for macroscopic numerical simulations. Such models can be classified into generalized Newtonian and rate-type models [7].

In the generalized Newtonian models, the blood viscosity μ is not constant but depends on the shear rate, which is a scalar measure of the strain tensor defined as follows:

$$\dot{\gamma} = \sqrt{2\gamma_{ik}\gamma_{kj}\delta_{ij}} \tag{3.4}$$

where δ_{ij} is Kronecker's delta. The stress tensor is thus given as follows:

$$\sigma_{ij} = 2\mu(\dot{\gamma})\gamma_{ij} \tag{3.5}$$

Generalized Newtonian models (e.g., power law, Powell–Eyring) differ from each other only in the specific form of the functional dependence of the viscosity to the shear ratesee [7] for review of several such models. A shortcoming of the generalized Newtonian models is that they are inherently incapable of describing the viscoelastic behavior of blood. This shortcoming can be mitigated by the so-called rate-type models, which rely on the Maxwell fluid constitutive equation [7]:

$$\sigma_{ij} + \lambda \overset{\nabla}{\sigma}_{ij} = 2\mu_0\gamma_{ij} \tag{3.6}$$

where $\overset{\nabla}{\sigma}_{ij}$ denotes the upper convected derivative operator of the stress tensor field defined as follows:

$$\overset{\nabla}{\sigma}_{ij} = \frac{D}{Dt}\left(\sigma_{ij}\right) - \sigma_{ik}\frac{\partial u_j}{\partial x_k} - \frac{\partial u_k}{\partial x_i}\sigma_{kj} \tag{3.7}$$

The constant λ is the stress relaxation time and μ_0 is the viscosity coefficient at zero shear rate. A more general model is the Oldroyd-type model:

$$\sigma_{ij} + \lambda_1 \overset{\nabla}{\sigma}_{ij} = 2\mu_0(\gamma_{ij} + \lambda_2 \overset{\nabla}{\gamma}_{ij}) \tag{3.8}$$

where λ_1 and λ_2 are relaxation times with $0 \leq \lambda_2 \leq \lambda_1$. Implementing such non-Newtonian constitutive models into the governing equations (3.1) gives rise to a highly non-linear system of equations and special non-linear numerical techniques are required to solve the resulting system of equations. Such methods are beyond the scope of this chapter and the interested reader is referred to [7] for more details.

In typical cardiovascular flow applications, the governing equations need to be solved in dynamically evolving domains bounded by compliant boundaries and containing arbitrarily complex 3D deformable bodies (see Fig. 3.1). Therefore, the governing equations (3.1) with a constitutive model for the stress tensor need to be supplemented with appropriate boundary conditions at the outer boundary of the flow domain—say the wall of the aorta—as well as at the inner moving immersed boundaries, e.g., the leaflets of a heart valve (see Fig. 3.1). In general the motion of the moving boundaries could either be prescribed (e.g., from imaging modalities) or be calculated through a fluid–structure interaction (FSI) scheme.

3.2.1.2 The Structural Equations

For FSI simulations the motion of structural components should be calculated in a coupled manner with the solution of the flow field. For the most general case, the

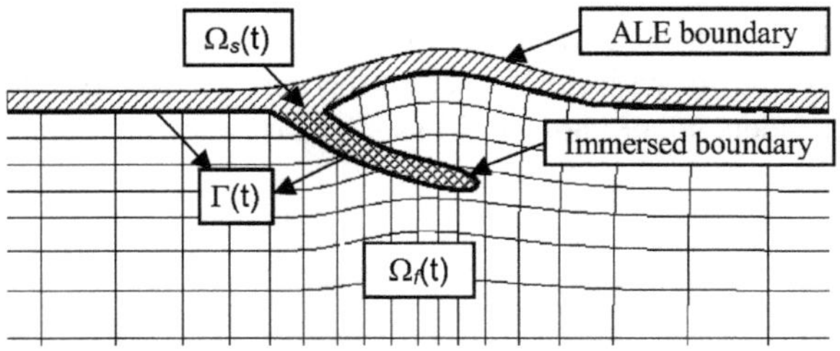

Fig. 3.1 A typical cardiovascular fluid–structure interaction problem with evolving fluid $\Omega_f(t)$ and the solid $\Omega_s(t)$ domains. The governing equations are solved on the fluid $\Omega_f(t)$ and solid $\Omega_s(t)$ domains with the boundary conditions from the other domain, i.e., the fluid domain provided the forces on the solid walls to calculate the motion of the solid walls, while the solid domain provides the location and velocity of the moving boundary $\Gamma(t)$

motion of a structural component can be divided into (1) a rigid body motion of the body's principal axes without changing its shape or size and (2) deformation of the body relative to its principal axis. However, in most cardiovascular applications such a general motion is not observed and only one of the components may be present. For example, in devices with rigid parts, such as bileaflet mechanical heart valves or cage-ball valves, only the rigid body motion is present, while in tissue valves or compliant arteries, only the deformation is present.

A general rigid body motion of the principal axes is governed by six equations: three linear momentum equations for the center of mass, describing translational motion, and three angular momentum equations, describing rotations (directional angles) around the principal axes [8]. These equations are formulated as follows:

$$m\frac{\mathrm{d}^2 r_i^{\mathrm{cm}}}{\mathrm{d}t^2} = F_{x_i}, \text{with } r_i^{\mathrm{cm}}(0) = r_{i0}^{\mathrm{cm}} \tag{3.9}$$

$$I_{ii}\frac{\mathrm{d}\omega_i}{\mathrm{d}t} + (I_{kk} - I_{jj})\omega_j\omega_k = M_i, \text{with } \omega_i(0) = \omega_{i0} \tag{3.10}$$

where m is the mass of the structure, F_{xi} is the total force imparted by the fluid in the x_i direction in the inertial frame, and I_{ii} and M_i are the moment of inertia and the total moment around the ith principal axis, respectively. Solving Equation (3.9) with the given initial condition yields the position of the center of mass. However, solving Equation (3.10) gives only the instantaneous angular velocities around each principal axis. Nevertheless, the orientation angles and the rotation matrix **A** can be calculated from the local angular velocities. There are several ways to define such orientation angles. In aerodynamics, the Euler angles ϕ, θ, and ψ are usually labeled as yaw, pitch, and roll angles, respectively. However, such definition can create a singularity at cos θ=0, which is sometimes referred to as "gimbal lock" [9]. To avoid gimbal lock, the quaternion representation (also known as the Euler parameters) of the rigid body rotation is used. The quaternion components are parameters similar to the Euler angles that define the orientation of the rigid body relative to the inertial frame but unlike Euler angles, there are only four components

$\varepsilon = \left[\varepsilon_1\ \varepsilon_2\ \varepsilon_3\ \varepsilon_4\right]^{\mathrm{T}}$. Since a general rotation can have only three degrees of freedom (DOF) and according to the Euler's rotation theorem [8] the four quaternion components are not independent and should satisfy the following constraint equation:

$$\varepsilon_1^2 + \varepsilon_2^2 + \varepsilon_3^2 + \varepsilon_4^2 = 1 \tag{3.11}$$

then the rotation matrix $\mathbf{A}(\varepsilon)$ is given by (see [8]):

$$\mathbf{A}(\varepsilon) = \begin{bmatrix} \varepsilon_1^2 - \varepsilon_2^2 - \varepsilon_3^2 + \varepsilon_4^2 & 2(\varepsilon_1\varepsilon_2 + \varepsilon_3\varepsilon_4) & 2(\varepsilon_1\varepsilon_3 - \varepsilon_2\varepsilon_4) \\ 2(\varepsilon_1\varepsilon_2 - \varepsilon_3\varepsilon_4) & -\varepsilon_1^2 + \varepsilon_2^2 - \varepsilon_3^2 + \varepsilon_4^2 & 2(\varepsilon_2\varepsilon_3 + \varepsilon_1\varepsilon_4) \\ 2(\varepsilon_1\varepsilon_3 + \varepsilon_2\varepsilon_4) & 2(\varepsilon_2\varepsilon_3 - \varepsilon_1\varepsilon_4) & -\varepsilon_1^2 - \varepsilon_2^2 + \varepsilon_3^2 + \varepsilon_4^2 \end{bmatrix} \tag{3.12}$$

and the time derivative of the quaternion is related to the local angular velocities:

$$\begin{bmatrix} \dot{\varepsilon}_1 \\ \dot{\varepsilon}_2 \\ \dot{\varepsilon}_3 \\ \dot{\varepsilon}_4 \end{bmatrix} = \frac{1}{2} \begin{bmatrix} 0 & \omega_3 & -\omega_2 & \omega_1 \\ -\omega_3 & 0 & \omega_1 & \omega_2 \\ \omega_2 & -\omega_1 & 0 & \omega_3 \\ -\omega_1 & -\omega_2 & -\omega_3 & 0 \end{bmatrix} \begin{bmatrix} \varepsilon_1 \\ \varepsilon_2 \\ \varepsilon_3 \\ \varepsilon_4 \end{bmatrix} \tag{3.13}$$

Therefore, with Equation (3.13) the quaternion and subsequently the rotation matrix $\mathbf{A}(\varepsilon)$ (Equation (3.12)) can be calculated. With $\mathbf{A}(\varepsilon)$ known, the body can now be rotated to the required orientation. For more details on the quaternion representation the reader is referred to [8, 10].

The solid deformation relative to the principal axes is governed by the conservation of linear momentum. Assuming an incompressible solid, the structural (solid) governing equations can be formulated in terms of the Cauchy stress tensor as follows [11]:

$$\begin{aligned} &\rho_{\mathrm{s}} \frac{\partial u_i^{\mathrm{s}}}{\partial \mathrm{t}} - \frac{\partial \sigma_{ji}^{\mathrm{s}}}{\partial x_j} = f_i \\ &\det\left\{F_{ij} = \partial x_i\,(\mathrm{t}) / \partial x_j\,(0)\right\} = 1 \end{aligned} \tag{3.14}$$

where ρ_{s}, u_i^{s}, and σ_{ji}^{s} are the solid density, velocity, and Cauchy stress tensor, respectively, F_{ij} is the deformation gradient tensor, and f_i the external force imparted by hemodynamic stresses. The $\det\left\{F_{ij}\right\}$ denotes the determinant of the tensor F_{ij} and expresses the incompressibility constraint. Equation (3.14) can also be written using the Piola–Kirchhoff stress tensors to express the stress relative to the reference configuration. This is in contrast to the Cauchy stress tensor, which expresses the stress relative to the present configuration. The Piola–Kirchhoff representation is formulated as follows [11]:

$$\begin{aligned} &\rho_{\mathrm{s}} \frac{\partial u_i^{\mathrm{s}}}{\partial \mathrm{t}} - \frac{\partial}{\partial x_j} F_{jk} S_{jk} = f_i \\ &\det\left\{F_{ij} = \partial x_i\,(\mathrm{t}) / \partial x_j\,(0)\right\} = 1 \end{aligned} \tag{3.15}$$

where S_{ij} is the second Piola–Kirchhoff stress tensor.

Unlike the fluid conservation laws cast in Eulerian coordinates (Equation (3.1)), Equation (3.15) is formulated in Lagrangian coordinates that move with the deforming structure. Similar to the fluid equations, constitutive equations are needed to close the structural deformation equations. For cardiovascular applications this step requires the development of tissue constitutive models, which can be radically different depending on the type of tissue and its location [12, 13].

Tissue undergoing large deformation (e.g., heart valves) is usually modeled as a hyperelastic material (neo-Hookean), an approach which is frequently used to model biomaterials, polymers, and rubbers (the Saint Venant–Kirchhoff model):

$$E_{ij} = (C_{ij} - \delta_{ij}),\ C_{ij} = F_{ki}F_{kj} \tag{3.16}$$

where the Lame moduli are denoted by λ and μ, and E is the Lagrange–Green strain, derived from the Cauchy–Green tensor C. Most biological tissues exhibit mechanical anisotropy and multi-axial anisotropic constitutive models should be used [14]. Different constitutive models such as pseudoelastic, randomly elastic, proelastic, and viscoelastic have been proposed for different blood vessels [13].

3.2.1.3 Boundary Conditions at the Fluid–Structure Interface

The fluid and the structural equations are coupled together at the fluid–structure interface with the kinematic (no-slip) and dynamic boundary conditions of the following form:

$$u_i = u_i^{\mathrm{s}} \tag{3.17}$$

$$f_i = (pn_i - \tau_{ij}n_j)\mathrm{d}A \tag{3.18}$$

where p is the hemodynamic pressure, dA is the surface area, n_j is the *jth* component of the normal to the surface, τ_{ij} is the hemodynamic shear stress tensor, and f_i is the fluid-imparted force on the right-hand side of structural equation (3.14). These conditions are valid even if the fluid–structure interface is moving.

3.2.2 Numerical Methods for Flows with Moving Boundaries

A major challenge in the numerical simulation of cardiovascular flow problems stems from the fact that the location of the fluid–structure interface, where the boundary conditions given by Equations (3.17) and (3.18) need to be satisfied, changes continuously in time. Consequently, special numerical methods are required that are capable of handling dynamically evolving domains with complex, deformable immersed boundaries. Such numerical techniques can be broadly classified into three categories: (a) boundary-conforming techniques in which the mesh conforms to the moving boundary at all times; (b) non-boundary-conforming techniques in which the mesh remains fixed; and (c) hybrid techniques that integrate both approaches (see Fig. 3.1 for an illustration of these general approaches). The various techniques for handling moving boundary problems are discussed in the following sections.

3.2.2.1 Boundary-Conforming Methods

As discussed above, boundary-conforming methods require constructing a computational mesh that always conforms to all solid boundaries in the flow domain (see Fig. 3.1). Such methods are classified as arbitrary Lagrangian–Eulerian (ALE) formulations and moving frame of reference (MFR) methods.

ALE Methods

In ALE methods the computational grid is fitted to and moves/deforms with the moving boundary. The grid movement is taken into account in the formulation of the Navier–Stokes by incorporating the so-called grid velocity terms as follows [15]:

$$\begin{aligned} &\frac{\partial(u_i - u_i^{g})}{\partial x_i} = 0 \\ &\frac{\partial u_i}{\partial t} + u_j \frac{\partial(u_i - u_i^{g})}{\partial x_j} = -\frac{\partial p}{\partial x_i} + \frac{1}{\mathrm{Re}}\frac{\partial \sigma_{ji}}{\partial x_j} - u_i \frac{\partial u_j^{g}}{\partial x_j} \end{aligned} \tag{3.19}$$

where u_i^{g} are the components of the grid velocity. Equations (3.19) are in weak conservation form, but often these equations are cast into the strong conservation form by employing either integral or differential formulations [16–18]. Regardless of the specific conservation form (strong vs. weak), however, a critical prerequisite for accurate solutions of the ALE equations is to satisfy the geometric conservation law (GCL) in discrete form [16, 17, 19]. For Equation (3.19), for instance, this can be accomplished by discretizing the convective terms and the term involving the divergence of the grid velocity vector to ensure that the GCL is explicitly satisfied and free stream is conserved (see [15] for details).

ALE formulation can be applied to both structured and unstructured grids [20]. The unstructured grid enables handling complex anatomical geometries [21]. ALE formulations have been used successfully to simulate a number of important flow problems, e.g., flows in a compliant curved tube as a model for the coronary artery [22]; flow in the ascending aorta with wall motion prescribed from MRI [23]; and blood flow through compliant carotid artery [24] and aorta [25], among others.

The ALE method works very well for problems with relatively simple geometries and moderate deformations such as FSI in compliant blood vessels. However, in problems with large structural deformations, obtaining smooth and well-conditioned computational meshes at every time step is far from trivial, if not impossible, and frequent remeshing may be the only option. For instance, Cheng et al. [26], who simulated the flow through a bileaflet mechanical heart valve using an ALE approach with structured grids, had to use interpolation between two previously generated meshes to obtain the intermediate mesh for a given leaflet angle and then apply an elliptic solver to the entire mesh to smooth the mesh. Because of these inherent difficulties, the ALE approach is not suitable for simulating cardiovascular flows involving complex geometries and large structural displacements, such as the flows arising in the study of native and/or prosthetic heart valves. Furthermore, simulating blood flow in deformable domains for large, realistic anatomic, and physiologic models of the cardiovascular system, even with moderate deformations, using the ALE method remains a formidable problem [27]. An alternative to the ALE method is the coupled momentum method (CCM), in which the equations of the deformation of the vessel wall are coupled at the variational level as a boundary condition for the fluid domain [27]. In this approach, wall motion is assumed to be small so that the fluid mesh is not updated, which makes the method efficient for large-scale FSI problems wherein underlying assumptions of small deformation and thin walls are valid [28].

Moving Frame of Reference Methods

If the domain in which the flow equations (3.1) are solved is not stationary (at rest), the equations should be solved in the non-inertial frame of reference. In the cardiovascular flows, such cases can arise to study blood flow in different cardiovascular parts if the subject (patient) is not at rest. The conventional formulation in the non-inertial reference frame, which produces source terms in the momentum equations, is as follows [29, 30]:

$$\begin{aligned}
&\frac{\partial u_i^{\mathrm{r}}}{\partial x_i} = 0\\
&\left(\frac{\partial u_i^{\mathrm{r}}}{\partial t}\right)_{\mathrm{r}} + u_j^{\mathrm{r}}\frac{\partial u_i^{\mathrm{r}}}{\partial x_j^{\mathrm{r}}} = -\frac{\partial p}{\partial x_j^{\mathrm{r}}} + \frac{1}{\mathrm{Re}}\frac{\partial \sigma_{ji}}{\partial x_j^{\mathrm{r}}} - \varepsilon_{ijk}\Omega_j\varepsilon_{klm}\Omega_l x_m^{\mathrm{r}}\\
&\qquad - 2\varepsilon_{ijk}\Omega_j u_k^{\mathrm{r}} - \varepsilon_{ijk}\frac{\mathrm{d}\Omega_j}{\mathrm{d}t}x_k^{\mathrm{r}} - A_{ij}\frac{\mathrm{d}^2 r_j^{\mathrm{cm}}}{\mathrm{d}t^2}
\end{aligned} \tag{3.20}$$

where x^{r} and u^{r} are the position and Cartesian relative velocity vector in the non-inertial reference frame, respectively, Ω_i is the angular velocity of the non-inertial frame about its origin, r^{cm} is the origin of the non-inertial reference frame with respect to the inertial reference frame, and A_{ij} are the components of the unitary rotation matrix, Equation (3.12), which transforms the coordinates from the inertial frame to the non-inertial frame as $x_i^{\mathrm{r}} = A_{ij}(x_i^{\mathrm{a}} - r_i^{\mathrm{cm}})$. The relation between the absolute velocity u^{a} and the relative velocity vector u^{r} is given as follows:

$$u_i^{\mathrm{a}} = A_{ji}(u_j^{\mathrm{r}} + \varepsilon_{jkl}\Omega_k x_l^{\mathrm{r}} + u_j^{\mathrm{cm}}) \tag{3.21}$$

where

$$u_i^{\mathrm{cn}} = A_{ij}\frac{\mathrm{d}^2 r_j^{\mathrm{cm}}}{\mathrm{d}t^2}$$

is the translational velocity of the center (origin) of the non-inertial frame. The time derivative in the non-inertial reference frame in Equation (3.20) is related to that of the inertial frame according to the following equation:

$$\left(\frac{\partial}{\partial t}\right)_{\mathrm{r}} = \left(\frac{\partial}{\partial t}\right)_{\mathrm{a}} + (\varepsilon_{jkl}\Omega_k x_l^{\mathrm{r}} + u_j^{\mathrm{cm}})\frac{\partial}{\partial x_j^{\mathrm{r}}} \tag{3.22}$$

The non-conservative formulation of Equation (3.20) contains four source terms: the centripetal force $\varepsilon_{ijk}\Omega_j\varepsilon_{klm}\Omega_l x_m^{\mathrm{r}} = \Omega \times \Omega \times \mathbf{x}^{\mathrm{r}}$; the Coriolis force $2\varepsilon_{ijk}\Omega_j u_k^{\mathrm{r}} = 2\Omega \times \mathbf{u}^{\mathrm{r}}$; and the force due to rotational and translational acceleration of the non-inertial reference frame

$$\varepsilon_{ijk}\frac{\mathrm{d}\Omega_j}{\mathrm{d}t}x_k^{\mathrm{r}} = \frac{\mathrm{d}\Omega}{\mathrm{d}t} \times \mathbf{x}^{\mathrm{r}} \text{and } A_{ij}\frac{\mathrm{d}^2 r_j^{\mathrm{cm}}}{\mathrm{d}t^2} = \mathbf{A}\frac{\mathrm{d}^2\mathbf{r}^{\mathrm{cm}}}{\mathrm{d}t^2}$$

respectively. As shown by Kim and Choi [30], such non-conservative formulations can lead to numerical instability. Instead, a conservative formulation for the non-inertial reference frame, developed by Beddhu et al. [31] and used by Kim and Choi [30], can be used:

$$\begin{aligned} &\frac{\partial u_i}{\partial x_i} = 0 \\ &\left(\frac{\partial u_i}{\partial t}\right)_{\mathrm{r}} + \frac{\partial}{\partial x_j^{\mathrm{r}}}\left[(u_i(u_j - v_j) + u_j w_i\right] = -\frac{\partial p}{\partial x_j^{\mathrm{r}}} + \frac{1}{\mathrm{Re}}\frac{\partial \sigma_{ji}}{\partial x_j^{\mathrm{r}}} \end{aligned} \tag{3.23}$$

where

$$u_i = u_i^{\mathrm{r}} + v_i = A_{ij}u_j^{\mathrm{a}}$$

$$v_i = w_i + u_i^{\mathrm{cm}}$$

$$w_i = \varepsilon_{ijk}\Omega_j x_k^{\mathrm{r}}$$

The equations in the non-inertial frame of reference have been successfully applied to solve a wide range of engineering and biological flows [29–35], among others, but have not been applied in cardiovascular flows. In cardiovascular simulations, these equations could be useful in simulations aimed at predicting hemodynamics under exercise (not at rest) conditions. In such a case, the governing equations should be formulated relative to the non-inertial frame of reference attached to the body of the patient and boundary-conforming or non-boundary-conforming techniques could be used to simulate the interaction of the blood flow with the various structural components within this moving frame of reference.

3.2.2.2 Non-boundary-Conforming Methods

Non-boundary-conforming methods are very attractive due to their enormous versatility in simulating problems involving large structural displacements and complex geometries [36–39]. In such methods the entire fluid computational domain is discretized with a single, fixed, non-boundary-conforming grid system (most commonly a Cartesian mesh is used as the fixed background mesh), while the structural domain is discretized with a separate (usually) unstructured grid, whose nodes comprise a set of Lagrangian points used to track its motion within the fluid domain. The effect of a moving immersed body on the fluid is accounted for by adding, either explicitly or implicitly, body forces to the governing equations of motion (Equation (3.1)) that are designed to account for the effect of a solid boundary. Since the grid used to discretize the fluid domain does not have to conform to the moving immersed boundaries, such methods are inherently applicable to moving boundary problems involving arbitrarily large structural displacements and are ideally suited for a wide range of cardiovascular flow applications.

Non-boundary-conforming methods treat complex solid surfaces as immersed boundaries and thus are collectively referred to as immersed boundary (IB) methods. Such methods are classified into two broad categories: diffused interface IB methods and sharp-interface IB methods. What distinguishes the two approaches is the manner in which the effect of the immersed boundary is taken into account into the governing equations.

Diffused Interface IB Methods: The Classical Peskin Formulation

This method was first introduced by Peskin [40] in the early 1970s, although some earlier work by Viecelli in the extension of the Marker and Cell method to cases with arbitrarily shaped and moving boundaries also falls into this category [41, 42]. Peskin and co-workers applied the immersed boundary (IB) methodology to study blood flow in the human heart [43–45]. In this method the vascular boundary is modeled as a set of elements linked by springs (elastic fibers) and a Lagrangian coordinate system is attached to track their location in space. The tracking information is then used to compute the spatial distribution of the external force field f_i^{IB} that is explicitly introduced in the governing equations (3.1) at the fixed Eulerian grid nodes, as follows:

$$\begin{aligned}
&\frac{\partial u_i}{\partial x_i} = 0\\
&\frac{Du_i}{Dt} = -\frac{\partial p}{\partial x_i} + \frac{1}{\mathrm{Re}}\frac{\partial \sigma_{ji}}{\partial x_j} + f_i^{\mathrm{IB}}
\end{aligned} \tag{3.24}$$

The body force is distributed on all nodes of the fixed background grid via a discrete delta function:

$$f_i^{\mathrm{IB}}(x_i, t) = \sum_K F_i^k(t)\delta(x_i - X_i^k) \tag{3.25}$$

where X_i^k and F_i^k are the position and the stress, respectively, on the *kth* Lagrangian massless point and δ is the Dirac delta function. Such body force formulation has the effect of smearing, or diffusing, the solid boundary over several fluid grid nodes in the vicinity of the boundary. Because of this inherent property of the method, Peskin's method [45] is known as a diffused interface method. The original IB method is only first-order accurate [40] but a variant of the method that is formally second-order accurate and combines adaptive mesh refinement to increase resolution in the vicinity of immersed boundaries has also been proposed [46].

The fictitious domain method [47] is another related fixed grid, diffused interface method that has also been applied to heart valve simulations [48–54]. In this method the immersed solid is, as in Peskin's classical IB method, free to move within the fluid mesh, but the two domains are coupled together at the solid–fluid interface through a Lagrange multiplier (or local body force) [54]. The fictitious domain method had been applied to simulate flow in a 2D model of the native valve [49], as well as in a 3D trileaflet heart valve at relatively low, non-physiological Reynolds number (peak systole Re=900) with symmetry assumption [48, 50, 51]. A major issue with the fictitious domain method is that it cannot yield accurate results for the viscous shear stresses on the solid boundary [54]. To remedy this major limitation, a combination of the fictitious domain method with adaptive mesh refinement has also been proposed [53, 54]. This enhanced fictitious domain method has been applied to carry out 3D FSI simulations assuming, as in previous studies, that the geometric symmetries of the valve will also be respected by the induced flow field [52]. Due to the high computational costs of this approach, however, it has yet to be used to carry out a full 3D simulation of heart valve flows at physiologic conditions.

Sharp-Interface Immersed Boundary Methods

As mentioned above, both the Peskin's and fictitious domain methods are diffused interface techniques causing the smearing of the solid–fluid interface. Because of this property, such methods typically require increased grid resolution in the vicinity of the boundary for accurate results, and their application to high Reynolds number flows can be impractical from a computational standpoint. To remedy this situation, a class of sharp-interface immersed boundary methods has recently been developed and is attracting increasing attention in the literaturesee for example [36, 55–60], among many others. In these methods the immersed boundary is treated as a sharp interface and its effect on the fluid can be accounted for in a variety of ways.

In the cut-cell methods [61] the shape of grid cells in the vicinity of the boundary is modified to produce a locally boundary-fitted mesh (see Fig. 3.2a). In immersed interface methods [59, 62] the jump conditions, caused by the discrete delta function in the classical IB method, are incorporated into the finite-difference scheme to avoid the approximation of the discrete delta function by a smooth function and to eliminate the smearing of the interface. In the hybrid Cartesian/immersed boundary (HCIB) method [56, 63] the boundary conditions are applied at the closest grid nodes to the immersed boundary (Fig. 3.2b). The various flow variables at such near-boundary nodes are calculated by interpolating linearly along an appropriate grid line between the nearest interior node, where flow variables are available from the solution of the governing equations, and the point where the grid line intersects the solid boundary, where physical boundary conditions are known [56, 58, 63].

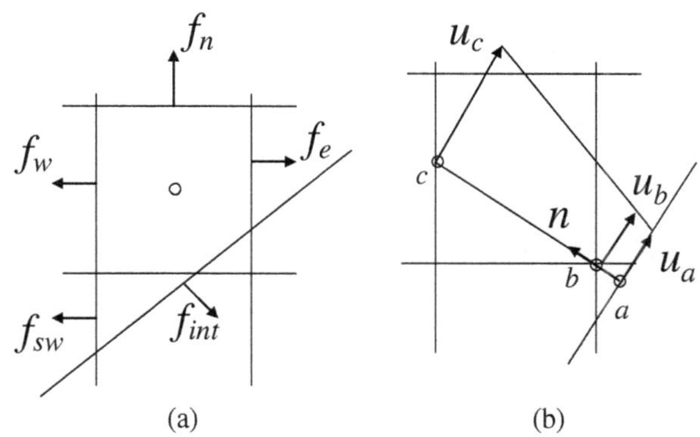

Fig. 3.2 Two types of immersed boundary methods: (**a**) cut-cell method, in which the local grid shape and fluxes in the vicinity of the immersed boundary are modified; (**b**) the reconstruction method, in which the boundary conditions on the node in the vicinity of the immersed boundary are reconstructed using an interpolation scheme

An important issue for the success of the HCIB method is the direction in which interpolation is carried out. Fadlun et al. [56] used a one-dimensional interpolation scheme along the gridline, which works well for simple geometries that are aligned with the gridline, but in complex geometries the choice of the interpolation direction can be ambiguous. Gilmanov et al. [64] and Gilmanov and Sotiropoulos [58] generalized the HCIB method such that the boundary conditions are reconstructed along the local normal to the boundary via an appropriate interpolation scheme, which works well for a general complex body shape. Gilmanov and Sotiropoulos [58] used a staged/non-staggered grid approach, which enhances the accuracy for flows with moving bodies, simplifies the implementation of velocity boundary conditions, and eliminates the need for the pressure boundary condition. Furthermore, the staggered/non-staggered approach enables satisfying the continuity equation to machine zero. For a detailed discussion of various sharp-interface methodologies, the reader is referred to [58] and the recent review paper by Mittal and Iaccarino [65].

A major issue for the efficient implementation of the immersed boundary method, and for that matter for other sharp-interface methods, is the efficiency of the algorithm for classifying at every time step the location of the nodes of the background grid relative to the moving immersed boundary. This issue was successfully addressed by Borazjani et al. [36], who incorporated a new and very efficient algorithm based on the ray-tracing algorithm [66] to classify the fluid domain grid nodes into fluid, solid, and immersed boundary.

A major shortcoming of all immersed boundary methods for which a Cartesian mesh is used to discretize the background domain arises from the fact that, depending on the complexity of the immersed boundary, a large number of Cartesian grid points could lie outside the fluid domain. Such unused nodes unnecessarily burden the computation since they incur a memory overhead without adding numerical resolution. This situation is especially exacerbated in complex multi-branch anatomic artery trees, as was recently highlighted in the work of Yokoi et al. [67], who employed an immersed boundary approach to simulate the flow in a very complex cerebral aneurysm anatomy. They immersed the aneurysm geometry in a Cartesian box discretized with $160 \times 50 \times 75$ grid nodes. Even though Yokoi et al. [67] did not provide any information about the number of nodes in the interior of the aneurysm model, a simple inspection of the geometry suggests that in their simulation the vast majority of the Cartesian nodes should reside in the exterior of the flow domain and only a few points are actually used to cover the vessel diameter.

To remedy this problem, which can significantly impact the memory efficiency of the code and make systematic grid refinement studies impractical, de Zélicourt et al. [68] recently proposed an unstructured Cartesian mesh paradigm. The approach is simple and straightforward to implement as it eliminates all unused Cartesian grid nodes from the grid, yielding a mesh that remains Cartesian but is now unstructured in the sense that the simple connectivity of the parent mesh has now been lost. Grid point re-numbering and indexing strategies used for unstructured mesh can be used to establish the connectivity of the mesh as described in [68]. This method has been used to successfully carry out grid refinement studies in very complex anatomic cases, as in a rather challenging clinical scenario of a single-ventricle patient with severe arterio-venous malformations [68]. The method has also been applied as part of a virtual surgery framework to demonstrate its potential as a surgical planning tool [69].

3.2.2.3 Hybrid Methods: Body-Fitted/Immersed Boundary Methods

The combination of body-fitted methods for some portion of the domain, together with the immersed boundary methods for another portion of the domain, provides great advantages, which are increasingly being exploited [57, 70]. Ge and Sotiropoulos [57] developed a novel paradigm integrating the HCIB method with body-fitted curvilinear grids. Their method, which was dubbed the curvilinear immersed boundary (CURVIB) method, is ideally suited for simulating heart valve flows in anatomic geometries. The empty aorta is discretized with a boundary-conforming curvilinear mesh and the valve leaflets are treated as sharp, immersed boundaries within the background curvilinear mesh (Fig. 3.3).

In the unstructured/immersed boundary method, similar to the CURVIB method, the background grid is body fitted, but with unstructured grids, and the immersed boundaries are treated with the sharp-interface method [70]. A similar hybrid framework, integrating an unstructured grid with a sharp-interface immersed boundary method, has also been applied to a patient-specific hemodynamic model of an aneurysm of the internal carotid artery constructed from a 3D rotational angiogram and stented with two different stent designs [71]. Such methods are thus ideally suited to be used for optimizing various medical devices, such as heart valves, ventricular assist devices, different stent designs, etc. in a patient-specific framework.

3.2.3 Fluid–Structure Interaction Algorithms

A practical way of simulating fluid–structure interaction (FSI) problems in cardiovascular flows is the partitioned approach, in which the system is partitioned into

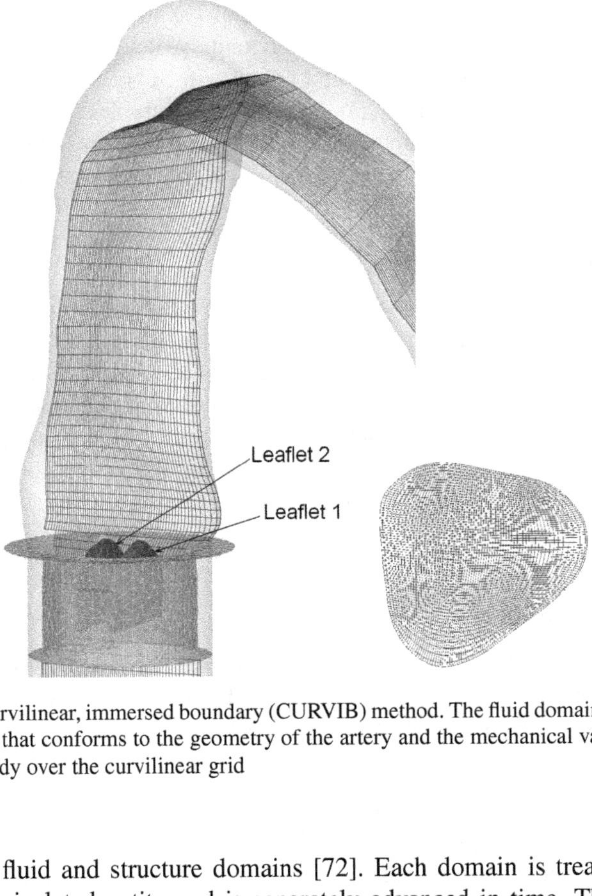

Fig. 3.3 The curvilinear, immersed boundary (CURVIB) method. The fluid domain is a curvilinear body-fitted grid that conforms to the geometry of the artery and the mechanical valve is placed as an immersed body over the curvilinear grid

two separate fluid and structure domains [72]. Each domain is treated computationally as an isolated entity and is separately advanced in time. The interaction effects are accounted for through boundary conditions (Equations (3.17) and (3.18)) at the fluid–structure interface [29, 36, 72]. Depending on how boundary conditions are applied (implicit vs. explicit), two types of coupling, loose and strong coupling (referred to as LC-FSI and SC-FSI, respectively), are possible, which will be discussed below.

Both LC-FSI and SC-FSI have been used to simulate cardiovascular flows. For example, Vierendeels et al. [73] used the SC-FSI method to simulate a 2D tissue valve and De Hart et al. [50] used the LC-FSI method to simulate a 3D tissue valve with the symmetry assumption. SC-FSI has been used to simulate the FSI between the flow and a compliant tube as well in [74], while LC-FSI has been used in [27].

3.2.3.1 Loose and Strong Coupling Strategies

In LC-FSI the boundary conditions at the interface are obtained from the domain solutions at the previous time level (explicit in time), while in SC-FSI the interfacial boundary conditions are obtained from the domain solutions at the current time level (implicit in time). In LC-FSI, the fluid and the structure domains are solved sequentially one after the other using the boundary conditions from the previous solution of the other domain. On the other hand, in SC-FSI the implicit boundary conditions are obtained by carrying out a number of sub-iterations at each physical time step and updating the boundary conditions based on the solution of the most recent subiterationsee [29, 36] for more details on coupling strategies.

3.2.3.2 Stability and Robustness Issues

As discussed previously the major difference between the two coupling strategies is in the treatment of the boundary condition, which leads to major differences in numerical properties. The LC-FSI approach is less stable and robust than the SC-FSI but requires many fewer computational resources per time step, as no sub-iterations are performed. Consequently, the decision to select one approach over the other should be based on a trade-off between efficiency and numerical stability [36]. In what follows we provide some simple guidelines for assessing the stability of different coupling methods.

The Added Mass and Flow Effects

In cases when the added mass is significant relative to the actual mass, the coupling strategies become unstable [36]. Borazjani et al. [36] showed by stability analysis of a simple FSI problem that when the stability ratio is large,

$$R_{\mathrm{st}} = \frac{\rho_{\mathrm{f}} H}{\rho_{\mathrm{s}} I} > 1$$

($\rho_{\mathrm{f}}, \rho_{\mathrm{s}}$ are fluid and structure density, respectively, H is the added mass coefficient, and I is the inertia or volume of the structure depending on the problem), both coupling strategies become unstable [36]. This FSI stability parameter is similar to the FSI stability parameter found in [74–76] for internal flow problems of flow in a domain with compliant walls. R_{st} can become large due either to high density ratio between fluid and structure or to geometry, e.g., low moment of inertia.

Nevertheless, Borazjani et al. [36] showed that the structural properties are not the only parameters affecting stability; local flow properties can also affect it. They found that in some cases—e.g., during closing of a mechanical heart valve—the flow has stabilizing effects and acts against the added mass effects. However, during the valve opening phase, the flow has a destabilizing effect and amplifies the added mass effects.

Non-linear Relaxation Strategies

When the stability ratio is larger than 1, both the loose- and strong coupling methods are unstable [36] and under-relaxation is required to stabilize the strong coupling iterations:

$$\bar{S}^{l+1} = (1-\alpha)\bar{S}^{l} + \alpha S^{l+1}$$

where S is the structural solution vector at iteration l. The upper bound of α for stability based on a simple FSI problem stability analysis has been found to be [36]:

$$0 < \alpha < \frac{2}{1 + R_{\text{st}}}$$

The speed of convergence and the efficiency of the under-relaxation directly depend on the under-relaxation coefficient α. The optimum value of α is problem dependent and might be found through trial and error. However, the optimum value of α might change within a given problem with time, especially if the problem is non-linear. A way to determine the optimum under-relaxation α coefficient dynamically is to use the Aitken acceleration technique for series [77] for the iterations of the non-linear system [78]. Following [36, 78], α can be computed as

$$\Delta\bar{S}^{l+1} = \bar{S}^{l} - S^{l+1}$$

$$\lambda^{l+1} = \lambda^{l} + (\lambda^{l} - 1)\frac{(\Delta\bar{S}^{l} - \Delta\bar{S}^{l+1})^{\mathrm{T}}\Delta\bar{S}^{l+1}}{(\Delta\bar{S}^{l} - \Delta\bar{S}^{l+1})^{2}}$$

$$\alpha = 1 - \lambda^{l+1}$$

For the first iteration (l=1), $\Delta\bar{S}^{l+1}$ can be calculated, while λ^{l+1} cannot, and as such a preset value should be used to calculate the first under-relaxation coefficient. Borazjani et al. [36] have shown the great effect of Aitken acceleration through numerical examples and found that the strong coupling converged within 4–5 iterations with Aitken acceleration, while 10–20 iterations were required with fixed under-relaxation coefficient α. Furthermore, the Aitken acceleration technique is extremely cheap from both memory and computational cost viewpoints, and is easy to implement.

3.2.4 Efficient Solvers for Physiologic Pulsatile Simulations

A major numerical challenge for solving the Navier–Stokes equation (3.1) under pulsatile physiologic conditions stems from the need to satisfy the discrete continuity equation to machine zero at every time step. For high-resolution simulations, this has to be accomplished on fine computational meshes, with temporal resolution typically requiring thousands of time steps per cardiac cycle. In this section we briefly discuss the issue of solver efficiency to emphasize its importance for high-resolution simulations of cardiovascular flows.

In the recent work of Ge and Sotiropoulos [57] the fractional step method is used to advance the governing equations in time. The performance of the fractional step method in unsteady flow simulations largely depends on the efficiency of the method used to solve the Poisson equation. For small problems the Poisson equation can be solved efficiently by direct methods. However, the cost of such methods greatly increases as the number of grid nodes is increased. Iterative solvers can provide efficient solutions for problems with large numbers of grid points. The most efficient iterative solvers available today include Krylov subspace methods, such as GMRES [79] and BICGSTAB [80], and multigrid methods [81]. Ge and Sotiropoulos [57] used a flexible GMRES (FGMRES) solver to solve the Poisson equation. However, the performance of the GMRES solver deteriorates as the mesh size is increased [82]. The use of multigrid as preconditioner can effectively restore the performance of the GMRES method and achieve grid-independent performance that is far superior to the multigrid method alone [82]. For grids with large aspect ratios, the deterioration of the performance of the multigrid can be effectively addressed using an appropriate grid-coarsening strategy that results in lower aspect ratio grids for coarser grids [57]. Algerbraic multigrid solvers (AMG) [39] are also efficient elliptic equation solvers that provide more flexibility than geometric multigrid solvers do since they do not use any geometric information from the grid.

3.2.5 High-Resolution Simulations of Cardiovascular Flow

In this section we review recent computational studies of selected cardiovascular flow problems and present a small sample of recent high-resolution simulations of these flows. The objective of this section is not to present an in-depth discussion of the various cases but rather to underscore the recent advances in our ability to perform high-resolution simulations of cardiovascular flows. We discuss simulations aimed at elucidating the hemodynamics of (1) mechanical bileaflet heart valves and (2) trileaflet heart valves.

3.2.5.1 Fluid–Structure Interaction Simulations of Mechanical Bileaflet Heart Valves

Computational advances in the area of bileaflet mechanical heart valves (BMHVs) have recently been reviewed in [83] and only a brief discussion is given here. The early simulations of BMHV flows were mostly with fixed leaflets and steady inflow conditions [84–88]. More recent simulations have been performed with moving leaflets and under pulsatile inflow conditions. A large number of such simulations have been two dimensional [89–93]. Only a few 3D simulations have been attempted. A 3D FSI simulation has been carried out by Cheng et al. [26] using only one-quarter of the valve (quadrant symmetry assumption) by an ALE method, while Tai et al. [94] used the immersed boundary method with overlapping grids to carry out FSI simulations of bileaflet MHV on an unstructured mesh (86,000 nodes, 450,000 elements in four zones). Tai et al. [94] preformed the simulations under physiologic conditions but the coarseness of the computational mesh did not allow them to obtain insights into the underlying physics of the flow. Nobili et al. [95] carried out FSI simulations of a BMHV using the commercial code Fluent on a relatively coarse mesh (1.2 million tetrahedral and 900,000 hexahedral), which could not capture the rich dynamics of the flow throughout the cardiac cycle as revealed by recent laboratory experiments [37]. Consequently, the simulations of Nobili et al. [95] yielded a simulated flow environment downstream of the valve leaflets that was significantly simpler and less rich dynamically than observed in experiments.

The CURVIB method of Ge and Sotiropoulos [57] was the first method to successfully capture all the flow features observed in experiments by performing high-resolution simulations (with about 10 million grid points) under pulsatile physiologic conditions [37]. The motion of the leaflets was prescribed from the experimental measurements in this work [37]. Borazjani et al. [36] extended the method to non-linear FSI problems and performed the first high-resolution 3D FSI simulation of a BMHV under pulsatile, physiologic conditions. The simulation results were in excellent agreement with the measurements of Dasi et al. [37] in terms of both leaflet kinematics and flow patterns.

The high-resolution numerical simulations reported in [36] along with the experimental measurements [37] have provided the first glimpse of the rich hemodynamics of the BMHV. The flow is fully described in [37] and here it suffices to mention that during early systole, the flow is dominated by organized laminar structures, while just before peak systole the flow breaks down into small-scale turbulent worm-like structures (see Fig. 3.4), which persist until the end of the cardiac cycle. As the new cardiac cycle begins, however, the small structures are washed out and new laminar organized structures are formed. The numerical simulations reported in [36] also captured the measured leaflet kinematics with remarkable accuracy (see Fig. 3.5).

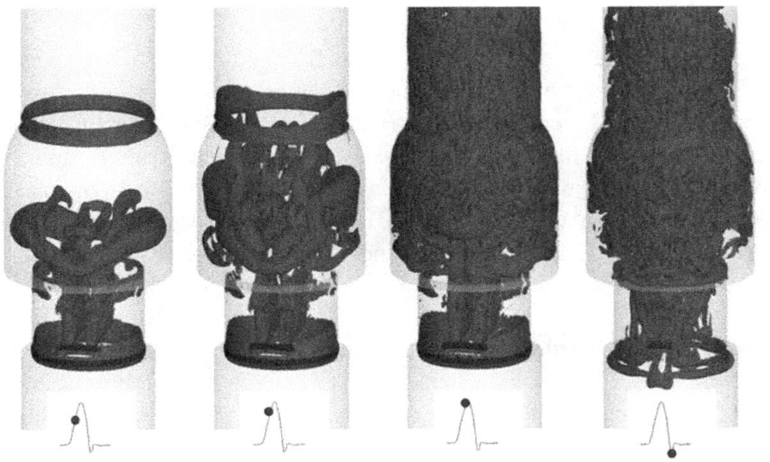

Fig. 3.4 The vortical structures of the flow through a bileaflet mechanical heart valve implanted in a straight aorta visualized by the iso-surfaces of q criteria at several time instants during the cardiac cycle. Adopted from Borazjani et al. [36]

The findings of Dasi et al. [37], Ge and Sotiropoulos [57], and Borazjani et al. [36] were also reproduced by De Tullio et al. [96], who performed FSI simulations of BMHV in a straight aorta.

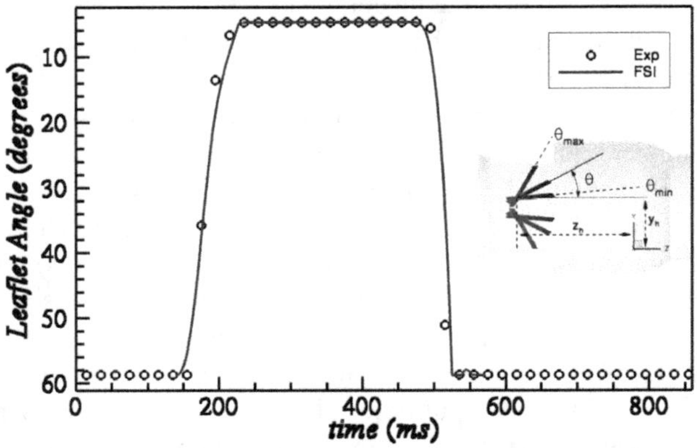

Fig. 3.5 The calculated leaflet angles for a mechanical heart valve based on FSI simulations of Borazjani et al. [36] (*solid lines*) and experimental measurements of Dasi et al. [37] (*open circles*)

To show the potential of the CURVIB-FSI solver as a viable tool for patient-specific simulations, Borazjani [29] and Borazjani et al. [97] applied the CURVIB-FSI method to carry out the first high-resolution 3D FSI simulations of a BMHV implanted in an anatomically realistic aorta obtained from MRI and compared the results with the straight aorta case (see Fig. 3.6). This study showed the significant effect of the aorta geometry on leaflet kinematics, flow patterns, and consequently the mechanical environment experienced by blood cells; it also underscored the need for patient-specific simulations [29, 97]. Such simulations can be used to guide valve implantation surgery in the future.

To illustrate the need for simulation-guided BMHV implantation, we carried out a set of FSI simulations by changing the orientation of the implanted valve in the anatomic aorta [98]. Figure 3.7 compares the calculated leaflet kinematics for two simulated orientations forming a 90° angle with each other. It can be observed that the opening phase for the two orientations is quite similar, but significant differences are observed in the closing phase. There is a larger asymmetry in the closing of the valves in the 90° orientation, i.e., one of the leaflets closes about 30 ms faster than the other. Furthermore, the rebound in the 90° orientation is larger than the base orientation. The difference in the valve kinematics is due to the difference in the hemodynamics induced by the moving leaflets and complex 3D anatomic geometry [98]. By comparing the hemodynamic environment of different valve orientations, the optimum hemodynamic orientation for a specific patient can be found beforehand to guide the valve replacement surgery.

3.2.5.2 Numerical Simulations of Trileaflet Heart Valve Hemodynamics

There are relatively few simulations of trileaflet heart valves. Two-dimensional simulations have been reported by van Loon et al. [53, 54] and De Hart et al. [49],

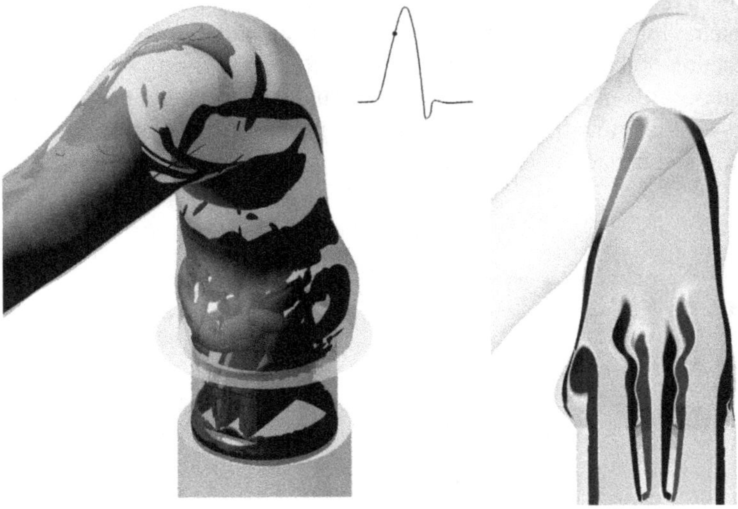

Fig. 3.6 The flow through a bileaflet mechanical heart valve implanted in an anatomic aorta visualized by the (**a**) iso-surfaces of q criteria and (**b**) out-of-plane vorticity in the mid-plane of the valve

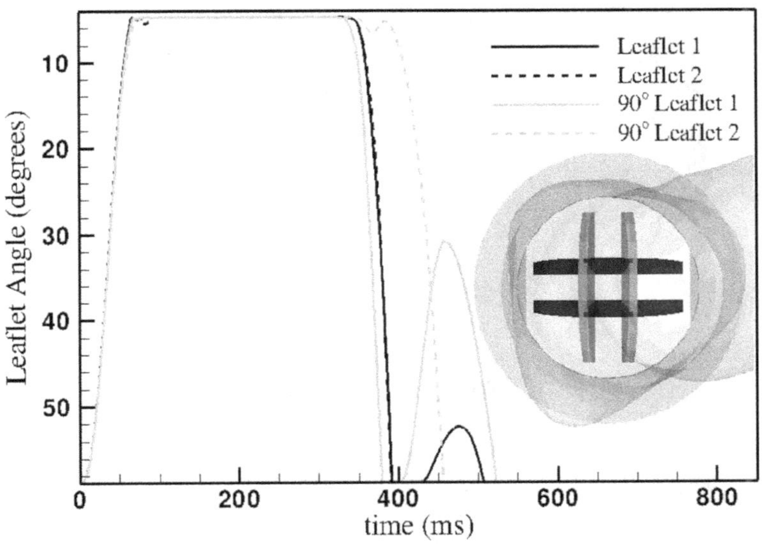

Fig. 3.7 The calculated leaflet angles for a mechanical heart valve based on FSI simulations in an anatomic aorta [97] compared to a mechanical valve implanted with 90° rotation under the same conditions. Adapted from [98]

who performed full FSI simulations with deformable slender structures, modeling the "leaflets" of a native valve, with a fictitious domain finite-element method. Most three-dimensional simulations have been performed by assuming that the mode-3 geometric symmetry of the valve is also preserved by the flow and considering only one of the three leaflets. For example, van Loon et al. [52] and De Hart et al. [53]

carried out three-dimensional FSI simulations of a tissue valve with imposed symmetry assumption and for a peak systole Reynolds number Re=900, a value that is only a fraction of the actual physiologic range (Re = 5,000–6,000).

Ge and Sotiropoulos [99] recently carried out high-resolution numerical simulations of the blood flow through a polymeric trileaflet valve without any symmetry assumption and under physiologic conditions. The motion of the trileaflet valve in

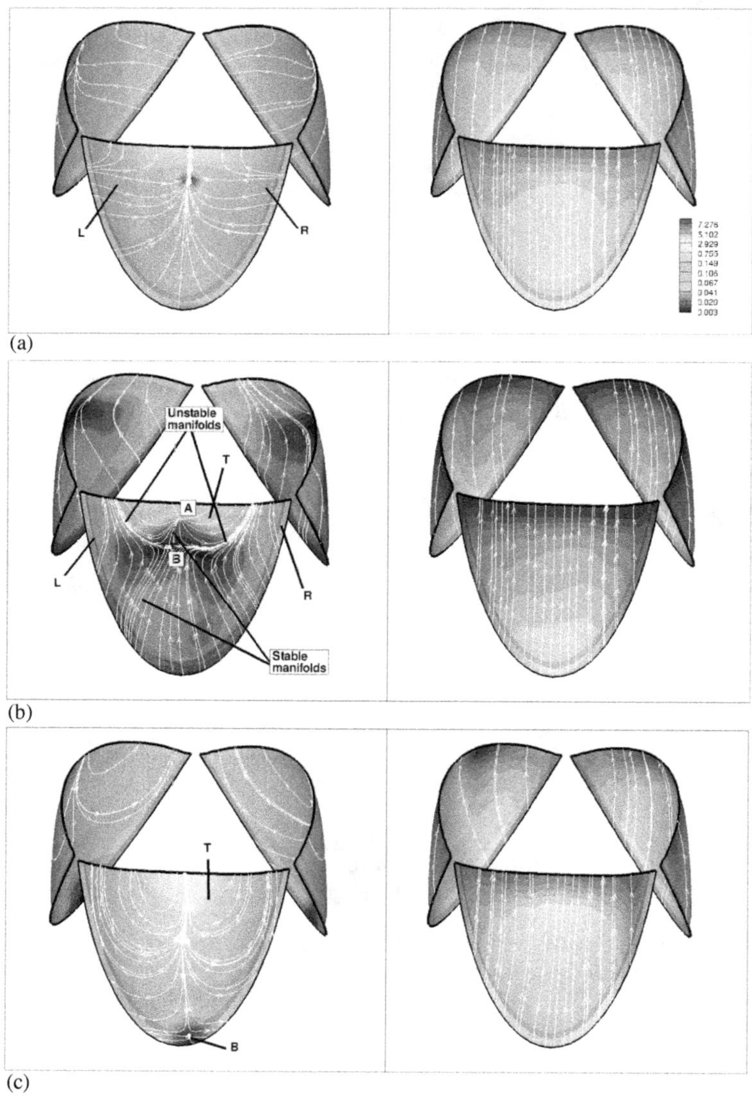

Fig. 3.8 Limiting streamlines and surface shear stress (in Pascal) contours on the aortic side (*left column*) and ventricle side (*right column*). Taken from Ge and Sotiropoulos [99]

this study, however, was not from a full FSI simulation and was prescribed based on the results in [100]. The results of Ge and Sotiropoulos [99] clarified for the first time the differential hemodynamics on the aortic and ventricular sides of the valve leaflets. Figure 3.8 shows calculated limiting streamlines and surface shear stress contours, which were employed by Ge and Sotiropoulos [99] to quantify the relative complexity of hemodynamics on the two sides. Complicated flow patterns were observed only on the aortic side of the valve near the region where focalized distribution of valve calcification is typically observed [101]. Even though these results are far from being conclusive, due to a number of simplifying assumptions, they do provide important evidence that there exists a link between the shear stress environment on the valve leaflets and the propensity of calcification lesions to occur on the aortic side of the valve.

3.3 Computational Methods for Blood Cell Scale Simulations

3.3.1 Background

In blood, a fundamental gap exists between cell-level biological processes and continuum-level function. This gap is filled by millions of RBCs which control the environment individual cells experience and determine continuum-level function. For example, the local stress environment influences stress-mediated platelet adhesion [102, 103], leukocyte adhesion [104], microvascular mechanics [105], artheriogenesis [106], and others. Furthermore, the local stress environment determines macroscopic two-phase rheological behavior of blood, which governs mass transport-mediated processes such as augmented mass transport—or an enhanced diffusivity 1,000 times greater than Brownian motion [107–109]—and platelet margination, which may cause increased numbers of platelets near the wall [110, 111].

Blood at arterial shear rates exhibits slight shear-thinning behavior where the viscosity of blood is shear dependent [112]. A portion of these shear-dependent properties is due to increased RBC deformation, as seen in the capillary number which is discussed below. As the shear rate increases, so does RBC deformation.

3.3.2 Review of Numerical Methods for Blood Cell-Resolving Simulations

Computational simulations of RBCs are prevalent in literature for the purposes of (1) modeling single-cell deformation for obtaining material properties from experimental results and (2) simulating RBCs suspended in fluid for the eventual study of micro- or macro-rheology. Due to the many different motivations in cardiovascular biology and the extreme complexity of simulating RBC suspensions, the level of sophistication via which RBCs are treated in these simulations is varied and not linear in either progression or complexity.

Mechanical stretching of RBCs by optical tweezers on attached beads has been simulated using the commercial software ABAQUS with a non-linear neo-Hookean membrane model for the RBC [113]. However, when fluid–solid interactions are present for a cell suspended in fluid, the problem formulation becomes much more difficult and problem-specific codes are required. Single and multiple RBCs in fluid have been simulated using the boundary-integral method [114–118]; immersed boundary method [3, 116–122]; particle methods [121–124]; and the lattice-Boltzmann method [125–128]. In the following sections, these simulation methods are discussed. Of special note is the large variation in treatment of simulated RBCs, from rigid two-dimensional (2D) spheres to three-dimensional (3D) non-linear deformable membranes. Furthermore, simulation methods are often limited to small numbers of red blood cells.

3.3.2.1 Boundary-Integral Methods for Cell-Level Simulation

The boundary-integral method solves the integral form of the Stokes equations on a computational mesh. Due to lack of inertia, RBC deformation is solved in equilibrium with the fluid stresses. The boundary-integral method is well-developed for the simulation of elastic capsules and RBCs in fluid. A few of the most relevant recent publications are discussed below. In general, the boundary-integral method is relatively intensive computationally and currently limited to simulations of one or two cells. Furthermore, the methodology is limited to Reynolds numbers identically equal to zero, whereas physiologic RBC suspensions occur at finite Reynolds number.

Breyiannis and Pozrikidis [115] present 2D suspensions of deformable fluid-filled circular and biconcave capsules in shear flow for rheologic purposes. However, instability in the biconcave shape precluded the measurement of normal stress differences in RBC suspensions. For single 3D particles, Ramanujan and Pozrikidis [114] investigated tank-treading behavior of elastic spherical and biconcave particles with varying internal viscosities. Results coincided well with experimental observations of RBC flipping, although simulation instability occurs at physiologic viscosity ratio, likely due to the lack of bending stiffness. Pozrikidis [116] furthered these simulations by including bending stiffness in 3D capsules and RBCs with bending stiffness, adding stability to the simulations. Pozrikidis [117] extended these results for a single 3D RBC with bending stiffness in shear flow where the RBC membrane properties are neo-Hookean. It was found that internal fluid viscosities less than two times the external viscosity cause the simulations to become unstable. Pozrikidis [118] further extended these simulations to a single RBC in tube flow but with equal internal and external viscosities. Issues arise due to fluid forces near the wall, causing increased cell curvature, and due to a lack of lubrication modeling. As can be seen in these simulations, treatment of single RBCs in boundary-integral simulations is generally very comprehensive, with 3D finite-element RBCs composed of linear or non-linear membranes with fluid inside. Bending stiffness may be included, as well as a higher fluid viscosity inside the cell due to hemoglobin (though the latter is more computationally intensive than $\hat{\mu} = 1$). The results presented in these boundary-integral simulations indicate the importance of including the bending stiffness of the RBC membrane and the viscosity of hemoglobin for stability purposes. Furthermore, comparative examples of deformed shapes of RBCs in simple shear are displayed for arterial shear rates (Ca_E=0.005–0.2) and are close to RBC properties.

3.3.2.2 Immersed Boundary Method

The immersed boundary method, as discussed earlier in Section 3.2.2.2, solves the fluid phase on a fixed grid, while the solid–fluid coupling is approximated by adding weighted forces to the fluid momentum near the boundary. As such, no limitations exist on fluid solution methods or solid modeling. Since the solid forces are generally applied within a given radius of a boundary node, the immersed boundary solid–fluid coupling fails as particles approach each other.

Eggleton and Popel [119] use the immersed boundary method to simulate 3D capsules and RBCs in shear flow using both neo-Hookean and Skalak models for membrane deformation. A finite Fourier transform method is used to solve the fluid phase. All simulations of RBCs become unstable due to high membrane curvature at the rim of the RBC. This is likely caused by a lack of bending stiffness and the high capillary number of the simulations (Ca_E =0.58). Interestingly, Eggleton and Popel find that the membrane area conservation constraint in the Skalak model is too restrictive for simulations with a reasonable time step. It is found that by reducing this restriction, changes in the surface area had only small effects on results. Bagchi et al. [3] present one- and two-particle interactions of 2D spherical capsules and RBCs using the immersed boundary method. The fluid phase is solved using finite differences, while the solid membrane is treated using a neo-Hookean model. The effective rigidity of interacting particles is found to be a function of their stiffness, bending stiffness, internal viscosity, and membrane viscoelastic properties. Liu and Liu [120] and Liu et al. [121, 122] present an extension of the immersed boundary method, the immersed finite-element method, for the simulation of 3D RBCs in venous rouleaux formation. RBC aggregation is modeled by curve fitting a depletion model for polymeric interactions. The fluid phase is solved using a Galerkin method, while the RBC membrane employs a Mooney–Rivlin model. In impressive simulations, Liu and Liu are able to simulate 10 3D RBCs and look at the Fahraeus–Lindqvist effect on blood viscosity in capillary-sized chambers. These simulations occur at venous shear rates where deformation is two to three orders of magnitude less than that found in arterial flows.

3.3.2.3 Particle Methods

Particle methods such as the semi-implicit or dissipative particle method simulate fluid as discrete particles which have some momentum associated with them. The position of these particles is tracked and fluid is simulated through the dynamics of particle collisions. Solids such as RBCs are simulated as collections of solid particles which behave differently in collision. Particle methods are computationally efficient, though quantitative modeling is method limited as discussed below. Dzwinel et al. [123] and Liu et al. [121, 122] simulate roughly 30 3D RBCs traveling through a capillary using an extension of dissipative particle dynamics. This particle method converges to a steady-state solution of the Fokker–Plank equation [129], which obeys Newton's laws, although the Newtonian fluid stress tensor is not produced by the method. The RBC models employed by Dzwinel et al. consist of a solid particle held together by conservative forces. No material properties similar to RBCs are given. Furthermore, as particles approach, hydrodynamic interactions are neglected and a Lennard–Jones repulsive potential keeps the particles apart. Tsubota et al. [124] use a semi-implicit particle method to simulate suspensions of 2D RBCs. The RBCs are composed of rings of solid particles held together by springs with no fluid inside the cell. Greater than 1,000 2D biconcave RBCs were simulated using 80 processors, though RBC properties were not similar to actual RBCs.

3.3.2.4 Lattice Boltzmann

The lattice-Boltzmann method (LBM) uses a fixed mesh to solve a discretized Boltzmann-like equation which converges to the Navier–Stokes equation. Solid boundaries are treated as moving through the fixed mesh and finite Reynolds numbers may be simulated. The LBM simulates mesoscopic timescales much smaller than the continuum timescale. The LBM approach has undergone much progress recently in simulating a large number of RBCs and capturing the rheology of blood flow. Therefore, this method is covered in more detail in the next section.

3.3.3 Lattice-Boltzmann Methodology

The lattice-Boltzmann technique is well-documented for the direct numerical simulation of particles suspended in fluid [130–135]. As initially described by Aidun et al. [130] and extended by Ding and Aidun [133], to incorporate lubrication forces at below-lattice length scales, the ALD method is a single-relaxation method which excludes the fluid inside a solid boundary. The background and formulation are presented below [136].

3.3.3.1 Lattice-Boltzmann BGK (LBGK) Model for Fluid Flow

The lattice-Boltzmann method discretizes the velocity space on a regular lattice, resulting in a lattice spacing based on the chosen set of discrete velocity vectors, $e_{\sigma i}$. A 3D 19-vector Cartesian discrete velocity set is chosen for $e_{\sigma i}$, where the subscripts σ and i denote the Cartesian directions. The time evolution of the fluid particle distribution function is calculated through a collision and streaming operator using the single-relaxation-time Bhatnagar–Gross–Krook (BGK) collision operator [137]:

$$f_{\sigma i}(r + e_{\sigma i}\Delta t, t + \Delta t) = f_{\sigma i}(r,t) - \frac{1}{\tau}(f_{\sigma i}(r,t) - f_{\sigma i}^{(0)}(r,t)) \tag{3.26}$$

where $f_{\sigma i}$ is the Boltzmann particle distribution function, $f_{\sigma i}^{(0)}$ is the equilibrium distribution function, subscripts σ and i determine the direction and magnitude of the particle velocity, respectively, r is the spatial location, and t is the time. The solution to the lattice-Boltzmann equation is obtained at low Mach number based on the lattice-Boltzmann pseudosound speed c_s. As demonstrated using a Chapman–Enskog expansion [138], the lattice-Boltzmann technique converges to the Navier–Stokes equations when the lattice spacing $c_{\sigma i}$ is much smaller than a characteristic length scale of the simulation. The equilibrium distribution function is defined as

$$f_{\sigma i}^{(0)} = \rho\,(r)\left[A_\sigma + B_\sigma\,(e_{\sigma i}\cdot u) + C_\sigma\,(e_{\sigma i}\cdot u)^2 + D_\sigma u^2 + O(u^3)\right] \tag{3.27}$$

where mass density, $\rho\,(r,t) = \sum\limits_{\sigma,i} f_{\sigma i}\,(r,t)$, and momentum, $\rho u\,(r,t) = \sum\limits_{\sigma,i} f_{\sigma i}\,(r,t)\,e_{\sigma i}$, must be conserved by the equilibrium distribution function, and coefficients A_σ to D_σ are given by Aidun et al. [130] and Ding and Aidun [133]. When the scaling for time and length is set to unity, the lattice-Boltzmann relaxation timescale τ is related to the fluid viscosity by $\tau = 3\nu + 0.5$. Fluid–particle boundaries are considered in a "link bounce-back" manner based on the lattice links crossing a solid boundary as shown in Fig. 3.9. The evaluation of rheological properties of the suspensions of deformable particles requires a high order of accuracy in computation of the stress distribution on the particle. The corrections required for accurate preservation of Galilean invariance in LB methods are discussed by MacMeccan et al. [127] and Clausen and Aidun [140]. A new LB method based on external boundary force (EBF) is presented [128] as an alternative method for LB analysis of deformable particles and capsules. The EBF method is more stable than the standard bounce-back method.

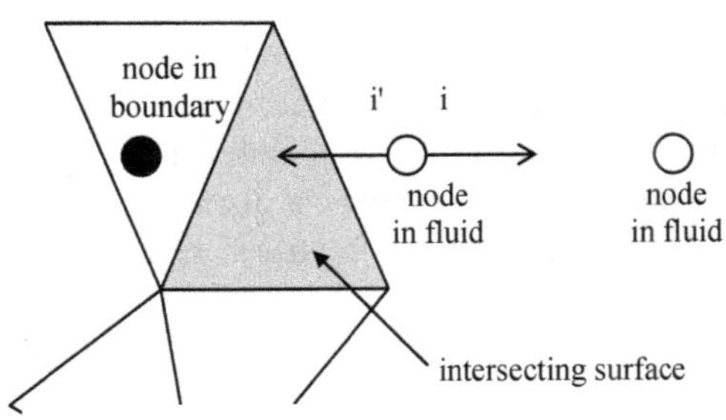

Fig. 3.9 Boundary treatment of finite-element solid in lattice fluid

3.3.3.2 Transient Finite-Element FSI Model

The trajectory and the deformation of an elastic deformable solid are governed by Cauchy's equation:

$$\frac{D\rho_s u_s}{Dt} = \nabla T_{\text{elastic}} \tag{3.28}$$

where

$$\frac{D\rho_s u_s}{Dt}$$

is the material derivative of solid momentum, and T_{elastic} is the elastic stress tensor. For deformable solid particles, the shearoelastic number $N_{\text{se}} = \mu\dot{\gamma}/G_S$ governs particle deformation, where μ is the fluid dynamic viscosity, $\dot{\gamma}$ is the fluid shear rate, and G_S is the particle's shear modulus. The capillary number

$$\text{Ca}_G = \frac{\mu\dot{\gamma}D}{G_M}$$

governs deformation of fluid-filled elastic capsules, where D is the particle diameter and G_M is the effective membrane's shear modulus ($G_M = G_S t_M$ for membranes with thickness t_M).

The transient finite-element method is chosen to calculate the time evolution of deformable particles in suspension (Equation (3.28)) because this method is well developed with great versatility and multiple commercial software applications. A general review of the transient finite-element method can be found in Bathe [11]. The transient finite-element method is derived by integrating virtual work over the volume of the element

$$\int \bar{\varepsilon}\, T_{\text{elastic}}\, \mathrm{d}V = \int \overline{X} F^{\text{traction}} \mathrm{d}A + \int \overline{X} F^{\text{body}} \mathrm{d}V \tag{3.29}$$

where $\bar{\varepsilon}$ is the virtual strain due to virtual displacements $\overline{X}$, F^{traction} are traction stresses on the surface, F^{body} are body stresses such as inertia, A is the surface area, and V is the volume. Equation (3.29) states that the amount of elastic work done on the solid (given by the area under the stress–strain curve and left-hand side of Equation (3.29)), is equal to the work done by the traction force on the boundary (area under the force–displacement curve) and solid inertia. A particle is discretized into elements consisting of nodes. The integrals in Equation (3.29) are evaluated for each element as functions of displacements and then summed over all elements in a solid object. The resulting transient finite-element equation

$$M\ddot{x} + C\dot{x} + Kx = F \tag{3.30}$$

determines the time evolution of the nodal displacement vector x and its time derivatives $\dot{x}$ and $\ddot{x}$, where the nodal displacement vector is defined as the deformed node location minus the undeformed node location. The global mass M, damping C, and stiffness K matrices are constructed from elemental matrices, while the force vector F is calculated from traction forces resulting from the coupling to the fluid as described later.

The damping matrix in the transient finite-element equation (Equation (3.30)) is chosen as a Rayleigh damping matrix:

$$C = \alpha_{\mathrm{D}} M + \beta_{\mathrm{D}} K \tag{3.31}$$

where the coefficients α_{D} and β_{D} are related to the solid body damping ratios via $\zeta(\omega_n) = 0.5\left(\alpha_{\mathrm{D}}\omega_n^{-1} + \beta_{\mathrm{D}}\omega_n\right)$, where ζ is the damping ratio for a given modal circular frequency ω_n. The Rayleigh damping coefficients are chosen based on the desired damping ratios for given solid material properties. To avoid influencing particle dynamics, damping ratios are chosen such that $\zeta(\omega_n) << 1$ for all simulations presented here. A finite-element object contains many discrete modes representing a spectrum of vibrational frequencies (ω_n), and these frequencies are calculated via a modal analysis in ANSYS using a block Lanczos routine. A review of finite-element modal analysis may be found in Bathe [11].

The transient finite-element equation is solved using a direct integration method, Newmark's method, where Newmark's equations

$$\ddot{x}_{t+1} = \beta_{\mathrm{N}}^{-1}\Delta t^{-2}\left[x_{t+1} - x_t - \Delta t\dot{x}_t - \Delta t^2(0.5 - \beta_{\mathrm{N}})\ddot{x}_t\right] \text{ and,} \tag{3.32}$$

$$\dot{x}_{t+1} = \dot{x}_t + \Delta t\left[(1-\gamma_{\mathrm{N}})\ddot{x}_t + \gamma_{\mathrm{N}}\ddot{x}_{t+1}\right] \tag{3.33}$$

combine with the transient finite-element equation (Equation (3.30)) to produce a solid-phase time evolution equation of the form

$$K'_{t+1}x_{t+1} = F'_{t+1} \tag{3.34}$$

where

$$K' = K + \frac{1}{\beta_{\mathrm{N}}\Delta t^2}M + \frac{\gamma_{\mathrm{N}}}{\beta_{\mathrm{N}}\Delta t}C, \text{ and} \tag{3.35}$$

$$\begin{aligned} F'_{t+1} =& F_{t+1} + \frac{M}{\beta_{\mathrm{N}}\Delta t^2}\left[x_t + \Delta t\dot{x}_t + \Delta t^2(0.5 - \beta_{\mathrm{N}})\ddot{x}_t\right] \\ &+ C\left[\frac{\gamma_{\mathrm{N}}}{\beta_{\mathrm{N}}\Delta t}x_t + \left(\frac{\gamma_{\mathrm{N}}}{\beta_{\mathrm{N}}} - 1\right)\dot{x}_t + \Delta t\left(0.5\frac{\gamma_{\mathrm{N}}}{\beta_{\mathrm{N}}} - 1\right)\ddot{x}_t\right] \end{aligned} \tag{3.36}$$

For convenience, Equation (3.36) is rewritten as $F'_{t+1} = F_{t+1} + M'_t + C'_t$. The choice of $\beta_{\mathrm{N}} = 1/6$ and $\gamma_{\mathrm{N}} = 1/2$ yields a constant acceleration method which is unconditionally stable. To guarantee convergence, the integration time step Δt is chosen as less than 0.1 of the smallest fundamental period as determined by the modal analysis of the finite-element particle. For particle shearoelastic numbers of the order 0.1, this constraint generally corresponds to time steps greater than the lattice time step. Thus, the finite-element integration is performed on a different timescale than the lattice-Boltzmann integration, separating the deformation and lattice timescales.

In order to simulate large numbers of deformable particles in suspension, the FEA must be incorporated efficiently, necessitating the assumption of small body-fixed deformations. To comply with this assumption, a particle coordinate system is fixed on a particle's center of mass and oriented using the average angular displacement of the finite-element nodes. Although elemental co-rotational procedures are typically utilized for large-rotation problems [141, 142], when averaged, the co-rotational procedure gives less consistent results than does a simple average angular displacement. The use of a body-fixed coordinate system for the solid particles results in invariant linear-elastic stiffness, mass, and damping matrices that may be determined a priori for each type of particle. Thus, the left-hand side of Equation (3.34) is constructed and inverted once for each particle type and applied to all particles of that type at all time steps. This simplification results in $O(n^2)$ operations for the time evolution of solid particles as opposed to $O(n^3)$ operations for inversion, where n is the size of the particle's finite-element matrices.

Coupling between fluids and finite-element solids is determined based on the lattice direction vectors crossing the solid boundary, forming links between the fluid and the solid [130]. Lattice links are found on the discretized finite-element surface using a ray-tracing algorithm commonly used in computer graphics. In this method, rays are projected along the lattice directions and tested for intersection with the triangles comprising the solid surface using a fast and minimum storage algorithm. The intersection is found through direct three-dimensional calculation using barycentric coordinates, which eliminates the need for two-dimensional projections or calculation of the plane equation for the triangle. For more details on this method, see implementation by Moller and Trumbore [143].

The fluid force on the moving solid boundary is determined by the following equation:

$$F_{\sigma i}^{(B)}\left(r+0.5e_{\sigma i'},t+0.5\Delta t\right)=2e_{\sigma i'}\left[f_{\sigma i'}^{t+}-\rho B_{\sigma}u_{\mathrm{b}}\cdot e_{\sigma i'}\right]+O(u^2) \qquad (3.37)$$

where $F_{\sigma i}^{(B)}$ is the force along the σith lattice vector ($e_{\sigma i}$), σi and σi' are lattice directions as defined in Fig. 3.9, and $f_{\sigma i'}^{t+}$ is the post-collision particle distribution function in the direction σi', $f_{\sigma i}^{t+}=f_{\sigma i}-\frac{1}{\tau}(f_{\sigma i}-f_{\sigma i}^{(0)})$. The boundary velocity u_{b} is determined by linear interpolation from the finite-element nodal velocity on the surface intersected by the link. The fluid particle distribution function along the σith lattice vector is modified by the presence of the solid boundary:

$$f_{\sigma i}(r,t+1)=f_{\sigma i'}^{t+}(r,t)+2\rho B_{\sigma}u_{\mathrm{b}}\cdot e_{\sigma i}+O(u^2) \qquad (3.38)$$

which imposes the correct velocity gradient in the fluid stress tensor. This boundary treatment assumes that the wall velocity is stationary with respect to the lattice timescale [135]. For a discussion of the Galilean invariance (GI) of the fluid–solid interaction force, see Clausen and Aidun [140]. A more stable method for computing the fluid–solid interaction force based on the external boundary force (EBF) is given by Wu and Aidun [128]. There is no need for GI correction with the EBF method.

At high-volume fractions, particles regularly approach within one lattice unit of each other. When no fluid node exists between two particles, sub-mesh modeling must be incorporated into the lattice-Boltzmann method. For the case of ideally smooth surfaces, Ding and Aidun [133] developed a lubrication model extending the analytic solution for approaching spheres to a model for curved surfaces that uses lattice links. This model performs well for a variety of different particle radii and fluid viscosities; however, the required integration time step becomes very small as particles approach contact. In the case of real particles, contact mechanics must be considered due to surface roughness. Furthermore, in the case of highly deformable particles such as RBCs, surface fluctuations due to Brownian motion will cause the cell membranes to collide at small separations, requiring the inclusion of below-lubrication contact modeling. Buxton et al. [144] study the impact of deformable lattice-spring particles with a wall and propose either an exponential repulsive force or a Lennard-Jones potential for link-wise contact mechanics. For the purposes of our method, we transition the lubrication model developed by Ding and Aidun [133] to contact, with special attention to the forces in suspension mechanics. Near-contact interactions along a link connecting approaching solid surfaces are calculated as follows:

$$\mathrm{d}F_{\sigma i}^{(B),t+1} = \begin{cases} 0, & \text{if} \quad c_{\sigma i} < g^{t+1} \\ \left[\frac{3\bar{q}}{2c_{\sigma i}^2\bar{\lambda}}\upsilon\rho\left(\frac{1}{(g^{t+1})^2} - \frac{1}{c_{\sigma i}^2}\right)\right]U_{\text{approach}}^{t+1}, & \text{if } g_{\mathrm{c}} < g^{t+1} < c_{\sigma i} \\ \left[\frac{3\bar{q}}{2c_{\sigma i}^2\bar{\lambda}}\upsilon\rho\left(\frac{1}{g_{\mathrm{c}}^2} - \frac{1}{c_{\sigma i}^2}\right)\right]U_{\text{approach}}^{t+1} & \\ \quad +A_{\mathrm{C}}\exp\left(\frac{-g^{t+1}+g_{\mathrm{c}}}{\sigma_{\mathrm{C}}}\right)\phi_{\sigma i} & \text{if} \quad g^{t+1} < g_{\mathrm{c}} \end{cases} \tag{3.39}$$

where $U_{\text{approach}}^{t+1}$ is the surface-approach velocity in link coordinates at time $t+1$, g^{t+1} is the link-wise gap between surfaces at time $t+1$, g_{c} is the contact cutoff distance, and $\overline{q} = 0.6$ [133]. These interactions are added to the lattice-Boltzmann, fluid–solid interactions for both particles' approaching surfaces when the gap between approaching surfaces is less than the link length $c_{\sigma i}$. Thus, the total force acting on a link connecting approaching surfaces is

$$F_{\sigma i}^{t+1} = F_{\sigma i}^{(B),t+0.5} + \mathrm{d}F_{\sigma i}^{(B),t+1} \tag{3.40}$$

where for the purposes of compactness, the link bounce-back force is written with a time superscript. The friction indicator is

$$\phi_{\sigma i} = \frac{(S_{\mathrm{N}} + \mu_{\mathrm{SF}}S_{\mathrm{T}}) \cdot e_{\sigma i}}{(S_{\mathrm{N}} + S_{\mathrm{T}}) \cdot e_{\sigma i}} \tag{3.41}$$

where S_N is the surface-normal vector, μ_{SF} is the coefficient of sliding friction, S_T is the projection of the surface-approach velocity on the surface tangent plane, and

$$S_T = \frac{S_N \times U_{\text{approach}} \times S_N}{\left|S_N \times U_{\text{approach}} \times S_N\right|} \tag{3.42}$$

The appropriate contact scale is $A_c = 6\pi\mu R U_{\text{Scale}}$, where U_{Scale} is the velocity scale of the problem in lattice units, e.g., $\dot{\gamma}D$ for particles in shear or the settling velocity for sedimentation problems. The contact cutoff distance g_c and the contact length scale σ_c are functions of the particle surface roughness and are determined a priori. The mean surface curvature $\bar{\lambda}$ in the lubrication model is the local mean curvature of the approaching finite-element surfaces. The local mean curvature of each individual surface is calculated as follows:

$$\lambda = \frac{\mathrm{d}T_{\text{surface}}}{\mathrm{d}s} \tag{3.43}$$

where T_{surface} is the tangent vector to the surface and s is a surface coordinate connecting finite-element surface centroids.

3.3.4 Membrane Models

RBC deformation is one of the most important aspects of blood rheology. RBC architecture consists of a cytoskeleton and a phospholipid membrane encapsulating a fluid solution of hemoglobin. The primary structural protein of the cytoskeleton, spectrin, is loosely coupled to the fluid membrane through proteins such as ankyrin. This composite structure gives the RBC both solid and fluid properties. At body temperature, the RBC membrane has a small but finite elastic shear modulus of $5.7 \times 10^{-3} \pm 1.8 \times 10^{-3}$ dyne/cm [145], a much larger area modulus of 288 ± 50 dyne/cm [146], and a bending stiffness of $2.2\times10^{-19}\pm0.3\times10^{-19}$ N m [147] caused by the spectrin skeleton and external negative charge. However, viscous behavior is observed in shear-thinning behavior of the membrane during micropipette aspiration [148], as well as the familiar tank-treading behavior [149] (Schmid-Schönbein et al. 1983). The fluid inside the RBC membrane is hemoglobin, which has a viscosity of 6 cP, approximately five times that of the surrounding plasma. In simple shear experiments, dimensionless parameters that contribute to RBC deformability are the capillary number

$$\mathrm{Ca_G} = 2\mu\dot{\gamma}R/G_S \tag{3.44}$$

where μ is the fluid viscosity, $\dot{\gamma}$ is the fluid shear rate, R is the particle radius, and G_S is the membrane shear modulus (or $\mathrm{Ca_E} = 2\mu\dot{\gamma}R/E_S$ based on Young's modulus E_S), non-dimensional bending stiffness

$$\hat{\kappa} = \frac{\mu \dot{\gamma} D^3}{\kappa_{\mathrm{B}}} \tag{3.45}$$

where κ_{B} is the membrane bending stiffness and D is the cell diameter membrane, Poisson ratio ν_{P}, which is related to membrane dilatation, and the viscosity ratio

$$\hat{\mu} = \frac{\mu_{\mathrm{f,inside}}}{\mu_{\mathrm{f,outside}}} \tag{3.46}$$

which relates the viscosity of fluid inside the membrane to the plasma viscosity outside the membrane.

The finite-element method has been used to describe single RBC deformation in the above-mentioned micropipette aspiration, stretching using beads [113], passing through capillaries [120], in shear flow [119], and elsewhere. In these simulations and others, red blood cells are often modeled using the non-linear Skalak model [150], which conserves RBC surface area, a neo-Hookean model [3, 113, 117], which does not conserve RBC surface area, or a Mooney–Rivlin model, which is an extension of the neo-Hookean model. While the effect of RBC membrane models is unknown for simulations of RBCs in high-volume-fraction suspensions, each model possesses unique properties. All three models are non-linear strain energy models, which treat the RBC membrane as an effective 2D analog characterized by principle stretches λ_1 and λ_2. At small capillary number, each model converges to a linear-elastic form. Using the strain energy formulation, the time evolution of the RBC membrane may be solved using the method of virtual work (left-hand side of Equation (3.29)):

$$\int \delta^o W \, \mathrm{d}V = \text{virtual work} \tag{3.47}$$

where δ^{o} is the virtual displacement operator, W is strain energy, and V is solid volume. Alternatively, the strain energy may be related to the stress by the following equation:

$$S_{ij}^{\mathrm{PK}} = \frac{\mathrm{d}W}{\mathrm{d}E_{ij}} \tag{3.48}$$

where S_{ij}^{PK} is the second Piola–Kirchhoff stress and the Green strain tensor is given by the following equation [11]:

$$E_{ij} = \frac{1}{2}\left(\frac{\mathrm{d}x_i}{\mathrm{d}r_j} + \frac{\mathrm{d}x_j}{\mathrm{d}r_i}\right) \tag{3.49}$$

where x is the deformation vector and r is the spatial location. The second Piola–Kirchhoff stress

$$S_{ij}^{\mathrm{PK}} = \frac{\mathrm{d}x_i}{\mathrm{d}r_m}\frac{\mathrm{d}x_j}{\mathrm{d}r_n} T_{mn} \tag{3.50}$$

expresses stress from strain in the undeformed elastic configuration, compared with the Cauchy stress T_{mn}, which expresses stress in the deformed configuration.

The strain energy equation for a neo-Hookean membrane may be given by the following equation [3]:

$$W = G_s t_M \left(\lambda_1^2 + \lambda_2^2 + \lambda_1^{-2}\lambda_2^{-2}\right) \tag{3.51}$$

where G_s is the shear modulus and t_M is the thickness of the membrane. When simulating RBCs, the Poisson ratio is set to 0.5, resulting in the preservation of the membrane volume. The material properties of the neo-Hookean membrane are easily matched to RBC properties through the shear modulus.

The Mooney–Rivlin model is an extension of the neo-Hookean model and has a strain energy of [151]

$$W = \frac{G_{MR}}{2}\left(\psi_{MR}\left(I_1 + 2 + \frac{1}{I_2 + 1}\right) + (1 - \psi_{MR})\left(\frac{I_1 + 2}{I_2 + 1} + I_2 + 1\right)\right) \tag{3.52}$$

where G_{MR} is an elastic modulus which, using the asymptotic limit at small deformations, is equal to the elastic membrane shear modulus G_s [151]. The choice of ψ_{MR} for RBCs has not been experimentally verified, although $\psi_{MR}.\approx 0.9$–1 is generally used [120, 150, 151]. The invariants are $I_1 = \lambda_1^2 + \lambda_2^2 - 2$ and $I_2 = \lambda_1^2\lambda_2^2 - 1$, which are related to area dilatation.

The Skalak models were developed specifically to model RBC deformation [150, 152] and generally have a strain energy function

$$W = \frac{B_{SK}}{4}\left(\frac{1}{2}I_1^2 + I_1 - I_2\right) + \frac{C_{SK}}{8}I_2^2 \tag{3.53}$$

where the coefficients B_{SK} and C_{SK} are material properties. To preserve membrane area, C_{SK} should be chosen much larger than B_{SK}, with suggested values of 5 and 0.005 dyne/cm, respectively. Also, viscoelasticity may be incorporated into this class of area-incompressible membranes [148].

The lattice-Boltzmann finite-element method is limited to simulations at small capillary number and models the RBC membrane using 504 linear-elastic, finite-element "shell" elements as described in MacMeccan et al. [127]. RBCs are modeled as three-dimensional biconcave elastic membranes encapsulating hemoglobin. The RBC membrane has an effective elastic shear modulus and bending stiffness with known material properties which are given above. A Poison ratio of 0.48 results in $E_Y/G_S = 2.96$ and similar behavior to the neo-Hookean model for RBC membrane deformation. The hemoglobin inside the RBC membrane is set to a viscosity of 6 cP, while the plasma surrounding the RBC has a viscosity of 1.2 cP at 37°C [153]. The RBCs have a major diameter of 7.8 μm and thickness of 2.2 μm at the flank and 0.9 μm at the dimple. The major RBC diameter is set to 24 lattice units. As will be seen in the following section, the linear-elastic membrane model is comparable to non-linear RBC membrane models for RBCs and suspensions of RBCs at physiologic viscosity ratio $\hat{\mu}$, as well as for small capillary numbers corresponding to physiologic low-arterial shear rates.

3.3.4.1 Comparison of Red Blood Cell Models

While the neo-Hookean, Skalak, and Mooney–Rivlin laws for membrane deformation have been successfully used to model RBC deformation in a variety of experimental applications, differences do exist between these models. For the purposes of differentiating between these models for RBC deformation, Barthes-Biesel et al. [151] and Eggleton and Popel [119] present comparative reviews of the models for spherical and biconcave RBC membranes in deformation. In the small deformation limit, all models converge to linear theory. At higher deformation values and under uniaxial or isotropic extension, Barthes-Biesel et al. [151] show that Mooney–Rivlin membranes tend to be strain softening, while Skalak membranes are always strain stiffening due to the near-conservation of membrane area. For the case of a sphere in axisymmetric straining flow, the area-extension modulus in the Skalak model C_{SK} is found to have a non-negligible effect on the deformed capsule shape. However, changes in the Skalak membrane shear modulus B_{SK} tend to dominate the deformed shape. For a thorough review of the membrane deformation models, see [139] MacMeccan (2007) and MacMeccan et al. [127].

3.3.5 Rheology, Stress, and Microstructure of Blood

Suspensions of RBCs and platelets at 40% hematocrit have been simulated using the coupled lattice-Boltzmann finite-element method to investigate the stress environment that platelets experience due to the two-phase nature of blood (MacMeccan, 2007) [127]. This stress environment is important in platelet deposition due to stress-mediated adhesion [102, 103]; augmented mass transport [107–108]; and platelet margination [108–110, 154]. Furthermore, the stress environment in blood is also important to areas such as leukocyte adhesion [104], microvascular mechanics [105], artheriogenesis [106, 155], and others. Lattice-Boltzmann, finite-element simulations of single RBCs in shear flow and pressure-driven flow accurately match experimental observations by Yao et al. [156] and Liu et al. [121, 122] in rheometers and in flow chambers as shown by MacMeccan et al. [127].

Simulations of blood suspensions consist of deformable 3D biconcave RBCs and ellipsoidal platelets with correct material properties at physiologic hematocrit. RBCs and platelets are initialized in an undeformed state at random location and orientations in unbounded shear flow where the flow and vorticity directions are periodic and the shear direction is shear periodic (MacMeccan, 2007). The effect of domain size is investigated by MacMeccan et al. [127] with no variation in bulk rheological properties between 77 and 847 RBCs in the domain. Furthermore, simulations show insensitivity to simulation particle Reynolds number between 0.03 and 0.2.

Following the method of Batchelor [157] and neglecting inertial terms, the particle bulk stress $\sum_{ij}$ can be separated into a Newtonian contribution due to the fluid and a potentially non-Newtonian contribution due to the particle shown as

$$\sum_{ij} = 2\mu E_{ij} + \frac{1}{V}\sum S_{ij}, \tag{3.54}$$

where E_{ij} is the strain rate tensor and S_{ij} is the contribution of an individual particle to the bulk stress known as the stresslet. The volume-averaged effect of all particle stresslets constitutes the contribution of the solid phase to the suspension stress. The stresslet for each particle is obtained via the integral

$$S_{ij} = \int_{A_0} \frac{1}{2}\left(\sigma_{ik} r_j + \sigma_{jk} r_i\right) n_k - \mu\left(u_i n_j + u_j n_i\right) \mathrm{d}A \tag{3.55}$$

where σ_{ij} is the stress in the suspending medium at the particle boundary, u is the velocity of the surface, n_i is the surface normal, and r is the position relative to the center of the particle. For the coupled lattice-Boltzmann, finite-element method, the surface integration is readily computed since the stress on the surface of the particles is known via the fluid–solid coupling. Also, the boundary velocity utilized in the second term of integral is known via the velocity of the finite-element deformation vector $\dot{x}$. The effective suspension viscosity is calculated from the particle bulk stress

$$\mu_{\text{eff}} = \frac{\langle \sum_{12} \rangle}{\dot{\gamma}} \tag{3.56}$$

where μ_{eff} is the effective suspension viscosity, and the ensemble-average particle bulk stress $\langle \sum_{12} \rangle$ is calculated using a time-integral average.

3.3.5.1 Bulk Rheology

To study the shear dependence of blood, 204 RBCs and 12 platelets are simulated in unbounded shear at 40.5% ha and shear rates ranging between 15 and 64 s^{-1}. Simulations are performed at a particle Reynolds number of 0.1 and a lattice-Boltzmann Mach number of 0.03.

Simulation parameters and simulation viscosity results are summarized in Table 3.1 with Ca_G based on the effective suspension viscosity μ_{eff} reported as a better indicator of RBC deformation in suspension compared to Ca_G. The linear-elastic, finite-element model performs well at the simulated capillary numbers (as described for a single RBC in Section 3.3.4 and RBCs in suspension as described later in Section 3.3.5.2); however, the higher Ca_G value based on μ_{eff} limits the range of shear rates investigated to only flows at low-arterial shear rates.

Table 3.1 Simulation parameters and simulation results for differing shear rates at 40.5% ha

$\dot{\gamma}$ (s^{-1})	RBC Ca_G	μ_{eff}/μ_{plasma}	RBC Ca_G based on μ_{eff}
15	0.011	4.55	0.05
22	0.016	4.35	0.07
31	0.023	4.19	0.09
51	0.037	4.02	0.15
64	0.047	3.94	0.18

Comparisons with experimental data that have been reported are often difficult because some experiments are performed at an in vitro temperature of 25°C, while others are performed at an in vivo temperature of 37°C. Both plasma viscosity and RBC membrane shear modulus are temperature dependent, with plasma viscosity and membrane shear modulus increasing to 1.6 cP [153] and 6.06 dyne/cm [145], respectively, at 25°C. To clarify the potential inconsistency in experimental results, simulations are run using material properties at 37°C as discussed in Section 3.3.4. In this section, results reported by Fung [158], which were performed at 25°C, are adjusted to an effective shear rate based on capillary number similarity.

Platelets are also included in simulations. Unlike RBCs, platelets have a complex internal structure consisting of a complex cytoskeleton which contains dense bodies and α-granules. Platelets are modeled as an effective solid using 711 linear-elastic, "brick-type" finite elements with a Young's modulus of $1.7 \pm 0.6 \times 10^3$ dyne/cm^2 and a shear modulus of $0.57 \pm 0.6 \times 10^3$ dyne/cm^2 [159]. Although platelet deformation is small compared with RBCs, the inclusion of a finite-element representation of platelets has negligible computational penalty due to their small numbers as compared with RBCs. The inactivated platelets simulated in this work have an approximately elliptical shape with a major diameter of 2.5 μm and a thickness of 0.7 μm (Haga et al.) [159], and they account for 0.1% of the total volume fraction.

3.3.5.2 Shear-Thinning Behavior

At the investigated shear rates and 40% ha, blood is often described [158] by Casson's equation:

$$\sqrt{\tau_{eff}} = \sqrt{\tau_{yield}} + \sqrt{\dot{\gamma}\,\mu_{eff}} \tag{3.57}$$

where τ_{eff} is the effective suspension shear stress and τ_{yield} is a constant which is the yield stress of the suspension in shear. A Casson plot of the suspension shear stress at different shear rates is shown in Fig. 3.10 with comparisons to experimental data. As seen in Fig. 3.10, the simulations display Casson behavior with a constant slope. The simulations are in very good agreement with experimental values reported by Merril et al. [5] for blood at 42.5% ha in a large-gap, Couette-type viscometer. Experimental values reported by Fung at 35.9 and 47.6% ha fall below and above simulations, respectively.

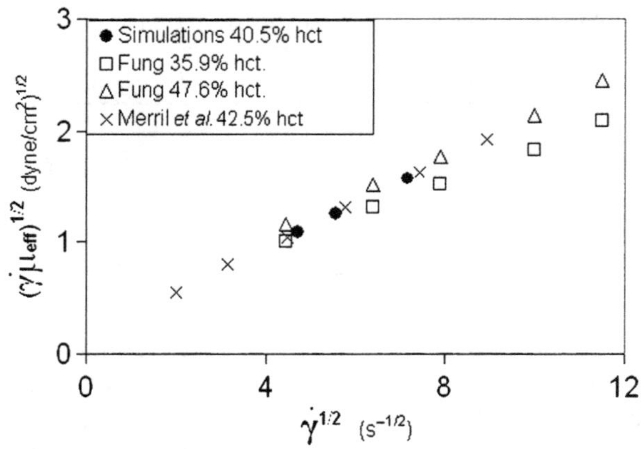

Fig. 3.10 Casson plot of shear stress as a function of shear rate for simulations of 204 RBCs and 12 platelets at 40.5% ha. Simulations are consistent with Casson behavior in experimental results using a Couette viscometer with a large gap [5, 158]

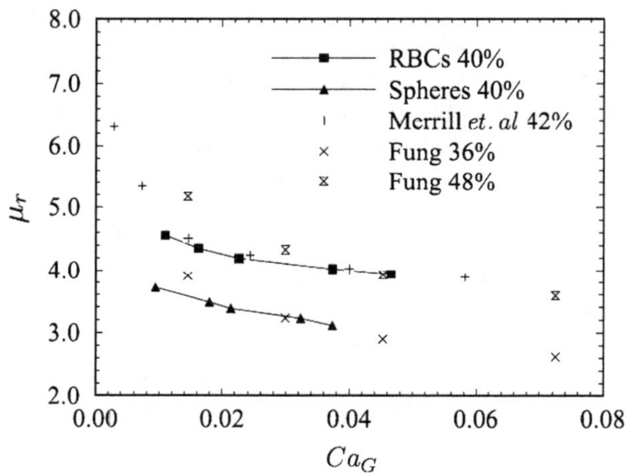

Fig. 3.11 Effective suspension viscosity of 204 RBCs and 12 platelets at 40.5% ha at shear rates ranging between 22 and 51 s^{-1}. Simulations display shear-thinning behavior consistent with experimental results in a Couette viscometer with large gap [158] and in large tubes [5]

A Casson fluid exhibits non-Newtonian and shear-thinning behavior. The viscosity of blood as a function of shear rate is shown in Fig. 3.11, with the effective viscosity in simulations decreasing from 5.46 cP at $\dot{\gamma} = 15\ s^{-1}$ to 4.73 cP at $\dot{\gamma} = 64\ s^{-1}$. Good agreement between simulations and experimental data indicates that the lattice-Boltzmann, finite-element method contains the necessary physics to predict blood behavior at physiologic hematocrit and shear rates. Furthermore, the linear-elastic RBC membrane model, which is consistent with published models, is appropriate for the simulation of RBC suspensions at the investigated shear rates.

In addition to shear-thinning rheology, the LBM-FEA method can probe the concentration dependence of blood viscosity as seen in Fig. 3.12. The RBCs in all simulations have a capillary number of 0.021 and the domain size is fixed, while

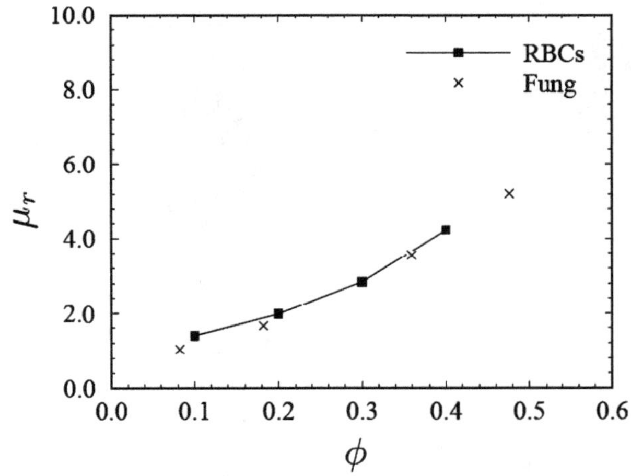

Fig. 3.12 Reduced suspension viscosity of simulations of blood at different volume fractions and $Ca_G = 0.021$ compared to experimental results by Fung [158] for blood in a Couette-type viscometer with a large gap

the number of particles varies from 54 to 216. The LBM-FEA method accurately predicts the concentration-dependent viscosity of blood at physiologic volume fraction, while slightly over-predicting viscosity at low volume fraction. The slight over-prediction at low volume fraction is most likely due to experimental and biological variations, in particular reported physical parameters for RBC membrane and plasma viscosity which are used to calculate experimental non-dimensional results. The accurate description of blood at varying concentrations is particularly important for the simulation of complex flows where RBC migration drives concentration gradients.

3.3.5.3 Microstructure

Snapshots of example simulations are shown in Fig. 3.13, with the initial locations shown on the left and the deformed cells at time $\dot{\gamma}\, t = 10$ shown on the right. The direction of shear flow is the same for all simulations and is shown by arrows. As seen in Fig. 3.13, many cells tend to align with the flow axis, consistent with the experimental observations of Goldsmith and Marlow [160]. Goldsmith and Marlow suspend normal RBCs among RBC ghosts that have been rendered transparent by having the hemoglobin removed. The suspensions flow through 65-μm tubes at 50% ha and $\dot{\gamma} = 66\ \text{s}^{-1}$, and RBC orientation is measured as defined in Fig. 3.14a. As seen in Fig. 3.14, RBC orientation is described using a cumulative φ_z orientation distribution which describes the probability of all RBCs orientated at less than φ_z. The range of φ_z is 0° to 90° due to cell symmetry ($\phi_z = 0° = 180°$) and the assumption in experiments that an RBC is equally likely to orient above and below the flow direction, e.g., $P_{\text{cumulative}}\,(+\phi_z) = P_{\text{cumulative}}\,(-\phi_z)$.

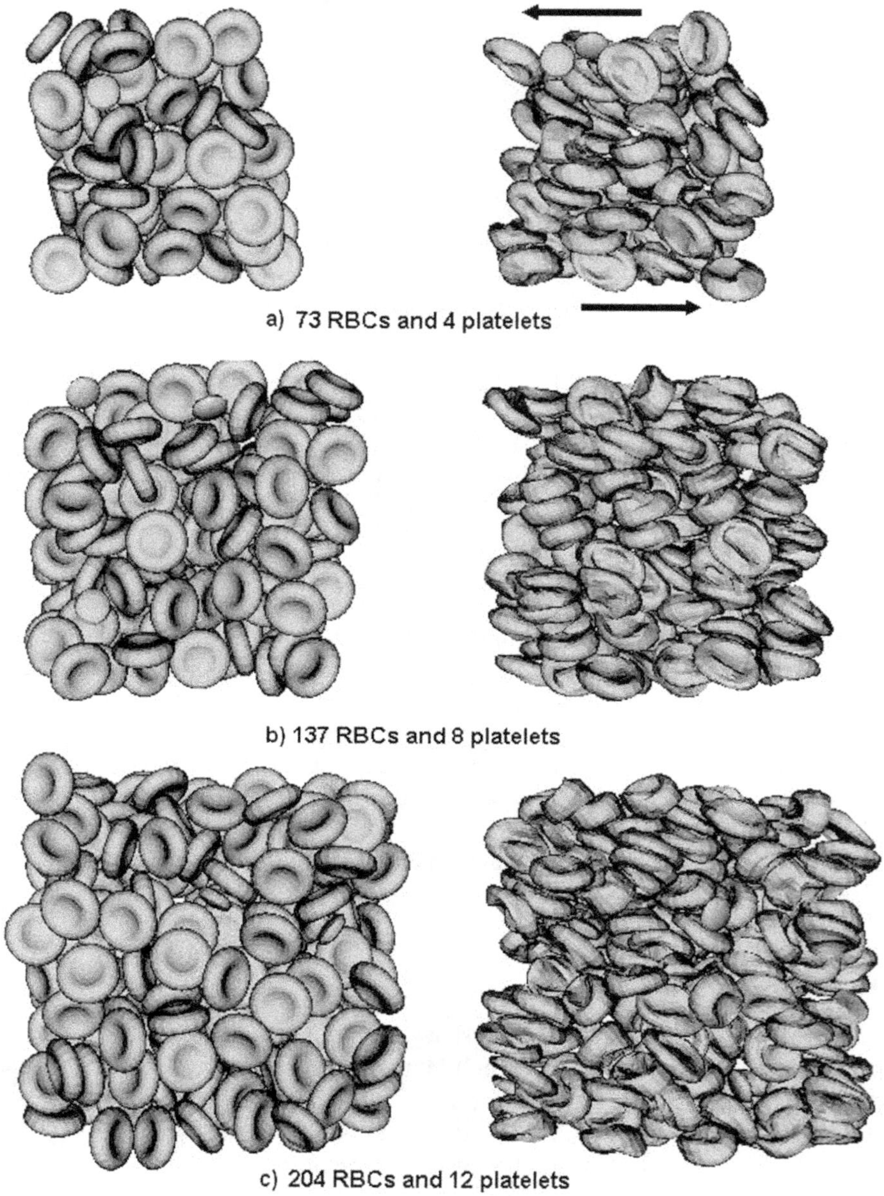

Fig. 3.13 Simulations of RBCs and platelets in unbounded shear at 40.5% ha and $Ca_G = 0.04$, drawn to equal scale. RBCs and platelets are initialized to random locations (*left side*) and deform in flow (*right side* at $\dot{\gamma}\, t = 10$). The direction of shear is shown by *arrows* in the first simulation

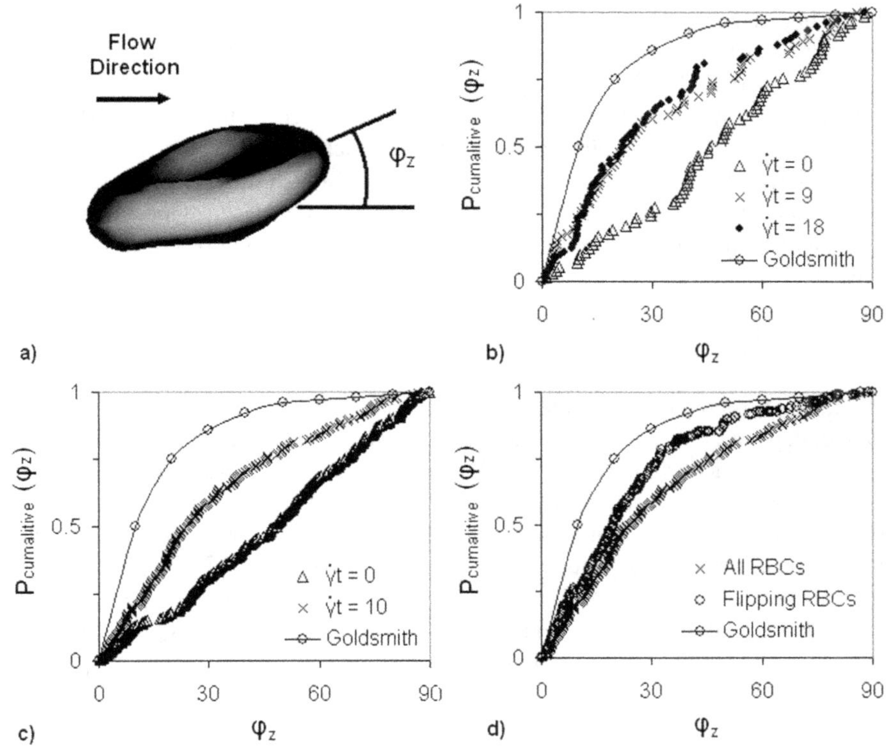

Fig. 3.14 Cumulative φ_z orientation distribution of RBCs in unbounded shear at 40.5% ha and $\dot{\gamma} = 55\ \text{s}^{-1}$. (**a**) RBC orientation, (**b**) simulation of 73 RBCs and 4 platelets, (**c**) simulation of 204 RBCs and 12 platelets, (**d**) 204 RBCs at $\dot{\gamma}t = 10$ with only RBCs within 20° of the flipping orientation (out-of-plane angle) considered. Experimental results are from Goldsmith and Marlow [160] in 65-μm tubes at 0.50% ha and $\dot{\gamma} = 66\ \text{s}^{-1}$

The cumulative φ_z orientation distribution of 73 RBCs in unbounded shear at 40.5% ha and $\dot{\gamma} = 55\ s^{-1}$ is shown in Fig. 3.14b at times $\dot{\gamma}t = 0,\ 9,\ \text{and}\ 18$. The linear cumulative distribution at $\dot{\gamma}t = 0$ indicates the random nature of initial particle orientations. At later times, the cumulative distribution reaches an approximate steady state ($\dot{\gamma}t = 9$ and 18 shown in Fig. 3.14b) with more RBCs aligned near the flow direction, $\phi_z = 0°$, than the shear direction. The cumulative φ_z orientation distribution of 204 RBCs is shown in Fig. 3.14c with identical results to 73 RBCs and 137 RBCs (not shown), indicating that the microstructure is invariant in domains larger than 70 particles in the case of unbounded shear flow. This result is consistent with invariant bulk viscosity measured in simulations with greater than 70 particles (MacMeccan, 2009) [127].

In Fig. 3.14b, c, the orientation of all RBCs is reported; however, Goldsmith and Marlow measured only orientation of RBCs which are clearly in the flipping orientation, not in the "wheel" configuration [160]. Thus, the cumulative distribution of RBCs which are only in the "flipping" orientation is shown in Fig. 3.14d with a higher percentage of RBCs reported aligned with the flow and better agreement with experimental results. The experimental results of Goldsmith and Marlow [160] show more RBCs aligned with the flow than simulations predict, likely from non-continuum effects and variations in shear due to the small tube diameter. The lack of tank-treading behavior in the linear finite-element model is not expected to influence simulation results because Goldsmith and Marlow do not report tank treading at 40% ha, although they do at 80% ha. Good qualitative agreement exists between experiments in tube flow and simulations in simple shear flow.

3.3.5.4 Local Stress Environment in Blood

The local stress environment that individual platelets experience causes increased platelet deposition through stress-mediated activation and adhesion. The local stress environment is determined by the local nature of blood microstructure and the presence of RBCs. One indicator of the local stress environment in blood flow is the existence of deformed shapes of RBCs in suspension. The successive deformation and orientation of an RBC suspended at 40% ha is shown in Fig. 3.15, along with experimental tracings by Goldsmith and Marlow [160] in RBC ghost suspensions and snapshots of a simulated RBC with shear times corresponding to experiments. The simulated RBC rotates in flow with the same period as the experimental RBC while

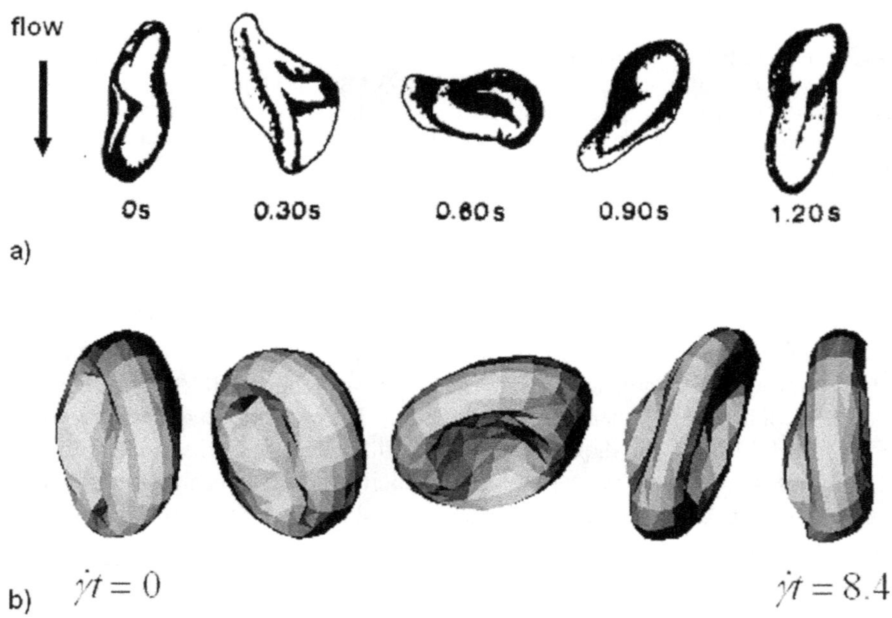

Fig. 3.15 Successive deformation and orientation of a single RBC suspended in 40% ha. **(a)** Experimental tracings from Goldsmith and Marlow [160, reprinted with permission from Elsevier]. **(b)** Example of RBC corresponding to the $\dot{\gamma}t$ shown above

continuously deforming into irregular shapes. Many characteristics of the deformed RBC shapes are visible in both experiments and simulations. As in simulation snapshots, the biconcave dimple on the RBC is visible in the experimental tracings at 0 and 1.2 s. Furthermore, the subtle folding of the RBC membrane and local deformation due to multi-particle interactions is visible in experiments (0.3 s) and simulations (0.9 s).

3.4 Future Directions

In this chapter, we reviewed state-of-the-art computational methods for macro- and micro-scale simulations of cardiovascular flows. At the macro-scale, major advances in recent years have made possible, among others, the first high-resolution, fluid–structure interaction simulations of flow in mechanical heart valves in anatomic geometries and at physiologic conditions. These simulations have provided for the first time a complete description of the instantaneous hemodynamic stresses experienced by blood elements, which have been linked to thromboembolic complications. In addition to heart valves, high-resolution direct numerical simulations have shed new insights into the hemodynamics of a wide range of cardiovascular flows, including aneurysms [67, 71, 161–164], arterial stenosis [20, 21], carotid artery bifurcation [22, 24, 155], and Fontan surgeries [68, 165]. Most of these simulations, however, have neglected the compliance of the arterial walls and a major computational challenge for the future is to develop macro-scale computational algorithms capable of performing high-resolution, FSI simulations in complex anatomic geometries with compliant vessels. At the micro-scale, significant progress has been made in our ability to simulate whole blood via direct numerical simulations of suspensions of blood cells, which account for flow–cell and cell–cell interactions.

In summary, the present-day computational methods have gone a long way toward accurately predicting cardiovascular hemodynamics at both the macro- and micro-scales. The next frontier for numerical models is to predict patient-specific hemodynamics at the micro-scale. This is a critical prerequisite for elucidating and quantifying the links between hemodynamic stresses and disease pathways, such as atherosclerosis, coronary thrombosis, and aortic valve calcification, and requires the development of multi-scale computational algorithms that couple patient-specific hemodynamic simulations at the macro-scale with micro-scale processes. Such a multi-scale approach is far from trivial, as it requires developing FSI algorithms that not only transition seamlessly from one scale to the next but also incorporate into the modeling framework the complex biophysics and biochemical processes at the cellular scale that lead to disease.

Acknowledgments This work was partially supported by a grant from the National Heart, Lung and Blood Institute (R01-HL-070262), the Minnesota Supercomputing Institute, and the National Science Foundation Graduate Research Fellowship.

References

1. Ge L, Dasi LP, Sotiropoulos F, Yoganathan AP (2008) Characterization of hemodynamic forces induced by mechanical heart valves: Reynolds vs. viscous stresses. Ann Biomed Eng 36:276–297
2. Ku DN (1997) Blood flow in arteries. Ann Rev Fluid Mech 29:399–434
3. Bagchi P, Johnson P, Popel A (2005) Computational fluid dynamic simulation of aggregation of deformable cells in a shear flow. J Biomech Eng 127:1070–1080
4. Baskurt OK, Meiselman HJ (2003) Blood rheology and hemodynamics. Semin Thromb Hemost 29:435–450
5. Merrill E, Cokelet G, Britten A, Wells R (1963) Non-Newtonian rheology of human blood – effect of fibrinogen deduced by "subtraction". Circ Res 13:48–55
6. Owens RG (2006) A new microstructure-based constitutive model for human blood. J Non-Newton Fluid Mech 140:57–70
7. Sequeira A, Janela J (2007) An overview of some mathematical models of blood rheology. In: Pereira MS (ed) A portrait of state-of-the-art research at the Technical University of Lisbon. Springer, Dordrecht, pp 65
8. Goldstein H, Poole CP, Safko JL (2002) Classical mechanics. Addison-Wesley, San Francisco, CA
9. Hoag D (1963) Apollo guidance and navigation, considerations of Apollo IMU Gimbal Lock. In: MIT Instrumentation Laboratory Document E-1344, MIT
10. Hughes PC (1986) Spacecraft attitude dynamics. Wiley, New York, NY
11. Bathe K-J (2003) Finite element procedures. Prentice Hall, Englewood Cliffs, NJ
12. Cowin SC, Doty SB (2007) Tissue mechanics. Springer, New York, NY
13. Vito RP, Dixon SA (2003) Blood vessel constitutive models?1995–2002. Ann Rev Biomed Eng 5:413–439
14. Kim H, Lu J, Sacks MS, Chandran KB (2008) Dynamic simulation of bioprosthetic heart valves using a stress resultant shell model. Ann Biomed Eng 36:262–275
15. Donea J, Giuliani S, Halleux JP (1982) An arbitrary Lagrangian–Eulerian finite element method for transient dynamic fluid–structure interactions. Comput Methods Appl Mech Eng 33:689–723
16. Morton SA, Melville RB, Visbal MR (1997) Accuracy and coupling issues of aeroelastic Navier–Stokes solutions on deforming meshes. AIAA paper 97-1085
17. Vinokur M (1989) An analysis of finite-difference and finite-volume formulations of conservation-laws. J Comput Phys 81:1
18. Warsi ZUA (2006) Fluid dynamics: theoretical and computational approaches. CRC Press, Boca Raton, FL
19. Yang Z, Mavriplis D (2006) Higher-order time integration schemes for aeroelastic applications on unstructured meshes. AIAA paper 2006-441
20. Taylor CA, Hughes TJR, Zarins CK (1998a) Finite element modeling of blood flow in arteries. Comput Methods Appl Mech Eng 158:155–196
21. Taylor CA, Hughes TJR, Zarins CK (1998b) Finite element modeling of three-dimensional pulsatile flow in the abdominal aorta: relevance to atherosclerosis. Ann Biomed Eng 26: 975–987
22. Qiu Y, Tarbell JM (2000) Numerical simulation of pulsatile flow in a compliant curved tube model of a coronary artery. J Biomech Eng 122:77
23. Jin S, Oshinski J, Giddens DP (2003) Effects of wall motion and compliance on flow patterns in the ascending aorta. J Biomech Eng 125:347
24. Perktold K, Hofer M, Rappitsch G, Loew M, Kuban BD, Freidman MH (1998) Validated computation of physiologic flow in a realistic coronary artery branch. J Biomech 31:217–228
25. Fernandez MA, Moubachir M (2005) A Newton method using exact Jacobians for solving fluid–structure coupling. Comput Struct 83:127–142
26. Cheng R, Lai YG, Chandran KB (2004) Three-dimensional fluid–structure interaction simulation of bileaflet mechanical heart valve flow dynamics. Ann Biomed Eng 32:1471

27. Figueroa CA, Vignon-Clementel IE, Jansen KE, Hughes TJR, Taylor CA (2006) A coupled momentum method for modeling blood flow in three-dimensional deformable arteries. Comput Meth Appl Mech Eng 195:5685–5706
28. Taylor CA, Humphrey JD (2009) Open problems in computational vascular biomechanics: hemodynamics and arterial wall mechanics. Comput Methods Appl Mech Eng 198: 3514–3523
29. Borazjani I (2008) Numerical simulations of fluid–structure interaction problems in biological flows. PhD thesis, University of Minnesota, Twin Cities.
30. Kim D, Choi H (2006) Immersed boundary method for flow around an arbitrarily moving body. J Comput Phys 212:662
31. Beddhu M, Taylor LK, Whitfield DL (1996) Strong conservative form of the incompressible Navier–Stokes equations in a rotating frame with a solution procedure. J Comput Phys 128:427–437
32. Dutsch H, Durst F, Becker S, Lienhart H (1998) Low-Reynolds-number flow around an oscillating circular cylinder at low Keulegan–Carpenter numbers. J Fluid Mech 360:249–271
33. Borazjani I, Sotiropoulos F (2009) Why don't mackerels swim like eels? The role of form and kinematics on the hydrodynamics of undulatory swimming. Phys Fluids 21:091109
34. Borazjani I, Sotiropoulos F (2010) On the role of form and kinematics on the hydrodynamics of body/caudal fin swimming. J Exp Biol 213:89–107
35. Vyšohlíd M, Mahesh K (2006) Large eddy simulation of crashback in marine propellers. AIAA paper 1415
36. Borazjani I, Ge L, Sotiropoulos F (2008) Curvilinear immersed boundary method for simulating fluid structure interaction with complex 3D rigid bodies. J Comput Phys 227:7587–7620
37. Dasi LP, Ge L, Simon HA, Sotiropoulos F, Yoganathan AP (2007) Vorticity dynamics of a bileaflet mechanical heart valve in an axisymmetric aorta. Phys Fluids 19:067105
38. Grigioni M, Daniele C, Del Gaudio C, Morbiducci U, Balducci A, D'Avenio G, Barbaro V (2005) Three-dimensional numeric simulation of flow through an aortic bileaflet valve in a realistic model of aortic root. ASAIO J 51:176
39. Ruge JW, Stuben K (1987) Algebraic multigrid. Multigrid Methods 3:73–130
40. Peskin CS (1972) Flow patterns around heart valves: a numerical method. J Comput Phys 10:252–271
41. Viecelli JA (1969) A method for including arbitrary external boundaries in the MAC incompressible fluid computing technique. J Comput Phys 4:543–551
42. Viecelli JA (1971) A computing method for incompressible flows bounded by moving walls. J Comput Phys 8:119–143
43. Peskin CS (1977) Numerical analysis of blood flow in the heart. J Comput Phys 25:220
44. Peskin CS, McQueen DM (1980) Modeling prosthetic heart valves for numerical analysis of blood flow in the heart. J Comput Phys 37:113–132
45. Peskin CS, McQueen DM (1989) A three-dimensional computational method for blood flow in the heart. 1. Immersed elastic fibers in a viscous incompressible fluid. J Comput Phys 81:372–405
46. Griffith BE, Hornung RD, McQueen DM, Peskin CS (2007) An adaptive, formally second order accurate version of the immersed boundary method. J Comput Phys 223:10–49
47. Glowinski R, Pan TW, Hesla TI, Joseph DD (1999) A distributed Lagrange multiplier/fictitious domain method for particulate flows. Int J Multiphas Flow 25:755–794
48. De Hart J, Baaijens FPT, Peters GWM, Schreurs PJG (2003a) A computational fluid–structure interaction analysis of a fiber-reinforced stentless aortic valve. J Biomech 36:699–712
49. De Hart J, Peters GWM, Schreurs PJG, Baaijens FPT (2000) A two-dimensional fluid–structure interaction model of the aortic value. J Biomech 33:1079–1088
50. De Hart J, Peters GWM, Schreurs PJG, Baaijens FPT (2003b) A three-dimensional computational analysis of fluid–structure interaction in the aortic valve. J Biomech 36:103–112

51. De Hart J, Peters GWM, Schreurs PJG, Baaijens FPT (2004) Collagen fibers reduce stresses and stabilize motion of aortic valve leaflets during systole. J Biomech 37: 303–311
52. van Loon R, Anderson PD, Baaijens FPT, van de Vosse FN (2005) A three-dimensional fluid–structure interaction method for heart valve modelling. Comptes Rendus-Mecanique 333:856–866
53. van Loon R, Anderson PD, de Hart J, Baaijens FPT (2004) A combined fictitious domain/adaptive meshing method for fluid–structure interaction in heart valves. Int J Numerical Methods Fluids 46:533
54. van Loon R, Anderson PD, van de Vosse FN (2006) A fluid–structure interaction method with solid-rigid contact for heart valve dynamics. J Comput Phys 217:806
55. Choi JI, Oberoi RC, Edwards JR, Rosati JA (2007) An immersed boundary method for complex incompressible flows. J Comput Phys 224:757–784
56. Fadlun EA, Verzicco R, Orlandi P, Mohd-Yusof J (2000) Combined immersed-boundary finite-difference methods for three-dimensional complex flow simulations. J Comput Phys 161:35–60
57. Ge L, Sotiropoulos F (2007) A numerical method for solving the 3D unsteady incompressible Navier–Stokes equations in curvilinear domains with complex immersed boundaries. J Comput Phys 225:1782–1809
58. Gilmanov A, Sotiropoulos F (2005) A hybrid Cartesian/immersed boundary method for simulating flows with 3D, geometrically complex, moving bodies. J Comput Phys 207: 457–492
59. Lee L, LeVeque RJ (2003) An immersed interface method for incompressible Navier-Stokes equations. SIAM J Sci Comput 25(3):832–856
60. Tseng YH, Ferziger JH (2003) A ghost-cell immersed boundary method for flow in complex geometry. J Comput Phys 192:593–623
61. Udaykumar HS, Mittal R, Shyy W (1999) Computation of solid–liquid phase fronts in the sharp interface limit on fixed grids. J Comput Phys 153:535–574
62. Leveque RJ, Li Z (1994) The immersed interface method for elliptic equations with discontinuous coefficients and singular sources. SIAM J Numer Anal 31:1019–1044
63. Mohd-Yosuf J (1997) Combined immersed-boundary/B-spline methods for simulations of flow in complex geometries. In: Annual research briefs, Center for Turbulence Research, Stanford, CA 94305-3035, USA pp 317–328
64. Gilmanov A, Sotiropoulos F, Balaras E (2003) A general reconstruction algorithm for simulating flows with complex 3D immersed boundaries on Cartesian grids. J Comput Phys 191:660–669
65. Mittal R, Iaccarino G (2005) Immersed boundary methods. Ann Rev Fluid Mech 37: 239–261
66. Haines E (1994) Point in polygon strategies, Academic Press Graphics Gems Series. Academic, Boston, MA, pp 24–46
67. Yokoi K, Feng X, Hao L, Fukasaku K (2005) Three-dimensional numerical simulation of flows with complex geometries in a regular Cartesian grid and its application to blood flow in cerebral artery with multiple aneurysms. J Comput Phys 202:1
68. de Zélicourt DA, Ge L, Wang C, Sotiropoulos F, Gilmanova A, Yoganathan A (2009) Flow simulations in arbitrarily complex cardiovascular anatomies – an unstructured Cartesian grid approach. Comput Fluids 38:1749–1762
69. Sundareswaran KS, de Zelicourt D, Sharma S, Kanter KR, Spray TL, Rossignac J, Sotiropoulos F, Fogel MA, Yoganathan AP (2009) Correction of pulmonary arteriovenous malformation using image-based surgical planning. JACC Cardiovasc Imaging 2: 1024–1030
70. Löhner R, Cebral JR, Camelli FE, Appanaboyina S, Baum JD, Mestreau EL, Soto OA (2008) Adaptive embedded and immersed unstructured grid techniques. Comput Methods Appl Mech Eng 197:2173–2197

71. Appanaboyina S, Mut F, Lohner R, Putman CM, Cebral JR (2008) Computational fluid dynamics of stented intracranial aneurysms using adaptive embedded unstructured grids. Int J Numer Meth Fluids 57(5):475–493
72. Felippa CA, Park KC, Farhat C (2001) Partitioned analysis of coupled mechanical systems. Comput Methods Appl Mech Eng 190:3247
73. Vierendeels J, Dumont K, Verdonck PR (2008) A partitioned strongly coupled fluid–structure interaction method to model heart valve dynamics. J Comput Appl Math 215:602–609
74. Causin P, Gerbeau JF, Nobile F (2005) Added-mass effect in the design of partitioned algorithms for fluid–structure problems. Comput Meth Appl Mech Eng 194:4506–4527
75. Conca C, Osses A, Planchard J (1997) Added mass and damping in fluid–structure interaction. Comput Methods Appl Mech Eng 146:387–405
76. Forster C, Wall WA, Ramm E (2007) Artificial added mass instabilities in sequential staggered coupling of nonlinear structures and incompressible viscous flows. Comput Methods Appl Mech Eng 196:1278–1293
77. Aitken AC (1926) On Bernoulli's numerical solution of algebraic equations. Proc R Soc Edinb 46:289–305
78. Irons BM, Tuck RC (1969) A version of the Aitken accelerator for computer iteration. Int J Numer Methods Eng 1:275–277
79. Saad Y, Schultz MH (1986) GMRES: a generalized minimal residual algorithm for solving nonsymmetric linear systems. SIAM J Sci Stat Comput 7:856
80. Sleijpen GLG, Fokkema DR (1993) BiCGStab (l) for linear equations involving unsymmetric matrices with complex spectrum. Electron Trans Numer Anal 1:2000
81. Trottenberg U, Oosterlee CW, Schüller A (2001) Multigrid: basics, parallelism and adaptivity. Academic, New York, NY
82. Oosterlee CW, Washio T (1998) An evaluation of parallel multigrid as a solver and a preconditioner for singularly perturbed problems. SIAM J Sci Comput 19:87–110
83. Sotiropoulos F, Borazjani I (2009) A review of the state-of-the-art numerical methods for simulating flow through mechanical heart valves. Med Biol Eng Comput 47:245–256
84. Ge L, Leo HL, Sotiropoulos F, Yoganathan AP (2005) Flow in a mechanical bileaflet heart valve at laminar and near-peak systole flow rates: CFD simulations and experiments. J Biomech Eng 127:782
85. Ge L, Jones SC, Sotiropoulos F, Healy TM, Yoganathan AP (2003) Numerical simulation of flow in mechanical heart valves: grid resolution and the assumption of flow symmetry. J Biomech Eng Trans ASME 125:709–718
86. Gotoh K, Minamino T, Katoh O, Hamano Y, Fukui S, Hori M, Kusuoka H, Mishima M, Inoue M, Kamada T (1988) The role of intracoronary thrombus in unstable angina: angiographic assessment and thrombolytic therapy during ongoing anginal attacks. Circulation 77:526–534
87. Kiris C, Kwak D, Rogers S (1997) Computational approach for probing the flow through artificial heart devices. J Biomech Eng 119:452
88. Mody N, Lomakin O, Doggett T, Diacovo T, King M (2005) Mechanics of transient platelet adhesion to von Willebrand factor under flow. Biophys J 88:1432–1443
89. Cheng R, Lai YG, Chandran KB (2003) Two-dimensional fluid–structure interaction simulation of bileaflet mechanical heart valve flow dynamics. J Heart Valve Dis 12:772
90. Rosenfeld M, Avrahami I, Einav S (2002) Unsteady effects on the flow across tilting disk valves. J Biomech Eng Trans ASME 124:21–29
91. Pedrizzetti G (2005) Kinematic characterization of valvular opening. Phys Rev Lett 94:194502
92. Pedrizzetti G, Domenichini F (2006) Flow-driven opening of a valvular leaflet. J Fluid Mech 569:321–330
93. Stijnen JMA, de Hart J, Bovendeerd PHM, van de Vosse FN (2004) Evaluation of a fictitious domain method for predicting dynamic response of mechanical heart valves. J Fluids Struct 19:835–850

94. Tai CH, Liew KM, Zhao Y (2007) Numerical simulation of 3D fluid-structure interaction flow using an immersed object method with overlapping grids. Comput Struct 85:749–762
95. Nobili M, Morbiducci U, Ponzini R, Del Gaudio C, Balducci A, Grigioni M, Maria Montevecchi F, Redaelli A (2008) Numerical simulation of the dynamics of a bileaflet prosthetic heart valve using a fluid–structure interaction approach. J Biomech 41:2539–2550
96. De Tullio MD, Cristallo A, Balaras E, Verzicco R (2009) Direct numerical simulation of the pulsatile flow through an aortic bileaflet mechanical heart valve. J Fluid Mech 622: 259–290
97. Borazjani I, Ge L, Sotiropoulos F (2010) High resolution fluid–structure interaction simulations of flow through a bi-leaflet mechanical heart valve in an anatomic aorta. Ann Biomed Eng 38(2):326–344. doi:10.1007/s10439-009-9807-x
98. Borazjani I, Sotiropoulos F (2010) The effect of implantation orientation of a bi-leaflet mechanical heart valve on kinematics and hemodynamics in an anatomic aorta. ASME J Biomech Eng. doi:10.1115/1.4002491
99. Ge L, Sotiropoulos F (2010) Direction and magnitude of hemodynamic stresses on the leaflets of aortic valves: is there a link with valve calcification? J Biomech Eng 131:0145051–014509
100. Haj-Ali R, Dasi LP, Kim HS, Choi J, Leo HW, Yoganathan AP (2008) Structural simulations of prosthetic tri-leaflet aortic heart valves. J Biomech 41:1510–1519
101. Davies PF, Shi C, DePaola N, Helmke BP, Polacek DC (2001) Hemodynamics and the focal origin of atherosclerosis a spatial approach to endothelial structure, gene expression, and function. Ann N Y Acad Sci 947:7–17
102. Alevriadou R, Moake J, Turner N, Ruggeri Z, Folie B, Phillips M, Schreiber A, Hrinda M, McIntire L (1993) Real-time analysis of shear-dependent thrombus formation and its blockade by inhibitors of von Willebrand factor binding to platelets. Blood 81:1263–1276
103. Kulkarni S, Dopheide S, Yap C, Ravanat R, Freund M, Mangin P, Heel K, Street A, Harper I, Lanza F et al (2000) A revised model of platelet aggregation. J Clin Invest 105:783–791
104. Munn L, Melder R, Jain R (1996) Role of erythrocytes in leukocyte–endothelial interactions: mathematical model and experimental validation. Biophys J 71:466–478
105. Sun C, Migliorini C, Munn LL (2003) Red blood cells initiate leukocyte rolling in postcapillary expansions: a lattice Boltzmann analysis. Biophys J 85:208–222
106. Sloop G (1998) Insights into the relationship of fatty streaks to raised atherosclerotic lesions provided by the hemorheologic–hemodynamic theory of atherogenesis. Med Hypotheses 51:385–388
107. Turitto V, Weiss H, Baumgartner H (1980) The effect of shear rate on platelet interaction with subendothelium exposed to citrated human blood. Microvasc Res 19:352
108. Goldsmith H, Bell D, Spain S, McIntosh F (1999) Effect of red blood cells and their aggregates on platelets and white cells in flowing blood. Biorheology 36:461–468
109. Goldsmith H, Kaufer E, McIntosh F (1995) Effect of hematocrit on adenine diphosphate-induced aggregation of human platelets in tube flow. Biorheology 32:537–552
110. Aarts P, van den Broek S, Prins G, Kuiken G, Sixma J, Heehaar R (1988) Blood platelets are concentrated near the wall and red blood cells, in the center in flowing blood. Arteriosclerosis 8:819–824
111. Wootton D, Markou C, Hanson S, Ku D (2001) A mechanistic model of acute platelet accumulation in thrombogenic stenoses. Ann Biomed Eng 29:321–329
112. Cha W, Beissinger R (1996) Augmented mass transport of macromolecules in sheared suspensions to surfaces B. Bovine serum albumin. J Colloid Interf Sci 178:1–9
113. Dao M, Limb CT, Suresh S (2003) Mechanics of the human red blood cell deformed by optical tweezers. J Mech Phys Solids 51:2259–2280
114. Ramanujan S, Pozrikidis C (1998) Deformation of liquid capsules enclosed by elastic membranes in simple shear flow: large deformations and the effect of fluid viscosities. J Fluid Mech 361:117–143
115. Breyiannis G, Pozrikidis C (2000) Simple shear flow of suspensions of elastic capsules. Theor Comput Fluid Dyn 13:327–347

116. Pozrikidis C (2001) Effect of membrane bending stiffness on the deformation of capsules in simple shear flow. J Fluid Mech 440:269–291
117. Pozrikidis C (2003) Numerical simulation of the flow-induced deformation of red blood cells. Ann Biomed Eng 31:1194–1205
118. Pozrikidis C (2005) Numerical simulation of cell motion in tube flow. Ann Biomed Eng 33:165–178
119. Eggleton C, Popel A (1998) Large deformation of red blood cell ghosts in a simple shear flow. Phys Fluids 10:1834–1845
120. Liu Y, Liu WK (2006) Rheology of red blood cell aggregation by computer simulation. J Comput Phys Arch 220(1):139–154. ISSN:0021-9991
121. Liu WK, Liu Y, Farrell D, Zhang L, Wang XS, Fukui Y, Patankar N, Zhang Y, Bajaj C, Lee J et al (2006a) Immersed finite element method and its applications to biological systems. Comput. Methods Appl Mech Eng 195:1722–1749
122. Liu X, Tang Z, Zeng Z, Chen X, Yao W, Yan Z, Shi Y, Shan H, Sun D, He D, Wen Z (2007) The measurement of shear modulus and membrane surface viscosity of RBC membrane with ektacytometry: a new technique. Math Biosci 209(1):190–204
123. Dzwinel W, Boryczko K, Yuen D (2003) A discrete-particle model of blood dynamics in capillary vessels. J Colloid Interf Sci 258(1):163–173
124. Tsubota K, Wada S, Yamaguchi T (2006) Particle method for computer simulation of red blood cell motion in blood flow. Comput Methods Programs Biomed 83:139–146
125. Dupin M, Halliday I, Care C (2006) A multi-component lattice Boltzmann scheme: towards the mesoscale simulation of blood flow. Med Eng Phys 8:3–18
126. Hyakutake T, Matsumoto T, Yanase S (2006) Lattice Boltzmann simulation of blood cell behavior at microvascular bifurcations. Math Comput Simul 72:134–140
127. MacMeccan R, Clausen J, Neitzel P, Aidun CK (2009) Simulating deformable particle suspensions using a coupled lattice-Boltzmann and finite-element method. J Fluid Mech 618:13–39
128. Wu J, Aidun CK (2009) Simulating 3D deformable particle suspensions using lattice Boltzmann method with discrete external boundary force. Int J Numer Methods Fluids 62(7):765–783. doi:10.1002/fld.2043
129. Dzwinel W, Yuen D (2002) Mesoscopic dispersion of colloidal agglomerate in a complex fluid modelled by a hybrid fluid–particle model. J Colloid Interf Sci 247(2):463–480
130. Aidun C, Lu Y, Ding EJ (1998) Direct analysis of particulate suspensions with inertia using the discrete Boltzmann equation. J Fluid Mech 373:287–311
131. Qi D (1999) Lattice-Boltzmann simulations of particles in non-zero-Reynolds-number flows. J Fluid Mech 385:41–62
132. Ding E, Aidun C (2000) The dynamics and scaling law for particles suspended in shear flow with inertia. J Fluid Mech 423:317–344
133. Ding E, Aidun C (2003) Extension of the lattice-Boltzmann method or direct simulation of suspended particles near contact. J Stat Phys 112:685–707
134. Ding E, Aidun C (2006) Cluster size distribution and scaling for spherical particles and red blood cells in pressure-driven flows at small Reynolds number. Phys Rev Lett 96:204502-1–204502-4
135. Ladd A, Verberg R (2001) Lattice-Boltzmann simulations of particle–fluid suspensions. J Stat Phys 104:1191–1251
136. Aidun CK, Clausen JR (2010) Lattice-Boltzmann method for Complex Flows. Annu Rev Fluid Mech 42:439–72
137. Chen S, Doolen G (1998) Lattice Boltzmann method for fluid flows. Ann Rev Fluid Mech 30:329–364
138. McNamara G, Zanetti G (1988) Use of the Boltzmann equation to simulate lattice-gas automata. Phys Rev Lett 61(20):2332–2335
139. MacMeccan R, Atlanta GA (2007) Mechanistic effects of erythrocytes on platelet deposition in coronary thrombosis. PhD Thesis, Georgia Institute of Technology, Atlanta, GA

140. Clausen J, Aidun CK (2009) Galilean invariance in the lattice-Boltzmann method and its effect on the calculation of rheological properties in suspensions. Int J Multiphas Flow 35:307–311
141. Rankin C, Brogan F (1986) An element independent corotational procedure for the treatments of large rotations. J Press Vessel Technol 108:165–174
142. Campanelli M, Berzeri M, Shabana A (2000) Performance of the incremental and non-incremental finite elements formulations in flexible multibody problems. J Mech Des 122:498–507
143. Moller T, Trumbore B (1977) Fast, minimum storage ray-triangle intersection. J Graph Tools 2:21–28
144. Buxton G, Verberg R, Jasnow D, Balazs A (2005) Newtonian fluid meets an elastic solid: coupling lattice Boltzmann and lattice-spring models. Phys Rev E 71:56707.
145. Waugh R, Evans E (1979) Thermoelasticity of red blood cell membrane. Biophys J 26: 115–132
146. Evans A, Waugh R, Melnik L (1976) Elastic area compressibility modulus of red cell membrane. Biophys J 16(6):585–595
147. Hwang W, Waugh R (1997) Energy of dissociation of lipid bilayer from the membrane skeleton of red blood cells. Biophys J 72:2669–2678
148. Tozeren A, Skalak R, Fedorcix K, Sung K, Chien S (1984) Constitutive equations of erythrocyte membrane incorporating evolving preferred configuration. Biophys J 45:541–549
149. Schmid-Schönbein H, Grebe R, Heidtmann H (1983) A new membrane concept for viscous RBC deformation in shear:spectrin oligomer complexes as a Bingham-fluid in shear and a dense periodic colloidal system in bending. Ann N Y Acad Sci 416:225–254
150. Skalak R, Tozeren S, Zarda R, Chien S (1973) Strain energy function of red blood cell membranes. Biophys J 13:245–264
151. Barthes-Biesel D, Diaz A, Dhenin E (2002) Effect of constitutive laws for two-dimensional membranes on flow-induced capsule deformation. J Fluid Mech 460:211–222
152. Evans E, Skalak R (1980) Mechanics and thermodynamics of biomembranes. CRC, Boca Raton, FL
153. Harkness J, Whittington R (1970) Blood-plasma viscosity: an approximate temperature-invariant arising from generalized concepts. Biorheology 6:169–187
154. Aarts P, Stendijk P, Sixma J, Heethaar R (1986) Fluid shear as a possible mechanism for platelet diffusivity in flowing blood. J Biomech 19:799–805
155. Jung J, Lyczkowski R, Panchal C, Hassanein A (2006) Multiphase hemodynamic simulation of pulsatile flow in a coronary artery. J Biomech 39:2064–2073
156. Yao W, Yan Z, Sun D, Ka W, Xie L, Chien S (2004) Low viscosity ektacytometry and its validation tested by flow chamber. J Biomech 34:1501–1509
157. Batchelor G (1970) The stress in a suspension of force-free particles. J Fluid Mech 43: 545–570
158. Fung Y (1993) Biomechanics mechanical properties of living tissues. Springer, New York, NY
159. Haga J, Beaudoin A, White J, Strony J (1998) Quantification of the passive mechanical properties of the resting platelet. Ann Biomed Eng 26:268–277
160. Goldsmith H, Marlow J (1979) Flow behavior of erythrocytes II. Particle motions in concentrated suspensions of ghost Cells. J Colloid Interf Sci 71:383–407
161. Le T, Borazjani I, Sotiropoulos F (2010) Vorticity dynamics in an intracranial aneurysm. ASME J Biomech Eng (In Press)
162. Shojima M, Oshima M, Takagi K, Torii R, Nagata K, Shirouzu I, Morita A, Kirino T (2005) Role of the bloodstream impacting force and the local pressure elevation in the rupture of cerebral aneurysms. Stroke 36:1933–1938
163. Steinman DA, Milner JS, Norley CJ, Lownie SP, Holdsworth DW (2003) Image-based computational simulation of flow dynamics in a giant intracranial aneurysm. Am J Neuroradiol 24:559–566

164. Wolters B, Rutten MCM, Schurink GWH, Kose U, de Hart J, van de Vosse FN (2005) A patient-specific computational model of fluid–structure interaction in abdominal aortic aneurysms. Med Eng Phys 27:871–883
165. Pekkan K, ZÃlicourt D, Ge L, Sotiropoulos F, Frakes D, Fogel MA, Yoganathan AP (2005) Physics-driven CFD modeling of complex anatomical cardiovascular flows: a TCPC case study. Ann Biomed Eng 33:284–300

Chapter 4
Formulation and Computational Implementation of Constitutive Models for Cardiovascular Soft Tissue Simulations

Michael S. Sacks and Jia Lu

Abstract Predictive computational modeling of the cardiovascular system has often been utilized as a powerful investigative tool. Motivated by the need for a deeper understanding of the underlying physiology, the identification of pathological initiators, as well as the development of bioprosthetic devices, a broad variety of modeling approaches have been introduced into the literature. Central to system- and organ-level functional simulations is the need for robust and physiologically meaningful constitutive models of the underlying soft tissue structures. While studied for many decades, Y.C. Fung popularized the field of soft tissue mechanics through a set of influential books which demonstrated the unique challenges involved in the mathematical characterization of living tissue mechanical behaviors. Overall, his main contribution was to establish constitutive relationships for the purpose of examining biological tissues in a continuum mechanics framework. Particular challenges in soft tissue constitutive modeling are encountered due to their complex mechanical behavior. For example, because of their oriented fibrous structures, they often exhibit pronounced mechanical anisotropy, nonlinear stress–strain relationships, large deformations, viscoelasticity, poroelasticity, and strong mechanical coupling. Taken as a whole, soft biological tissues defy simple material models. The focus of this chapter is the description and computational application of relevant biomechanical constitutive theories. Throughout this chapter, we will utilize the assumption of hyperelastic behavior as fundamental to soft tissue biomechanics, utilizing the concept of pseudo-elasticity, so that the loading response is modeled only.

M.S. Sacks (✉)
Department of Bioengineering, Swanson School of Engineering, University of Pittsburgh, Pittsburgh, PA 15219, USA; School of Medicine, The McGowan Institute, University of Pittsburgh, Pittsburgh, PA 15219, USA
e-mail: msacks@pitt.edu

K.B. Chandran et al. (eds.), *Image-Based Computational Modeling of the Human Circulatory and Pulmonary Systems*, DOI 10.1007/978-1-4419-7350-4_4,

4.1 Introduction

Predictive computational modeling of the cardiovascular system has often been utilized as a powerful investigative tool. Motivated by the need for a deeper understanding of the underlying physiology, the identification of pathological initiators, as well as the development of bioprosthetic devices, a broad variety of modeling approaches have been introduced into the literature. Central to system- and organ-level functional simulations is the need for robust and physiologically meaningful constitutive models of the underlying soft tissue structures. While studied for many decades, Y.C. Fung popularized the field of soft tissue mechanics through a range of influential books [1, 2] which demonstrated the unique challenges involved in the mathematical characterization of living tissue mechanical behaviors. Overall, his main contribution was to establish constitutive relationships for the purpose of examining biological tissues in a continuum mechanics framework. Particular challenges in soft tissue constitutive modeling are encountered due to their complex mechanical behavior. For example, because of their oriented fibrous structures, they often exhibit pronounced mechanical anisotropy, nonlinear stress–strain relationships, large deformations, viscoelasticity, poroelasticity, and strong mechanical coupling. Taken as a whole, then, soft biological tissues defy simple material models.

At present, soft tissue biomechanical modeling in the cardiovascular area remains quite an active area of research as investigators struggle to bridge the gap between the matured field of mechanics and evolving physiological understandings. In particular, there is a recent trend in increasing focus models based on tissue microscale features and function, as well as in investigating mechanisms contributing to mechanical responses associated with varied length scales. In particular, the complex function and microstructural architecture of the extracellular matrix fibrous constituents are of primary interest. The collagen fiber network is of interest because it is the primary load-bearing tissue constituent and exhibits multi-scale characteristics that influence organ-level anisotropy, as well as strongly interacting with the cellular constituents. In a predictive model framework, relating these characteristics to scale-relevant tissue behavior can produce a great deal of realism. This includes efforts to address pathology and engineered tissue development which look toward microscale-based models for insight and guidance. Relating the observed mechanical response to tissue structure is perhaps more paramount than in other traditional material applications, where the continuum scale is usually at most of the size of large polymer molecules. In contrast, biological soft tissues are composed of a dense network of primarily collagen and various elastin fibers, which indicates a continuum scale at the fiber scale (typically $\sim$1 μm). In addition, the fibers can undergo large rotations and exhibit nonlinear stress–strain behavior that can induce complex behaviors at the macro-specimen scale not easily accounted for in classic material models. Accounting for these behaviors in both experimental evaluation and formulation of appropriate constitutive models continues to be challenging.

In order to either determine the constitutive model form or estimate model parameters, multi-axial mechanical experiments on soft biological tissues are required [3], which are generally difficult to perform. Just a few of the experimental problems include small specimen sizes, structural and compositional heterogeneity, difficulty in gripping (without doing damage), dramatic effects of different gripping techniques (St. Venant-like effects), difficulty in precisely identifying material axes, difficulty in assuring constant forces along specimen edges, large specimen-to-specimen variability, and time-dependent changes due to biological degradation. In addition, a question of homogeneity of deformation within the specimen is paramount. These issues can often frustrate the application of even the most straightforward attempts to develop a constitutive model. Most early biomechanical investigations of cardiovascular tissues were confined to uniaxial studies because of the difficulties in controlling two- or even three-dimensional boundary conditions. Due to the presence of mechanical anisotropy and related theoretical considerations, uniaxial mechanical data are insufficient for parameter estimation for generalized three-dimensional constitutive equations, even if multi-dimensional strain data from the uniaxial experiment are available. There have also been investigations using inflation of circular membranes, which under assumption of isotropy can provide the necessary experimental data [4, 5]. Virtually all tissues are mechanically anisotropic; hence this method must be modified for a general application. Further, when attempting to determine material constants for complex nonlinear constitutive models, testing methods are required that include comprehensive testing protocols that allow large variations in stress and strain states for accurate material parameter estimation [6, 7].

The focus of this chapter is the description and computational application of relevant biomechanical constitutive theories. Throughout this chapter, we will utilize the assumption of hyperelastic behavior as fundamental to soft tissue biomechanics, utilizing the concept of pseudo-elasticity [1] so that the loading response is modeled only. Proper formulation for a discretized solution procedure is rooted in a linearization of the equilibrium equations in a Lagrangian framework.

4.2 Constitutive Models for Cardiovascular Soft Tissues

Originally introduced to describe the behavior of skin tissue, the strain energy function proposed by Tong and Fung [8] has provided a reliable foundation for soft tissue mechanics. Researchers have successfully used variations of this model to describe a number of soft tissue applications including skin, blood vessels, myocardium, and heart valvular tissue [9–11]. Although originally formulated to couple quadratic and exponential forms, a basic quadratic form was later found to be sufficient. The Fung-type strain energy function Ψ is given as

$$W(\mathbf{E}) = \frac{c}{2} e^{Q}, \quad \text{where } Q = \mathbf{a}_{ijkl}\mathbf{E}_{ij}\mathbf{E}_{kl} \tag{4.1}$$

where $\mathbf{E}$ is the Green–Lagrange strain tensor, c is the scalar parameter in units of stress, and $\mathbf{a}$ is the nondimensional material parameter. Both two- and three-dimensional forms have been proposed. It is known that most native and biologically derived tissues exhibit incompressibility because of their high water content and low permeability. The incompressibility condition requires J=det(**F**)=1 so that the stress statement for a hyperelastic constitutive equation is expressed as [12]

$$\mathbf{S} = -p'\mathbf{C}^{-1} + \frac{\partial W}{\partial \mathbf{E}} \tag{4.2}$$

where p' is a Lagrangian multiplier enforcing the incompressibility constraint.

As an example of a computational implementation, we utilized a comprehensive experimental biaxial mechanical dataset that included high in-plane shear stresses using glutaraldehyde-treated bovine pericardium (GLBP), a critical biomaterial utilized in heart valve bioprostheses [9]. Compared to our previous study [13], GLBP demonstrated a substantially different response under high-shear strains. This finding was underscored by the inability of the standard Fung model, applied successfully in [13], to fit the high-shear data. To develop an appropriate constitutive model, we utilized an interpolation technique for the pseudo-elastic response to guide modification of the final model form. Ultimately, we developed a full expansion of quadric terms of Q to account for in-plane shear strains [13]

$$Q = A_1E_{11}^2 + A_2E_{22}^2 + 2A_3E_{11}E_{22} + A_4E_{12}^2 + 2A_5E_{12}E_{11} + 2A_6E_{12}E_{22} \tag{4.3}$$

and c and A_i are material constants. Note that Equation (4.3) represents a generalized pseudo-elastic Fung constitutive model and other variants [8, 9, 13–15] could be easily treated as a subset or an expansion of this model. This eight-parameter modified Fung model, utilizing additional quadratic terms, was found to fit the complete dataset well. Model parameters were also constrained to satisfy physical plausibility of the strain energy function. Specifically, the derived constraints were sufficient for physically realistic strain energy functions and extend similar relationships derived by Humphrey [16] to include the presence of shear strains. The derived restrictions were applied to simpler forms used in the current study by eliminating the additional terms. Following Equations (4.2) and (4.3), we have

$$S_{11} = \frac{c}{2}(2A_1E_{11} + 2A_3E_{22} + 2A_5E_{12})e^Q \tag{4.4}$$

$$S_{12} = \frac{c}{2}(2A_4E_{12} + 2A_5E_{11} + 2A_6E_{22} + 2B_1E_{12}E_{22}^2)e^Q \tag{4.5}$$

$$S_{22} = \frac{c}{2}(2A_2E_{22} + 2A_3E_{11} + 2A_6E_{12} + 2B_1E_{12}^2E_{22})e^Q \tag{4.6}$$

We first require that Ψ vanishes in the reference configuration and increases with deformation, i.e., $\Psi \geq 0$ so that $c \geq 0$. Next, the parameters used in the current model must also satisfy the following simple deformations:

Case 1: $S_{11} > 0$ when $E_{11} > 0$ when $E_{12} = E_{22} = 0$, leading to $A_1E_{11} + A_3E_{22} + A_5E_{12} > 0$, or $A_1 > 0$.

Case 2: $S_{22} > 0$ when $E_{22} > 0$ $E_{12} = E_{11} = 0$, leading to $A_2E_{22} + A_3E_{11} + A_6E_{12} + B_1E_{12}^2E_{22} > 0$, or $A_1 > 0$.

Case 3: $S_{12} > 0$, which is sub-divided into the following two sub-cases:

(a) $E_{11} = 0,\ E_{12} > 0$, and $E_{22} \neq 0$ (i.e., simple shear in X_1-direction) so that $A_4E_{12} + A_6E_{22} + B_1E_{12}E_{22}^2 > 0$.

(b) $S_{12} > 0$ when $E_{22} = 0,\ E_{12} > 0$, and $E_{11} \neq 0$ (i.e., simple shear in X_2-direction) so that $A_4E_{12} + A_5E_{11} > 0$.

To enforce these conditions, they were simultaneously implemented during the regression procedure.

Conventionally, experimentally measured stresses are fit to the predicted stress by the constitutive model, with parameter estimates obtained using nonlinear regression techniques to obtain model parameter values [3]. However, the resulting model may be numerically unstable for computational implementation and thus will generally not lead to convergent solutions. Moreover, physical plausibility of the resultant constitutive model must be enforced [9, 17, 18]. In most commercial FE codes, the Newton–Raphson method is usually used to evaluate the material stiffness matrix, which is a function of fourth-order elasticity tensor $\mathbb{C}$, which in material description is obtained as

$$\mathbb{C} = \frac{\partial \mathbf{S}}{\partial \mathbf{E}} \tag{4.7}$$

If $\mathbf{C}$ is positive definite, full rank, and well conditioned, then the numerical solution of the equilibrium equations will be stable and accurate. From Equation (4.7) it can be seen that elasticity tensor $\mathbf{C}$ is the second partial derivative (or Hessian) of Ψ. It is known that whenever the Hessian of a function is positive definite, the corresponding function will be convex [19]. Thus, by requiring the elasticity tensor $\mathbf{C}$ to be positive definite, the convexity of the generalized Fung pseudo-elastic constitutive model is thus enforced.

As originally proposed, a priori restrictions on the material parameters to ensure convexity were not presented [20]. However, to generate a convex anisotropic strain energy function, the material parameters must be developed with additional constraints. Several authors have proposed schemes to address the convexity condition [21–23]. Additionally, Fung-type models are not commonly presented in an invariant form. Criscione et al. [24] presented a Fung model formulation in terms of invariants for the generalized case of three orthonormal directions of material symmetry. The complexity of this form has been a barrier to widespread use. Instead, we utilized a more straightforward approach by expressing $\mathbf{C}$ as the Hessian $\mathbf{D}$ expressed in matrix form as

$$\mathbf{D} = \frac{c}{2}e^{Q}\begin{bmatrix} 2A_1 + \zeta_1^2 & 2A_3 + \zeta_1\zeta_2 & 2A_5 + \zeta_1\zeta_3 \\ & 2A_2 + \zeta_2^2 & 2A_6 + \zeta_2\zeta_3 \\ \text{sym} & & 2A_4 + \zeta_3^2 \end{bmatrix} \tag{4.8}$$

where

$$\zeta_1 = 2A_1E_{11} + 2A_3E_{22} + 2A_5E_{12}$$

$$\zeta_2 = 2A_2E_{22} + 2A_3E_{11} + 2A_6E_{12}$$

$$\zeta_3 = 2A_4E_{12} + 2A_5E_{11} + 2A_6E_{22}$$

The corresponding Cauchy stress and the elasticity tensors in the spatial description can be easily obtained by standard push-forward operations (the interested reader is referred to [17] for details). For planar biaxial loading of biomembranes, strict convexity physically implies that the projections of the contours of Ψ on the E_{11}–E_{22}, E_{11}–E_{12}, and E_{22}–E_{12} planes form convex surfaces [17]. To impose a convexity condition on the current strain energy function (Equation (4.5)), we examined the matrix **D** of Equation (4.15) in the reference configuration (where $E_{11} = E_{22} = E_{12} = 0$):

$$\mathrm{D} = \frac{c}{2}\begin{bmatrix} 2A_1 & 2A_3 & 2A_5 \\ 2A_3 & 2A_2 & 2A_6 \\ 2A_5 & 2A_6 & 2A_4 \end{bmatrix} \tag{4.9}$$

This definition of positive definiteness is equivalent to the requirement that the determinants associated with all upper-left submatrices are positive. From Equation (4.9), the parameter constraints that satisfy positive definiteness for **D** in the reference configuration (i.e., $E_{11} = E_{22} = E_{12} = 0$) are as follows:

$$c > 0,\ A_1 > |A_3|,\ A_2 > |A_3| \text{ and } A_1A_2A_4 + 2A_3A_6A_5 - A_5^2A_2 - A_6^2A_1 - A_3^2A_4 > 0 \tag{4.10}$$

However, these constraints are not sufficient to guarantee **D** is positive definite over the entire strain range. Rather, we consider these to be a necessary condition that if parameter constraints fail to satisfy these constraints, convexity will be violated. It is therefore necessary that **D** be evaluated at every point in the measured strain range to verify that positive definiteness is always satisfied. Practically, it is usually sufficient to examine the convexity of **D** along projections of W against two strain components with the other component set to zero, such as the E_{11}–E_{22} plane with $E_{12} = 0.0$.

4.2.1 Condition Number of D

The condition number of **D** is defined as

$$R = \|\mathbf{D}\| \cdot \left\|\mathbf{D}^{-1}\right\| \tag{4.11}$$

where the norm of $\mathbf{D}$ is defined as $\|\mathbf{D}\| = \max_{1 \leq j \leq n} \sum_{i=1}^{n} |\mathbf{D}_{ij}|$. We evaluated R over the experimental strain ranges. Theoretically, the lower the value of R, the better the numerical stability. In our experience with Fung pseudo-elastic constitutive models, we have found that $R \leq 200$ was acceptable for numerical convergence. However, we note that we do not have analytic bounds similar to Equation (4.10) for the condition number inequality. This is because solving the condition number inequality through matrix manipulation leads to a complex formulation, complicating the parameter bounds on the nonlinear regression. Thus, for the present work, we only tested whether $R \leq 200$ was satisfied once the model parameters were obtained.

4.3 Structural Constitutive Models

Although the Fung-type phenomenological constitutive models discussed above were successful in the above applications, they were unable to elucidate the underlying mechanisms of tissue behavior. Structural constitutive models attempt to integrate information on tissue composition and structure to avoid ambiguities in material characterization and offer insight into the function, the structure, and the mechanics of tissue components. Structural constitutive models have been developed for a variety of intact tissues and tissue components including lung [25], collagen [26, 27], cartilage [28], passive myocardium [29], heart valves [30], and maturing skin [31]. Perhaps the most complete approach for structural constitutive modeling of soft tissues has been developed by Lanir et al. [32–34]. In this approach, the tissue's total strain energy is assumed to be the sum of the individual fiber strain energies, linked through appropriate tensor transformation from the fiber coordinates to the global tissue coordinates.

However, the description of fiber-scale properties such as orientation and crimp is cast in terms of statistical distributions due to a microstructure complexity that prohibits individual representation. This stochastic description is based on data homogenized at a representative element scale of approximately 100 μm. At this scale, characterization of the fiber microstructure is relative to the fiber ensemble. For example, the fiber angular distribution function describes the dominant fiber alignment direction in an ensemble framework. In this sense, structural models have a meso-scale focus, as they characterize an organ-scale response in terms of fiber structure attributes at a scale in-between cellular and organ. However, critical structural information (such as fiber orientations) modeled using assumed statistical distributions (usually Gaussian) with the distribution parameters numerically estimated fits to the mechanical testing data. The framework established by the early work of Lanir was given practical relevance to soft tissue modeling with the introduction of an experimentally derived form for fiber orientation by Sacks [7]. This work used data collected using SALS techniques to develop analytical functions suitable for direct implementation in the context of a structural model. Additionally, Sacks presented a distribution for fiber recruitment stemming from crimp periods observed in bovine pericardium. Note that the term fiber ensemble refers to a collection of fibers sharing a common orientation. This modeling technique treats the fibrous component by homogenizing the fiber ensemble response.

We have extended this approach by demonstrating that the complete planar biaxial mechanical response of the tissue could be simulated by combining the fiber ensemble stress–strain response, derived from a single equi-biaxial test, with the experimentally determined fiber angular distribution [7]. It is this latter method that we exploit as follows: It is assumed that a representative volume element (RVE) can be identified that is large enough to represent the processes associated with the microstructure of the material in some average sense, yet small compared to the characteristic length scale of the microstructure, i.e., the tissue thickness. The RVE is treated as a three-dimensional continuum and it is assumed that the material can be modeled as a hyperelastic solid. Within the RVE, the following assumptions are made:

1. The tissue can be idealized as a planar (i.e., network of fibers) embedded in a compliant ground substance (i.e., the matrix).
2. The collagen fibers are undulated, which gradually disappear with stretch. The load required to straighten the collagen fiber is considered negligible compared to the load transmitted by the stretched fibers. Hence, collagen fibers transmit load only if stretched beyond the point where all the undulations have disappeared and are assumed to be linearly elastic.
3. The degree of fiber undulation can vary considerably. At the tissue level, the gradual straightening of the linear elastic collagen fibers with variable undulations produces the classic nonlinear stress–strain relationship [1].
4. The fiber strain can be computed from the tensorial transformation of the global strain tensor referenced to fiber coordinates (i.e., the affine transformation assumptions).
5. The strain energy function of the tissue is the sum of the individual fiber strain energies.

Thus, the complete tissue-level strain energy density function Ψ assumes the form

$$\Psi(\mathbf{E}) = \int_{\theta} R(\theta)\Psi_{\text{ens}}(\theta, \mathbf{E})d\theta = \int_{\theta} R(\theta)\Psi_{\text{ens}}(\mathbf{E}_{\text{ens}})d\theta \tag{4.12}$$

where $R(\theta)$ is the fiber ensemble angular density function, Ψ_{ens} is the strain energy associated with an individual fiber ensemble, which in turn equals the sum of individual fiber strain energies of the ensemble, and $E_{\text{ens}} = \mathbf{N}^T\mathbf{E}\mathbf{N}$ is the uniaxial Green–Lagrange strain acting in the ensemble direction. Assuming a pseudo-elastic, hyperelastic model, the tissue level stress $\mathbf{S}$ is given as

$$\mathbf{S}(\mathbf{E}) = \frac{\partial\Psi(\mathbf{E})}{\partial\mathbf{E}} - p'\mathbf{C}^{-1} \tag{4.13}$$

where the Lagrange multiplier p accounts for the incompressible nature associated with nonfibrillar components of the ECM. In the present study, p was eliminated as a result of the configuration of planar biaxial testing. Expanding Equation (4.13) results in

$$\mathbf{S}(\mathbf{E}) = \int_{\theta} R(\theta) \frac{\partial \Psi}{\partial \mathbf{E}_{\text{ens}}} \frac{\partial \mathbf{E}_{\text{ens}}}{\partial \mathbf{E}} d\theta = \int_{\theta} R(\theta) \mathbf{S}_{\text{ens}}(\mathbf{E}_{\text{ens}}) \mathbf{N} \otimes \mathbf{N} \, d\theta \tag{4.14}$$

where $\mathbf{S}_{\text{ens}}$ represents the fiber ensemble stress.

The formulation for collagen fibers assumes that the fibers are crimped and transmit load only if stretched beyond the point where all the crimps have disappeared. Once in straightened condition, fibers are assumed to act in a linear manner. The fiber stress–strain relation was assumed to be $S = \eta E_{\text{t}}$, where E_{t} is the true fiber strain which is related to the ensemble or fiber strain by

$$E_{\text{t}} = \frac{E_{\text{ens}} - E_{\text{s}}}{1 + 2E_{\text{s}}}$$

where E_{s} is the fiber slack strain. In terms of E_{t}, the fiber stress–strain relation is

$$S_{\text{f}} = \frac{\partial \Psi_{\text{f}}}{\partial E_{\text{f}}} = \frac{\partial \Psi_{\text{f}}}{v E_{\text{t}}} \frac{\partial E_{\text{t}}}{\partial E_{\text{f}}} = \eta^* \times \frac{E_{\text{f}} - E_{\text{s}}}{1 + 2E_{\text{s}}} \frac{1}{1 + 2E_{\text{s}}} = \eta^* \times \frac{E_{\text{f}} - E_{\text{s}}}{(1 + 2E_{\text{s}})^2} \tag{4.15}$$

The ensuing fiber ensemble stress–strain relation is then described in terms of a fiber recruitment function D:

$$S_{\text{ens}}(E_{\text{ens}}) = \int_0^{E_{\text{ens}}} D(x) \, S_{\text{f}}(x) \, \mathrm{d}x = \eta^* \cdot \int_0^{E_{\text{ens}}} D(x) \left[\frac{E_{\text{ens}} - x}{(1 + 2x)^2} \right] dx \tag{4.16}$$

$D(x)$ describes the fraction of fibers realizing a fully straightened state at a given strain magnitude and is defined over $E_{\text{s}} \in [0, E_{\text{ub}}]$, where E_{ub} is the upper bound strain defining an ensemble strain state where all fibers are straight.

Combining Equations (4.14) and (4.16) leads to a complete formulation for the tissue stress

$$\mathbf{S}(\mathbf{E}) = \eta \int_{-\frac{\pi}{2}}^{\frac{\pi}{2}} R(\theta) \left\{ \int_0^{E_{\text{ens}}(\theta)} D(x) \left[\frac{E_{\text{ens}}(\theta) - x}{(1 + 2x)^2} \right] \mathrm{d}x \right\} \mathbf{N} \otimes \mathbf{N} \, \mathrm{d}\theta \tag{4.17}$$

Here, for convenience the fiber volume fraction ϕ is absorbed into the fiber effective modulus such that $\eta = \phi \cdot \eta^*$. Under equi-biaxial strain conditions, $E_{11} = E_{22}$ and $E_{12} = 0$. Under the assumption that $D(x)$ does not depend on θ, the fiber ensemble stress–strain relationship can be obtained directly from the experimental data using $S_{\text{ens}} = S_{11} + S_{22}$, and $E_{\text{ens}} = E_{11} = E_{22}$ [7, 34]. Hence, the parameters of the fiber ensemble model were experimentally determined directly from the equi-biaxial stretch protocol.

Forms and functions describing fiber recruitment, a process in which fibers in an individual ensemble engage into load transmission as they become fully straightened, have been presented in the literature [7]. As an example, the recruitment function $D(E_s)$ was represented by means of a unimodal modified beta distribution with mean μ and standard deviation σ [35]. The beta distribution $B(y)$ is defined on the interval $y \in [0,\ 1]$, which was modified to be defined on the interval $[0,\ E_{ub}]$, with $y = E_{ens}/E_{ub}$, so that

$$D(E_{ens}, \mu, \sigma) = \begin{cases} \frac{B(y, s_1, s_2)}{E_{ub}} & 0 \leq E_{ens} \leq E_{ub} \\ 0 & \text{otherwise} \end{cases} \tag{4.18}$$

where s_1 and s_2 are shape parameters computed from μ and σ using

$$s_1 = \frac{\mu^2 - \mu^3 - \mu\sigma^2}{\sigma^2}, \quad s_2 = s_1 \frac{1 - \mu}{\mu} \tag{4.19}$$

The beta function was restricted to a unimodal shape with a lower bound of zero and to prevent the upper bound from going to infinity, by enforcing $s_1,\ s_2 > 1$. Note that for an unbound function $\mu, \sigma \in [0, 1]$. To estimate the fiber-effective modulus η, we note that if the tissue would consist of all straight fibers, the maximum tangent modulus (MTM) would equal η. At strain values $E_{ens} > E_{ub}$, all fibers are recruited, and MTM is determined using the equation

$$\text{MTM} = \frac{\partial S_{ens}(E_{ens})}{\partial E_{ens}} = \eta \int_0^{E_{ub}} \frac{D(x)}{(1 + 2x)^2} dx \tag{4.20}$$

The greater the degree of slacked fibers that exist in the tissue (either a longer slack length per fiber, or more slacked fibers), the lower the MTM as compared to η. Hence, MTM provides a *lower bound* value for η, because the integral on the right-hand side of Equation (4.20) is lower than 1 because of the presence of fiber slack.

For actual implementation, it is necessary to determine the statistical distribution function of the angular distribution of the collagen fibers, $R(\theta)$. As $R(\theta)d\theta$ is defined as the fraction of fibers oriented between θ and $\theta + d\theta$ and subjected to the normalization constraint

$$\int_{-\pi/2}^{\pi/2} R(\theta)\, \mathrm{d}\theta = 1$$

it can be determined directly from experimentally measured discrete angular fiber distribution $P_f(\theta)$ using

$$R(i) = \frac{R_{\mathrm{f}}(i)}{\sum\limits_{j} R_{\mathrm{f}}(j)} \tag{4.21}$$

such as done using SALS in [36]. In that study, it was demonstrated that a structural constitutive modeling approach was able to accurately predict equi- and nonequi-biaxial test protocols for pericardium. *An important aspect of our approach is that only a single equi-biaxial test to determine the fiber stress–strain response and $R(\theta)$ determined by SALS is required to determine the complete planar biaxial mechanical response.* Studies of collagen fiber crimp suggest that the local fiber strains follow closely the externally applied tissue strains so that the affine transformation assumption appears to be valid. Note that extension of this approach to 3D is a straightforward extension of the above approach.

4.4 Finite-Element Implementation

The constitutive models presented in the preceding sections can be implemented into a computational framework suitable for finite-element analysis [37]. The process most commonly utilized is the updated Lagrangian formulation, which requires constitutive relationships to be cast in terms of a work balance expressed in the weak form. Implementation of models requires a generalized treatment for the stress and material Jacobian. Proper formulation for a discretized solution procedure is rooted in a linearization of the equilibrium equation in a Lagrangian framework. In brief, assuming a Newton–Raphson procedure based on internal work terms contributing to the material Jacobian

$$\mathrm{d}(\delta W_{\mathrm{int}}) = \int_V [d\underline{\sigma} : \delta\underline{\mathbf{D}} + \underline{\sigma} : \mathrm{d}(\delta\underline{\mathbf{D}})]\mathrm{d}V \tag{4.22}$$

A key component of this process involves the formulation of a consistent material Jacobian. The method presented will be compatible for both Fung and structural-based models cast in terms of the undeformed reference state. Separating the volumetric and deviatoric components of the deformation provides advantages in the treatment of incompressible hyperelastic materials. The multiplicative decompositions of the deformation gradient $\underline{\mathbf{F}}$ and the right Cauchy–Green tensors are given as

$$\underline{\mathbf{F}} = J^{1/3}\hat{\underline{\mathbf{F}}} \quad \underline{\mathbf{C}} = J^{2/3}\hat{\underline{\mathbf{C}}} \tag{4.23}$$

where J is the Jacobian describing volume change defined as the determinant of the deformation gradient. The tensors $\hat{\underline{\mathbf{F}}}$ and $\hat{\underline{\mathbf{C}}}$ are associated with the volume-preserving deformation of the material. This decomposition leads to representing the strain energy function in a decoupled form:

$$\psi(\underline{\mathbf{C}}) = \psi_{\mathrm{vol}}(J) + \psi_{\mathrm{dev}}(\hat{\underline{\mathbf{C}}}) \tag{4.24}$$

The second Piola–Kirchhoff stress can be written as

$$\underline{\mathbf{S}} = 2\frac{\partial\,\psi\,(\underline{\mathbf{C}})}{\partial\,\underline{\mathbf{C}}} = \underline{\mathbf{S}}_{\text{vol}} + \underline{\mathbf{S}}_{\text{dev}} \tag{4.25}$$

where $\underline{\mathbf{S}}_{\text{vol}}$ and $\underline{\mathbf{S}}_{\text{dev}}$ are defined as

$$\underline{\mathbf{S}}_{\text{vol}} = 2\frac{\partial\,\psi_{\text{vol}}(J)}{\partial\,\underline{\mathbf{C}}} = pJ\underline{\mathbf{C}}^{-1} \tag{4.26}$$

$$\underline{\mathbf{S}}_{\text{dev}} = 2\frac{\partial\,\psi_{\text{dev}}(\hat{\underline{\mathbf{C}}})}{\partial\,\underline{\mathbf{C}}} = J^{-2/3}\text{dev}\left[\hat{\underline{\mathbf{S}}}\right] \tag{4.27}$$

The deviatoric operator is defined as $\text{dev}(\bullet) = (\bullet) = 1/3[(\bullet) : \hat{\underline{\mathbf{C}}}]\hat{\underline{\mathbf{C}}}^{-1}$. The Lagrange multiplier p characterizes the hydrostatic pressure and is commonly defined as $p = \kappa(J - 1)$. In an incompressible context, κ is the bulk modulus and can be determined using the condition of plane stress. Associated with the deviatoric component of the deformation gradient, $\hat{\underline{\mathbf{S}}}$ is given as

$$\hat{\underline{\mathbf{S}}} = 2\frac{\partial\,\psi_{\text{dev}}(\hat{\underline{\mathbf{C}}})}{\partial\,\hat{\underline{\mathbf{C}}}} \tag{4.28}$$

Therefore, the second Piola–Kirchhoff stress tensor in terms of the strain energy is

$$\underline{\mathbf{S}} = pJ\mathbf{C}^{-1} + 2J^{-2/3}\text{dev}\left[\frac{\partial\,\psi(\hat{\underline{\mathbf{C}}})}{\partial\hat{\underline{\mathbf{C}}}}\right] \tag{4.29}$$

As used in most codes, an updated Lagrangian formulation approach stems from the statement of virtual power capturing the weak form of the momentum equation [38]. Expressed in terms of the virtual velocity, the work balance can be expressed as

$$\delta W(\phi, \delta V) = \int_V \underline{\boldsymbol{\sigma}} : \delta\underline{\mathbf{D}}\,\mathrm{d}V - \int_V \underset{\sim}{\mathbf{f}} \cdot \delta\underset{\sim}{\mathbf{v}}\,\mathrm{d}V - \int_{\partial v} \underset{\sim}{\mathbf{t}} \cdot \delta\underset{\sim}{\mathbf{v}}\,\mathrm{d}A = 0 \tag{4.30}$$

The pair $\boldsymbol{\sigma}$ and $\boldsymbol{D}$ is the work-conjugate with respect to the current deformed volume and ϕ represents a trial solution. This relationship includes both the internal and external components as defined by the equation

$$\begin{aligned} \delta W_{\text{int}} &= \textstyle\int_V \underline{\boldsymbol{\sigma}} : \delta\underline{\mathbf{D}}\,\mathrm{d}V \\ \delta W_{\text{ext}} &= \textstyle\int_V \underset{\sim}{\mathbf{f}} \cdot \delta\underset{\sim}{\mathbf{v}}\,\mathrm{d}V - \int_{\partial v} \underset{\sim}{\mathbf{t}} \cdot \delta\underset{\sim}{\mathbf{v}}\,\mathrm{d}A \end{aligned} \tag{4.31}$$

The following cast this relationship into a form suitable for the Newton–Raphson procedure. Linearizing in the direction of the increment $\underset{\sim}{\mathbf{u}}$ leads to

$$\mathrm{d}(\delta W)[\underset{\sim}{\mathbf{u}}] = \mathrm{d}(\delta W_{\mathrm{int}})[\underset{\sim}{\mathbf{u}}] - \mathrm{d}(\delta W_{\mathrm{ext}})[\underset{\sim}{\mathbf{u}}] \tag{4.32}$$

Expressing the work statement in a Lagrangian framework requires an alternate work-conjugate pair. The Jacobian (J) can be used to relate the current volume element to the initial reference state, $\mathrm{d}V = J\,\mathrm{d}V_0$. Internal virtual work expressed on the current and reference configuration is given by

$$\delta W_{\mathrm{int}} = \int_V \underline{\sigma} : \delta\underline{\mathbf{D}}\,\mathrm{d}V = \int_{V_0} \underline{\mathbf{S}} : \delta\underline{\mathbf{E}}\,\mathrm{d}V_0 \tag{4.33}$$

From Equation (4.33), the rate of change of internal virtual work is expressed as

$$\mathrm{d}\delta W_{\mathrm{int}} = \mathrm{d}\left(\int_{V_0} \underline{\mathbf{S}} : \delta\underline{\mathbf{E}}\,\mathrm{d}V_0\right) = \int_{V_0} \mathrm{d}\underline{\mathbf{S}} : \delta\underline{\mathbf{E}}\,\mathrm{d}V_0 + \int_{V_0} \underline{\mathbf{S}} : \mathrm{d}\delta\underline{\mathbf{E}}\,\mathrm{d}V_0 \tag{4.34}$$

Making use of the relation

$$\mathrm{d}(\delta\underline{\mathbf{E}}) = (\mathrm{d}\underline{\mathbf{F}}^{\mathrm{T}}\delta\underline{\mathbf{F}} + \delta\underline{\mathbf{F}}^{\mathrm{T}}\mathrm{d}\underline{\mathbf{F}})/2 \tag{4.35}$$

and recalling $\mathrm{d}V = J\,\mathrm{d}V_0$, the second term in (4.34) can be written as

$$\int_{V_0} \mathbf{S} : \mathrm{d}(\delta\mathbf{E})\mathrm{d}V_0 = \int_{V_0} \underline{\mathbf{F}}\,\underline{\mathbf{S}}\,\underline{\mathbf{F}}^{\mathrm{T}} : \mathrm{d}\underline{\mathbf{F}}^{\mathrm{T}}\delta\underline{\mathbf{F}}\,\mathrm{d}V_0 = \int_V \underline{\sigma} : \mathrm{d}\underline{\mathbf{F}}^{\mathrm{T}}\delta\underline{\mathbf{F}}\,\mathrm{d}V \tag{4.36}$$

The second Piola–Kirchhoff stress is defined as

$$\underline{\mathbf{S}} = 2\frac{\partial\,\psi(\underline{\mathbf{C}})}{\partial\,\underline{\mathbf{C}}} \tag{4.37}$$

From Equations (4.4)–(4.18), $d\underline{\mathbf{S}}$ can be written as

$$\mathrm{d}\underline{\mathbf{S}} = 2\frac{\partial\,\psi^2(\underline{\mathbf{C}})}{\partial\,\underline{\mathbf{C}}^2} : \mathrm{d}\underline{\mathbf{C}} = 4\frac{\partial\,\psi^2(\underline{\mathbf{C}})}{\partial\,\underline{\mathbf{C}}^2} : \underline{\mathbf{F}}^{\mathrm{T}}\,\mathrm{d}\,\underline{\mathbf{D}}\,\underline{\mathbf{F}} \tag{4.38}$$

Deriving Equation (4.38) requires the relation

$$\mathrm{d}\underline{\mathbf{C}} = 2\underline{\mathbf{F}}^{\mathrm{T}}\,\mathrm{d}\,\underline{\mathbf{D}}\,\underline{\mathbf{F}} \tag{4.39}$$

Combining the above gives

$$\mathrm{d}\delta W_{\mathrm{int}} = \int_{V_0} \delta\underline{\mathbf{D}} : \left(4\underline{\mathbf{F}}\underline{\mathbf{F}}\frac{\partial\psi^2(\underline{\mathbf{C}})}{\partial\underline{\mathbf{C}}^2}\underline{\mathbf{F}}^{\mathrm{T}}\underline{\mathbf{F}}^{\mathrm{T}}\right) : \mathrm{d}\underline{\mathbf{D}}\,\mathrm{d}V_0 + \int_V \underline{\sigma} : \mathrm{d}\underline{\mathbf{F}}^{\mathrm{T}}\delta\underline{\mathbf{F}}\,\mathrm{d}V \tag{4.40}$$

The material Jacobian, which provides the co-rotational stress rate per reference volume, is expressed as

$$\underline{\mathbb{C}} = \underline{\mathbf{F}} \cdot \left[\frac{\partial^2 \psi(\underline{\mathbf{C}})}{\partial \underline{\mathbf{C}}^2} : \left(2\underline{\mathbf{F}}^{\mathrm{T}}\, \mathrm{d}\underline{\mathbf{D}}\, \underline{\mathbf{F}} \right) \right] \cdot \underline{\mathbf{F}}^{\mathrm{T}} \tag{4.41}$$

In the above form, the fourth-order elasticity tensor can be defined in terms of the strain energy expressed in the reference frame. This provides a convenient framework to reduce the Fung-type and structural-based models as presented previously to forms containing derivatives of the strain energy with respect to the Green strain and Right Cauchy–Green tensors.

4.4.1 Fung Model Implementation Example

As an example, we present an implementation of the Fung model incorporated into the FE software package ABAQUS through its user subroutine UMAT [23, 39]. An eight-node biquadratic, reduced integration plane stress element (ABAQUS element-type CPS8R) was used for all simulations, using the updated Lagrangian formulation. Static simulations allowing for nonlinearity arising from both the constitutive law and the large geometric deformations were performed and the default convergence criteria set by ABAQUS were kept unchanged for all simulations. To ensure that the stress update and the tangent stiffness were properly implemented into ABAQUS, displacement-controlled, single-element tests were performed. Briefly, nodal displacements of the single element were prescribed as boundary conditions to control the element deformation. The resulting FE-updated strains were used as input to the constitutive model to calculate the theoretical updated stresses. The theoretical updated stresses were then compared with the FE-updated stresses to validate the correctness of the model implementation. Note that for anisotropic materials, the resulting asymmetry of the model requires that full element simulations be used, precluding the use of partial models (i.e., quarter models) that require material and geometric symmetries.

4.4.2 Biaxial Testing Simulations

The biaxial testing simulation model consisted of 400 plane stress CPS8R elements describing a specimen geometry of 25 mm $\times$ 25 mm $\times$ 0.4 mm (Fig. 4.1). Four evenly spaced node forces, with 5 mm between two adjacent nodes and 2.5 mm inside the specimen edge, were imposed on each side (Fig. 4.1). Each node force was 2.5 N, imposing a net total of 1 MPa Lagrangian stress on each edge. Similar to the actual biaxial testing setup, only the central region was used for stress–strain measurements. This was accomplished by averaging the stress and strain tensor components for 16 elements located in the center of the FE model, delimiting a 5-mm $\times$ 5-mm region.

Nonlinear regression results for the Fung model, with and without enforcing Equation (4.19), are illustrated in Fig. 4.2 and parameter values are taken from

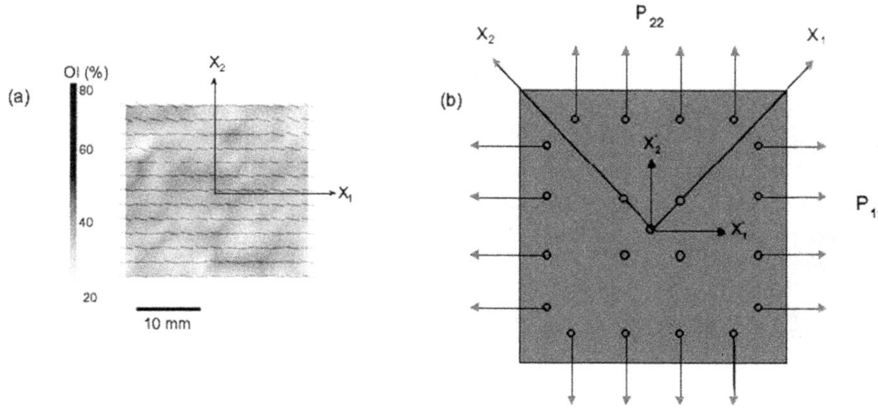

Fig. 4.1 (**a**) Biaxial test specimen overlaid on a grayscale collagen fiber orientation index (OI) values demonstrating high uniformity of both fiber-preferred directions and OI, along with the definition of the material axes X_1–X_2. (**b**) A schematic of the biaxial testing specimen, with P_{11} and P_{22} the Lagrangian stresses imposed on each side of the sample by the four evenly spaced suture attachments. Specimens were tested with material axes (X'_1–X'_2) inclined to a 45° angle with respect to the biaxial device stretching axes (X_1–X_2) to generate a combined state of normal and shearing stresses

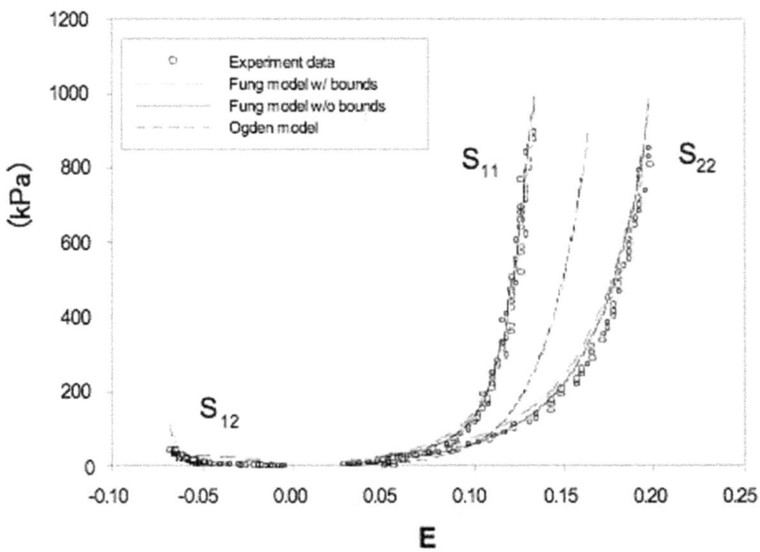

Fig. 4.2 Nonlinear regression fitting results for the Fung elastic model (Equation (4.5)) with and without the bounds of Equation (4.10). Also shown for comparison is an isotropic Ogden model [23]

[23, 39]. Further, with the bounds imposed, the matrix **D** satisfied the positive-definiteness requirement over the experimental ranges and Ψ satisfied the convexity condition (Fig. 4.3d, e, f). As a comparison (Fig. 4.3), without the bounds, the matrix **D** failed to be positive definite and the strain energy function was not convex (Fig. 4.3a–c). The results of the single-element test of the equi-biaxial stress protocol demonstrate exact agreement between the theoretical and FE solutions. They indicated that the material model was successfully incorporated into the FE analysis. Results of biaxial test FE simulations exhibited symmetric deformations for the isotropic model and prominent shear deformation for the orthotropic model. Overall, there was excellent agreement between the FE-predicted and experimentally measured stress responses. The small discrepancies between FE output and experimental data were attributable to idealized model conditions, such as the use of point loads that are exactly spaced and have exactly equal values. While it is possible to attempt to achieve this experimentally, the current simulations suggest that this is not absolutely necessary for accurate simulations. This is fortunate since multi-axial soft tissue experimentation is difficult and prone to error resulting from the inherent nonlinearity of the tissue. It should be noted that our finite-element analysis was conducted using material axes oriented at 45° with respect to biaxial stretching axes (Fig. 4.1a) and point load boundary conditions. Our findings should

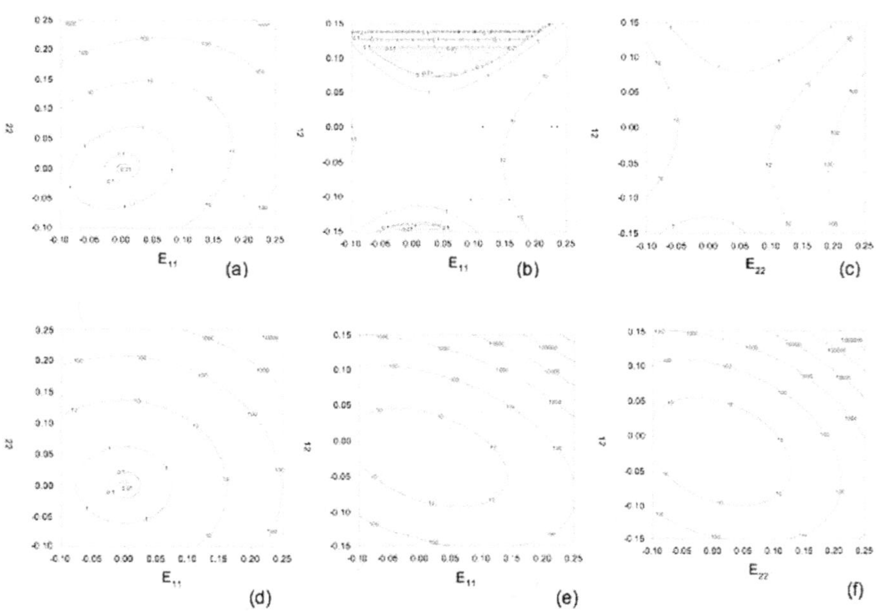

Fig. 4.3 Contour plots representing the projections of the log of the strain energy function potential onto the (**a**) E_{11}–E_{22}, (**b**) E_{11}–E_{12}, and (**c**) E_{22}–E_{12} strain planes *without* the convexity condition imposed. Also shown are the same projections onto the (**d**) E_{11}–E_{22}, (**e**) E_{11}–E_{12}, and (**f**) E_{22}–E_{12} strain planes *with* the convexity condition imposed. Note especially the presence of hyperbolic (*saddle-shaped*) strain energy function in (**b**) and (**c**)

be interpreted in light of these restrictions. When varying the orientations of tissue material axes and tissue clamping methods, the region of uniform stress and strain would certainly be changed. Those effects are addressed in another study [39].

For proper computational implementation, the Fung model requires explicit expression of shear behavior; therefore the E_{12} term has to be included. In the literature, a four-parameter Fung model (e.g., $Q = A_1E_{11}^2 + A_2E_{22}^2 + 2A_3E_{11}E_{22}$) is commonly used to describe and quantify tissue mechanical properties. However, it is not adequate for any realistic numerical simulation in which shearing is expected. This also suggests that mechanical testing methods that utilize extensional strains alone will not provide sufficient information to develop a strain energy function for computational applications. Wilber and Walton [40] studied sufficient and necessary conditions for the Fung model to satisfy the Legendre–Hadamard and strong ellipticity conditions. They showed that to satisfy either of these conditions, the material parameters in the Q term (i.e., the A_i) of the Fung model needed to be equal. This is an unrealistic constraint, since by satisfying it the Fung model can only have two parameters (c and A), and thus consequently becomes an isotropic material model. Obviously, it cannot fit with the anisotropic biaxial experimental data utilized in the present study.

Construction of a constitutive model with a physical plausibility property a priori is ideal. There have been several attempts [22, 41–44] in this direction. For example, recently Holzapfel and Gasser [22] proposed a constitutive model for arterial layers. The model was constructed with physically meaningful material parameters and demonstrated physical plausibility (convexity). Structural approaches have been shown to be physically plausible [43, 45]. Moreover, when quantitative tissue structural information is available, they have been demonstrated to be robust and substantially simplify parameter information while simultaneously providing insight into the underlying physical basis for the observed mechanical behavior [7]. Taken as a whole, our previous [9, 13, 46] and current work has begun to address the complete process of computational implementation. This process included the following:

(1) The acquisition of appropriate experimental data
(2) Formulating the appropriate constitutive model
(3) Obtaining and refining material parameters
(4) FE implementation and validation

Additionally, the approach outlined in this chapter was formulated within a general commercially available finite-element code and can thus be straightforwardly adopted.

4.4.3 Prosthetic Valve Simulations

As an example of the above approach, we present simulations of pericardial BHV under quasi-static loading that directly incorporate experimentally measured leaflet structures and the Fung elastic material models for bovine pericardium, which include the effects of in-plane shear as well as additional parameter constraints to facilitate numerical convergence. Moreover, simulations were validated by comparing simulation results with leaflet strain fields from the same BHV leaflets used for leaflet material property determination. The effects of variations in individual leaflet mechanical properties on the resulting leaflet stress field were also explored. In order to obtain accurate BHV FE quasi-static simulations, the following key study aspects need to be addressed:

1. Experimental validation of the FE-predicted in-surface strain tensor at different quasi-static pressures over the entire leaflet surface
2. Collagen fiber structural information to determine local material axes and constitutive model of the nonlinear anisotropic leaflet material properties for each leaflet
3. Material parameter constraints to facilitate convergence of the numerical solutions
4. Well-defined valve leaflet geometry and proper prescription of boundary conditions

In addition to the above aspects of the model, there are two additional parameters that need to be defined. The first is the out-of-plane transverse shear stiffness $K_{ij}(i = 1,2, j = 3)$ for the leaflet, and the other is the coefficient of friction f at the contact interface between the leaflets. The experimental data for accurately defining these two parameters are not currently available. We therefore conducted a study of these two parameters to investigate the sensitivity of the model solution. We assumed that the transverse shear stiffness $K_{ij}(i = 1,2, j = 3)$ was of the same value for both K_{13} and K_{23}, and varied it using values of 1, 10, 30, 50, and 100 kPa. For the friction coefficient, the value was varied using values of 0.0 (i.e., no friction), 0.1, 0.3, and 0.5. To investigate the sensitivity of the BHV to variations in leaflet mechanical properties, we also modified the model by choosing individual leaflet material parameter sets and using them for all three leaflets.

Details of the approach have been presented in detail in [47]. Briefly, we simulated quasi-static BHV leaflet deformation under 40, 80, and 120 mmHg quasi-static transvalvular pressures. The Fung elastic material model utilizing material parameters and axes was derived from actual leaflet biaxial tests and measured leaflet collagen fiber structure (Fig. 4.4). Rigorous experimental validation of the predicted leaflet strain field was used to validate the model results. An overall discrepancy of 2.36% strain between the FE results and experimental measurements was obtained, indicating good agreement between computed and measured major principal strains (Fig. 4.5a). Parametric studies utilizing the material parameter set from one leaflet for all three leaflets resulted in substantial variations in leaflet stress and strain distributions (Fig. 4.5b). This result suggests that utilization of actual leaflet material properties is essential for accurate BHV FE simulations. The present study also underscores the need for rigorous experimentation and accurate constitutive models in simulating BHV function and design.

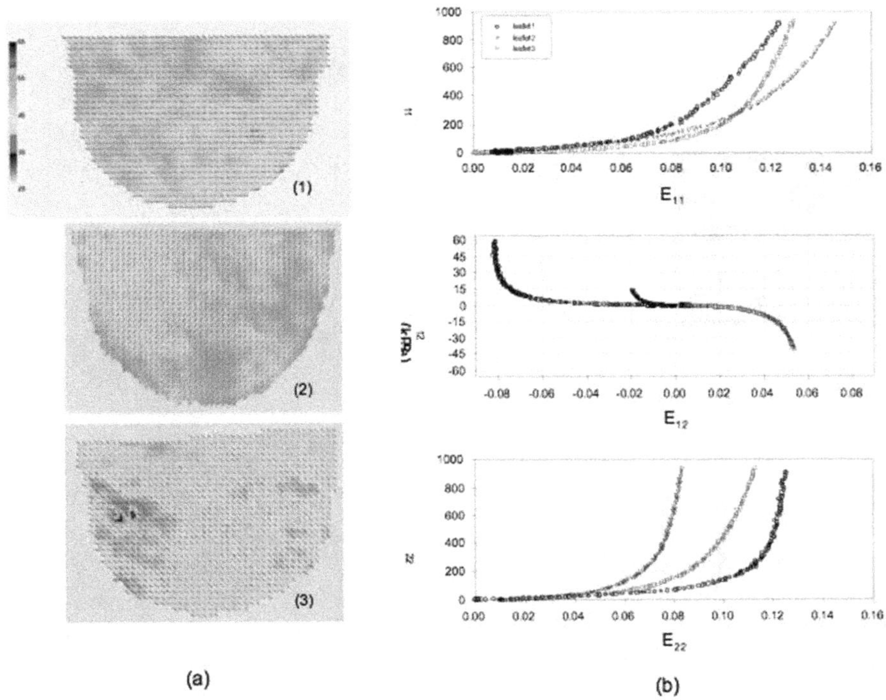

Fig. 4.4 (**a**) Representative collagen fiber architectural data for leaflets 1–3, where the vectors represent the local preferred fiber orientations, and the color scale is described in [36]. Most leaflets had a $\pm 45°$ preferred orientation and relatively uniform degree of orientation throughout the leaflet. (**b**) Corresponding biaxial mechanical data for the 1 MPa equi-biaxial Lagrangian stress test protocol for the same leaflets, showing a moderate degree of variability

4.4.4 Engineered Heart Valve Leaflet Tissue Simulations

Regardless of the design specifics of current prosthetic valve devices, none offers any potential for growth, and therefore pediatric patients requiring valve replacement will require reoperations to place larger devices to accommodate the growth of the patient. It is believed that a tissue-engineered heart valve can accommodate many of these requirements, especially those pertaining to somatic growth. Tissue engineering (TE) describes the process of combining engineered materials with living cells to produce viable structures for the replacement of diseased or deficient native tissues [48]. The challenges facing TE researchers involve the contrasting requirements between favorable tissue growth conditions and functional in vivo properties. The initial stage of the process involves the use of a synthetic scaffold to serve as host to an integrated cell seed population. Tissue growth follows and is stimulated using strain-based conditioning techniques, a process in which the evolving tissue is exposed to dynamic biomechanical effects [49, 50]. As the tissue growth advances the scaffold, material biodegrades and is completely absent

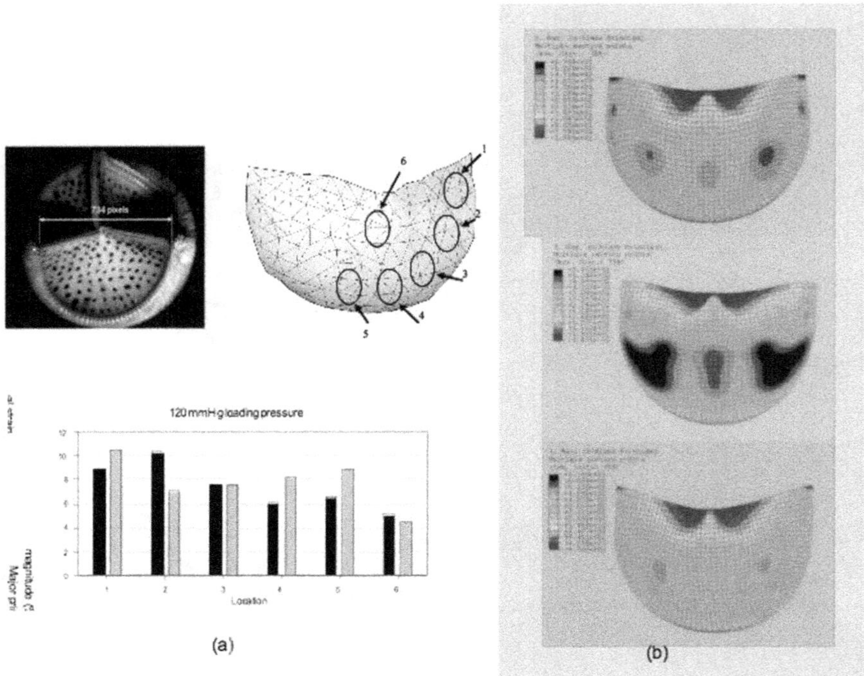

Fig. 4.5 (**a**) One camera view of valve 1 at 0 mmHg transvalvular pressure, showing the marker array used for strain measurements, and the width (in pixels) of the valve, a typical triangular element mesh constructed from the digitized markers positions, where the numbered regions indicate areas used for strain tensor component comparisons as shown in the lower figure. Overall there was excellent agreement with the model with errors less than 2.5% strain. (**b**) Simulation results with the same material properties for all three leaflets on aortic side of leaflet surface and on the ventricular side for the same simulations. These simulations clearly show that even moderate variations in mechanical properties as shown in Fig. 4.4 can produce substantial variations in actual leaflet stress distributions

in the matured tissue. Therefore, in addition to the broad range of biocompatible requirements, scaffold materials must support the strain-based conditioning process. A long-term research aim is the definition of these requirements and the development of supporting guidelines for scaffold design. Further details can be found in [51].

Scaffolds fabricated by electrospinning natural polymers, synthetic polymers, or polymer blends have received widespread attention. Beyond the relative affordability and simplicity of electrospun natural polymers, their popularity is largely a result of a versatile manufacturing process in which slight alterations during fabrication enable the production of scaffolds with a wide array of fiber morphologies (i.e., fiber diameter, porosity, packing density, and orientation) directly influencing bulk mechanical properties. Controlled mechanical anisotropy, for example, is attained by using a rotating collection surface that induces a preferred fiber direction as the rotational speed of the collector increases (Fig. 4.6a). This ability is

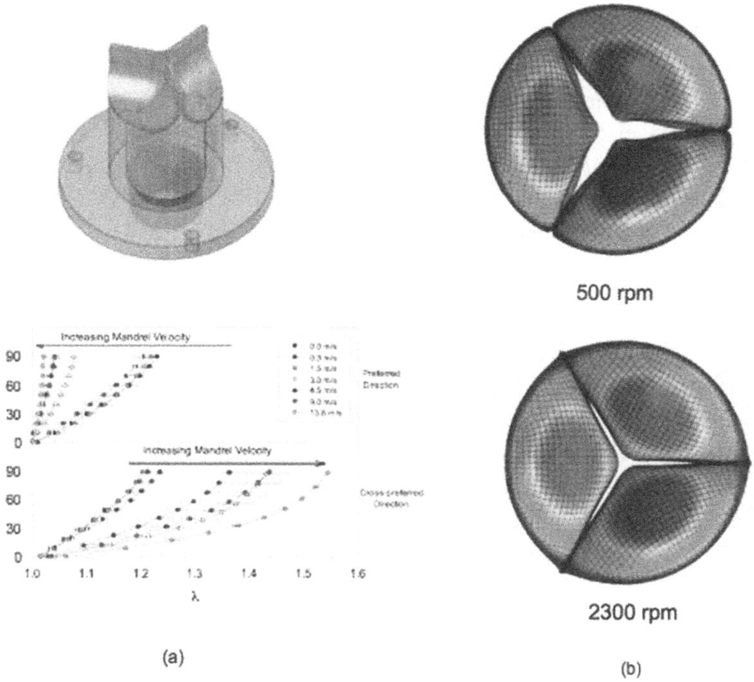

Fig. 4.6 (**a**) TEHV in vitro bioreactor stent design [52, 53], along with stress–strain data of ES-PEUU scaffold [54] showing the influence of fabrication mandrel speed on the resulting biaxial mechanical properties. (**b**) The influence on the resulting leaflet coaptation is demonstrated by the deformed shaped of scaffolds prepared with mandrel speeds of 500 and 2,300 rpm

extremely beneficial in mimicking native tissue architecture and has even been shown to approximate the highly nonlinear biaxial mechanical response of collagenous soft tissues. In particular, our laboratory has utilized electrospun PEUU (ES-PEUU) scaffolds with tensile biaxial mechanical properties remarkably similar to the native pulmonary valve, including the ability to undergo large physiological strains and exhibit pronounced mechanical anisotropy (Fig. 4.6a). Computational simulations need to be directed at determining the ES-PEUU properties that can provide the maximum benefit from a biomechanical point of view. If these properties can be controlled by the fabrication process, what properties should be targeted to mimic the homogenous strain field of native tissue? And, to what practical extent can this information guide the fabrication process to produce TE material with improved qualities? In addition to material anisotropy, the initial undeformed scaffold shape can be manipulated.

To begin to address these questions, finite-element simulations [37] using a stented leaflet design intended for organ-level bioreactor studies [52, 53] were conducted under 80 mmHg transvalvular pressure for isotropic and anisotropic scaffolds (Fig. 4.6b). Note that the circumferential direction was taken as the preferred material direction for all simulations. The biaxial mechanical properties of ES-PEUU scaffolds (Fig. 4.6) [54] were modeled on the Fung model, as for pericardium in ABAQUS, through custom-written UMAT subroutines along with the material parameters of the scaffold type [2]. Quadratic hexahedral elements were used to model the leaflet. Overall geometric characteristics indicated that without sufficient mechanical anisotropy, leaflet response in the radial direction cannot develop the strain magnitude required to permit free edge engagement with adjacent leaflets. Note that the initial scaffold geometry was constant for each case and was representative of a "near"-closed condition.

In addition to basic functional characteristics, our laboratory has utilized a unique ability to investigate cellular deformations within 3D elastomeric fibrous scaffolds [55]. Scaffold specimens microintegrated with vascular smooth muscle cells were subjected to controlled biaxial stretch with 3D cellular deformations and local fiber microarchitecture simultaneously quantified [56, 57] (Fig. 4.7a). We demonstrated that the local fiber geometry followed an affine behavior so that it could be predicted by macroscaffold deformations. However, local cellular deformations depended nonlinearly on changes in fiber microarchitecture and ceased at large strains where the scaffold fibers completely straightened (Fig. 4.7a). *Thus, local scaffold microstructural changes induced by macrolevel applied strain dominated*

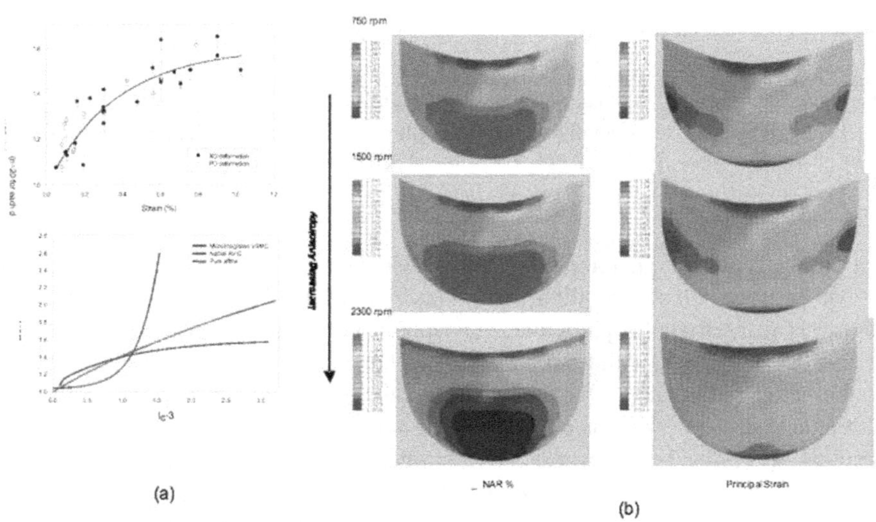

Fig. 4.7 (**a**) A composite of all NAR measurements (mean ± s.e.m) demonstrated a rapid increase to ~60% strain, after which a plateau was observed with further strain increases, indicating that nuclei deformations are dominated by local fiber straightening. A composite cell-scaffold deformation response (*bottom*) is provided for native porcine aortic valve leaflet, cell-integrated electrospun PEUU, and a theoretical purely affine cell deformation response to macroscopic strain. Figure reproduced with permission from [88]. (**b**) Resulting simulations of the NAR magnitude and homogeneity influenced by material anisotropy as characterized by mandrel speed. As the strength of anisotropy increases, NAR magnitude increases along with improved homogeneity of NAR and principal strain

cellular deformations so that monotonic increases in scaffold strain do not necessitate similar levels of cellular deformation. This result has fundamental implications when attempting to elucidate the events of de novo tissue development and remodeling in engineered tissues, which are thought to depend substantially on cellular deformations.

To attempt to capture these characteristics, we performed additional simulations. For the mechanically isotropic, substantial regional variations were observed. In contrast, use of a mechanically anisotropic scaffold (9 m/s mandrel spin rate) resulted in a more uniform principal strain distribution (Fig. 4.7b). Thus, tailoring the fabrication process, by way of mandrel spin rate, may provide an effective and practical method to introduce a moderate degree of anisotropy into scaffold material. These qualities can be controlled to provide a biomechanical response that more closely mimics the native tissue and therefore provide a physiologically relevant deformation state for the microintegrated cell population. Beyond aspects of strain homogeneity, the change in nuclear aspect ratio (NAR) over the leaflet was also predicted and shown to vary with mechanical anisotropy, but was overall less sensitive than the principal strain (Fig. 4.7b).

4.5 Finite-Element Models of Heart Valve Leaflets

Advanced heart valve simulations will require the development of advanced elements. At the structural level, heart valve leaflets are best described as thin shell structures. Although theoretically it is possible to model a leaflet using 3D solid elements, displacement-based elements such as the eight-node bricks perform poorly for thin structures and numerical remedies must be implemented. The modified thin solid elements are often categorized as continuum shells. The analysis of shell structures has been of interest for several decades and continues to be an area of novel activities. Tremendous efforts have been devoted to the development and improvement of shell formulations in the computational mechanics. As a result, a large number of shell elements are available in literature and in commercial packages. Yet, despite these rich resources, leaflet modeling is still challenging. Overall, finite-element shell elements have been mainly derived along three avenues of development:

(1) The degenerated approach
(2) The stress-resultant or direct (Cosserat) shell theory
(3) The continuum shell concept

The kinematical underpinning common to all approaches is summarized in Appendix. The first two are based on the same kinematic assumptions; they differ primarily in the manner the kinetic variables (i.e., stresses) are treated. The degenerated approach utilizes the 3D continuum equilibrium equations, while the direct theory replaces the 3D equilibrium equations with a set of, in some sense, equivalent equations in terms of stress resultants. The resultant approach needs a set of constitutive equations different from a 3D constitutive law. The third approach treats a shell structure as a 3D body, as it really is, without introducing additional kinematic and/or kinetic assumptions. But the low-order finite elements are not suitable for thin structures due to numerical locking, and thus the element must be properly modified.

4.5.1 Degenerate Solid Shell

The degenerated shell formulation may be traced back to the early contributions by Ahmad et al. [58], Hughes and Liu [59], Dvorkin and Bathe [60], and Hallquist et al. [61]. This family of elements was initially designed for "small strain, large rotation" problems, although later developments carried the methods to fully nonlinear regions. In the degenerated formulation, the position vectors and the displacement field are constructed by interpolating the nodal values, viz.

$$\begin{aligned}\mathbf{X} &= \sum_{i=1}^{\text{nen}} N_{\text{I}}\hat{\mathbf{X}}_i + \theta^3 \sum_{i=1}^{\text{nen}} N_{\text{I}}\mathbf{D}_i \\ \mathbf{x} &= \sum_{i=1}^{\text{nen}} N_{\text{I}}\hat{\mathbf{x}}_i + \theta^3 \sum_{i=1}^{\text{nen}} N_{\text{I}}\mathbf{d}_i \\ \mathbf{u} &= \sum_{i=1}^{\text{nen}} N_{\text{I}}\hat{\mathbf{u}}_i + \theta^3 \sum_{i=1}^{\text{nen}} N_{\text{I}}\Delta\mathbf{d}_i\end{aligned} \tag{4.42}$$

Despite the 3D nature, the element is considered as a surface element with nodes placed on the reference surface. The primary nodal unknowns consist of $\mathbf{u}_i$ and $\Delta\mathbf{d}_i$ (or equivalently $\hat{\mathbf{x}}_i$ and $\mathbf{d}_i$). The strain field in an element is computed from the finite-element displacement field. The strains need to be properly modified to remedy numerical pathologies discussed later. The director $\mathbf{d}_i$ can be either inextensible (no thickness strain) or extensible; in the former case, the directors undergo rigid rotations only and thus $\Delta\mathbf{d}_i$ is related to an incremental rotation which in general is not a linear map. In [59], a linearized update algorithm was implemented.

Despite being a "surface element," integrals in the weak form should be carried out over the 3D volume the element occupies:

$$\int_V (\cdot)\mathrm{d}V = \int_{\square}\int_{h^-}^{h^+} (\cdot)j\,\mathrm{d}\theta^1\,\mathrm{d}\theta^2\,\mathrm{d}\theta^2 \tag{4.43}$$

where j is the Jacobian of the geometric mapping. Although reduced (surface) quadratures are admissible for linear material, multiple layers of quadratures in the thickness direction are necessary for nonlinear constitutive equations. Plane stress condition is normally imposed. For linear elastic material this can be done simply using the plane stress constitutive equation, but for fully nonlinear materials the thickness stretch needs to be locally solved from the plane stress condition. The degenerated formulation admits hyperelastic materials. A finite strain hyperelastic degenerated shell was presented in [62]. The formation utilized the geometrically exact rotation updates developed in [63, 64].

A distinct feature of the degenerated shell is that it allows a 3D constitutive equation to be directly implemented. This feature is attractive for traditional engineering materials for which the constitutive laws are readily known, but for heart valve or other thin biological organs, this "3D" feature is actually unfavorable. As we alluded to before, thin tissues are normally characterized in a 2D biaxial tension test because it is difficult to determine the mechanical properties in the thickness direction. Starting from a 2D constitutive equation, one must make appropriate assumptions about the mechanical properties in the thickness direction in order to extend the constitutive into a 3D form for a degenerated shell formulation. Only for linear material and small strain formulation can the constitutive equation be equivalently recast into a resultant form and thus bypass a full 3D constitutive description.

4.5.2 Element Pathology

A direct implementation of the generated shell formulation is susceptible to numerical locking. For thin structural elements, a prominent type of locking is the so-called shear locking, a numerical pathology arising from the presence of artificial transverse shear strain in the finite-element kinematics. Recalling Equation (4.54), the transverse shear strain *i*

$$2E_{\alpha 3} = \mathbf{u}_{\alpha} \cdot \mathbf{d} + \hat{\mathbf{G}}_{\alpha} \cdot \Delta\mathbf{d} \tag{4.44}$$

Consider as an example a flat four-node element undergoing a pure bending, in which in the analytical sense the directors are supposed to remain normal to the shell surface. Numerically, however, the director displacement $\Delta\mathbf{d}$ is a bilinear function of coordinates and thus $\hat{\mathbf{G}}_{\alpha} \cdot \Delta\mathbf{d}$ is nonzero except for at some special locations. This gives rise to an artificial shear strain known as the parasite shear. The contribution to the strain energy depends inversely on the shell thickness. Thus, the element pathologically stiffens as the element thickness decreases.

Shear locking is well understood and effective remedies have been developed. Methods for eliminating shear locking include the assumed natural strain (ANS) method [60, 65] and enhanced assumed strain (EAS) method [66, 67]. In the ANS approach, the pathological shear is modified but still linked to nodal values and therefore no additional kinematic variables are introduced. A widely accepted ANS modification for the four-node shell element was proposed in [60]. In the EAS concept, additional strain terms are introduced to enrich the finite-element kinematics. This approach is more general and has been extensively used in the design of low-order elements and in particular of thin shell elements [68–71].

4.5.3 Stress-Resultant Shell

An alternative approach to shell mechanics is to formulate the equilibrium equations in terms of stress resultants. The degenerated shell in some sense also admits resultants, but the reduction is carried out numerically. In the resultant approach the reduction is carried out analytically prior to a numerical realization.

The stress resultant $\mathbf{n}^\alpha$, the stress couple $\mathbf{m}^\alpha$, and the across-the-thickness stress resultant $\mathbf{l}$ are defined as follows:

$$\mathbf{n}^\alpha = \frac{1}{\bar{j}} \int_{h^-}^{h^+} \sigma g^\alpha j \, \mathrm{d}\theta^3, \tilde{\mathbf{m}}^\alpha = \frac{1}{\bar{j}} \int_{h^-}^{h^+} \xi \sigma g^\alpha j \, \mathrm{d}\theta^3, \mathbf{l} = \frac{1}{\bar{j}} \int_{h^-}^{h^+} \sigma g^3 j \, \mathrm{d}\theta^3 \tag{4.45}$$

where $\boldsymbol{\sigma}$ is the Cauchy stress, j is the Jacobian of the 3D deformation gradient, and $\bar{j} = \|\hat{\mathbf{g}}_1 \times \hat{\mathbf{g}}_2\|$ refers to the surface Jacobian. All relevant variables are defined in Appendix. With these resultant variables the equilibrium equations can be expressed as

$$\frac{1}{\bar{j}} \left(\bar{j}\mathbf{n}^\alpha\right)_{,\alpha} + \bar{\mathbf{n}} = \bar{\rho}\ddot{\varphi}, \quad \frac{1}{\bar{j}} \left(\bar{j}\tilde{\mathbf{m}}^\alpha\right)_{,\alpha} - \bar{\mathbf{l}} + \tilde{\bar{\mathbf{m}}} = \bar{I}_\rho \ddot{\mathbf{t}} \tag{4.46}$$

where $\bar{\mathbf{n}}$ and $\tilde{\bar{\mathbf{m}}}$ are the resultants of applied loads, and $\bar{\mathbf{l}} = \lambda \mathbf{t} + \mathbf{l}$, where λ is an undetermined director tension.

The stress resultants nevertheless do not directly work-conjugate the strain variables defined in Equation (4.52). For example, the resultant $\mathbf{n}$ contains a transverse term, which obviously is not work-conjugate to the in-plane strain $\varepsilon_{\alpha\beta}$. In the direct (Cosserat) shell theory [72], these stress variables are properly combined to yield three work-conjugate stress measures:

$$\tilde{\mathbf{n}} = \tilde{n}^{\alpha\beta} \hat{\mathbf{g}}_\alpha \otimes \hat{\mathbf{g}}_\beta, \tilde{\mathbf{q}} = \tilde{q}^\alpha \hat{\mathbf{g}}_\alpha, \tilde{\mathbf{m}} = \tilde{m}^{\alpha\beta} \hat{\mathbf{g}}_\alpha \otimes \hat{\mathbf{g}}_\beta \tag{4.47}$$

The relations between the conjugate and nonconjugate stress variables are omitted. In terms of the conjugate stress variables, the weak form of the momentum balance equations can be written as

$$\int_A \left[\tilde{n}^{\alpha\beta} \delta\varepsilon_{\alpha\beta} + \tilde{m}^{\alpha\beta} \delta\rho_{\alpha\beta} + \tilde{q}^\alpha \delta\gamma_\alpha\right] \mathrm{d}\mu - \Pi_{\text{ext}}(\delta\mathbf{u}) = 0 \tag{4.48}$$

where $\mathrm{d}\mu = \bar{j}\, \mathrm{d}\theta^1 \, \mathrm{d}\theta^2$ is the current surface area element.

The system is closed with the introduction of a set of constitutive equations that relate the three stress resultants in Equation (4.38) to the strain variables in Equation (4.52). In a properly invariant formulation, these constitutive equations are described in terms of reference resultants, viz.

$$\begin{aligned} \tilde{N}^{\alpha\beta} &= \tilde{N}^{\alpha\beta}(\varepsilon_{\alpha\beta}, \rho_{\alpha\beta}, \delta_\alpha, \theta^\alpha) \\ \tilde{M}^{\alpha\beta} &= \tilde{M}^{\alpha\beta}(\varepsilon_{\alpha\beta}, \rho_{\alpha\beta}, \delta_\alpha, \theta^\alpha) \\ \tilde{Q}^\alpha &= \tilde{Q}^\alpha(\varepsilon_{\alpha\beta}, \rho_{\alpha\beta}, \delta_\alpha, \theta^\alpha) \end{aligned} \tag{4.49}$$

For example, the following constitutive model inspired by the linear shell theory is commonly used:

$$\begin{aligned}\tilde{N}^{\alpha\beta} &= HD^{\alpha\beta\delta\gamma}\varepsilon_{\delta\gamma} \\ \tilde{M}^{\alpha\beta} &= \tfrac{H^3}{12}D^{\alpha\beta\delta\gamma}\rho_{\delta\gamma} \\ \tilde{Q}^{\alpha} &= G\delta_{\alpha}\end{aligned} \tag{4.50}$$

Here $D^{\alpha\beta\delta\gamma}$ is a (plane stress) elastic tensor and G is a "shear modulus." The "shear modulus" is really a penalty parameter introduced to approximately enforce the Kirchhoff–Love assumption. At the end, the Cauchy stress resultants defined in Equation (4.38) are related to the reference invariants by

$$\tilde{n}^{\alpha\beta} = \bar{j}^{-1}\tilde{N}^{\alpha\beta},\ \tilde{m}^{\alpha\beta} = \bar{j}^{-1}\tilde{N}^{\alpha\beta},\ \tilde{q}^{\alpha} = \bar{j}^{-1}\tilde{Q}^{\alpha} \tag{4.51}$$

This framework allows one to treat a shell as a deformable surface equipped with a director field. The weak form involves only surface integrals and no through-thickness integration is necessary. The contributions by Simo and co-workers constituted a body of seminal work on stress-resultant shells [63, 73, 74]. They introduced a geometrically exact description for rotation updates which was then widely adapted in computational shell mechanics. The shell elements by Simo's group allow for large rotation updates for any rotation increment less than 180°. The elements thus are well suited to large rotation applications. A summary of the major developments on the stress-resultant shell is contained in [75].

The stress-resultant shell is well suited to small to moderately large deformations. However, the implementation of hyperelastic constitutive models is not easy. Constitutive equations such as Equation (4.41) are based on the premise that the laminar stress varies linearly over the thickness, and in this case the moment can be explicitly derived. It goes without saying that the assumption of linear stress distribution is not valid for general hyperelastic materials even if the laminar strain varies linearly. To implement a hyperelastic model, one must establish the bending-curvature relation either analytically or experimentally. In addition, if the material is incompressible, thickness stretch must be properly considered. A significant amount of research has been devoted to hyperelastic stress-resultant shells. Among others, Simmonds [76] and Schieck et al. [77] submitted methods to consistently derive approximated bending-curve relations from hyperelastic functions.

Nevertheless, the resultant formulation offers the flexibility of separately considering the in-plane and bending functions. The present authors have followed this concept to derive a fully experiment-based, stress-resultant shell model for prosthetic heart valves [78]. For the in-plane part we have used the Fung-type in-plane constitutive equation:

$$\begin{aligned}\tilde{N}^{11} &= Hce^{Q}(A_1E_{11} + A_3E_{22} + A_5E_{12}) \\ \tilde{N}^{22} &= Hce^{Q}(A_3E_{11} + A_2E_{22} + A_6E_{12}) \\ \tilde{N}^{12} &= Hce^{Q}(A_5E_{11} + A_6E_{22} + A_4E_{12})\end{aligned} \tag{4.52}$$

See Section 4.2 for an in-depth discussion of the experimental determination of the material parameters. The bending part was described by a polynomial form

$$\begin{aligned}\tilde{M}^{11} &= \tfrac{H^3}{12}[a_1\rho_{11} + b_1(\mathrm{sign}(\rho_{11}))\rho_{11}^2 + 0.25(a_1 + a_2)\rho_{22}]\\ \tilde{M}^{22} &= \tfrac{H^3}{12}[0.25(a_1 + a_2)\rho_{11} + a_2\rho_{22} + b_2(\mathrm{sign}(\rho_{22}))\rho_{22}^2]\\ \tilde{M}^{12} &= \tfrac{H^3}{12}[0.25(a_1 + a_2)\rho_{12}]\end{aligned} \tag{4.53}$$

The parameters in Equation (4.53) in the preferred direction (1-direction) were characterized experimentally using a three-point bending test [79, 80]. The parameters in the transverse direction as well as the coupling terms were assigned based on the anisotropic ratio in the in-plane model. This shell model was utilized to simulate the opening and closing dynamics [81, 82].

4.5.4 Continuum Shell

Perhaps the most straightforward approach to thin-shell modeling is to treat shells as three-dimensional solids, as they should be treated. The advantage of this direct approach is that a 3D constitutive model can be applied without any modification and that the zero stress condition in the thickness direction can be naturally imposed. Existing technologies for enforcing the incompressibility constraint can be readily utilized. However, low-order displacement elements, in particular the tri-linear eight-node brick, behave poorly in thin-shell analysis due to numerical locking. In addition to the afore-discussed shear locking, the continuum element is prone to the so-called Poisson thickness locking. This is a pathology resulting from insufficient displacement resolution in the thickness direction. A continuum element employing linear displacement in the thickness direction gives only a constant thickness strain, whereas owing to the Poisson effect the thickness strain should vary linearly, or at least approximately so, in the thickness direction. The constant thickness strain leads to an artificial stiffening effect, albeit one that is less severe than the shear locking. Fortunately, both the shear locking and the Poisson thickness locking can be resolved within the framework of EAS. Continuum shell elements with ANS and EAS modifications are presented by several authors [71, 83, 84]. The solid element in [83] is currently used by the present authors in a fluid–structure interaction analysis of heart valves.

4.6 Summary

In this chapter we presented approaches toward FE implementation of general constitutive models, including in-plane shear effects, with restrictions necessary to achieve numerical stability. These results suggest that accurate design and functional simulations of medical devices using realistic nonlinear anisotropic material models are both feasible and practical. The successful FE implementation of the anisotropic Fung elastic model will eventually facilitate the medical design process. The specific aims of this work focus on the development of soft tissue constitutive models implemented in a computational framework. Levering the models within the context of finite-element analysis is the primary aim. Although the semi-lunar heart valves served as a modeling focus, the modeling techniques involved are applicable to all cardiovascular soft biological tissue applications. Both phenomenological and physiologically guided approaches can be employed. For engineered heart valve tissues, the modeling of such approaches is still in its infancy. A prudent approach is to first develop a general model for the virgin scaffold of interest and then to extend it for the particular cell source utilized. Moreover, a key component of any future work will involve the development of a high-fidelity physiologically based model based on in vivo imaging techniques.

Acknowledgments We would like to acknowledge funding by NIH/NHBLBI grants HL68816 and HL071814, as well as National Science Foundation Grant BES-9978858.

Appendix: Shell Kinematics

In a shell theory the position $\mathbf{X}$ of a generic material point is described as

$$\mathbf{X}(\theta^1,\theta^2,\theta^3) = \hat{\mathbf{X}}(\theta^1,\theta^2) + \theta^3\mathbf{D}(\theta^1,\theta^2),\ \theta^3 \in [H^-,H^+] \tag{4.54}$$

where $\hat{\mathbf{X}}$ denotes a point in a reference surface, and $\mathbf{D}$ is a vector based at the surface point that defines the surface director (or the "fiber direction"). In the classical shell kinematics, $\mathbf{D}$ is the outward surface normal, although in general it does not have to be so. Note in Equation (4.54) the vector $\mathbf{D}$ is a unit vector, where $H := H^+ - H^-$ is the initial thickness of the shell. Alternatively, one can redefine the parametric range of θ^3 while allowing $\mathbf{D}$ to have its physical length.

In a deformed configuration the geometry is described as

$$\mathbf{x}(\theta^1,\theta^2,\theta^3) = \hat{\mathbf{x}}(\theta^1,\theta^2) + \theta^3\mathbf{d}(\theta^1,\theta^2) \tag{4.55}$$

The difference $\mathbf{u} = \mathbf{x} - \mathbf{X}$ gives the deformation from the reference to the current configuration. Following (1) and (2), the displacement is

$$\mathbf{u}(\theta^1,\theta^2,\theta^3) = \hat{\mathbf{u}}(\theta^1,\theta^2) + \theta^3\Delta\mathbf{d}(\theta^1,\theta^2) \tag{4.56}$$

where $\hat{\mathbf{u}} = \hat{\mathbf{x}} - \hat{\mathbf{X}}$ is the displacement of the reference surface and $\Delta\mathbf{d}$ denotes the director displacement.

For further development it proves to be convenient to write the strain variables in the curvilinear coordinate system. The basis vectors are

$$\begin{aligned} \mathbf{G}_\alpha &= \tfrac{\partial \mathbf{x}}{\partial \theta^\alpha} = \hat{\mathbf{G}}_\alpha + \theta^3\mathbf{D}_{,\alpha}\ \alpha = 1,2 \\ \mathbf{G}_3 &= \tfrac{\partial \mathbf{x}}{\partial \theta^3} = \mathbf{D} \end{aligned} \tag{4.57}$$

where $\hat{\mathbf{G}}_\alpha = \hat{\mathbf{X}}_{,\alpha}$ are the tangent vectors in the reference surface. The corresponding basis vectors in the current configuration are denoted by lower case letters, as $\mathbf{g}_\alpha$ $(\alpha = 1, 2)$ and $\mathbf{g}_3$. Let $\mathbf{G}^i$ denote the contravariant basis such that $\mathbf{G}^i \cdot \mathbf{G}_j = \delta^i_j$; we can write the deformation gradient as

$$\mathbf{F} = \mathbf{g}_\alpha \otimes \mathbf{G}^\alpha + \mathbf{g}_3 \otimes \mathbf{G}^3 \tag{4.58}$$

The Green–Lagrangian strain tensor is

$$\mathbf{E} = \frac{1}{2}(\mathbf{F}^T\mathbf{F} - \mathbf{I})$$

Let us denote the components by $\mathbf{E} = E_{\alpha\beta}\mathbf{G}^\alpha \otimes \mathbf{G}^\beta + E_{\alpha 3}(\mathbf{G}^\alpha \otimes \mathbf{G}^3 + \mathbf{G}^3 \otimes \mathbf{G}^\alpha) + E_{33}\mathbf{G}^3 \otimes \mathbf{G}^3$, then

$$\begin{aligned}
E_{\alpha\beta} &= \tfrac{1}{2}(\hat{\mathbf{g}}_\alpha \cdot \hat{\mathbf{g}}_\beta - \hat{\mathbf{G}}_\alpha \cdot \hat{\mathbf{G}}_\beta) + \tfrac{\theta^3}{2}(\hat{\mathbf{g}}_\alpha \cdot \mathbf{d}_{,\beta} + \hat{\mathbf{g}}_\beta \cdot \mathbf{d}_{,\alpha} - \hat{\mathbf{G}}_\alpha \cdot \mathbf{D}_{,\beta} - \hat{\mathbf{G}}_\beta \cdot \mathbf{D}_{,\alpha}) \\
&\quad + \tfrac{(\theta^3)^2}{2}(\mathbf{d}_{,\alpha} \cdot \mathbf{d}_{,\beta} - \mathbf{D}_{,\alpha} \cdot \mathbf{D}_{,\beta}) \\
E_{\alpha 3} &= \tfrac{1}{2}(\hat{\mathbf{g}}_\alpha \cdot \mathbf{d} - \hat{\mathbf{G}}_\alpha \cdot \mathbf{D}) + \tfrac{\theta^3}{2}(\mathbf{d} \cdot \mathbf{d}_{,\alpha} - \mathbf{D} \cdot \mathbf{D}_{,\alpha}) \\
E_{33} &= \tfrac{1}{2}(\mathbf{d} \cdot \mathbf{d} - \mathbf{D} \cdot \mathbf{D})
\end{aligned} \tag{4.59}$$

Here $E_{\alpha\beta}$ is the surface strain in a laminar plane (i.e., θ^3 = const.), $E_{\alpha 3}$ is the transverse shear, and E_{33} is the thickness strain. These strains can be further reduced if additional kinematic assumptions are incorporated. If the director remains normal to the surface (the Kirchhoff shell), one has

$$\begin{aligned}
E_{\alpha\beta} &= \tfrac{1}{2}(\hat{\mathbf{g}}_\alpha \cdot \hat{\mathbf{g}}_\beta - \hat{\mathbf{G}}_\alpha \cdot \hat{\mathbf{G}}_\beta) + \theta^3(\hat{\mathbf{g}}_{,\alpha} \cdot \mathbf{d}_{,\beta} - \hat{\mathbf{G}}_{,\alpha} \cdot \mathbf{D}_{,\beta}) + \tfrac{(\theta^3)^2}{2}\mathbf{D}_{,\alpha} \cdot \mathbf{D}_{,\beta} \\
E_{\alpha 3} &= \tfrac{1}{2}(\hat{\mathbf{g}}_\alpha \cdot \mathbf{d} - \hat{\mathbf{G}}_\alpha \cdot \mathbf{D})
\end{aligned} \tag{4.60}$$

Here $a_{\alpha\beta} := \hat{\mathbf{g}}_\alpha \cdot \hat{\mathbf{g}}_\beta$ is the component of the surface metric tensor. Likewise, $\kappa_{\alpha\beta} := \hat{\mathbf{g}}_{,\alpha} \cdot \mathbf{d}_{,\beta}$ is the curvature tensor, and $\gamma_\alpha := \hat{\mathbf{g}}\alpha \cdot \mathbf{d}$ the transverse shear strain. The corresponding relative deformation measures are defined as the difference between those of the current and the reference configuration

$$\varepsilon_{\alpha\beta} = \frac{1}{2}(a_{\alpha\beta} - A_{\alpha\beta}),\ \rho_{\alpha\beta} = \kappa_{\alpha\beta} - K_{\alpha\beta},\ \delta_\alpha = \gamma_\alpha - \Gamma_\alpha \tag{4.61}$$

In light of the variables in Equation (4.52) and further ignoring the quadratic term in Equation (4.51), the strain variables reduce to

$$\begin{aligned}
E_{\alpha\beta} &= \varepsilon_{\alpha\beta} + \theta^3\kappa_{\alpha\beta} \\
2E_{\alpha 3} &= \delta_\alpha
\end{aligned} \tag{4.62}$$

In addition, if the director is inextensible, the thickness strain $E_{33} = 0$. Equation (4.52) corresponds to the classical assumption that the laminar strain varies linearly with the shell thickness. It should be noted that this assumption is valid only in the small strain limit.

References

1. Fung YC (1993) Biomechanics: mechanical properties of living tissues, 2nd edn. Springer, New York, NY. p 568
2. Fung YC (1990) Biomechanics: motion, flow, stress, and growth. Springer, New York, NY, p 569
3. Sacks MS (2000) Biaxial mechanical evaluation of planar biological materials. J Elast 61: 199–246
4. Yang WH, Lu CH (1973 March) General deformations of neo-Hookean membranes. J Appl Mech 40:1–12
5. Wineman AS. (1976) Large axisymmetric inflation of a nonlinear viscoelastic membrane by lateral pressure. Trans Soc Rheol 20(2):203–225
6. Choi HS, Vito RP (1990) Two-dimensional stress–strain relationship for canine pericardium. J Biomech Eng 112(2):153–159
7. Sacks MS (2003) Incorporation of experimentally-derived fiber orientation into a structural constitutive model for planar collagenous tissues. J Biomech Eng 125(2):280–287
8. Tong P, Fung YC (1976) The stress–strain relationship for the skin. J Biomech 9(10):649–657
9. Sun W et al (2003) Biaxial mechanical response of bioprosthetic heart valve biomaterials to high in-plane shear. J Biomech Eng 125:372–380
10. Yin FC (1981) Ventricular wall stress. Circ Res 49(4):829–842
11. Chuong CJ, Fung YC (1983) Three-dimensional stress distribution in arteries. J Biomech Eng 105(3):268–274
12. Fung Y, Tong P (eds) (2001) Classical and computational solid mechanics. Advanced series in engineering science. World Scientific, Hackensack, NJ, p 952
13. Sacks MS (1999) A method for planar biaxial mechanical testing that includes in-plane shear. J Biomech Eng 121(5):551–555
14. Humphrey JD, Vawter DL, Vito RP (1987) Pseudoelasticity of excised visceral pleura. J Biomech Eng 109:115–120
15. Chew PH, Yin FC, Zeger SL (1986) Biaxial stress–strain properties of canine pericardium. J Mol Cell Cardiol 18(6):567–578
16. Humphrey JD, Strumpf RK, Yin FC (1992) A constitutive theory for biomembranes: application to epicardial mechanics. J Biomech Eng 114(4):461–466
17. Holzapfel GA (2000) Nonlinear solid mechanics : a continuum approach for engineering. Wiley, Chichester, pp xiv, 455
18. Humphrey JD (2002) Cardiovascular solid mechanics: cells, tissues, and organs. Springer, New York, NY, pp xvi, 757
19. Hildebrand F (1980) Advanced calculus for applications. Prentice-Hall, Englewood Cliffs, NJ
20. Fung YC, Fronek K, Patitucci P (1979) Pseudoelasticity of arteries and the choice of its mathematical expression. Am J Physiol 237(5):H620–H631
21. Humphrey JD (1999) An evaluation of pseudoelastic descriptors used in arterial mechanics. J Biomech Eng 121(2):259–262
22. Holzapfel GA, Gasser TC (2000) A new constitutive framework for arterial wall mechanics and a comparative study of material models. J Elast 61:1–48
23. Sun W, Sacks MS (2005) Finite element implementation of a generalized Fung-elastic constitutive model for planar soft tissues. Biomech Model Mechanobiol 4(2–3):190–199
24. Criscione JC, Douglas AS, Hunter WC (2001) Physically based strain invariant set for materials exhibiting transversely isotropic behavior. J Mech Phys Solids 49(4):871–897
25. Mijailovich SM, Stamenovic D, Fredberg JJ (1993) Toward a kinetic theory of connective tissue micromechanics. J Appl Physiol 74(2):665–681

26. Comninou M, Yannas IV (1976) Dependence of stress–strain nonlinearity of connective tissues on the geometry of collagen fibers. J Biomech 9:427–433
27. Buckley CP, Lloyd DW, Konopasek M (1980) On the deformation of slender filaments with planar crimp: theory, numerical solution and applications to tendon collagen and textile materials. Proc R Soc Lond A372:33–64
28. Farquhar T, Dawson PR, Torzilli PA (1990) A microstructural model for the anisotropic drained stiffness of articular cartilage. J Biomech Eng 112(4):414–425
29. Horowitz A et al (1988) Structural three dimensional constitutive law for the passive myocardium. J Biomech Eng 110:200–207
30. Billiar KL, Sacks MS (2000b) Biaxial mechanical properties of the native and glutaraldehyde-treated aortic valve cusp: Part II – a structural constitutive model. J Biomech Eng 122(4): 327–335
31. Belkoff SM, Haut RC (1991) A structural model used to evaluate the changing microstructure of maturing rat skin. J Biomech 24(8):711–720
32. Lanir Y (1983) Constitutive equations for fibrous connective tissues. J Biomech 16:1–12
33. Lanir Y (1979) A structural theory for the homogeneous biaxial stress–strain relationships in flat collagenous tissues. J Biomech 12:423–436
34. Lanir Y, Lichtenstein O, Imanuel O (1996) Optimal design of biaxial tests for structural material characterization of flat tissues. J Biomech Eng 118:41–47
35. Mendenhall W, Sincich T (1988) Statistics for the engineering and computer sciences. Dellen, San Francisco, CA, p 1036
36. Sacks MS, Smith DB, Hiester ED (1997) A small angle light scattering device for planar connective tissue microstructural analysis. Ann Biomed Eng 25(4):678–689
37. Schmidt DE (2009) Multi-scale biomechanical modeling of heart valve tissue. In: Civil and environmental engineering. Carnegie Mellon University, Pittsburgh, PA, p 156
38. Bonet J, Wood RD (1997) Nonlinear continuum mechanics for finite element analysis. Cambridge University Press, Cambridge, MA
39. Sun W, Sacks MS, Scott MJ (2005) Effects of boundary conditions on the estimation of the planar biaxial mechanical properties of soft tissues. J Biomech Eng 127(4):709–715
40. Wilber JP, Walton JR (2002) The convexity properties of a class of constitutive models for biological soft tissues. Math Mech Solids 7:217–236
41. Ogden R (2003) Nonlinear elasticity, anisotropy, material stability, and residual stresses in soft tissue. In: Ogden R (ed) Biomechanics of soft tissue in cardiovascular system. Springer, New York, NY
42. Schroder J, Neff P (2003) Aspects of formulation of hyperelastic soft tissues with polyconvex anisotropic strain energies. In: Bathe KJ (ed) Second MIT conference on computational fluid and solid mechanics. MIT, Cambridge, MA
43. Lanir Y (1994) Plausibility of structural constitutive-equations for isotropic soft-tissues in finite static deformations. J Appl Mech Trans Asme 61(3):695–702
44. Lanir Y (1996) Plausibility of structural constitutive equations for swelling tissues – implications of the C–N and S–E conditions. J Biomech Eng 118(1):10–6
45. Abovsky M, Lanir Y, Nevo E (1996) Tethering affects the mechanics of coronary capillaries. J Biomech 29(5):597–607
46. Sacks MS, Chuong CJ (1998) Orthotropic mechanical properties of chemically treated bovine pericardium. Ann Biomed Eng 26(5):892–902
47. Sun W, Abad A, Sacks MS (2005) Simulated bioprosthetic heart valve deformation under quasi-static loading. J Biomech Eng 127(6):905–914
48. Hutmacher DW, Goh JC, Teoh SH (2001) An introduction to biodegradable materials for tissue engineering applications. Ann Acad Med Singapore 30(2):183–191
49. Hutmacher DW (2000) Scaffolds in tissue engineering bone and cartilage. Biomaterials 21(24):2529–2543
50. Hutmacher DW (2001) Scaffold design and fabrication technologies for engineering tissues – state of the art and future perspectives. J Biomater Sci Polym Ed 12(1):107–124

51. Sacks MS, Schoen FJ, Mayer JE Jr (2009) Bioengineering challenges for heart valve tissue engineering. Ann Rev Biomed Eng 11:289–313
52. Hildebrand DK et al (2004) Design and hydrodynamic evaluation of a novel pulsatile bioreactor for biologically active heart valves. Ann Biomed Eng 32(8):1039–1049
53. Ramaswamy S et al (2009) The role of organ level conditioning on the promotion of engineered heart valve tissue development in-vitro using mesenchymal stem cells. Biomaterials 31(6):1114–1125
54. Courtney T et al (2006) Design and analysis of tissue engineering scaffolds that mimic soft tissue mechanical anisotropy. Biomaterials 27(19):3631–3638
55. Stankus JJ et al (2006) Microintegrating smooth muscle cells into a biodegradable, elastomeric fiber matrix. Biomaterials 27(5):735–744
56. Stella JA et al (2008) Tissue-to-cellular level deformation coupling in cell micro-integrated elastomeric scaffolds. Biomaterials 29(22):3228–3236
57. Stella JA, Wagner WR, Sacks MS (2010) Scale-dependent fiber kinematics of elastomeric electrospun scaffolds for soft tissueengineering. J Biomed Mater Res A, 93(3):1032–1042
58. Ahmad S, Irons BM, Zienkiewicz OC (1970) Analysis of thick and shell structures by curved finite element. Int J Numer Methods Eng 2:419–459
59. Thomas J, Hughes R, Liu WK (1981) Nonlinear finite element analysis of shells: Part I. Three-dimensional shells. Comput Methods Appl Mech Eng, 26(3):331–362
60. Dvorkin EN, Bathe KJ (1984) A continuum mechanics based four-node shell element for general nonlinear analysis. Eng Comput 1:77–88
61. Hallquist JO, Benson DJ, Goudreau GL (1986) Implementation of a modified Hughes–Liu shell into a fully vectorized explicit finite element code. In: Bergan P (ed) Finite element methods for nonlinear problems. Springer, Berlin, pp 238–297
62. Betsch P, Gruttmann E, Stein E (1996) A 4-node finite shell element for the implementation of general hyperelastic 3D-elasticity at finite strains. Comput Methods Appl Mech Eng 130: 57–79
63. Simo JC, Fox DD (1989) On a stress resultant geometrically exact shell model. Part I: formulation and optimal parameterization. Comput Methods Appl Mech Eng 72: 267–304
64. Simo JC, Fox DD (1989) On a stress resultant geometrically exact shell model. Part II: the linear theory; computational aspects. Comput Methods Appl Mech Eng 73:53–92
65. Parks KC, Stanley GM (1986) A curved C^0 shell element based on assumed natural-coordinate strains. J Appl Mech 53:278–290
66. Simo JC, Fox DD, Rifai MS (1990) On a stress resultant geometrically exact shell model. Part III: computational aspects of the nonlinear-theory. Comput Methods Appl Mech Eng 79:21–70
67. Simo JC, Armero F (1992) Geometrically non-linear enhanced strain mixed methods and the method of incompatible modes. Int J Numer Methods Eng 33:1413–1449
68. Bischoff AM, Ramm AE (1997) Shear deformable shell elements for large strains and rotations. Int J Numer Methods Eng 40(23):4427–4449
69. Andelfinger U, Ramm E (1993) EAS-elements for two-dimensional, three dimensional, plate and shell structures and their equivalence to HR-elements. Int J Numer Methods Eng 36: 1311–1337
70. Betsch P, Stein E (1995) An assumed strain approach avoiding artificial thickness straining for a non-linear 4-node shell element. Commun Numer Methods Eng 11:899–909
71. Betsch P, Stein E (1991) A non-linear extensible 4-node shell element based on continuum theory and assumed strain interpolations. J Nonlinear Sci 6:169–199
72. Naghdi PM (1972) The theory of plates and shells. In: Truesdell C (ed) Handbuch der Physik, vol VIa/2. Springer, Berlin, pp 425–640
73. Simo JC, Rifai MS, Fox DD (1990) On a stress resultant geometrically exact shell model. Part IV: variable thickness shells with through-the-thickness stretching. Comput Methods Appl Mech Eng 81:91–126

74. Simo JC (1993) On a stress resultant geometrically exact shell model. Part VII: shell intersections with 5/6-dof finite element formulations. Comput Methods Appl Mech Eng 108:319–339
75. Ibrahimbegovic A (1997) Stress resultant geometrically exact shell theory for finite rotations and its finite element implementation. Appl Mech Rev 50:199–226
76. Simmonds JG (1985) The strain energy density of rubber-like shells. Int J Solids Struct 21: 67–77
77. Schieck B, Pietraszkicwiez W, Stumpf H (1992) Theory and numerical analysis of shells undergoing large elastic strains. Int J Solids Struct 29:689–709
78. Kim H et al (2007) An experimentally derived stress resultant shell model for heart valve dynamic simulations. Ann Biomed Eng 35(1):30–44
79. Mirnajafi A et al (2005) The effects of collagen fiber orientation on the flexural properties of pericardial heterograft biomaterials. Biomaterials 26(7):795–804
80. Engelmayr GC Jr et al (2003) A novel bioreactor for the dynamic flexural stimulation of tissue engineered heart valve biomaterials. Biomaterials 24(14):2523–2532
81. Kim H et al (2008) Dynamic simulation of bioprosthetic heart valves using a stress resultant shell model. Ann Biomed Eng 36(2):262–275
82. Kim H et al (2006) Dynamic simulation pericardial bioprosthetic heart valve function. J Biomech Eng 128(5):717–724
83. Vu-Quoc L, Tan XG (2003) Optimal solid shells for non-linear analyses of multilayer composites. I. Statics. Comput Methods Appl Mech Eng 192:975–1016
84. Parisch H (1995) A continuum-based shell theory for non-linear applications. Int J Numer Methods Eng 38:1855–1883

Chapter 5
Algorithms for Fluid–Structure Interaction

Sarah C. Vigmostad and H.S. Udaykumar

Abstract The human body presents several fluid–structure interaction (FSI) problems, such as the operation of the heart and its valves, motion of blood cells in the circulation, peristaltic contractions in the gut, vibration of vocal cords, operation of the lungs during breathing, contraction of the urinary bladder, and a host of others. Modeling such problems and devising computational techniques to solve the governing equations is an increasingly popular and powerful way to understand the behavior of these systems in the healthy and pathological states. Fluid–structure interaction problems in different organ systems can present different challenges. In the presence of blood (i.e., in the cardiovascular system), FSI problems are plagued by numerical stiffness arising from added mass effects, while such constraints are absent in the presence of air (as in the respiratory and voice systems). In addition, at large scales, fluid inertia can play a significant role, leading to unsteady, transitional, and weakly turbulent flows. At small scales (as in blood cells), viscous effects assume importance. This chapter provides a survey of some of the important issues that arise in the cardiovascular system when FSI problems are tackled. Three primary techniques are discussed, viz., the immersed boundary approach, the immersed interface approach, and the sharp interface approach. The suitability of these approaches to specific problems is addressed and example calculations are shown to illustrate the state of the art of FSI in the cardiovascular system.

H.S. Udaykumar (✉)
Department of Mechanical and Industrial Engineering, 2408 Seamans Center, College of Engineering, The University of Iowa, 52242, Iowa City, IA, USA
e-mail: hs-kumar@uiowa.edu

K.B. Chandran et al. (eds.), *Image-Based Computational Modeling of the Human Circulatory and Pulmonary Systems*, DOI 10.1007/978-1-4419-7350-4_5,

5.1 Introduction

5.1.1 Key Aspects of Fluid–Structure Interaction Problems

Fluid–structure interaction (FSI) problems [1] arise when fluids interact with immersed bodies or enclosing vessels in such a way that the overall dynamics of the system is described by the coupled solution of the laws of motion governing the solid and fluid subsystems. In essence, the fluid influences the motion and deformation of the solid, and in turn the solid influences the fluid flow. Such situations arise in many fields of engineering and are abound in natural systems. The fluid flows may involve gases in the compressible or incompressible regime with rather weak viscous effects (as in the aerodynamics of a jet wing or the opening behavior of a parachute) or incompressible liquids with significant viscous effects (as in many biomechanics applications including blood flow in a deforming artery, the movement of bacteria using flagella, and the motion of the heart, blood vessels, blood cells, the GI tract, lungs, and prosthetics such as heart valves). Solving fluid–structure interaction problems remains a challenge to computational physicists due to several peculiarities associated with the coupling of the fluid and solid subsystems. The purpose of this chapter is to present the key issues involved, briefly describe some of the choices of methods available, characterize these methods according to their regime of applicability, and present some typical computational results that illustrate important aspects of some of the available methods.

Any approach to solving FSI problems requires the simultaneous solution of the governing equations of the fluid and the solid structure along with the treatment of the surface dividing the two subdomains. Most often this dividing surface itself is to be determined as part of the solution process since its position and configuration are determined by the coupled interaction of the fluid flowfield and the solid deformation. Typically, in the dynamics of membranes, or rigid/deformable bodies of substantial thickness, equilibrium at the solid–fluid interface requires the enforcement of kinematic compatibility (i.e., no-slip and no-penetration for viscous fluids) and continuity of tractions at the fluid–structure interface. Therefore, a complete solution of the FSI problem must evolve the fields in the fluid and solid subdomains in concert with the motion of the interface separating the subdomains. While the fluid and solid mechanical equations individually are nonlinear, the coupling at the interface introduces yet another source of nonlinearity, leading to a problem that is quite challenging to solve. Frequently, the constraints posed for the solution process, e.g., time step sizes for stable computations, can be more strongly imposed by the interfacial dynamics than by the dynamics of the bulk fluid flow or solid stress fields. Therefore, the presence of the immersed boundary necessitates substantial modifications to the computational techniques used for handling the mechanics of the bulk media.

Computational FSI approaches in large part differ on three primary grounds:

1. The way in which the fluid and solid subdomains are discretized and solved. Under this broad issue the various classes of body-conforming, moving mesh methods [2, 3] (finite volume, finite element, ALE) and fixed grid (immersed boundary methods [4, 5]) can be identified.
2. The methods used to keep track of and advance the solid–fluid interface. Here, one can think of methods that explicitly track the interface using points and surface meshes in a Lagrangian framework or the alternative approach of using implicit surface representations (such as level sets [6]) in an Eulerian framework.
3. The methods used to enforce the coupling of the fluid and solid solutions. Here, the choices are to formulate loosely coupled algorithms [7] (where the subsystems are evolved in a sequence during each time step) or strongly coupled algorithms (where the subsystems are evolved simultaneously in some form of iterative process).

In this chapter, we will discuss and illustrate with some example calculations that the choice of method in its three essential aspects listed above is dependent on the nature of the problem, its inherent (numerical) "stiffness," the type of solver chosen for the fluid and for the structure, as well as on other limitations [8, 9]. Choosing a discretization scheme is usually dictated by the particular expertise of the researcher, although depending on the problem some choices are perhaps better than others. While some flexibility is available regarding the choice of methods to solve the solid and fluid equations in their individual subdomains and how to keep track of the interface deformation and location, as will be shown, the coupling method has a considerable impact on the convergence and stability of the FSI solution. Therefore, a careful choice of the coupling of fluid and solid becomes necessary. The key issues underlying the choice of fluid–solid coupling technique are presented in this chapter.

5.2 Governing Equations and Important Parameters

Throughout this chapter, because we are focused on FSI problems in biomedical systems, fluid governing equations are assumed to be described by the incompressible Navier–Stokes equations for fluid flow:

$$\vec{\nabla} \cdot \vec{u} = 0 \tag{5.1}$$

$$\rho \left(\frac{\partial \vec{u}}{\partial t} + \vec{u} \cdot \vec{\nabla} \vec{u} \right) = -\vec{\nabla} p + \mu \nabla^2 \vec{u} + \vec{F} \tag{5.2}$$

Here, $\vec{u}$ is the velocity field, ρ and μ are the density and viscosity of the fluid, p represents the fluid pressure, and $\vec{F}$ is a volumetric force (body force) that provides a local source of momentum to the fluid. Nondimensionalization is performed using the scales that apply to the particular system being investigated. In effect, choosing representative scales for the variables that appear in the above equations, one can then write in nondimensional form (dropping the customary asterisks on nondimensional quantities):

$$\vec{\nabla} \cdot \vec{u} = 0 \tag{5.3}$$

$$\frac{\partial \vec{u}}{\partial t} + \vec{u} \cdot \vec{\nabla} \vec{u} = -\vec{\nabla} p + \frac{1}{\mathrm{Re}} \nabla^2 \vec{u} + \vec{f} \tag{5.4}$$

where Re is the Reynolds number, computed from the characteristic velocity (U), length scale (L), density ρ_f, and viscosity μ: ($\mathrm{Re} = \rho_f UL/\mu$).

The governing equations for a solid capture the deformation of the solid under the action of fluid forces. Again, due to the focus on biomedical systems, the solid in the present work is modeled in the form of a nonlinear elastic material and is governed by the force balance:

$$\frac{\rho_s}{\rho_f} \ddot{\vec{x}} - \mathrm{Div} \underline{\boldsymbol{\sigma}}_{\mathrm{s}} = \vec{b} \tag{5.5}$$

where ρ_{s} is the density of the solid, $\ddot{\vec{x}}$ is acceleration of the solid, $\underline{\boldsymbol{\sigma}}_{\mathrm{s}}$ is the stress tensor in the structure, and $\vec{b}$ describes the body forces acting on the structure. Nondimensionalization of the solid equations can be performed with the reference values listed above for the fluid variables. The material properties of the solid material (especially deformable materials such as tissue) are required as inputs to the constitutive model for the solid; one example is the popular hyperelastic (Fung-type) solid model [10] that will be discussed later in this chapter.

The coupling between the fluid and solid subdomain solutions is enforced at the interface Γ_{sf} (see Fig. 5.1c) through the kinematic and dynamic matching conditions specifying the position of the interface on the fluid side, the continuity of velocities at the interface, and the continuity of surface tractions at the interface. Therefore,

Interface position:

$$\varphi(\vec{x}, t) = 0 \equiv \vec{x}_{\mathrm{s}}|_{\Gamma_{\mathrm{sf}}} \tag{5.6}$$

where $\varphi(\vec{x}, t)$ is the level-set field that defines the fluid–solid interface for the fluid flow calculations.

Interface velocity and acceleration:

$$\vec{u}_{\mathrm{f}}|_{\Gamma_{\mathrm{sf}}} = \dot{\vec{x}}_{\mathrm{s}}\Big|_{\Gamma_{\mathrm{sf}}}, \quad \vec{a}_{\mathrm{f}}|_{\Gamma_{\mathrm{sf}}} = \ddot{\vec{x}}_{\mathrm{s}}\Big|_{\Gamma_{\mathrm{sf}}} \tag{5.7}$$

Continuity of tractions:

$$\underline{\sigma}_{\mathrm{s}}\Big|_{\Gamma_{\mathrm{sf}}} \cdot \vec{n} = \underline{\sigma}_{\mathrm{f}}\Big|_{\Gamma_{\mathrm{sf}}} \cdot \vec{n} \tag{5.8}$$

In Equation (5.8), $\underline{\sigma}$ is the stress tensor, $\vec{n}$ is the local normal to the interface, subscripts s and f indicate the solid and fluid, respectively.

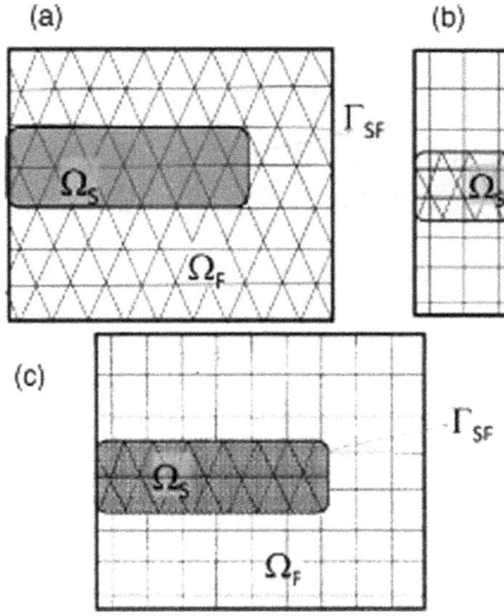

Fig. 5.1 Comparison between alternative approaches for the treatment of fluid flows interacting with embedded solid boundaries. (**a**) A body-fitted grid system where the grid moves with the deforming boundary. (**b**) A fixed grid immersed boundary approach where the fluid flow is solved in the interior of the embedded boundary. (**c**) A sharp interface approach where the flowfield is computed in the fluid subdomain on a fixed Cartesian grid, the solid solver operates in a finite element setting in the solid subdomain, and appropriate interface conditions are used on the solid surface to connect the two subdomains

The previous equations may appear to represent a rather limited setting for fluid–structure interaction problems, but in reality encompass a wide range of biologically relevant FSI problems. Examples include the deformation of cells [11–13], the pumping action of the heart [14], deformation of blood vessels [15], the operation of heart valves [16, 17], locomotion of cells due to flagellar motion [18, 19], and a host of others. The previous system of equations also provides a platform to examine general concepts and challenges associated with solving coupled fluid–structure interaction problems.

5.3 Spatial Discretization to Couple Fluid and Solid Dynamics

While many researchers have investigated various approaches to solving fluid–structure interaction problems, including conditions for stability, energy conservation issues, and the related convergence of matrix solvers, the majority of these investigations and techniques are applied to aeroelastic (the coupling between aerodynamics and flexible structures) problems. These methods relied on boundary-fitted moving meshes, whether structured or unstructured [3, 20, 21]. Many of the lessons learned in the aeroelastic problems can, in part, translate into viscous, incompressible flow problems interacting with highly deformable, soft structures. However, the latter category of problems introduces its own set of unique challenges. For example, in aerodynamics problems, the flow speeds are large enough that the flow lies in the transonic regime, and the structural motion and/or deformation of interest, such as in the phenomenon of flutter, is quite modest under normal conditions. In most biomedical FSI problems, the flow speeds are quite low, viscous effects may be significant, and the structures may be highly deformable. Therefore, the demands placed on the FSI calculations in such systems can be quite different from those in aeroelasticity. Numerous researchers have analyzed flows interacting with highly deformable structures in the setting relevant to biological systems. A discussion of several approaches follows, including advantages and shortcomings of the different approaches.

5.4 ALE-Type Methods

In Arbitrary Lagrangian–Eulerian (ALE) methods, a body-fitted mesh is employed, which moves according to the computed or prescribed deformation of the interface. While the fluid mesh motion is dependent upon the structural deformation, the structural dynamics is computed independently using a Lagrangian framework. While this method has been used extensively for moving rigid bodies, applications involving large deformation, particularly deformable structures and complex geometries, are quite difficult with the ALE formulation [3, 20, 22, 23]. As the boundary moves and the mesh deforms accordingly, high grid velocities can lead to skewing of the mesh, which has been shown to adversely affect accuracy [24, 25]. To avoid mesh distortion, frequent remeshing is necessary, which leads to increased computation times and can introduce artificial diffusivity. Particularly in 3D, this approach is not attractive for high deformation simulations since the task of generating and maintaining a "nice" mesh can occupy a considerable amount of computational resources.

5.5 Immersed Boundary Method

In contrast to the moving mesh approach, *fixed grid* methods can circumvent the problem of generating meshes to conform to the deforming structure. In this context, the immersed boundary (IB) method was first introduced by Peskin in [14] as a numerical method to study fluid dynamics in the heart and its valves. In this approach, moving elastic or viscoelastic interfaces can be embedded within a viscous incompressible fluid domain. The fluid variables are defined in an Eulerian framework, while the interface is described using a Lagrangian reference frame. The interfacial effects are transmitted to the fluid through the source term $\vec{f}$ in the momentum equation.

Equations (5.7) and (5.8) describe the Navier–Stokes equations for a viscous, incompressible fluid. The effect of the immersed boundary is communicated to the flow as a "body force":

$$\mathbf{f}(\mathbf{x},t)=\int_{\sigma}\mathbf{F}(s,t)\delta(\mathbf{x}-\mathbf{X}(s,t)\mathrm{d}s \tag{5.9}$$

Equation (5.9) uses a numerical delta function to describe the interaction between the membrane and the fluid. Once the fluid velocity is found, kinematic compatibility is applied so that the boundary moves with the fluid velocity:

$$\frac{\partial \mathbf{X}(s,t)}{\partial t}=\mathbf{u}(\mathbf{X}(s,t),t)=\int_{\Omega}\mathbf{u}(\mathbf{x},t)\delta(\mathbf{x}-\mathbf{X}(s,t))\mathrm{d}\mathbf{x} \tag{5.10}$$

In these equations, $\mathbf{u}(\mathbf{x},t)$ is the fluid velocity, $p(\mathbf{x},t)$ the fluid pressure, and ρ and μ are the density and viscosity of the fluid, respectively. The force per unit volume exerted onto the fluid by the membrane is described as $\mathbf{f}(\mathbf{x},t)$, while the membrane geometry is represented by parameterization, viz., $\mathbf{X}(s,t)$ using arc-length parameter s, where $0 \le s \le L_s$.

The fluid domain is typically described using a fixed Cartesian grid (recent papers have investigated the use of an adaptive grid [26] as well as triangulations and prismatic grids), while the immersed structure is represented using finite element meshes. By employing a smoothed Dirac delta function to represent the effects of the deforming elastic interface, the immersed boundary method leads to a smearing of the interface region. The elastic forces are spread over several mesh points [4].

Peskin and colleagues have employed the IB method to study the FSI solutions for native and prosthetic heart valves [27–29]. This study was extended to the whole heart [30], as well as to other cardiovascular regions [31, 32]. The limitations of this method lie primarily in the treatment of the interfacial forces. The interface is smeared, and discontinuities cannot be captured. In this way, accuracy at the interface can be adversely affected, which can influence the fluid solution as well as the movement of the interface [4, 33, 34]. Furthermore, the method is restricted to treating immersed interfaces that occupy no volume in the fluid domain [35]. The majority of work done with the IB method treats the structural motion explicitly, leading to high stiffness and time step limitations. Stockie and Wetton examined the numerical stiffness of the IB method [36], comparing time-stepping schemes in an attempt to address the time step limitations encountered by typical IB approaches. The stiffness of the method is explained primarily by the relatively high structural force combined with low fluid viscosity, characteristic of the type of simulations utilized by the IB method. Tu and Peskin [37] implemented a fully implicit solver to overcome these limitations. While they achieved stability, their computations were too expensive for practical use. Stockie and Wetton compared iterative schemes with the explicit approaches and found that while iterative schemes relieved the time step restrictions, they led to increased volume loss when studying closed-boundary (i.e., an ellipse relaxing to a circle at steady state) simulations. This volume loss can be attributed to the lack of a divergence-free velocity field near the interface [36]. Later versions of the immersed boundary method have attempted to alleviate some of these difficulties associated with the classical Peskin approach with some success [38–40].

5.6 Immersed Interface Method

To overcome the limitations of the IB method, Leveque and Li [41, 42] developed the immersed interface method (IIM), which was later applied to the Navier–Stokes equations by Lai and Li [43, 44]. This method begins with Equations (5.9) and (5.10) where the interfacial effects are incorporated into the momentum equation as a source term, **f**. However, the goal of the approach is to preserve discontinuities and treat the interface sharply (referred to as a *sharp interface model*). To achieve this, the interfacial force is incorporated into a series of jump conditions applied on the interface. These jump conditions preserve the discontinuities in velocity and pressure and their derivatives across the interface.

Equations (5.9) and (5.10) in the immersed boundary method are described in terms of "mollified" delta functions, resulting effectively in "smearing" of the interface and may be classified as a "diffuse interface approach." The desire, in the IIM context, is to remove these smoothed delta functions in order to retain jump discontinuities as such, thereby conforming to a "sharp interface approach." This is performed using jump conditions that apply at the precise interface locations. A jump $[\varphi]$ describes the difference of the value of the variable φ from one side (+) of the interface to the other (−):

$$[\varphi] = \varphi^+ - \varphi^- \tag{5.11}$$

The first jump condition simply describes the no-slip condition—the membrane moves at the local fluid velocity and thus:

$$[\mathbf{u}] = 0 \tag{5.12}$$

A pressure jump across a boundary must be balanced with internal forces within the interface itself. This is described in the following equations:

$$[p] = f_\mathrm{n} \tag{5.13}$$

$$[p_\mathrm{n}] = \frac{\partial f_\mathrm{t}}{\partial s} \tag{5.14}$$

Here, f_n is the strength of the normal force residing on the interface, and f_t is the corresponding tangential force strength.

Finally, the momentum flux jump balances the viscous shear stress on the interface with the tangential forces at the boundary.

$$\mu\left[\frac{\partial \mathbf{u}}{\partial \mathbf{n}}\right] = -f_t\tau \tag{5.15}$$

where $\tau = (-\sin\theta, \cos\theta)$ is the unit tangent direction. The IIM has been applied to simulate moving membranes and other zero-thickness applications where it is reasonable to assume a constant viscosity and negligible inertial effects. However, issues of stability and stiffness remain in the case of the IIM approach as well. When the jumps in the interfacial momentum and energy balances become large the explicit time-stepping schemes typically employed with the immersed boundary methods face increasingly severe restrictions on allowable spatial and temporal step sizes. Another type of method that relies on application of jump conditions in the discretized equations of fluid flow that is similar to the IIM is the ghost fluid method (GFM [34, 45]). This method has also been applied to simulate the dynamics of two-fluid systems (droplets, bubbles, etc.) and can be employed for computing the dynamics of membranes. While the sharp treatment of the immersed boundaries (membranes) is certainly a hallmark of such methods, the presence of highly localized large source terms on a small set of mesh points may lead to spurious flows and stability issues with such methods. This becomes particularly evident when the fluid viscosity decreases and cannot suppress the growth of perturbations at the interface.

5.7 Sharp Interface Method

In fixed grid methods, it is desirable to retain a sharp interface description (in contrast to the immersed boundary method) while avoiding the placement of jump-capturing source terms on the mesh. This can be done if the fluid calculation is performed separately from the solid deformation calculations and the two fields are matched at the solid–fluid boundary through compatibility conditions (Equation 5.7). In this way, the stresses in the solid are not placed in the fluid momentum equations as in the IBM or IIM approaches. This type of technique is traditional in fluid–structure interactions in aeroelasticity and in general in boundary conforming mesh or ALE-type methods. In the context of fixed grid methods, such a sharp interface method has been presented in several papers by Udaykumar and coworkers. The sharp interface method has the following features:

1. Fluid and solid fields are entirely separated, with a fixed Cartesian-grid finite volume approach for the fluid and a finite element approach for the solid. The two fields are matched at the solid–fluid interface. This leads to the so-called partitioned or segregated FSI model.
2. No source terms are placed on the fluid mesh.
3. The immersed boundary does not have to be thin.

A straightforward implementation of the sharp interface method for a variety of moving boundary problems has been described in several papers by Udaykumar and coworkers [17, 33, 46–50]. In the sharp interface method, the interfaces are represented by level sets [51]. The level-set field ϕ represents the normal distance from the interface and is positive outside and negative inside the immersed object. The interface location is implicitly embedded in the ϕ-field since the $\varphi = 0$ contour represents the immersed boundary. This method allows a natural representation and computation of complex objects and geometrical characteristics such as normal and curvature on Eulerian mesh in 2D or 3D [33]. In the case of moving interfaces, the motion of the boundary is tracked by advecting the level set using:

$$\varphi_t + \vec{V}.\vec{\nabla}\varphi = 0 \tag{5.16}$$

where $\vec{V}$ is the level-set velocity field calculated from the physics of the problem at hand. It is noted that while use of the level-set interface representation facilitates the implementation of the sharp interface approach, other ways of representing and tracking interfaces (such as through markers and curves/surfaces) can also be employed.

The basic idea in the sharp interface approach is to obtain discrete forms of the fluid flow equations in a "redefined" fluid domain that does not conform to the Cartesian mesh, but instead is defined by the zero-level contour of the level-set field. Thus, as seen in Fig. 5.1, the flow geometry is irregular in shape and the flowfield is computed in this irregularly shaped domain. In the presence of moving boundaries, the computational mesh is fixed in space and therefore the issue of generating a suitable mesh as the boundary moves is side-stepped. The challenge in this framework is to convey to the discretization scheme the shape of the embedded solid boundaries at which boundary conditions on the flowfield should be applied. A simplifying feature is that, in a finite-difference context, the computational nodes where these boundary effects must be included form a lower-dimensional set in the overall flow domain. At points located away from the embedded boundaries, the discretization remains simple. Figure 5.1a shows the demarcation of "bulk" nodes (i.e., those away from the interface) and "interface" nodes (i.e., points located adjacent to the interface) based on the level-set field ϕ. Figure 5.1c shows details of discretization about an interface point i. As seen in Fig. 5.1c, point $i+1$ is on the other side of the interface and will not be included in the discretization at point i. Instead, boundary conditions at the interface point I will be included.

These basic ideas can be illustrated for the general transport equation:

$$\frac{\partial \psi}{\partial t} + \vec{u} \cdot \vec{\nabla}\psi = \alpha \nabla^2 \psi \tag{5.17}$$

where ψ is a variable that may represent velocity, temperature, or species and α is a diffusion coefficient. A second-order central difference scheme is used for discretization of bulk points, i.e., those that do not lie next to an interface (Fig. 5.1b). The standard central difference scheme in 1D for the discretization of the Laplace operator for a function ψ at point i is

$$\frac{\partial^2 \psi}{\partial x^2} = \frac{\psi_{i+1,j} - \psi_{i,j}}{\Delta x^2} - \frac{\psi_{i,j} - \psi_{i-1,j}}{\Delta x^2} \tag{5.18}$$

The discretization for an interfacial point (with reference to Fig. 5.1c) is

$$\frac{\partial^2 \psi}{\partial x^2} = \frac{2}{\chi(1+\chi)} \frac{(\psi_I - \psi_{i,j})}{\Delta x^2} - \frac{2}{(1+\chi)} \frac{(\psi_{i,j} - \psi_{i-1,j})}{\Delta x^2} \tag{5.19}$$

$$\chi = \frac{\Delta x_I}{\Delta x} \cong \left| \frac{(\phi)_{I_x} - (\phi)_{i,j}}{(\phi)_{i+1,j} - (\phi)_{i,j}} \right| = \left| \frac{0 - (\phi)_{i,j}}{(\phi)_{i+1,j} - (\phi)_{i,j}} \right| \tag{5.20}$$

The discrete form at the interfacial points differs from the bulk points through the introduction of parameter χ into the discretization (Fig. 5.2). This parameter is dependent on the level-set field and represents the distance of the point i from the interface as shown in Equation (5.19). ψ_I represents the interfacial value obtained from the interface boundary conditions. The relevant boundary conditions are applied at the interface depending on the type of interface and the physics of the problem. While Dirichlet conditions are straightforward to apply, Neumann conditions (as apply for the pressure Poisson equation) need to be judiciously constructed. One way of imposing Neumann condition was presented in Marella et al. [33]. An improved approach is presented in Marella et al. [50].

The sharp interface method allows boundary conditions to be applied in a sharp fashion at the interfaces without any smearing of the interface or sacrificing solution accuracy. Since the interface conditions for the fluid subdomain amount to the specification of position and velocity of the solid surface only, this allows for flexibility in

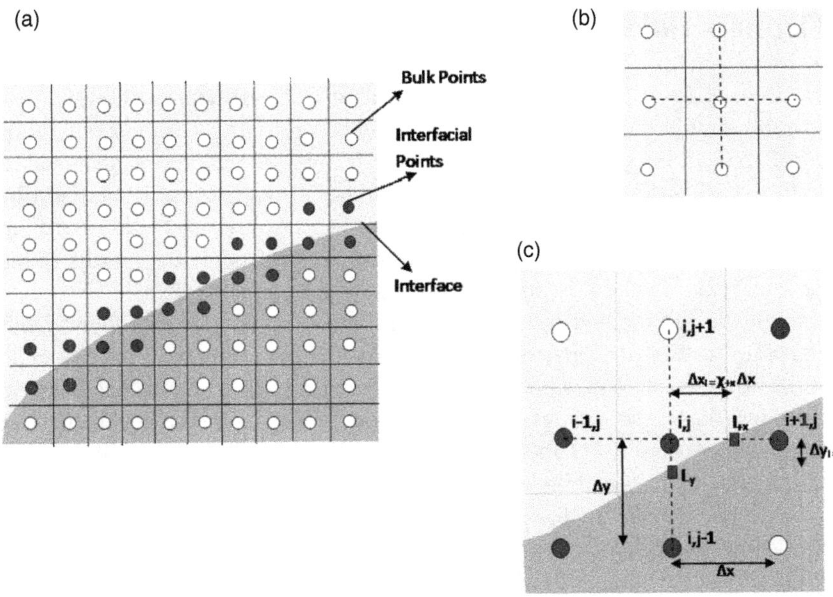

Fig. 5.2 (**a**) Definition of the bulk (*clear circles*) and interfacial (*filled circles*) points. The interface is given by the zero contour of the level-set field. (**b**) Standard 5-point bulk stencil in two dimensions. (**c**) The configuration of a typical interfacial point

treating the solid subdomain as a totally separate entity. Once the solid deformation is computed, the rate of displacement of the solid provides the boundary conditions to be applied on the fluid domain. Furthermore, the fluid side solution is obtained independently of the nature of the solid, i.e., whether it is rigid or deformable. Therefore, in the sharp interface treatment the facility to use well-developed stand-alone solvers on the fluid and solid side exists, a feature that will be exploited in the simulations of heart valve dynamics that will be shown as examples later in this chapter.

5.8 Finite Element Methods

Several numerical techniques have been developed for fluid–structure interaction applications in which the fluid and structural equations are discretized using finite elements. In these methods, the interfacial effects are not incorporated in the momentum equation, but rather are implemented in the discrete form of the Navier–Stokes equations. Some of the frequently adopted methods are discussed below.

5.9 Fictitious Domain Method

The fictitious domain (FD) method is a monolithic finite element method that has been used to simulate large deformation structures interacting with viscous fluids [16, 22, 52, 53]. In the FD approach, a fixed mesh Eulerian reference frame is used to describe the fluid domain, while a Lagrangian mesh is used to track and describe the structure. In this approach, Lagrange multipliers are used to couple the fluid and structure, enforcing the no-slip condition on the interface. Due to limits of the approach, inertial effects of the structure are assumed to be negligible.

Although the FD method has been used to approximate a bioprosthetic heart valve in both 2D and 3D, physiologic Reynolds numbers (Re) have not yet been achieved. A finer mesh than was computationally reasonable would be necessary to perform physiologic simulations [54]. This shortcoming, combined with the fact that the accuracy of the FD method is lower at the interface, limits its usefulness in understanding the complex stresses experienced by the valve leaflet [55]. Furthermore, because this method treats the FSI solution monolithically, it does not have the desired flexibility that partitioned methods provide. De Hart et al. described high variability in the magnitude of the matrix components caused by the differences in material properties between the fluid and the solid. As noted, a poorly conditioned matrix can lead to difficulties in convergence of the system [53].

5.10 Immersed Finite Element Methods

In an effort to simulate immersed, deformable bodies such as red blood cells (RBCs), proteins, and other geometries occupying a nonnegligible volume, the immersed finite element method (IFEM) was developed as an extension of the IB method. Like the fictitious domain method, this approach operates using finite element discretization, where the fluid mesh is a fixed Eulerian grid framework, and the structure is represented using an overlying Lagrangian mesh. Inertial effects due to the structural motion can be incorporated, and fluid–structure coupling is performed in a fashion similar to that of the IB method, where smoothing functions transmit the structural forces to the fluid [56–58].

In this approach, the Eulerian mesh occupies the entire domain, regardless of the structural volume. Due to the limitations of the interpolation and coupling techniques, the structural mesh is required to be at least twice as dense as that of the surrounding fluid mesh. The fluid–structure coupling is performed explicitly, such that the structural deformation is computed based upon the fluid solution at the previous time step [58]. As previously discussed, this type of coupling severely limits the allowable time step size [9].

This method has been used to model deformable, volumetric structures in viscous fluids, including RBC aggregation, cell–cell interaction, and the motion of deformable spheres [56–58]. While the IFEM approach improves upon some of the limitations of the IB method, the use of smoothing functions at the interface, combined with time step limitations, are both limitations that are similar to the immersed boundary method.

5.11 Issues Related to the Temporal Update of the Coupled Fluid–Solid System

The development of a computer model capable of simulating interactions of tissue, organs, or engineered prostheses with fluids (such as blood) must contend with a unique set of challenges depending on the physical problem being solved. For example, a method that may be suitable to study the deformation of membrane-enclosed fluid sacs (cells) with low to modest inertial effects may not be suitable or may need to be revamped considerably to handle the deformation of a rather thick, multilayered arterial wall in the presence of near-transitional blood flow. To judiciously choose the appropriate solution technique, a researcher needs to understand the difficulties involved in discretizing and numerically solving the governing equations. In particular, it is important to understand the particular set of challenges entailed by the combined solution of the fluid and solid subsystems (i.e., as distinguished from the standard solution procedures for each subsystem in isolation). Some of these peculiarities of FSI modeling are outlined below.

5.12 Numerical Stiffness

Obtaining solutions to FSI problems can be thought of as solving a system of coupled differential equations. The general form of the discrete forms for solving the equations in each subdomain can be cast in matrix form as

$$A\varphi = b \tag{5.21}$$

Here, A is the matrix of coefficients that determine the evolution of the field φ that is defined at the discrete finite difference/finite element nodes and b contains source terms and boundary contributions. Numerical solutions obtained by solving the above discrete system are inevitably accompanied by errors in the solution. Such errors are unavoidable in any approximation of the continuous fields in the solid and fluid subdomains by a finite set of computational nodes at which the continuous functions may be approximated. These errors may be due to round-off (resulting from finite computer precision) or may be particular to the numerical method chosen (for example, due to truncation errors). If these errors grow over the course of the solution process, the result is said to be unstable [59]. While stability of numerical schemes is of concern in solving the governing equations in a single field system (i.e., solid/fluid alone), a particular challenge in developing stable numerical techniques for FSI problems is that of "stiffness." Numerical stiffness accompanies differential equations associated with two or more disparate time scales. In numerically stiff problems, stability is only maintained when the time step size chosen to solve the problem is such that the fastest responses can be captured during the time evolution of the system. The shortest time scale dictates the time step size permissible. Typically, phenomena at the long time scales are of interest as well, and therefore the number of time steps required depends on the ratio of the slow to fast response times, leading to large computation times when the two time scales are disparate.

The fluid response time scale can be thought of as

$$t_{\mathrm{f}} = \frac{L}{U_0} = \sqrt{\frac{L^2 \rho_{\mathrm{f}}}{\Delta p}} \tag{5.22}$$

where L is the length scale, U_0 the characteristic velocity, ρ_{f} the fluid density, and Δp the pressure difference. The structure responds in time scales of the order:

$$t_{\mathrm{s}} = \sqrt{\frac{hL\rho_{\mathrm{s}}}{E}} \tag{5.23}$$

where h is the thickness of the structure, ρ_{s} the structural density, and E the elastic modulus of the structure. From here, the ratio of fluid response time to structural response time can be considered:

$$\frac{t_f}{t_s} = \sqrt{\frac{L}{h}}\sqrt{\frac{E}{\Delta p}}\sqrt{\frac{\rho_f}{\rho_s}} \tag{5.24}$$

As an example, the ratio of time scales for simulating a bioprosthetic heart valve can be examined by plugging physiologic values into Equation (5.24). When these values [60] are used, the ratio of time scales is roughly 40, implying that the structure responds 40 times more rapidly than the fluid. For this reason, even very small changes in the fluid can lead to significant changes in the structure, reflecting the numerical stiffness problem. In fact, there are two situations in which the time scale ratio above can be large. The first is for a system for which the material stiffness is large, i.e., E has a large value; the second is when the ratio $\frac{\rho_f L}{\rho_s h}$ is large. In this latter case, the numerical stiffness is caused by the so-called strong added mass effect; the fluid density is comparable to the solid density or the solid is very slender (h/L is very small) so that during the course of its motion the solid has to displace a large mass of fluid, i.e., the added mass of the fluid set in motion by the solid boundary is large.

Alternately, from a strictly mathematical viewpoint, stiffness is identified by examining the Jacobian of the coupled partial differential equation system governing the FSI problem. The overall coupled solution pertains to a system of algebraic equations obtained from the discretization:

$$\begin{pmatrix} A_f & A_{fI} \\ A_{sI} & A_s \end{pmatrix} \begin{bmatrix} \varphi_f \\ \varphi_s \end{bmatrix} = \begin{bmatrix} b_f \\ b_s \end{bmatrix} \tag{5.25}$$

Here, A is a submatrix of coefficients, φ is the field variable, and b is the source vector, with subscripts f, s, and I implying the fluid, solid, and interface, respectively. Therefore, in analogy with Equation (5.21), which applies to a single field, in Equation (5.25) for a coupled FSI problem the overall coefficient matrix is comprised of submatrices pertaining to the solid alone (A_s), fluid alone (A_f), and transfer terms from solid to fluid (A_{fI}) and fluid to solid (A_{fI}) at the interface. If the eigenvalues of the overall coefficient matrix differ greatly in magnitude, the system is said to be stiff. Clearly, when dealing with stiff systems, the method used for temporal discretization becomes very important and has a significant impact on the stability of the solution. If an inappropriate discretization scheme is chosen, even small perturbations can lead to catastrophic instabilities [61]. Typically, for stiff systems of equations, stability constraints can be surmounted by adopting an implicit-in-time discretization or by solving the coupled system in a block-wise (*monolithic*) fashion, i.e., by solving the coupled fluid–solid equation system simultaneously [9, 62, 63]. In practice however, devising a fully implicit approach for fluid–solid coupling may not be an attractive proposition, because the modeler may desire to use well-developed fluid and solid solvers that have been developed independently as mature pieces of software. In this case, the issue of stiffness must be addressed as it will limit the allowable time step for computation of the coupled system. In addition, even if a monolithic solver were developed, under certain conditions (arising from material properties of the solid and fluid), the matrix A can be ill-conditioned and therefore solution of the system is severely hindered. Unfortunately, the conditions under which such problems arise apply quite frequently to problems of biomedical interest and are described below.

5.13 Material Density and Slenderness

The behavior of an FSI system can depend on the geometric characteristics of the structure and fluid domain. For example, the solid structure may be rather slender as in the case of a tissue heart valve, or it may possess substantial thickness and material strength as in the case of large organs such as the heart or stomach. The fluid domain in turn may be a narrow convoluted conduit as in the case of minor blood vessels, or it may be spacious as in the case of the stomach. The considerations pertaining to handling the interaction between the fluid and the solid are determined by these factors as well. Causin et al. [64] examined explicit versus implicit partitioned FSI algorithms by studying a simplified 2D model of the interaction between blood and an artery wall. The propagation of a pressure wave through an incompressible fluid with thin, deformable walls was examined, and it was found that a loosely coupled method exhibited numerical instabilities triggered by two factors: the density of the structure and the aspect ratio of the geometry. For a given geometry, instabilities arose when the density of the structure decreased below a certain threshold value. Likewise, while keeping the density constant, if the length of the domain (i.e., the artery) was extended beyond some threshold, instabilities were observed. The same experiments were then performed using a strongly coupled algorithm in which at every time step, subiterations of the fluid and structure solutions were performed in an effort to achieve a balance of energy at the interface. During the subiteration process, relaxation was employed to limit fluctuations of the solution and help enhance stability of the implicit scheme. Similar to the results of the loosely coupled scheme, it was found that heavier underrelaxation was necessary when the structural density decreased, and likewise when the domain was slender.

Ultimately, a relationship was established between the fluid and structural densities (ρ_f and ρ_s, respectively), a geometrical component related to the aspect ratio (μ_{max}, which increases as R/L, the ratio of the radius to the length, decreases), and the structural thickness (h_s), which stated that the loosely coupled system was *unconditionally unstable* if

$$\rho_f \mu_{max} > \rho_s h_s \tag{5.26}$$

This was true regardless of the time step size chosen. A similar, if less restrictive, criterion was found for the iterative (strongly coupled) approach to solving the FSI problem. Again, the solution's stability was dictated by the ratio of the fluid to structural densities, and by the geometry of the domain. It was found that an underrelaxation parameter was necessary to ensure convergence, and that parameter was again dependent upon these properties [64]. In some cases, the underrelaxation factor necessary to maintain stability can become quite large and therefore the computations can become fairly onerous.

5.14 Rapidity of Motion and Deformation

In addition to the spatial distribution of the fluid and solid subdomains, the time scales of the interaction are important to consider when devising a simulation procedure. As an example, the heart valves possess highly deformable leaflets, which open and close due to pressure gradients, typically undergoing approximately 75 full open/close cycles each minute. These rapidly rotating leaflets are also continuously undergoing complex deformation, with the shape and profile of the valve constantly changing. The properties of the valve leaflet are highly nonlinear, anisotropic, and require a complex material model to accurately capture the in-plane, bending, and shear stress components attributed to the normal motion of a BHV [65]. Moreover, the leaflets themselves are very thin, occupying a negligible volume and contributing very little inertial effects to the system. Geometrically, this is a challenge to render, and numerically, this further contributes to the highly deformable nature of the valve.

In the initial stages of valve leaflet motion the acceleration of the valve leaflet can be rather large, implying that the added mass effect is significant in the initial stages of opening. Inclusion of the added mass effect has been shown to exacerbate the stiffness problem in FSI problems [64]. To include the added mass effect, when computing the displacement of a structure immersed in a viscous fluid, an accurate computation must take into consideration the inertia of the surrounding fluid. In this way, as an immersed structure moves, it displaces some volume of fluid in doing so. This characteristic can be incorporated numerically as a modification when computing the interfacial pressure. Without the added mass term, the pressure at the interface is generally computed by enforcing:

$$\frac{\mathrm{d}p}{\mathrm{d}\mathbf{n}} = 0 \tag{5.27}$$

To incorporate the effects of added mass, the interfacial pressure is subject to the following condition:

$$\frac{\mathrm{dp}}{\mathrm{d}\mathbf{n}} = -\rho \mathbf{a_n} \tag{5.28}$$

Inclusion of this condition into the pressure Poisson equation which is solved in the case of incompressible flows is challenging, particularly for cases where strong added mass effects can confer instability to the fluid–solid coupling.

5.15 Techniques for Coupling of the Temporal Update of the Fluid and Solid Subsystems

Broadly speaking, from the standpoint of temporal update of the coupled system, FSI algorithms generally fall into one of two categories: monolithic systems or partitioned systems. The distinction lies in the approach to solving the fluid and structure governing equations, and the coupling of the fluid and structure at the interface.

Monolithic approaches solve the fluid and structure simultaneously, operating on the entire, aggregate system of fluid and solid governing equations directly, choosing the same discretization method for both sets of governing equations (be it finite element, finite volume, finite difference, etc.). In this way, the coupling of the fluid and structure and enforcement of interface conditions is directly performed during the solution of the system of equations. While this uniform approach has its appeal, this method has some shortcomings. The potential for ill-conditioned matrices (resulting from disparate physical responses of the solid and fluid, i.e., the stiffness problem) hinders the solution process [66]. Furthermore, this type of approach prohibits the use of preexisting codes for fluid dynamics and structural mechanics solutions, rendering it impractical for many realistic fluid–structure interaction problems [7, 21, 67].

Partitioned approaches take advantage of preexisting solid and fluid mechanics computer codes by solving the governing equations for the fluid and structure independently, subject to appropriate interface conditions. As a result, the solution process of the fluid and structure is asynchronous, i.e., a lag is introduced between the two domains. This lag can be reduced or even eliminated by updating the interface conditions and repeating the solution of the fluid and structure multiple times per time step until a converged solution is achieved. The process of iterating these solutions is referred to as *subiterating*, and partitioned methods utilizing subiterations are termed *strongly coupled* (or *implicit*) *partitioned procedures*. Conversely, those algorithms which maintain a lag in the solution by limiting the computations to a single fluid and a single solid solution per time step are referred to as *weakly* (or *loosely*) *coupled partitioned procedures* [7, 21, 63, 67, 68].

5.16 Weak and Strong Coupling Algorithms

The major drawback of weakly coupled partitioned algorithms is the inherent time lag between the fluid and structure solutions. Weakly coupled systems follow an algorithm similar to the one described here.

New Time Step (t^{n+1}):

Move the structure based on solution at t^n

Solve fluid governing equations, using appropriate compatibility conditions from structure at t^n (typically enforcing the kinematic conditions)

Solve structural equations based on fluid results at time t^{n+1} (typically using fluid tractions to compute structural displacement)

Note that in this algorithm, at any given time step the fluid solution is being computed based on structural information (displacement field, loads, velocity) from the previous time step. This illustrates the inherent asynchronous behavior of the weakly coupled partitioning algorithms, which can prevent conservation of energy and momentum. Due to the lack of energy conservation, particularly for large displacement computations, severe limits must be placed on the allowable time step size required to maintain stability, and in many cases stable computations are infeasible [9, 63, 68, 69].

In aerodynamic applications, however, loosely coupled methods have been used extensively [3, 20] due in part to the significant difference in mass between the fluid and the structure. Because the mass of the structure (building, airfoil, bridge, etc.) is typically substantially larger than that of the surrounding air, the structure is only minimally affected by the dynamics of the air. The effect of the solid (which is far stiffer than the surrounding air) on the air is therefore far greater than that of the air on the structure. This leads to a predominantly one-way influence, i.e., that of the structure on the air. As a result, in this scenario the structure dominates the FSI system, making this application suitable (even ideal) for a loosely coupled algorithm [7].

In thinking about strongly coupled partitioned systems, one can consider a scenario opposite to that in aerodynamics. In the case of a system where the mass of the fluid and structure are comparable, each aspect makes comparable contributions to the overall dynamics of the system. These strong interactions mean that perturbations by the fluid are felt immediately by the structure, while the converse is also true. In this way, a staggered approach (i.e., using loosely coupled partitioning) would not be appropriate, as any changes would only be felt (and responded to) in the following time step [70]. This, combined with the other limitations of the loosely coupled partition methods, motivates the development of strong (or implicit) coupling methods using partitioned solvers. In these types of approaches, subiterations (i.e., repeatedly solving both the fluid and the structure) are performed throughout the time step, updating the interface conditions as the solution continues, until convergence is declared. Van Brummelen and de Borst describe a general algorithm for this approach, which is summarized below with modifications to describe an incompressible fluid [66].

New Time Step (t^{n+1}):
Move the structure based on solution at t^n
If necessary, update mesh
Iterate (k+1) until converged:

Solve kinematic condition ($\mathbf{u}_f^{t+1,k+1} = \mathbf{u}_s^{t+1,k}$)
Solve fluid governing equations
Compute fluid forces on the interface and transfer to the solid solver

Solve structural equations
Check for convergence

It is worth noting that the above algorithm reduces to the weakly coupled algorithm if the number of subiterations is limited to 1. Additionally, the subiterations couple a stand-alone fluid dynamics problem (assuming boundary conditions are available) and an independent structural mechanics problem. In this way, the behavior of the approach is independent of the particular fluid dynamics or structural mechanics solver utilized, assuming that appropriately fine spatial and time step sizes are used [70]. Convergence of the subiterations can be declared by various means, but is typically based upon the examination of the error or residual of interface parameters, such as the displacement, velocity, and/or stresses at the interface over successive subiterations [71].

5.17 Three Different Approaches to FSI Modeling in Biomedical Applications

In the following, we show examples of computations performed with three different approaches for fluid–structure interaction problems. In Approach 1, the embedded object, a model for a white blood cell, is treated as an immersed boundary and the standard IBM approach of Peskin is applied. In Approach 2, the immersed interface approach as in Li and Lai [44] is used for the study of the dynamics of a capsule containing a fluid. The embedded object here is a hyperelastic membrane that is driven to deform by the surrounding fluid forces. In this case, boundary conditions on the fluid are used to enforce no-slip and no-penetration on the solid surface. In Approach 3, we illustrate the use of the sharp interface method to study the dynamics of a mechanical (rigid) heart valve and the dynamics of a bioprosthetic (flexible tissue) heart valve. In Approach 3, we contrast the two problems, i.e., the MHV and BHV dynamics, in terms of their demands on the FSI approach (in particular with regard to the coupling strategy) employed for their solution.

5.17.1 FSI Approach 1

An example of the first approach is the deformation of a white blood cell during aspiration into a micropipette. This system is analyzed using a mixed Eulerian–Lagrangian numerical technique. The cell (Fig. 5.3) is modeled as a compound drop comprised of fluid layers of density ρ_i and viscosity $\nu_{i,}$ occupying regions Ω_i $(i = 1, 2, 3)$. The regions Ω_1, Ω_2, and Ω_3 represent the suspending fluid, the cytoplasm, and the nucleus. The interface Γ_{12} (cell membrane) is modeled as a membrane with a constant tension or an elastic structure with a nonlinear constitutive relation, while the interface Γ_{23} (nuclear membrane) is always ascribed a

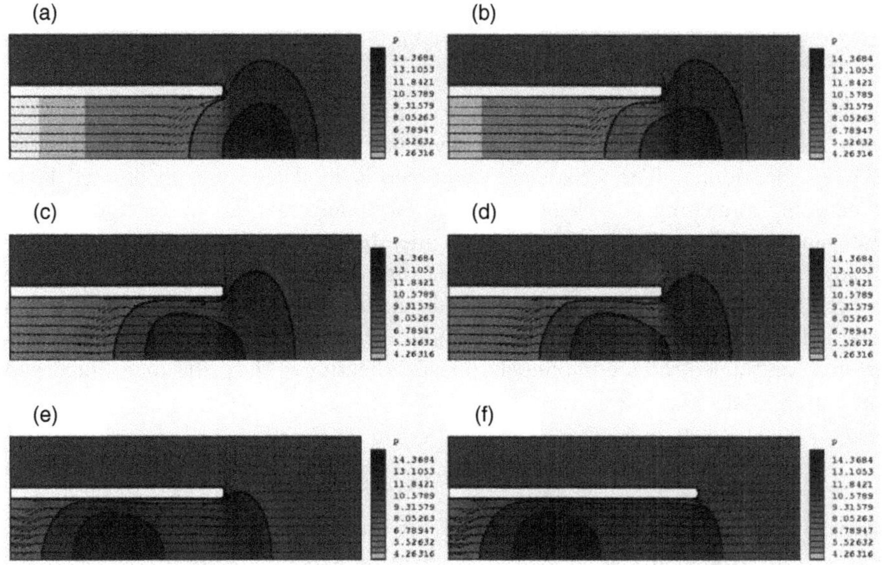

Fig. 5.3 Simulation results for the entry of a modeled leukocyte during aspiration into a micropipette. **a–f** Sequence of entry of the leukocyte as it is aspirated into the micropipette

constant tension. R_0 is the undeformed radius of the cell. For micropipette aspiration simulations, the system is set up as shown in Fig. 5.3. By suitable pressure boundary conditions at the boundaries of the domain, including the portion of the domain boundary that lies within the micropipette, the cell is aspirated [13]. The surface of the micropipette is a no-slip solid boundary that is enforced sharply, while the cell membrane is treated with the immersed boundary method.

The force contribution is from both the tangential and the normal forces acting on the membrane. The following equation identifies the individual components of the force:

$$\begin{aligned} \vec{f} &= \vec{f_{\mathrm{t}}} + \vec{f_{\mathrm{n}}} \\ &= \int \left(\frac{\partial \sigma_{\mathrm{s}}}{\partial s}\right)\hat{t}\delta(x - x_{\mathrm{I}})\mathrm{d}s + \int (\sigma_s \kappa_s + \sigma_\varphi \kappa_\varphi)\hat{n}\delta(x - x_{\mathrm{I}})\mathrm{d}s \end{aligned} \tag{5.29}$$

where $\sigma_{\mathrm{s}/\varphi}$ is calculated using the constitutive relation which is discussed below, x and x_{I} are the locations of points in the flowfield and on the interface, respectively, $\hat{t}$ and $\hat{n}$ are unit vectors in the tangential and normal directions to the interface, and $\delta(\mathrm{x} - \mathrm{x_I})$ is the numerical delta function (centered around x_{I}) used in the immersed boundary technique [4] to distribute the singular surface forces to the underlying mesh. The numerical implementation and other details can be found in Marella and Udaykumar [13]. The velocity at an interface is continuous across it and is obtained at an interfacial point from bilinear interpolation of the fluid velocities stored at grid points around it. The membrane is advected with the interface velocity using the following kinematic relation:

$$\frac{d\vec{X}_k}{dt} = \vec{V}_{\Gamma_{ij}} \tag{5.30}$$

where $\vec{X}_k$ is the position of the kth marker on the interface Γ_{ij}.

In the example of a leukocyte, an axisymmetric model proposed by Dong et al. [72, 73] is adopted. The leukocyte membrane is modeled as a prestressed layer undergoing stretching and shearing during large deformation. In the undeformed state, an isotropic tension σ_0 resides in the membrane which persists and is additive to any stresses due to deformation. The constitutive relationship for the membrane is expressed in terms of the principal membrane layer tensions and principal stretch ratios. These tensions are comprised of both isotropic (elastic stretching and pre-tension) and anisotropic components (elastic shearing). Thus, the meridional and circumferential stresses are given by

$$\sigma_{s/\varphi}(s) = \sigma_0 + E_a A(\lambda_s, \lambda_\varphi) + E_b B_{s/\varphi}(\lambda_s, \lambda_\varphi) \tag{5.31}$$

where E_a and E_b are the elastic moduli for area dilation and membrane shear, and A, B give the isotropic and anisotropic elastic tensions. Subscript φ stands for meridional direction and s for circumferential direction. λ_s, λ_ϕ are the principal stretch ratios in the meridional and circumferential directions. Due to unfolding of the leukocyte surface we expect a nonlinear behavior of the membrane. The critical point at which the membrane is unfolded corresponds to the condition $(\lambda_s\lambda_\phi - 1) = 1.0$ (i.e., a maximum 100% excess area is provided). At this point the membrane stresses increase rapidly and the smooth portion of the membrane offers large resistance to area dilatation. The set of functions for A and $B_{S/\varphi}$ suggested to express the nonlinear membrane response are

$$A = (\lambda_s\lambda_\phi - 1)^\rho \tag{5.32}$$

$$B_{s/\phi}(\lambda_s, \lambda_\phi) = \frac{1}{2\lambda_s\lambda_\phi}(\lambda_{s/\phi}^2 - \frac{1}{\lambda_{s/\phi}^2}) \tag{5.33}$$

where the exponent $\rho >> 1$ implies rapid stiffening of the membrane when the folds are smoothed. In the following example, $\rho = 30$ is chosen to represent the steep change in the tension after the cell membrane is unfolded. This model is the same as used in Dong et al. [72, 73].

The stretch ratios λ_s and λ_φ are determined from the strains at the interfacial points using the following expressions:

$$\lambda_s = \sqrt{1 + 2\varepsilon_{ss}} \ , \ \ \lambda_\varphi = \sqrt{1 + 2\varepsilon_{\phi\phi}} \tag{5.34}$$

where $\varepsilon_{ss}, \varepsilon_{\phi\phi}$ are the membrane strains in the circumferential and meridional directions. The strain distribution along the interface can be calculated using various approaches. The strain rate at the markers is computed from the gradients of the underlying velocity field. The strain rate tensor ($\dot{\varepsilon}$) in an axisymmetric coordinate system is given by Equation (5.35):

$$\dot{\varepsilon} = \begin{bmatrix} \dot{\varepsilon}_{xx} & \dot{\varepsilon}_{xy} & 0 \\ \dot{\varepsilon}_{yx} & \dot{\varepsilon}_{yy} & 0 \\ 0 & 0 & \dot{\varepsilon}_{\phi\phi} \end{bmatrix} = \frac{1}{2}(\vec{\nabla}V + V\overleftarrow{\nabla}) = \begin{bmatrix} \frac{\partial u}{\partial x} & \frac{1}{2}(\frac{\partial u}{\partial y} + \frac{\partial v}{\partial x}) & 0 \\ \frac{1}{2}(\frac{\partial v}{\partial x} + \frac{\partial u}{\partial y}) & \frac{\partial v}{\partial y} & 0 \\ 0 & 0 & \frac{v}{y} \end{bmatrix} \quad (5.35)$$

In Equation (5.35), x and y denote the axial and radial coordinate directions, respectively. The strain rates at the markers are obtained by a bilinear interpolation of the strain rates at the surrounding grid points. The circumferential strain rate $\dot{\varepsilon}_{ss}$ is calculated from $\dot{\varepsilon}_{xx}, \dot{\varepsilon}_{xy}, \dot{\varepsilon}_{yy}$ by transforming to axes tangential and normal to the interface.

$$\dot{\varepsilon}_{ss} = \frac{1}{2}(\dot{\varepsilon}_{xx} + \dot{\varepsilon}_{yy}) + \frac{1}{2}(\dot{\varepsilon}_{xx} - \dot{\varepsilon}_{yy}) \cos 2\theta + \dot{\varepsilon}_{xy} \sin 2\theta \quad (5.36)$$

where θ is the angle made by the tangent with the positive x-axis. The strain at the marker points is interpolated from the strain rates and updated using

$$\varepsilon_{ss}^{k+1} = \varepsilon_{ss}^{k} + \dot{\varepsilon}_{ss}^{k}\Delta t, \quad \varepsilon_{\phi\phi}^{k+1} = \varepsilon_{\phi\phi}^{k} + \dot{\varepsilon}_{\phi\phi}^{k}\Delta t \quad (5.37)$$

The interfacial stresses are evaluated from these strains using the constitutive equations (Equation 5.31) and are transferred to the flow through the source terms (Equation 5.29) in the momentum equation (Equation 5.4) and the new flowfield is calculated.

5.17.1.1 Results

A realistic portrayal of a leukocyte deforming in suspension should account for further elements in the internal structure of the cell. As indicated by Dong et al. [73], the cytoplasm is highly viscous compared to plasma, while the viscosity of the nuclear fluid (2,323 poise) is higher than the cytoplasm (200 poise). The above method can be used to study the deformation and recovery behavior of a compound cell with large viscosity jumps between the suspending fluid, cytoplasm, and nucleus. In this example, the parameters chosen are $\mu_1 = 1.0$, $\mu_2 = 40.0$, and $\mu_3 = 400.0$. In comparison to the case with equal viscosities in cytoplasm and nucleus, the discontinuous viscosity case provides markedly different behavior. Figure 5.3 shows the aspiration into a micropipette of a compound cell with constant, uniform membrane tension. The aspiration is effected by applying a suction pressure at the left end of the modeled pipette. The highly viscous nucleus slows the entry of the cell into the pipette. The rather stiff, viscous nucleus also deforms modestly as the cell flows into the pipette. Once the cell is in the pipette, the nucleus is able to partly recover from the deformation. In the simulation shown, the forces acting on the cell as well as on the nuclear membrane are spread over a finite area of the underlying mesh using the numerical delta function inherent in the immersed boundary approach. Therefore, these simulations do not treat the membranes as sharp entities.

5.17.2 FSI Approach 2

One can treat a membrane separating two fluid regions as a sharp entity (i.e., without spreading its influence over a region of the mesh) using the immersed interface method, the ghost fluid method, or the sharp interface method. In all these methods, the displacement of the membrane is computed based on fluid flow conditions (i.e., the massless membrane moves with the fluid velocity consistent with the kinematic compatibility constraint), and jump conditions are imposed on the fluid based on the leaflet stresses. The force balance can be described conceptually by examining the forces acting on a control volume. This scenario is illustrated in Fig. 5.4.

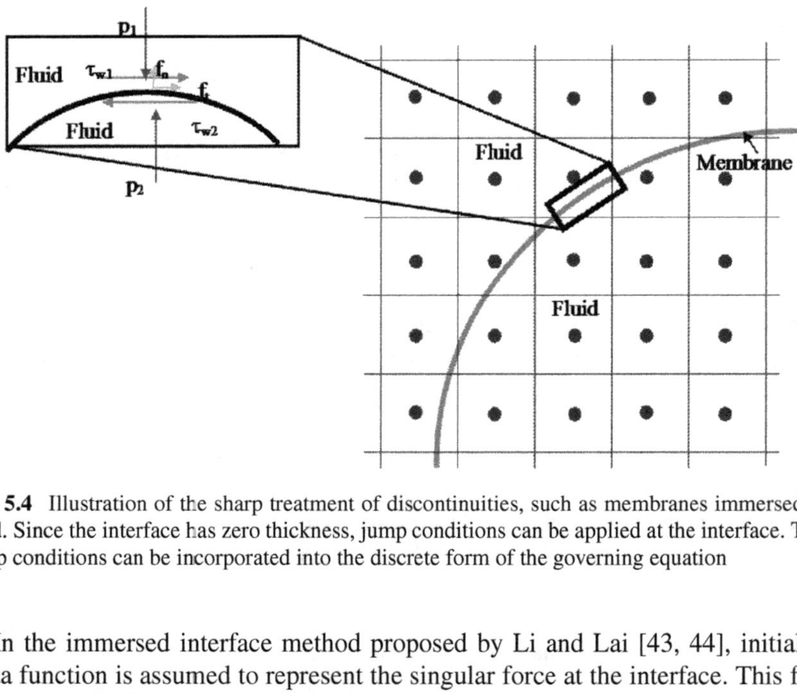

Fig. 5.4 Illustration of the sharp treatment of discontinuities, such as membranes immersed in a fluid. Since the interface has zero thickness, jump conditions can be applied at the interface. These jump conditions can be incorporated into the discrete form of the governing equation

In the immersed interface method proposed by Li and Lai [43, 44], initially a delta function is assumed to represent the singular force at the interface. This force is shown in the following equation:

$$\mathbf{f}(\mathbf{x}, t) = \int_{\sigma} \mathbf{F}(r, s, t)\, \delta\left(\mathbf{x} - \mathbf{X}(r, s, t)\right) \mathrm{d}r\mathrm{d}s \tag{5.38}$$

where $\mathbf{f}(\mathbf{x}, t)$ is the force density exerted by the membrane, and the membrane itself, σ, is represented by $\mathbf{X}(r, s, t)$ where r, s are parameters of a reference configuration where $0 \le r \le L_r$ and $0 \le s \le L_s$. The Dirac delta function is three dimensional, and the membrane force $\mathbf{F}(r, s, t)$ is a function of its configuration, where $\mathbf{F}(r, s, t) = \mathbf{S}(\mathbf{X}(r, s, t), t)$. In other words, at any time, t, a region on the surface of the membrane is mapped onto a patch with area $L_r \times L_s$ and all calculations are performed in this reference space. These calculations could also be performed in the current space with $\mathbf{F}(\sigma,t)$ instead of mapping it back to a reference configuration. In this way, Equation (5.38) would change to

$$\mathbf{f}(\mathbf{x},t) = \int_{\sigma} \mathbf{F}(l,m,t)\,\delta\left(\mathbf{x} - \mathbf{X}(\sigma,t)\right) \mathrm{d}l \mathrm{d}m \tag{5.39}$$

where l, m are surface parameterizations in the two *local* tangent directions,τ_1, τ_2. The unit normal can be calculated with respect to these two tangent directions as

$$\mathbf{n} = \frac{\tau_1 \times \tau_2}{|\tau_1 \times \tau_2|} \tag{5.40}$$

It should be noted that in the following derivations, the unit normal, $\mathbf{n}$, can also be described using information about the level sets:

$$\mathbf{n} = \left\langle n^x, n^y, n^z \right\rangle = \frac{\nabla \phi}{|\nabla \phi|} \tag{5.41}$$

Further, the first tangent direction is obtained by finding the smallest component of $\mathbf{n}$, i.e., $\min(n^x, n^y, n^z)$. If the first (x) direction is the smallest, for example, then choose the unit vector in that direction $\langle 1,0,0\rangle$ and take the cross-product of the normal vector and this new vector to obtain the first tangent:

$$\boldsymbol{\tau}_1 = \frac{\mathbf{n} \times \langle 1,0,0\rangle}{|\mathbf{n} \times \langle 1,0,0\rangle|} = \left\langle 0, \frac{n^z}{\sqrt{(n^y)^2 + (n^z)^2}}, \frac{-n^y}{\sqrt{(n^y)^2 + (n^z)^2}} \right\rangle \tag{5.42}$$

Finally, in 3D, a second tangent will be taken as the cross-product of the normal and first tangent.

$$\boldsymbol{\tau}_2 = \mathbf{n} \times \boldsymbol{\tau}_1 \tag{5.43}$$

At the interface the jumps in the variables will be written in the form $[\psi] = \psi_I^+ - \psi_I^-$ so that the jump of variable ψ is the difference in the value ψ on either side of the interface. Detailed derivations of the jump conditions are given in [74]. A summary of the jumps in pressure and velocities, and their derivatives, in the normal and two tangential directions are shown below.

$$[u] = [v] = [w] = 0 \tag{5.44a}$$

$$[u_x + v_y + w_z] = 0 \tag{5.44b}$$

$$\left[\frac{\partial \mathbf{u}}{\partial \mathbf{n}}\right] \cdot \mathbf{n} = 0 \tag{5.44c}$$

$$[\mathbf{u}_t] + [\nabla \mathbf{u}] \cdot \mathbf{u} = \mathbf{0} \tag{5.44d}$$

$$[\nabla u \cdot \tau_1] = [\nabla v \cdot \tau_1] = [\nabla w \cdot \tau_1] = 0 \tag{5.44e}$$

$$[\nabla u \cdot \tau_2] = [\nabla v \cdot \tau_2] = [\nabla w \cdot \tau_2] = 0 \tag{5.44f}$$

$$[p] = [\mu]\frac{\partial \hat{\mathbf{u}}}{\partial \mathbf{n}} \cdot \mathbf{n} + \mathbf{F} \cdot \mathbf{n} \tag{5.44g}$$

$$\left[\mu \frac{\partial \mathbf{u}}{\partial \mathbf{n}}\right] = [\mu]\left(\frac{\partial \hat{u}}{\partial \mathbf{n}} \cdot \mathbf{n}\right)\mathbf{n} + (\mathbf{F} \cdot \mathbf{n})\mathbf{n} - \mathbf{F} \tag{5.44h}$$

$$\left[\frac{\partial p}{\partial \mathbf{n}}\right] = \frac{\partial}{\partial \boldsymbol{\tau}_1}\mathbf{F} \cdot \boldsymbol{\tau}_1 + \frac{\partial}{\partial \boldsymbol{\tau}_2}\mathbf{F} \cdot \boldsymbol{\tau}_2 \tag{5.44i}$$

$$\left[\frac{\partial p}{\partial \boldsymbol{\tau}_1}\right] = \frac{\partial}{\partial \boldsymbol{\tau}_1}\left([\mu]\left(\frac{\partial \hat{u}}{\partial \mathbf{n}} \cdot \mathbf{n}\right) + (\mathbf{F} \cdot \mathbf{n})\right) \tag{5.44j}$$

$$\left[\frac{\partial p}{\partial \boldsymbol{\tau}_2}\right] = \frac{\partial}{\partial \boldsymbol{\tau}_2}\left([\mu]\left(\frac{\partial \hat{u}}{\partial \mathbf{n}} \cdot \mathbf{n}\right) + (\mathbf{F} \cdot \mathbf{n})\right) \tag{5.44k}$$

The right-hand side quantities in the above equations can be computed for grid points that lie adjacent to the interface. In terms of coupling the fluid and solid, the jump conditions above need to be introduced into the discrete form of the governing equations at those grid points that lie adjacent to the interface. This is done by employing the ghost fluid method [75]; higher-order approaches for incorporating the jump conditions are also available as in immersed interface methods [43]. The details of the implementation of jump conditions are provided in Marella et al. [76]. Once the fluid flowfield is computed the fluid velocity field is (bilinearly) interpolated to the points on the membrane, the solid nodes are advanced based on this velocity to their new positions, and the resulting strains and stresses are computed for the membrane.

5.17.2.1 Results

Approach 2 has been implemented into the sharp interface flow solver using the formulation described above. In 2D, this approach has been used to simulate several basic membrane problems and has been validated against benchmark cases.

To demonstrate the capabilities of Approach 2, the method was used to simulate a membrane that is circular in its equilibrium state (i.e., in the undeformed condition). Initially, the membrane is in a stretched configuration so that it is elliptical in shape. Subsequently, it is allowed to recover back to its unstressed state. This is illustrated in Fig. 5.5. The fluid and membrane are fully coupled, so that the stress in the membrane affects the fluid, and in turn, the fluid motion affects the membrane behavior. No slip is enforced at the interface implicitly, since the membrane moves with the fluid velocity. Membrane stress is computed assuming a hyperelastic material model. Results are shown at three stages: early in the simulation, midway through the simulation, and as the solution reaches steady state. Pressure and U–velocity contours are shown in Fig. 5.5. As can be observed, the fluid has a high

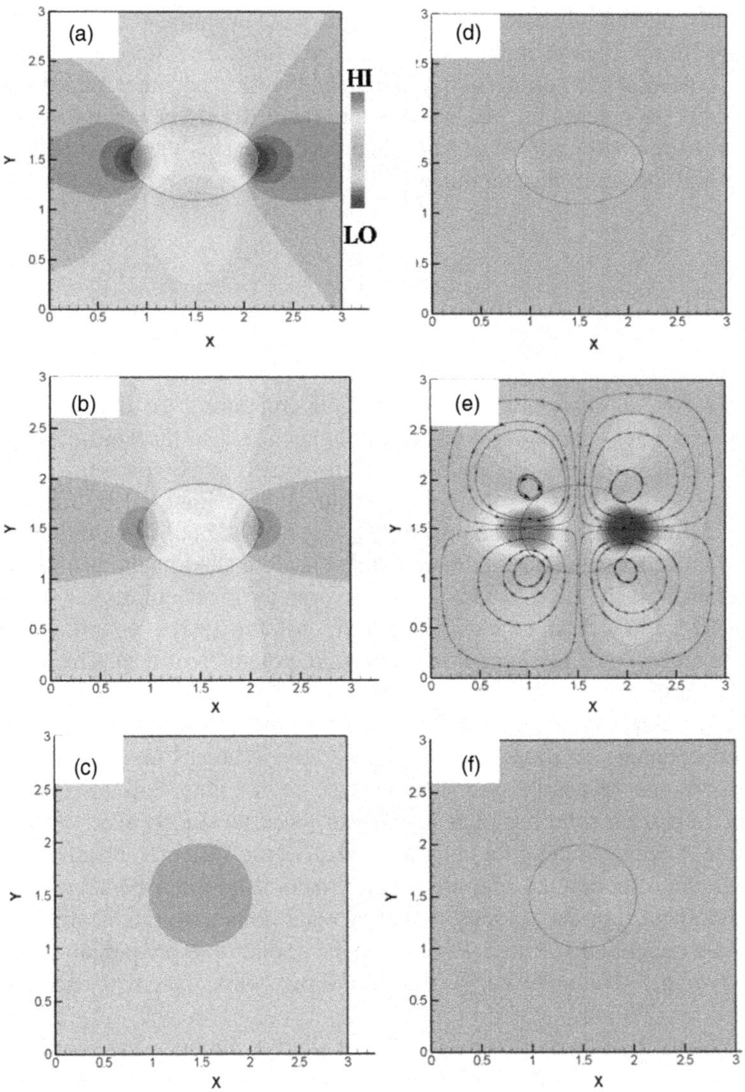

Fig. 5.5 Results of calculations with the immersed interface method. An initially stretched capsule is allowed to relax to the final equilibrium position. **a–c** Sequence of *shapes* in time of the capsule along with pressure contours. The final equilibrium *circular shape* is shown in (**c**). **d–f** The velocity field generated due to the membrane forces during the shape recovery. The membrane is initially at rest and has zero velocity in the final equilibrium position

pressure jump due to the high tension in the membrane early on (Fig. 5.5a). As the ellipse relaxes back to its equilibrium state, the pressure field becomes more uniform, until at the final stage, the ellipse returns to a circle with a constant pressure difference across the membrane.

Similarly, the initially quiescent fluid in which the membrane is immersed begins to flow due to the stresses in the membrane. Over time, the ellipse and surrounding fluid move to establish equilibrium. Ultimately, the fluid becomes stationary again, as the membrane reaches its final, steady-state shape. It is to be noted that, as shown by the pressure contours in Fig. 5.5a, Approach 2 retains the sharp nature of the interface and the jump discontinuities are preserved as such by the discretization scheme.

5.17.3 FSI Approach 3

The above two approaches are well suited for computing the dynamics of flexible, thin structures interacting with flows; in such a case the inertia of the solid is small and the mass of the solid can be neglected in comparison to the fluid. Then, the coupling between the solid and the fluid is imposed by means of the so-called Neumann–Dirichlet approach [9], i.e., the fluid subdomain obtains force information from the solid subdomain, while the solid subdomain receives velocity information from the fluid subdomain. This strategy for coupling has two significant problems. First, from a practical point of view, this type of coupling is suitable only for membranes (i.e., thin deformable solid structures of negligible mass). For solid boundaries with finite thickness, the notion of singular forces supplied by the solids to the fluids does not hold and therefore the idea underlying the original immersed boundary or immersed interface methods breaks down and alternative methods become necessary. Second, this type of coupling fails to remain stable when the immersed solid has high stiffness or when the added mass of the fluid is significant. A better strategy for coupling then is of the Dirichlet–Neumann [9] type (where the fluid subdomain supplies surface tractions to the solid solver, while the solid subdomain supplies velocity boundary conditions to the fluid). In fact, it has been shown recently that Robin–Robin boundary conditions are perhaps even better than the Dirichlet–Neumann conditions for FSI problems of the type encountered in biomechanics [77, 78].

Apart from the immersed interface/ghost fluid techniques for handling membranes, another approach is also possible in a sharp interface Cartesian grid setting; this is one where the fluid and solid dynamics are computed in a segregated fashion with specialized solvers for each subdomain. Then the strategy of Dirichlet–Neumann coupling for FSI treatment can be employed, and surface tractions (from the fluid side) and solid boundary displacements (from the solid side) are exchanged between the two subdomains at the surface separating them. This approach is the norm in conventional body-fitted mesh-based approaches; here we adapt the same to a non-body-fitted (Cartesian, fixed mesh) system while still maintaining a sharp interface. In effect, the sharp interface approach follows the main ideas that have been used in FSI computations with body-fitted moving meshes. However, in the present approach the fluid mesh is freed from conforming to the solid motion. Instead, the fixed Cartesian mesh allows the large deformations of the solid to occur through (or over) it. This is a significant advantage, in that problems that plague Lagrangian or ALE-type methods, viz., those relating to mesh generation and management, are avoided altogether. The key aspects of this second approach to the solution of the FSI problem are then as follows:

1. The solid dynamics is computed entirely independently of the fluid solver using standard FE methods. The fluid and solid solutions are segregated.
2. The fluid flow computations receive as inputs the position of the solid surfaces, and the velocities and accelerations thereof. These are applied as boundary conditions on the embedded solid surface using the methods described in Marella et al. [33].
3. The fluid flow solver relies on level-set representations of the embedded surfaces to facilitate the sharp interface treatment in the discretization process. Therefore, algorithms are required to translate the information on solid surface position and other variables that are supplied from the solid surface (to the fluid) or supplied to the solid surface (from the fluid) into the level-set-based representation on the fluid mesh.
4. The fluid flow solution provides the forces that act on the solid and result in its deformation. These forces include all the tensor components that contribute to the surface traction, i.e., the pressure and viscous stresses on the solid surface.

The above approach to coupling the fluid and solid mechanics is called a partitioned solution process and has the advantage of allowing specialized and mature solvers to be used for the solid and fluid field solutions. For instance, in the present case, finite element discretization is employed for the solid while a finite difference approach is used for the fluid flow solution. However, it is critical that these independent solutions are performed in a tightly coupled fashion or stability of the aggregate calculations will be impossible to ensure.

To devise an implicit technique for fluid–solid coupling, the subiteration approach is adopted, as illustrated in Fig. 5.6. In this approach, the fluid flow equations (Equations 5.3 and 5.4) and structural equations (Equation 5.5) are solved in an iterative fashion over each time step, until convergence of the overall system is achieved. Communication between the solid and the fluid domain comes primarily from the conditions that have to be satisfied on the surface of the embedded solid. These include the kinematic compatibility defining the continuity of the fluid and solid positions, velocities, and accelerations (amounting to no-slip and no-penetration conditions, i.e., the Dirichlet conditions):

$$\vec{X}_{\mathrm{f}} = \vec{X}_{\mathrm{s}} \tag{5.45a}$$

$$\vec{u}_{\mathrm{f}} = \vec{u}_{\mathrm{s}} \tag{5.45b}$$

$$\vec{a}_{\mathrm{f}} = \vec{a}_{\mathrm{s}} \tag{5.45c}$$

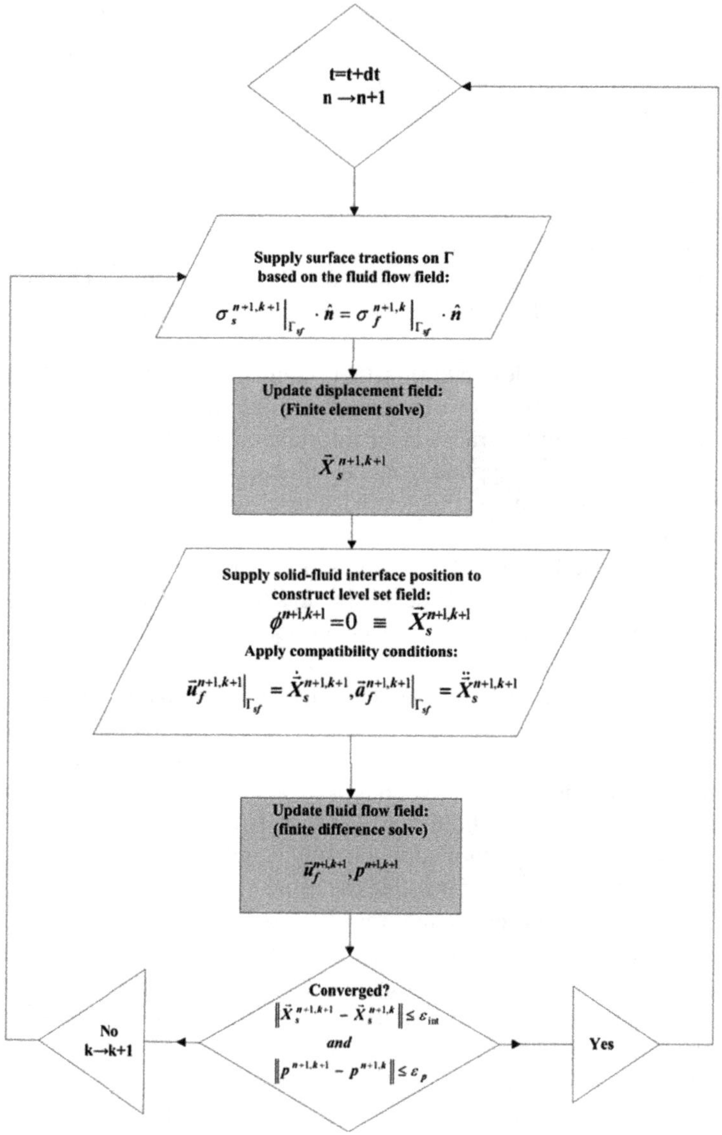

Fig. 5.6 Algorithm for the implicit (strongly coupled) calculation of the FSI problem

The added mass effect comes into play through the boundary condition on the pressure:

$$\left(\frac{\partial p}{\partial n}\right)_{\mathrm{f}} = -\rho \vec{a}_{\mathrm{s}} \cdot \hat{n} \tag{5.46}$$

where $\hat{n}$ indicates the local unit normal direction.

The conditions that apply on the structure in the course of solution of the solid field equations using the finite element approach are that the tractions on the solid surface are supplied by the fluid loading as follows:

$$\sigma_f \cdot \hat{n} = \Sigma_s \cdot \hat{n} \tag{5.47}$$

Incorporation of these conditions into the fluid solver as well as the solid solver is straightforward and no significant modifications to apply such conditions are necessary in either subdomain. The application of the approach to two problems is described below, i.e., to compute the dynamics of rigid and flexible heart valves. The behavior of these two systems in terms of the FSI coupling is very different. In the case of the rigid valve, since the density of the valve material is much higher than that of the fluid, a straightforward loose-coupling approach suffices. In the case of a flexible heart valve, however, since the density of the tissue is comparable with that of blood, strong added mass effects arise and the FSI computations must be of the strong-coupling type.

5.18 Modeling of Mechanical Heart Valves

Mechanical heart valves have rigid leaflets and come in several designs. One particular case is that of a bileaflet valve. A two-dimensional representation of this system is shown in Fig. 5.7. Due to the symmetry of this system, the centerline (left boundary) represents a line of symmetry, while the right boundary represents the valve housing and is a nonslip wall. Between the housing and the leaflets, at the point of the closure of the valve, there exists a clearance gap of about 100 μm in width. This gap and the leakage jet that issues through it are of critical importance in the tendency of the valve to induce thrombotic events. The model therefore seeks to compute the FSI problem of valve motion under the imposed pressure loading while also resolving the dynamics of blood carrying platelets (modeled as point particles) as it flows through the narrow leakage gap at closure. In addition, there is a rebound event as the valve closes and makes contact with the hinge. This rebound may also have implications for thrombosis by influencing flow patterns distal to the valve.

The rigid leaflet is pivoted by means of hinges attached to the valve housing, so that leaflet rotation can be described by the relationship [79–81]

$$\frac{\mathrm{d}^2\theta}{\mathrm{d}t^2} = \frac{M}{I_o} \tag{5.48}$$

In the equation, $\theta(t)$ is the opening angle, indicating the leaflet position at any instant t, I_o is the moment of inertia of the leaflet about the pivot (3.3 e^{-9} kg/m^2), and M is the total momentum applied on the leaflet from the external forces inducing the leaflet motion. The external momentum can be calculated as

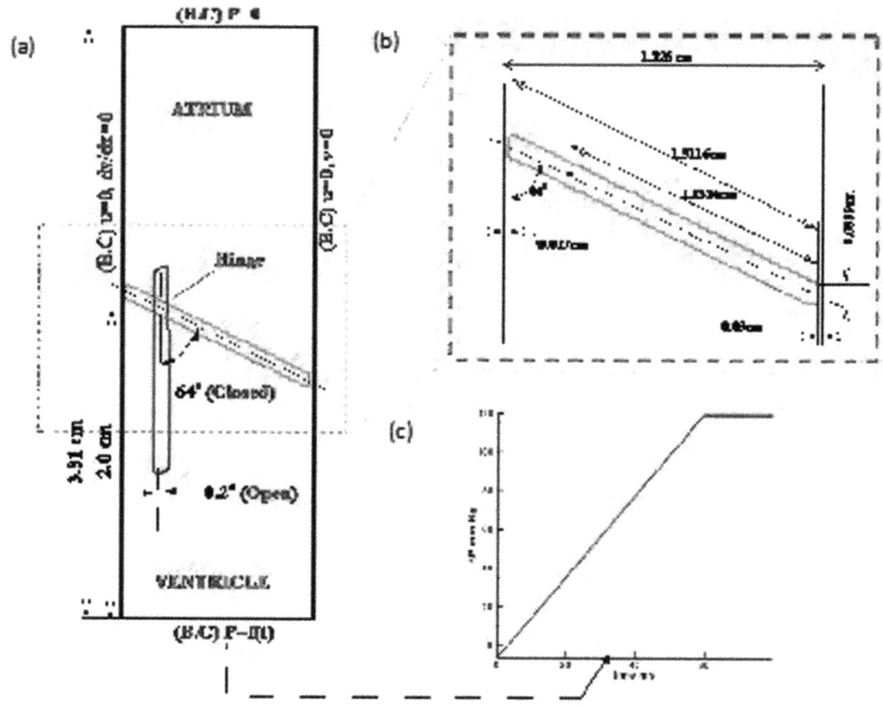

Fig. 5.7 Setup of the bileaflet mechanical heart valve simulation. (**a**) Overall computational setup. (**b**) View of the geometry at instant of closure. (**c**) The pressure versus time profile supplied on the ventricular side

$$M = M_{\mathrm{G}} + M_{\mathrm{P}} + M_{\mathrm{F}} \tag{5.49}$$

M_{G} is the momentum resulting from the buoyancy and the gravitational force and is given by

$$M_{\mathrm{G}} = gV\left(\rho_l - \rho_{\mathrm{f}}\right)\left(\frac{l}{2} - a\right)\cos(\theta) \tag{5.50}$$

Here, g is the acceleration due to gravity, V is the leaflet volume, ρ_{l} $(2{,}000\ \mathrm{kg/m^3})$ and ρ_{f} $(1{,}056\ \mathrm{kg/m^3})$ are the leaflet and fluid densities, respectively, and l (13.16 mm) and a are the leaflet radius and pivot length (distance between the hinge location and the left-side edge of the valve, 1.79 mm), respectively. The leaflet thickness e is 0.899 mm. M_{P} is the momentum resulting from the blood pressure and M_{F} is the momentum resulting from shear forces. These quantities are calculated from the flowfield by integrating the normal (pressure) and tangential (shear stresses) fluid forces acting on the valve surface.

5.19 Leaflet Rebound

The leaflet impacts against the valve seating lip at the instant of valve closure. After impact, the leaflet bounces back from the housing. The governing equation of leaflet dynamics during impact can be expressed as [80]

$$\omega_2 = -\sigma\omega_1 \tag{5.51}$$

where σ is the coefficient of resilience that depends upon the material of the leaflet and the valve housing. ω_1 and ω_2 are the angular velocities before and after impact, respectively. The coefficient of resilience σ is specified as 0.5 [80].

5.19.1 Results

The nominal local Reynolds number based on properties of blood is calculated to be around 1,800 based on the gap dimensions and the maximum velocity computed in the gap after closure. The above Reynolds number indicates that the flow through the valve is expected to be laminar for the bulk of the flow. However, the disparate length scales (i.e., between the overall valve dimensions and the leakage region) pose a challenge in adequately meshing the leakage gap ($\sim$15 cells at least are required to resolve the jet flows in these gaps) to obtain reliable solutions in reasonable time making optimal use of computational resources. The local mesh refinement algorithm solves this problem and refines meshes only in regions that need to resolve high solution gradients or curvature.

5.20 Effect of Flow During Closure and Rebound Phases

Figure 5.8 shows vorticity contours, absolute values of shear stress, and the simulated platelet activation in the region of the gap between the leaflet and the housing at time instants near the valve closure and rebound phases. The most significant feature observed in the figure is the interaction between the boundary layers separating the wall and the leaflet edge. As seen in Fig. 5.8a, the boundary layers from the wall and the leaflet edge separate and come closer. Due to the mutual interaction of these vortex sheets of opposite sign, secondary vortex layers are formed below the main vortex structures, both on the wall and on the valve. Instability of these vortex sheets causes the sheets to roll up, thus leading to a self-sustaining regeneration of pockets of vorticity that lead to local recirculating flows in the region distal to the valve. However, the interaction of the vortices in this region does not allow vortices to shed periodically and to be carried away downstream as in the central jet on the left edge of the valve. Figure 5.8a–f shows the various stages of vortex interaction in the valve closure and rebound stages with the corresponding instantaneous shear stress and activation parameter. The activation parameter was computed using particle tracking techniques and accumulating data on the shear stress and residence

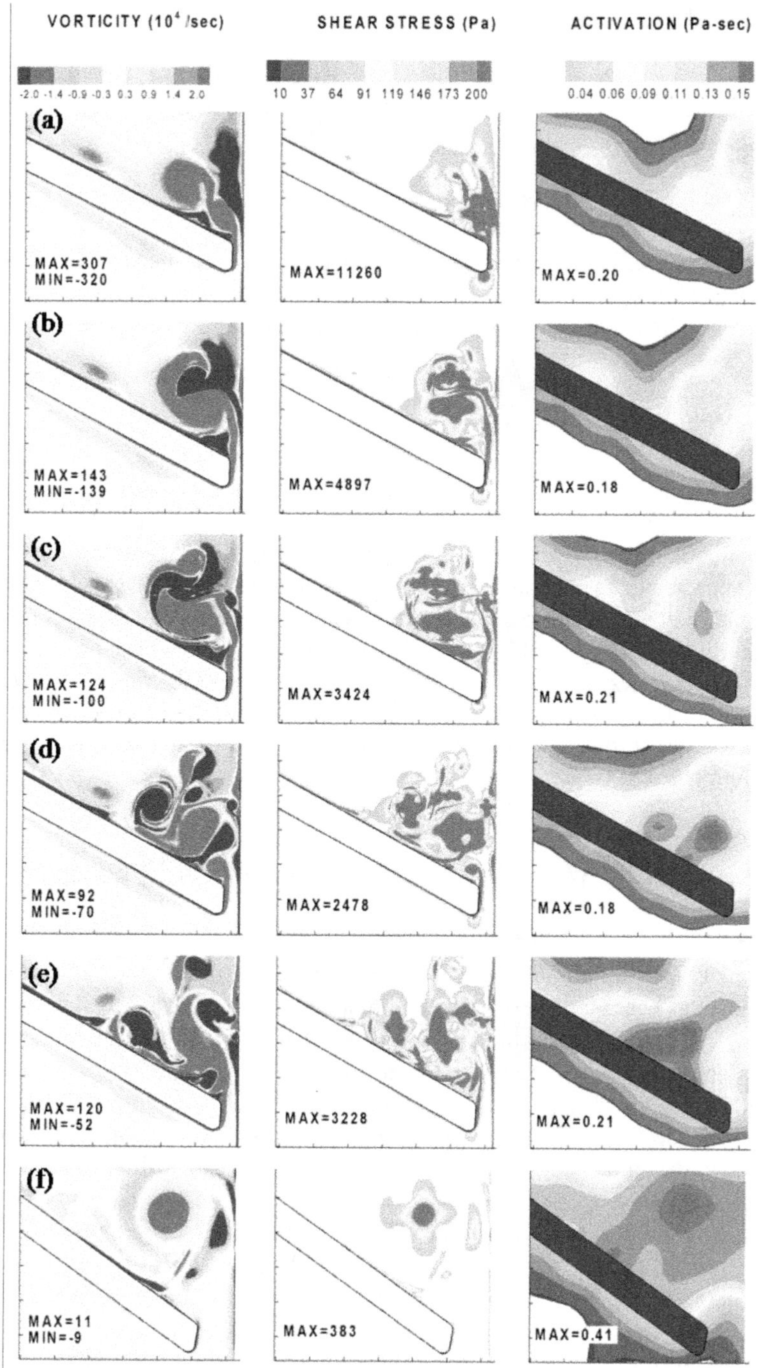

Fig. 5.8 Evolution of the vorticity, shear stress, and platelet activation fields during closure and rebound of the valve leaflet. The specific region shown is in the leakage gap between the leaflet edge and valve housing. **a–f** shows the sequence in time

time experienced by individual platelets (see [17] for details). Thus, high activation parameter contours indicate the presence of platelets that have experienced high shear stresses over significant times in that region of the flow domain.

In terms of energy budget, the pressure differential across the valve at the time of closure results in the creation of recirculating fluid pockets rather than a strong jet with a vertically directed velocity field. The separated boundary layer from the leaflet edge draws in the vorticity from the housing wall and the two structures roll around each other. As the flow evolves, these pockets of vorticity break up but remain in the same region. The dynamics of flow in this region causes shear stresses to be high and also causes recirculations that are localized in the region near the gap distal to the valve. This high shear coupled with high residence time makes the region immediately distal to the valve leaflet, a potential site for platelet activation and deposition (see Fig. 5.8f). Corresponding contours of the activation parameter indicate the distal (atrial) side of the right edge of the leaflet as a high-risk region for platelet activation and deposition. This is because particles entrained in the vortex interaction zone are subject to high shears from the circulating flow, but are unable to escape quickly from the region. It may be noted that if the interaction between vorticity generated at the housing wall and valve edge does not occur, the particles are likely to get activated by the high shear region in the small gap but may rapidly advect out of the housing region with the leakage jet and would then be unlikely to cause significant deposition. The resolution of the flow in the gap is therefore of crucial importance. Eventually, the strong vortex structures with high shear rates are carried downstream by the flow, away from the valve leaflet. However, weak vortex structures with recirculating flow on the downstream edge of the leaflet persist even after 5–10 ms after valve closure. Figure 5.8e,f shows the flow characteristics and corresponding shear rates and activation parameters a few milliseconds after the initial impact of the valve against the housing. At this stage, though the high shear region has moved downstream with the vortex structure, the downstream edge of the valve continues to show significant activation levels of platelets.

The periodically shedding vortices in the central orifice are likely regions of free emboli formation as reported by other researchers [82–84] but are not likely to cause deposition on the valve itself. But interaction of vorticity from the housing wall and valve develops a potential thrombogenic region on the valve leaflet edge and housing. The high vorticity strength in the recirculating flow region also has significance as potential sites for cavitation because of highly negative pressures that develop at the vortex core [79, 81, 85]. Thus, well-resolved calculations of the flow in this FSI problem, particularly in the narrow leakage gap region, are crucial from the standpoint of improving the performance of the implant.

5.21 Modeling of Tissue Heart Valves

5.21.1 Challenges in Modeling Tissue Heart Valves

Bioprosthetic heart valves (BHVs) possess highly deformable leaflets, which open and close due to pressure gradients, typically undergoing approximately 75 full open/close cycles each minute. These rapidly rotating leaflets are also undergoing complex deformation, with constantly changing shape and profile of the valve. The properties of the valve leaflet are highly nonlinear, anisotropic, and require a complex material model to accurately capture the in-plane, bending, and shear stress components attributed to the normal motion of a BHV [65]. Moreover, the leaflets themselves are very thin, occupying a negligible volume. Geometrically, this is a challenge to render, and numerically, this further contributes to the highly deformable nature of the valve. In simulating the dynamics of the valve, the time scales of solid response relative to the fluid time scale play an important role in determining the stability of the numerical solution procedure. The ratio of solid to fluid time scale determines the type of coupling procedure, i.e., explicit (i.e., loosely coupled [21, 86, 87]), subcycling [2], or fully implicit (i.e., strongly coupled [62, 69, 71]), that is best suited to solve the FSI problem. The ratio of time scales for simulating a tissue heart valve can be examined by using physiologic values in Equation (5.24). When these values are used, the ratio of time scales is roughly 40. Therefore, the structural responses are rapid when compared to the fluid and in fact the time step restriction posed by the fluid flow is no longer suitable to employ in the computations. This is in contrast to the situation in aeroelasticity, where the solid time scale is larger than the fluid scale [68]. For this reason, in the computation of tissue valves, even very small changes in the fluid can lead to significant changes in the structure, underscoring the need for a strong coupling method. The computation is also complicated because the fluid inflow condition is constantly changing. In addition, physiological inflow conditions dictate that the Reynolds number for flow through the valve can exceed 2,000 for a certain duration in one pulse. This implies that the inertia of the fluid is significant. Thus, features such as shear and boundary layer, flow separation, and vortical structures can be expected in the flow and must be captured if the effects of the blood flow on the valve leaflet are to be adequately treated. Thus, the mechanical heart valve simulations shown in the previous section were performed with a loosely coupled algorithm where one iteration per time step suffices to obtain stable computations of valve motion. However, in the following calculation, the subiterations required for convergence are fairly large in the beginning of the calculation (order of 100) and settle down as the calculation proceeds to order of 10 iterations. This is because at the beginning of the calculation, the acceleration of the valve is large and the added mass moved corresponds to the fluid enclosed within the entire domain. Thus, the added mass effect is particularly strong in the initial stages of the calculation.

5.21.1.1 Results of Simulations

An illustration of the conditions employed for the present case is shown in Fig. 5.9 along with key material parameters. A physiologically realistic time-varying pressure is applied on the left ventricle (inlet) side and a zero pressure is applied at the aortic (outlet) side of the geometry to represent the pressure drop experienced by the aortic valve. Standard pressure boundary condition treatment is employed on the inlet side while the outlet is treated via extrapolation of velocities and pressure.

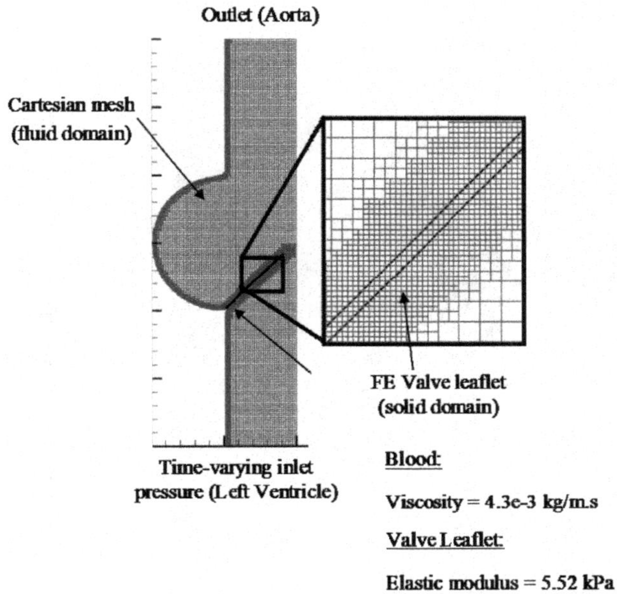

Fig. 5.9 Setup for the computation of the dynamics of a bioprosthetic (tissue) aortic heart valve. The locally refined mesh around the tissue surface is shown in the inset and the material properties used in the FSI modeling are listed

The pressure waveform shown in Fig. 5.9 is a standard physiologic pressure variation measured across the aortic valve [88]. Only the opening phase of valve motion is computed here to demonstrate the methodology. Material properties for the valve leaflet are specified based the experiments performed by Sun et al. [65] to simulate physiologic conditions as closely as possible.

A simplified (monotonically increasing) version of the inlet pressure waveform is used as the inlet boundary condition for simulation since it captures the essential features of the flow and leaflet deformation behavior during the opening phase of valve motion. Fig. 5.9 shows a magnified view of the mesh employed for the calculation of the flow and the finite element mesh employed in the solution of the solid displacement field. The finite element mesh was initialized based on the base (coarse) mesh that was used in the flow calculations before the local refinement process became activated by the developing flow gradients. This mesh is maintained in the solid during the computations.

Figure 5.10 shows the shape of the valve leaflet and the streamlines corresponding to the four instants of time. As the pressure rises in the left ventricle (the chamber of the heart responsible for pumping freshly oxygenated blood to the body's organs), the valve begins to open with a convex (upward) curvature. At this stage (Fig. 5.10a), the applied pressure is primarily working to push the valve leaflet open. The flow through the domain is primarily unidirectional, including in the sinus. This is in contrast to the flow results presented in de Hart and coworkers [16] where recirculations

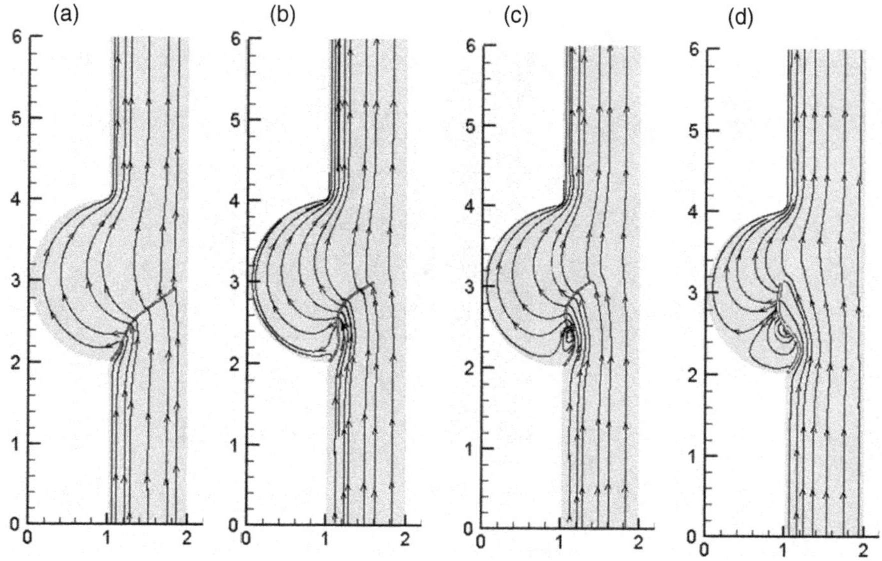

Fig. 5.10 Valve leaflet shapes and flow patterns (via streamlines) at four instants during the opening phase of a tissue heart valve

in the sinus were indicated. However, as shown in [89], the type of flow observed by de Hart et al. is due to the rather large stiffness of the valve material assumed in that work (corresponding to the elastic modulus of a rubber-like material). In the current computations, weak recirculating flows are observed in the later stages of valve opening (corresponding to Fig. 5.10c, d) in parts of the sinus.

While quantitative comparison of the opening behavior of the leaflet is not possible at this stage, qualitative comparisons can be made with experimental data as well as limited numerical simulation data [90]. The leaflet motion during opening is shown in Fig. 5.11, which shows the motion obtained experimentally during physiologic flow loop simulations, as well as results computed using 3D FEM simulations applying a spatially constant (but time-varying) pressure difference as the boundary condition [90]. It is important to note that the initial geometry of the current study (employing a 2D simulation using one leaflet in the analysis) does not incorporate the coaptated shape (the portion of the leaflet area in contact with and resting against the other leaflets when the valve is in the fully closed position) visible in the initial 3D geometries. Therefore, the key items for comparison are the rate of opening of the valve as defined by the average position of the valve during the course of opening as well as the shape of the valve in the mid-to-later stages of opening when the effects of the initial coaptation of the valve have diminished. As is clear from the figures, the shape assumed by the deformed valve is very different from that of the stiffer valve used by [16]. In particular, when the tissue valve opens, a change in curvature can be observed. The change in curvature of the present valve comes about in the early stage of valve opening. At this stage, the valve bulges outward

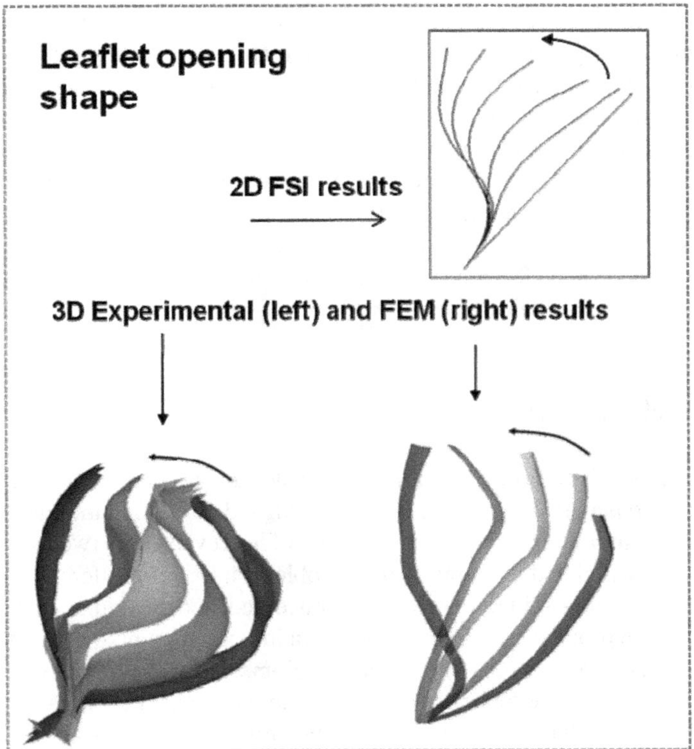

Fig. 5.11 Comparison of the valve opening characteristics obtained from two-dimensional FSI calculations with those obtained from experiments and 3D FEM computations of valve mechanics with an imposed pressure difference across the valve

under the action of the imposed pressure. This gives the valve a convex (upward) curvature in the early part of the opening phase. As the valve opens and the blood flows with increasing velocity through the central orifice of the valve, the leaflet edges gain in velocity and the concave upward curvature develops in the region close to the edge of the leaflet. The resulting s-shaped profile is then maintained as the valve opens fully. This type of change in curvature along the valve surface has also been observed in three-dimensional finite element calculations of valve motion under the action of applied pressures [90], as well as during physiologic flow loop simulations [91]. Additionally, the appearance of the change in signs of curvature during valve opening is made possible by the dynamic nature of the present calculations. Quasi-static calculations for leaflet deformation, as performed by de Hart et al. where valve inertia is neglected, or the rigid valve calculations performed by Dumont and coworkers [55], would not capture this type of shape characteristic and therefore would not properly capture the realistic response of tissue heart valves.

The results obtained using the current method and the comparison of the results with the previous studies underscores an important point. The details of valve shape, surface stresses experienced, and other fluid effects that impact the leaflet performance and pathologies are dependent upon the ability to accurately compute and fully resolve the flowfield, including the shear and boundary layers. This is only possible when the interfaces embedded in the flow are captured accurately (in a sharp manner) and the grid employed is fine enough that the flow features are adequately captured. Similarly, the flowfield and valve dynamics are highly dependent upon the use of realistic material properties. However, this type of computation faces challenges in numerical stiffness and requires a highly resolved grid. Simulating physiologically realistic Reynolds numbers with realistic material properties has not been possible in past studies.

5.22 Concluding Remarks

There are a host of FSI problems in the human system and they have extremely significant roles to play in determining human health and disease. In the cardiovascular system, the dynamics of blood cells and motion of heart valves are two examples that have been addressed in this chapter. These problems have very different characteristics as far as the fluid–solid interactions are concerned, and the numerical techniques suitable for each problem are quite different. An important point to note is that while computational techniques for FSI problems in biomechanics have drawn extensively from FSI techniques in aeroelasticity and other applications, there are some key elements of the biomechanical problems that have prompted changes in the numerical techniques. Chief among these is the issue of strong added mass effects in some biomedical problems. An example is the tissue heart valve dynamics problem. Due to the challenges presented by FSI problems of the type shown in this chapter, viz., large deformations, unsteady and pulsatile flows, strong added mass effects, and highly flexible, soft structures, the development of efficient, robust FSI methods for biomedical systems continues to be a matter of ongoing research.

References

1. Morand HJP, Ohayon R (1995) Fluid structure interaction: applied numerical methods. Wiley; Masson, Chichester; New York; Paris. viii, 212p
2. Piperno S (1997) Explicit/implicit fluid/structure staggered procedures with a structural predictor and fluid subcycling for 2D inviscid aeroelastic simulations. Int J Numerical Methods Fluids 25(10):1207–1226
3. Piperno S, Farhat C, Larrouturou B (1995) Partitioned procedures for the transient solution of coupled aroelastic problems. 1. Model problem, theory and 2-dimensional application. Comput Methods Appl Mech Eng 124(1–2):79–112
4. Peskin CS (2002) The immersed boundary method. Acta Numerica 11:479–517
5. Mittal R, Iaccarino G (2005) Immersed boundary methods. Annu Rev Fluid Mech 37:239–261
6. Osher S, Sethian JA (1988) Fronts propagating with curvature-dependent speed – algorithms based on Hamilton-Jacobi formulations. J Comput Phys 79(1):12–49
7. Michler C (2005) Efficient numerical methods for fluid-structure interaction. In: Aerospace engineering, Delft University of Technology, Delft, p 161

8. Michler C, van Brummelen EH, Hulshoff SJ, de Borst R (2003) The relevance of conservation for stability and accuracy of numerical methods for fluid-structure interaction. Comput Methods Appl Mech Eng 192(37–38):4195–4215
9. Le Tallec P, Mouro J (2001) Fluid structure interaction with large structural displacements. Comput Methods Appl Mech Eng 190(24–25):3039–3067
10. Sun W, Sacks MS (2005) Finite element implementation of a generalized Fung-elastic constitutive model for planar soft tissues. Biomech Model Mechanobiol 4(2–3): 190–199
11. Kan HC, Shyy M, Udaykumar HS, Vigneron P, Tran-Son-Tay R (1999) Effects of nucleus on leukocyte recovery. Ann Biomed Eng 27(5):648–655
12. Tran-Son-Tay R, Kan HC, Udaykumar HS, Damay E, Shyy W (1998) Rheological modelling of leukocytes. Med Biol Eng Comput 36(2):246–250
13. Marella SV, Udaykumar HS (2004) Computational analysis of the deformability of leukocytes modeled with viscous and elastic structural components. Phys Fluids 16(2):244–264
14. Peskin CS (1977) Numerical analysis of blood flow in the heart. J Comput Phys 25(3): 220–252
15. Arthurs KM, Moore LC, Peskin CS, Pitman EB, Layton HE (1998) Modeling arteriolar flow and mass transport using the immersed boundary method. J Comput Phys 147(2): 402–440
16. De Hart J, Peters GW, Schreurs PJ, Baaijens FP (2000) A two-dimensional fluid-structure interaction model of the aortic valve. J Biomech 33(9):1079–1088
17. Krishnan S, Udaykumar HS, Marshall JS, Chandran KB (2006) Two-dimensional dynamic simulation of platelet activation during mechanical heart valve closure. Ann Biomed Eng 34(10):1519–1534
18. Dillon RH, Fauci LJ, Yang XZ (2006) Sperm motility and multiciliary beating: an integrative mechanical model. Comput Math Appl 52(5):749–758
19. Dillon RH, Fauci LJ, Omoto C (2003) Mathematical modeling of axoneme mechanics and fluid dynamics in ciliary and sperm motility. Dyn Continuous Dis Impulsive Syst A Math Anal 10(5):745–757
20. Farhat C, Lesoinne M, LeTallec P (1998) Load and motion transfer algorithms for fluid/structure interaction problems with non-matching discrete interfaces: momentum and energy conservation, optimal discretization and application to aeroelasticity. Comput Methods Appl Mech Eng 157(1–2):95–114
21. Felippa CA, Park KC, Farhat C (2001) Partitioned analysis of coupled mechanical systems. Comput Methods Appl Mech Eng 190(24–25):3247–3270
22. Baaijens FPT (2001) A fictitious domain/mortar element method for fluid-structure interaction. Int J Numerical Methods Fluids 35(7):743–761
23. Bhardwaj MK, Kapania RK, Reichenbach E, Guruswamy GP (1998) Computational fluid dynamics/computational structural dynamics interaction methodology for aircraft wings. AIAA 36(12):2179–2186
24. Wagner GJ, Moes N, Liu WK, Belytschko T (2001) The extended finite element method for rigid particles in Stokes flow. Int J Numerical Methods Eng 51:293–313
25. Fedkiw RP (2002) Coupling an Eulerian fluid calculation to a Lagrangian solid calculation with the ghost fluid method. J Comput Phys 175:200–224
26. Griffith BE, Hornung RE, Mcqueen D, Peskin CS (2006) An adaptive, formally second order accurate version of the immersed boundary method. J Comput Phys 223(1):10–49
27. Peskin CS (1982) The fluid-dynamics of heart-valves – experimental, theoretical, and computational methods. Annu Rev Fluid Mech 14:235–259
28. Mcqueen DM, Peskin CS, Yellin EL (1982) Fluid-dynamics of the mitral-valve – physiological-aspects of a mathematical-model. Am J Physiol 242(6):1095–1110
29. McQueen DM, Peskin CS (1991) Computational studies of blood flow in the heart in two and three dimensions. In: Proceedings of the 1991 IEEE 17th Annual Northeast Bioengineering conference (Cat. No.91CH2997-5), 4–5 April 1991, IEEE, Hartford, CT

30. Peskin CS, Mcqueen DM (1989) A 3-dimensional computational method for blood-flow in the heart. 1. Immersed elastic fibers in a viscous incompressible fluid. J Comput Phys 81(2): 372–405
31. Fogelson AL, Guy RD (2004) Platelet-wall interactions in continuum models of platelet thrombosis: formulation and numerical solution. Math Med Biol J IMA 21(4):293–334
32. Wang NT, Fogelson AL (1999) Computational methods for continuum models of platelet aggregation. J Comput Phys 151(2):649–675
33. Marella S, Krishnan S, Liu H, Udaykumar HS (2005) Sharp interface Cartesian grid method I: an easily implemented technique for 3D moving boundary computations. J Comput Phys 210(1):1–31
34. Kang MJ, Fedkiw RP, Liu X-D (2000) A boundary condition capturing method for multiphase incompressible flow. J Sci Comput 15(3):323–359
35. Zhang L, Gerstenberger A, Wang XD, Liu WK (2004) Immersed finite element method. Comput Methods Appl Mech Eng 193(21–22):2051–2067
36. Stockie JM, Wetton BR (1999) Analysis of stiffness in the immersed boundary method and implications for time-stepping schemes. J Comput Phys 154(1):41–64
37. Tu C, Peskin CS (1992) Stability and instability in the computation of flows with moving immersed boundaries: a comparison of three methods. SIAM J Sci Stat Comput 13(6):1361–1376
38. Griffith BE, Hornung RD, McQueen DM, Peskin CS (2007) An adaptive, formally second order accurate version of the immersed boundary method. J Comput Phys 223(1):10–49
39. Griffith BE, Peskin CS (2005) On the order of accuracy of the immersed boundary method: higher order convergence rates for sufficiently smooth problems. J Comput Phys 208(1): 75–105
40. Peskin CS, Printz BF (1993) Improved volume conservation in the computation of flows with immersed elastic boundaries. J Comput Phys 105(1):33–46
41. Leveque RJ, Li ZL (1997) Immersed interface methods for Stokes flow with elastic boundaries or surface tension. SIAM J Sci Comput 18(3):709–735
42. Leveque RJ, Li ZL (1994) The immersed interface method for elliptic-equations with discontinuous coefficients and singular sources. SIAM J Numerical Anal 31(4):1019–1044
43. Lai M-C, Li Z (2001) A remark on jump conditions for the three-dimensional Navier-Stokes equations involving an immersed moving membrane. Appl Math Lett 14:149–154
44. Li Z, Lai M-C (2001) The immersed interface method for the Navier-Stokes equations with singular forces. J Comput Phys 171:822–842
45. Fedkiw RP, Aslam T, Merriman B, Osher S (1999) A non-oscillatory Eulerian approach to interfaces in multimaterial flows (the ghost fluid method). J Comput Phys 152(2):457–492
46. Udaykumar HS, Krishnan S, Marella S (2009) Adaptively refined, parallelised sharp interface Cartesian grid method for three-dimensional moving boundary problems. Int J Comput Fluid Dyn 23(1):1–24
47. Udaykumar HS, Mittal R, Rampunggoon P (2002) Interface tracking finite volume method for complex solid-fluid interactions on fixed meshes. Commun Numerical Methods Eng 18(2):89–97
48. Udaykumar HS, Krishnan S, Marella S (2008) Adaptively refined, parallelized sharp interface Cartesian grid method for 3-D moving boundary problems. Int J Comput Fluid Dyn 23(1): 1–24
49. Liu H, Marella S, Krishnan S, Udaykumar HS (2004) Sharp interface cartesian grid method II: a technique for simulating droplet interactions with surfaces of arbitrary shape 210(1): 32–54
50. Marella S (2006) A parallelized sharp-interface fixed grid method for moving boundary problems. In: Mechanical engineering. The University of Iowa, Iowa City, IA
51. Sethian JA (1999) Level set methods and fast marching methods: evolving interfaces in computational geometry, fluid mechanics, computer vision, and materials science. Cambridge monographs on applied and computational mathematics, 2nd edn, vol 3. Cambridge University Press, Cambridge, UK; New York, pp xx, 378

52. Bertrand F, Tanguy PA, Thibault F (1997) A three-dimensional fictitious domain method for incompressible fluid flow problems. Int J Numerical Methods Fluids 25(6):719–736
53. De Hart J, Peters GW, Schreurs PJ, Baaijens FP (2003) A three-dimensional computational analysis of fluid-structure interaction in the aortic valve. J Biomech 36(1):103–112
54. De Hart J, Baaijens FP, Peters GW, Schreurs PJ (2003) A computational fluid-structure interaction analysis of a fiber-reinforced stentless aortic valve. J Biomech 36(5):699–712
55. Dumont K, Stijnen JM, Vierendeels J, van de Vosse FN, Verdonck PR (2004) Validation of a fluid-structure interaction model of a heart valve using the dynamic mesh method in fluent. Comput Methods Biomech Biomed Eng 7(3):139–146
56. Liu Y, Zhang L, Wang X, Liu WK (2004) Coupling of Navier-Stokes equations with protein molecular dynamics and its application to hemodynamics. In: Proceedings of the 12th international conference on Finite Element Methods in Fluids, 2–4 April 2003, International Journal for Numerical Methods in Fluids, Wiley, Nagoya
57. Zhang L, Gerstenberger A, Wang X, Liu WK (2004) Immersed finite element method. Comput Methods Appl Mech Eng 193(21–22):2051–2067
58. Liu WK, Liu Y, Farrell D, Zhang L, Wang XS, Fukui Y, Patankar N, Zhang Y, Bajaj C, Lee J, Hong J, Chen X, Hsu H (2006) Immersed finite element method and its applications to biological systems. Comput Methods Appl Mech Eng 195(13–16):1722–1749
59. Hoffman K, Chiang S (2004) Computational fluid dynamics, 4th edn, vol 1. Engineering Education System, Wichita
60. Vigmostad SC, Udaykumar HS, Lu J, Chandran KB Fluid-structure interaction methods in biological flows with special emphasis on heart valve dynamics. Int J Numer Methods Biomed Eng 26(3–4):435–470
61. Heath M (2001) Scientific computing: an introductory survey, 2 edn. McGraw-Hill, New York, NY
62. Matthies HG, Niekamp R, Steindorf J (2006) Algorithms for strong coupling procedures. Comput Methods Appl Mech Eng 195(17–18):2028–2049
63. Michler C, Hulshoff SJ, van Brummelen EH, de Borst R (2004) A monolithic approach to fluid-structure interaction. Comput Fluids 33(5–6):839–848
64. Causin P, Gerbeau JF, Nobile F (2005) Added-mass effect in the design of partitioned algorithms for fluid-structure problems. Comput Methods Appl Mech Eng 194(42–44): 4506–4527
65. Sun W, Sacks MS, Sellaro TL, Slaughter WS, Scott MJ (2003) Biaxial mechanical response of bioprosthetic heart valve biomaterials to high in-plane shear. J Biomech Eng 125(3): 372–380
66. Van Brummelen EH, De Borst R (2005) On the nonnormality of subiteration for a fluid-structure-interaction problem. SIAM J Sci Comput 27(2):599–621
67. Jaiman RK, Jiao X, Geubelle PH, Loth E (2005) Assessment of conservative load transfer for fluid-solid interface with non-matching meshes. Int J Numerical Methods Eng 64(15): 2014–2038
68. Michler C, van Brummelen EH, de Borst R (2006) Error-amplification analysis of subiteration-preconditioned GMRES for fluid-structure interaction. Comput Methods Appl Mech Eng 195(17–18):2124–2148
69. Matthies HG, Steindorf J (2003) Partitioned strong coupling algorithms for fluid-structure interaction. Comput Struct 81(8–11):805–812
70. Michler C (2006) Michler Thesis – Edit! In: Aerospace engineering? pp 4195–4215
71. Matthies HG, Steindorf J (2002) Partitioned but strongly coupled iteration schemes for nonlinear fluid-structure interaction. Comput Struct 80(27–30):1991–1999
72. Dong C, Skalak R (1992) Leukocyte deformability – finite-element modeling of large viscoelastic deformation. J Theor Biol 158(2):173–193
73. Dong C, Skalak R, Sung KL (1991) Cytoplasmic rheology of passive neutrophils. Biorheology 28(6):557–567
74. Vigmostad S (2007) A sharp interface fluid-structure interaction for bioprosthetic heart valves. The University of Iowa, Iowa City, IA, p 169

75. Fedkiw RP (2002) Coupling an Eulerian fluid calculation to a Lagrangian solid calculation with the ghost fluid method. J Comput Phys 175(1):200–224
76. Liu H, Krishnan S, Marella S, Udaykumar HS (2005) Sharp interface Cartesian grid method II: a technique for simulating droplet interactions with surfaces of arbitrary shape. J Comput Phys 210(1):32–54
77. Badia S, Nobile F, Vergara C (2008) Fluid-structure partitioned procedures based on Robin transmission conditions. J Comput Phys 227(14):7027–7051
78. Badia S, Nobile F, Vergara C (2009) Robin-Robin preconditioned Krylov methods for fluid-structure interaction problems. Comput Methods Appl Mech Eng 198(33–36):2768–2784
79. Cheng R, Lai YG, Chandran KB (2004) Three-dimensional fluid-structure interaction simulation of bileaflet mechanical heart valve flow dynamics. Ann Biomed Eng 32(11):1471–1483
80. Cheng R, Lai YG, Chandran KB (2003) Two-dimensional fluid-structure interaction simulation of bileaflet mechanical heart valve flow dynamics. J Heart Valve Dis 12(6):772–780
81. Lai YG, Chandran KB, Lemmon J (2002) A numerical simulation of mechanical heart valve closure fluid dynamics. J Biomech 35(7):881–892
82. Bluestein D, Yin W, Affeld K, Jesty J (2004) Flow-induced platelet activation in mechanical heart valves. J Heart Valve Dis 13(3):501–508
83. Bluestein D, Li YM, Krukenkamp IB (2002) Free emboli formation in the wake of bi-leaflet mechanical heart valves and the effects of implantation techniques. J Biomech 35(12): 1533–1540
84. Bluestein D, Rambod E, Gharib M (2000) Vortex shedding as a mechanism for free emboli formation in mechanical heart valves. J Biomech Eng Trans Asme 122(2):125–134
85. Avrahami I, Rosenfeld M, Einav S, Eichler M, Reul H (2000) Can vortices in the flow across mechanical heart valves contribute to cavitation? Med Biol Eng Comput 38(1):93–97
86. Farhat C, Lesoinne M (2000) Two efficient staggered algorithms for the serial and parallel solution of three-dimensional nonlinear transient aeroelastic problems. Comput Methods Appl Mech Eng 182(3–4):499–515
87. Farhat C, Lesoinne M, Stern P, Lanteri S (1997) High performance solution of three-dimensional nonlinear aeroelastic problems via parallel partitioned algorithms: methodology and preliminary results. Adv Eng Softw 28(1):43–61
88. Hole JW (1996) Hole's human anatomy & physiology, 7th edn. Wm. C. Brown, Dubuque, IA, pp ix, 238
89. Vigmostad S, Udaykumar HS, Lu J, Sacks MS, Chandran KB (2009) A fluid-structure interaction model for 3D heart valve dynamics. In: Proceedings of the ASME summer bioengineering conference, SBC2008, n PART B, pp 1101–1102
90. Kim H, Lu J, Sacks MS, Chandran KB (2008) Dynamic simulation of bioprosthetic heart valves using a stress resultant shell model. Ann Biomed Eng 36(2):262–275
91. Iyengar AKS, Sugimoto H, Smith DB, Sacks MS (2001) Dynamic in vitro quantification of bioprosthetic heart valve leaflet motion using structured light projection. Ann Biomed Eng 29(11):963–973

Chapter 6
Mesoscale Analysis of Blood Flow

Jeffrey S. Marshall, Jennifer K.W. Chesnutt, and H.S. Udaykumar

Abstract Blood flow in the cardiovascular system and its interaction with the vessel walls plays a crucial role in health and disease. Individual blood cells play varied and vital roles in the circulation, including transport of nutrients and dissolved gases (red blood cells), fighting infections and disease (white blood cells), and healing of wounds (platelets). Malfunctioning of blood cells can result in pathologies such as sickle cell disease (red blood cells), ischemia (white blood cells), atherosclerosis (white blood cells and platelets) and thrombosis (platelets and red blood cells). To better understand the behavior of blood cells and their role in health and disease, microscale models that capture the dynamics of individual cells and their interactions with other cells/vessel walls can be very useful. However, since even micro-volumes of blood contain extremely large numbers of cells, connecting blood flow phenomena to cell dynamics and cell–cell/cell–wall interactions limits the usefulness of micro-scale models. Mesoscale models that do not model individual cells in detail, but allow for the treatment of large numbers of cells can provide important insights into the impact of the particulate nature of blood; such mesoscale models can represent transport phenomena, aggregation/disaggregation of cell clusters, collisional interactions of cells with each other and with walls and other phenomena important to healthy and pathological states in the circulation. This chapter describes the important features of such mesoscale models of blood; the treatment of the particulate nature of blood and the modeling and simulation of cell–cell and cell–surface interactions are covered. The examples presented illustrate the state of the art in mesoscale modeling of blood flow.

J.S. Marshall (✉)
School of Engineering, University of Vermont, Burlington, VT 05405, USA
e-mail: jeffm@cems.uvm.edu

K.B. Chandran et al. (eds.), *Image-Based Computational Modeling of the Human Circulatory and Pulmonary Systems*, DOI 10.1007/978-1-4419-7350-4_6,

6.1 Introduction

Part III of this volume discusses a wide variety of physiological problems that occur in various parts of the cardiovascular and pulmonary systems, and the role of fluid, vesicle, and cellular dynamics in both producing and mitigating these problems. In some cases, these problems are addressable by consideration of fluid–solid systems occurring within a narrow range of length scales. As an example of such a problem, recall that red blood cells (RBC) deform from a biconcave shape with no flow into a parachute-type shape within the capillaries and that this shape change increases the contact area between the cell and the capillary wall, hence increasing the rate of oxygen transport to the surrounding tissues. Certain diseases, such as diabetes, can result in a reduction in RBC deformability [1], leading to conditions such as tissue hypoxia. For this and similar problems within the capillaries, it is reasonable to construct a computational model that fully captures the flow and cell deformations within a sufficiently large region so as to be reasonably representative of the phenomenon.

On the other hand, a great many of the biomedical phenomena discussed in this volume occur in the larger scale vessels of the cardiovascular systems, such as the small arteries and veins, the arterioles, and the venules, as well as within regions of the valves separating the different chambers of the heart, particularly in cases where artificial valves have replaced natural valves. Phenomena such as thrombosis formation within stenotic arteries or mechanical heart valves, clot formation, plasma skimming at bifurcations, and leukocyte margination and endothelial attachment all entail flows consisting of tens to hundreds of thousands of blood cells. Mechanical processes such as RBC aggregation and aggregate breakup, RBC deformation and migration, leukocyte adhesion and rolling, and platelet activation all play important roles in the development of such phenomena. These processes can change rapidly within the flow field and are often interdependent, both with each other and with the imposed fluid flow. For instance, the presence of RBCs and the formation of rouleaux (RBC aggregates) are known to have a strong effect on rolling of leukocytes [2–4] and on onset of platelet activation in regions of high shear stress [5–9]. The formation and breakup of rouleaux have a strong influence on the margination of leukocytes and platelets [10–13]. In this process, RBCs migrate to the central region of a blood vessel and platelets and leukocytes collect preferentially near the vessel walls, which strongly influences the ability of leukocytes and platelets to adhere to endothelial cells lining the vessel walls.

The traditional approach for solution of blood flow problems is to select an appropriate continuum model that displays shear-thinning behavior similar to that observed for blood. For instance, Buchanan et al. [14] examine predictions of three different stress constitutive models—a Newtonian fluid, a power law fluid, and a Quemada fluid—for pulsatile flow through a stenosed tube, as well as the consequences of each constitutive model on particle residence time within the tube. To account for the nonuniform distribution of RBCs over the cross section of narrower blood vessels, other investigators use a two-fluid multiphase continuum model [15–17], in which one fluid represents the RBC-free layer near the vessel walls and the other fluid represents the RBC-rich region within the central part of the vessel. For prediction of RBC aggregation, the continuum fluid models are supplemented by kinetic-theory-based population balance models for aggregate formation rate and breakup, following along the lines of the classical Smoluchowski equation [18–23]. A similar population balance approach has also been used for other blood elements, such as activated platelets [24]. While continuum models of this type can readily be developed, calibrated, and tested for slowly varying flows in which the microstructure does not differ much from the equilibrium state, most of the various phenomena discussed earlier in this chapter occur in regions of blood flow which are distinctly nonequilibrium. These include a sudden branching of a blood vessel, a stent, or other type of arterial constriction, or the high-shear region that occurs where a bileaflet heart valve closes [25]. In such cases, the fluid microstructure can deviate strongly from equilibrium values, leading to breakdown of the continuum models at precisely the locations in which one is most interested in modeling.

A number of researchers have explored development of direct computational models for blood cells and their interactions, consisting of detailed computation of both the flow around each blood cell and the deformation of the cell and its membrane. Microscale models of this type have proven very useful in understanding the local behavior of blood cells in different situations, such as the deformation of an RBC in a shear flow [26–28], the transition from tank-treading to flip-flopping motions of RBCs [29], the dynamics of two RBCs colliding under a shear flow [30, 31], and the impact of RBCs with leukocytes and their effect on leukocyte rolling [3, 4]. However, microscale models are limited to treatment of a relatively small number of blood cells. Many microscale models deal with only 1–2 cells, and even the most recent models of this type can only handle on the order of 20–30 cells [32]. A typical arteriole has a width of about 80 μm, so that at physiological hematocrit values, 60–80 RBCs would lie on any given cross section of the vessel. Assuming a computational domain length of about 10 times the vessel width, which would be minimal for any simulation, yields over 70,000 RBCs in even a fairly mild computation. Problems involving phenomena such as the effect of RBCs on platelet activation in artificial heart valves would necessitate inclusion of many times more RBCs than even this estimate. While microscale models provide valuable information on the behavior of blood microstructure in different situations, their use for direct solution of many of the most pressing blood flow problems is well out of reach of model computational abilities for all but the smallest capillaries.

This chapter focuses on computational modeling of blood flow phenomena for which macroscale models are generally not adequate but complete computation via microscale models is not feasible. For such problems, some type of *mesoscale* model is required which incorporates some of the dynamics of the microscale models, but at the same time makes sufficient simplifications so as to treat the large numbers of blood cells characteristic of the physiological phenomenon. Two types of mesoscale models are explored in this chapter. The first of these, commonly known as *dissipative particle dynamics* (DPD), is an offshoot of the molecular dynamics approach, with the difference that in DPD each computational particle represents a large group of molecules. The second type of model, commonly referred to as the *discrete element method* (DEM), is based on an outgrowth of the method of the same name used for prediction of granular flows, as well as the various approaches developed in the fluid mechanics community for simulation of dense particulate flows. While both the DEM and DPD methods are based on a particle type of discretization, there are many differences in the formulation and philosophy of the two approaches.

The chapter first presents an overview of the different length and time scales involved in different blood flow phenomena, which serves to quantify the challenges in computationally modeling such phenomena. Next, we explore computational models for cell adhesion, which is of central importance in many of the phenomena discussed above. We then provide background on microscale models of blood cell interaction in shear flows, discussing both the contributions and limitations of such models. The development of mesoscale models using the discrete element method for colliding and adhesive blood cells is examined, followed by a survey of the use of dissipative particle dynamics for both polymeric and colloidal fluids. For both methods, we emphasize the various approximations that must be made in the development of the mesoscale model, as well as the advantages of the mesoscale modeling approach in enabling consideration of large ensembles of aggregating blood cells under different flow conditions. The chapter concludes with a discussion of the potential for bridging the various scales of modeling involved in blood flow analysis.

6.2 Scaling Estimates

Blood flow phenomena occur over a large range of length and time scales, as indicated in Table 6.1. In this table, we have restricted attention to the primary elements involved in mechanical mesoscale and microscale models and have thus either omitted or grossly oversimplified features involved with the complex series of chemical reactions that trigger blood clotting, cell signaling, etc. All subcellular processes are also omitted, except for certain processes occurring on the cell membrane, which are important for cell adhesion. We have also not included time scales typically associated with development of diseases, such as thrombus formation, which are typically measured in years. Even with these provisos, the range of length scales that govern essential blood flow processes is enormous, ranging from the 10 nm length scale of the ligands that control bonding of colliding RBCs to the 3 cm size of the aorta leading into and out of the heart—a variation of over six orders of magnitude.

The range of time scale variation is similarly large, but the time scale of a process does not vary in proportion to the length scales. For instance, the fluid advection time scale is similar throughout the cardiovascular system, from the aorta to the smallest capillaries, due to the fact that the flow rate within the different blood vessels varies in rough proportion to the vessel size. On the other hand, a single element on this table can exhibit processes with widely different time scales. For instance, simulation of the motion of a red blood cell with effective diameter d exhibits the

Table 6.1 Length and time scales for different components of the cardiovascular system

Feature	Actions	Length scales	Time scales
Ligands and receptors	Embedded in cell membrane, allow cells to adhere following collision	10 nm	Bonding $\sim 10^2$–10^{-14} s
Blood plasma proteins involved in hemostasis and thrombosis (fibrinogen, thrombin, etc.)	Form protein network that binds together blood clots, influence blood viscosity	50 nm	Activation $\sim$ 1–10 s
Platelets	Activate and anchor protein network in blood clots	2 μm	Particle advection $\sim 10^{-6}$–10^{-2} s Activation $\sim$ 0.001–0.1 s
Red blood cells (erythrocytes)	Carry oxygen to tissues, influence blood viscosity, aggregate to form rouleaux	6 μm	Particle advection $\sim 10^{-5}$–10^{-2} s Collision $\sim 10^{-3}$–10^{-2} s
White blood cells (leukocytes, etc.)	Adhere to endothelial cells and migrate into surrounding tissues to fight disease	8 μm	Particle advection $\sim 10^{-5}$–10^{-2} s Collision $\sim 10^{-3}$–10^{-2} s
Capillaries	Smallest blood vessels, over which oxygen exchange occurs	10 μm	Fluid advection $\sim$ 0.01 s
Arterioles, venules	Blood vessels leading to and from capillaries	80 μm	Fluid advection $\sim$ 0.015 s
Small arteries and veins	Blood vessels connecting arterioles/venules and arteries/veins	300 μm	Fluid advection $\sim$ 0.015 s
Large arteries and veins	Blood vessels transporting blood to and from heart and organs/muscles	5 mm	Fluid advection $\sim$ 0.025 s
Ascending aorta	Vessel transporting blood from the heart to the arteries	30 mm	Fluid advection $\sim$ 0.05 s

advective time scale $T_P = d/U$, where U is the fluid characteristic velocity. If the time step does not resolve this scale, then collision of RBCs with each other will be missed by the simulation. However, there is also an inherent time scale for drift of a particle suspended in a fluid flow, given by $T_S = \mathrm{St}\,L/U$, where L is the characteristic fluid length scale (the vessel diameter) and St is the Stokes number, $\mathrm{St} \equiv \rho_P d^2 U/18\mu L$. Here, μ is the blood plasma viscosity and ρ_P is the density of the RBC. If an explicit method is used for time stepping, the computation will become unstable if the time step exceeds T_S. This instability can be avoided by use of an implicit time evolution algorithm, but the solution will not resolve the RBC drift relative to the fluid flow for time steps above T_S. The collision of two RBCs introduces elastic deformation of the blood cells, leading to a repulsive response force. This collision occurs over a characteristic time scale $T_C = d(\rho_P^2/E^2 v_0)^{1/5}$, where E is the effective Young's modulus of the RBC and $v_0 \approx U\,\mathrm{St}$ is the typical particle impact velocity. Dulinska et al. [33] measured the effective elastic modulus E of RBCs to be about 2.6×10^5 dyne/cm^2, and the plasma viscosity is typically about $\mu = 1.2$ cP. Two colliding RBCs are attracted to each other by the attachment of proteins, termed ligands and receptors, attached to the cell membranes, which have a bonding time scale T_B that can be expressed in terms of reaction rate of the ligands and number density of receptors and ligands.

For flow in an arteriole, with $L \sim 80\ \mu\mathrm{m}$ and $U \sim 0.8$ mm/s, we obtain the fluid advection time $T_F \cong 0.1\,s$, the cell advection time scale $T_P \cong 0.006$ s, the collision time scale $T_C \cong 0.00005$ s, and the cell drift time scale $T_S \cong 10^{-6}$ s. The bonding time of the ligands and receptors exhibits broad variation, from about 10^2–10^{-14} s depending on the values selected for the various reaction rates, but a nominal value might be selected as roughly $T_B \sim 10^{-9}$ s. Thus, there are over eight orders of magnitude difference between the fluid time scale and the smallest scales associated with the RBC motion. Similarly, adhesion of RBCs occurs when the cells are separated by a gap on the order of the ligand length scale (10 nm), which is three orders of magnitude smaller than the RBC diameter, which in turn is three orders of magnitude smaller than the arteriole width. Solution of the flow field using a level-set method on a Cartesian grid over a region of volume L^3 over a time interval of order T_F would thus require $O(10^5)$ time steps on a grid with over $O(10^{18})$ grid points. A good desktop workstation can solve computational fluid dynamic problems with thousands of time steps and $O(10^6)$ grid points, over a span of several weeks per run, and parallel networks routinely solve problems with about $O(10^7)$ grid points. Clearly, substantial simplifications are necessary to make such problems tractable.

6.3 Modeling Adhesion Force Between Blood Cells

All three of the major types of blood entities—red blood cells, white blood cells, and platelets—exhibit adhesion to each other and to the surrounding endothelial cells at different stages in their life cycles. Cell adhesion is the dominant mechanism underlying clotting, thrombosis, rouleaux formation, and numerous other blood processes. There are two distinct physical mechanisms that are believed to give rise to adhesion, illustrated in Fig. 6.1. The first of these mechanisms, which we will refer to simply as *clotting*, involves the activation of platelets and the activation and subsequent polymerization of large-scale proteins (e.g., fibrinogen) suspended in the blood plasma. Polymerization of these proteins results in the formation of a dense protein mesh that entraps red blood cells and other blood constituents, and along with aggregation of platelets, ultimately forms a clot. The second mechanism for cell adhesion occurs due to bonding of two classes of proteins embedded in the cell membrane, termed ligands and receptors, when two circulating cells collide with

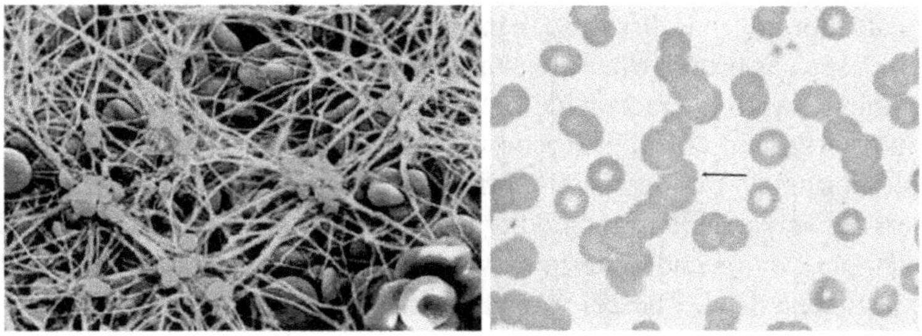

(a) (b)

Fig. 6.1 Illustration of the two major mechanisms for blood cell adhesion: (**a**) red blood cells trapped in a fibrin network forming a blood clot and (**b**) ligand–receptor binding of red blood cells to form rouleaux (reproduced with permission from (**a**) Yuri Veklich and John Weisel, University of Pennsylvania and (**b**) © 2009 Rector and Visitors of the University of Virginia, Charles E. Hess, M.D. and Lindsey Krstic, B.A.)

each other, or when a cell collides with a cell on a surface, such as the endothelial wall. This *ligand–receptor binding* process is primarily responsible for adhesion of leukocytes and platelets to endothelial cells lining the vessel walls. Red blood cells possess ligands and receptors, although the mechanism of red blood cell adhesion is unclear, with the major theories being either the bridging theory whereby the adsorption of macromolecules onto adjacent cell surfaces bridges cells together [34] or the depletion theory whereby a reduced concentration of macromolecules near red blood cell surfaces lowers osmotic forces near the cells which pulls cells together [35]. Nevertheless, models of ligand–receptor binding of red blood cells have been used in computational studies [30, 36, 37], and ligand–receptor binding is presumed for red blood cell adhesion here. Both of the adhesion processes of clotting and ligand–receptor binding involve complex chains of chemical reactions (e.g., [38, 39]), which are beyond the scope of this review. Rather, we give only a brief mechanistic review of the basic aspects of these adhesion processes, which are of primary importance for the development of mesoscale blood flow models.

Clotting begins upon injury to the endothelial wall with two processes: (1) platelet adhesion, activation, and aggregation and (2) creation of a fibrin network gel. First, platelets adhere to the endothelium via specific proteins on the platelet surfaces, which bind to specific proteins in the subendothelium. Binding of particular proteins results in the activation of bound platelets. Activated platelets release platelet agonists and the nucleotide adenosine diphosphate, which activate additional platelets that are recruited to the injury site, resulting in platelet aggregation at the site. Activated platelets also catalyze thrombin formation from the protein prothrombin, necessary for the production of the fibrin network gel. Thrombin is a strong platelet agonist, thus enhancing platelet activation and engaging new platelets in the clotting process [40, 41]. Creation of the fibrin network gel is controlled by the solution of large-scale proteins that are suspended in the blood, typically constituting about 7% of the blood plasma. The major protein involved in this process is fibrinogen, which is a composite structure consisting of six protein chains. In its normal state, fibrinogen is dissolved in the blood plasma. When thrombin generation is triggered, such as occurs when a wound results in bleeding from the blood vessel, the thrombin enzyme interacts with the fibrinogen and removes small pieces from the fibrinogen protein chain. The removal of these pieces leads to a degeneration of soluble fibrinogen into insoluble fibrin monomers. Unlike fibrinogen, the fibrin monomers actively adhere to each other as well as to activated platelets, leading to the formation of an extended fibrin network gel that entraps blood cells and other blood constituents [41]. The gel formation is limited by production and transport of thrombin molecules, so that complete formation of the resulting clot occurs over a time scale of several minutes. A mesoscale model of clot formation is presented by Boryczko et al. [42] using the dissipative particle dynamics method, which is discussed further in Section 6.5.

While clotting leads to protein meshes with length scales much larger than the size of the blood cells, other adhesion forces operate on length scales much smaller than the cell dimensions, and therefore are of importance only in the event of cell collision. These short-range adhesion forces are of two types—specific and nonspecific. Nonspecific adhesion forces, which are independent of the identity of the colliding cells, include forces such as electrostatic interaction, steric stabilization, and van der Waals forces. The electrostatic force between cells is repulsive since cells have a slight negative charge. Steric stabilization provides an even larger repulsive force between the surfaces [43, 44], but both of these forces have very short length scales (on the order of a few nanometers). The van der Waals force is generally attractive and has a somewhat longer length scale than the repulsive forces, but still on the scale of about 10 nm. The combination of these nonspecific forces leads to a curve for energy variation with separation distance that exhibits a large repulsive peak a few nanometers from the cell surface followed by a smaller attractive trough (Fig. 6.2). The equilibrium separation distance between the cell surfaces, coinciding with the minimum energy location, occurs at about 7 nm [45]. This length scale is three orders of magnitude smaller than the diameter of a red blood cell.

As the cell surfaces approach this equilibrium separation distance, proteins attached to the two cell membranes begin to intermingle and attach to each other (Fig. 6.3). The binding force of these proteins decreases somewhat the equilibrium separation distance of the cell surfaces, but the energy maximum at smaller distances

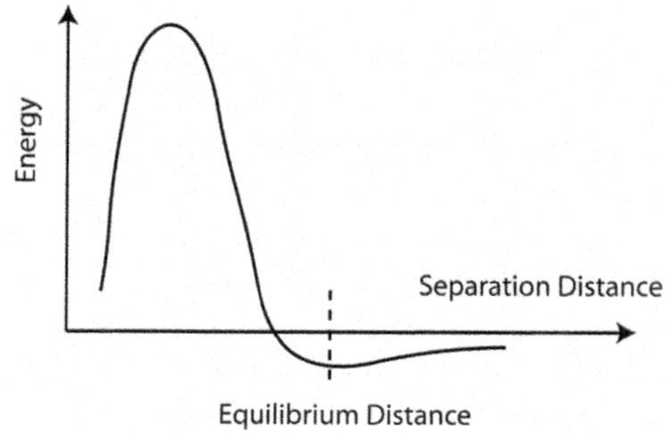

Fig. 6.2 Energy minimum due to nonspecific adhesion forces

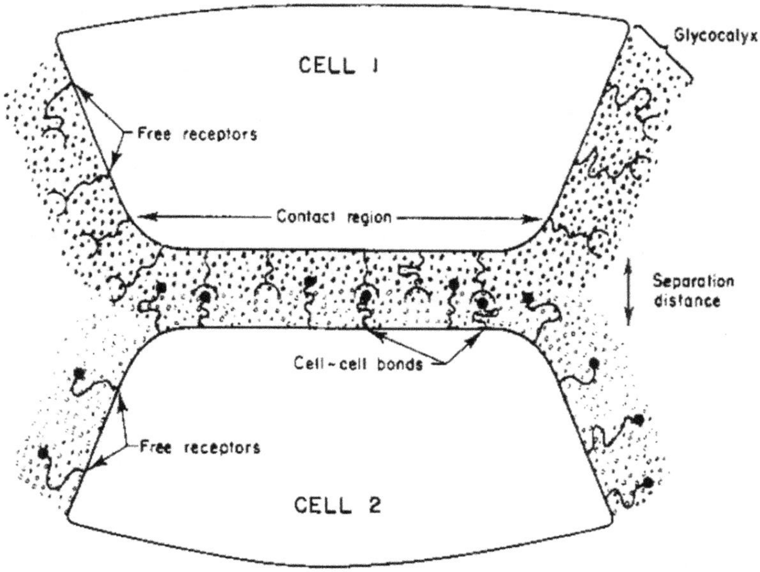

Fig. 6.3 Model for the adhesion of two cells through bonding of receptors and ligands. Bonding occurs in the contact region where the separation distance is constant (sketch of contact region is greatly enlarged for detail). Repulsion due to interaction of glycocalyces and negative surface charges is balanced by the bonds and the van der Waals attractive force (reproduced with permission from Bell et al. [47])

is sufficiently large that it cannot be overcome and the cell surface separation remains on the order of 10 nm. The protein binding serves to substantially strengthen the bond between the surfaces, resisting any attempt to separate the surfaces from the equilibrium location. There are a great many different types of molecules involved with surface adhesion of cells of different types [46], but for the purposes of the current review it suffices to refer to these proteins generically simply as ligands and receptors, and the adhesion process itself as ligand–receptor binding. Unlike the electrostatic and van der Waals forces, the protein connections involved in ligand–receptor binding are highly specific to cell types involved in the collision.

A kinetic model of ligand–receptor binding was proposed by Bell [45] and Bell et al. [47], who assumed that the force exerted by each protein bond could be modeled as a spring with spring constant σ and equilibrium length λ. The energy released by compressing or stretching a bond to a length x_b is

$$E_s = -\frac{\sigma}{2}(x_b - \lambda)^2 \tag{6.1}$$

Other studies [3, 48, 49] assume that bonds break at some maximum bond length, thereby releasing the load. The total energy associated with receptor–ligand bonds per unit area of the membrane is the product $N_b E_s$ of the single-bond energy E_s and the number density $N_b(t)$ of bonds joining the cell membranes. A kinetic equation governs the formation and elimination of bonds such that the bond number density can be determined by a balance between bond formation and elimination, given by

$$\frac{dN_b}{dt} = k_f(N_{l0} - N_b)(N_{r0} - N_b) - k_r N_b \tag{6.2}$$

Here, N_{l0} and N_{r0} are the initial receptor and ligand densities on the membrane, and k_f and k_r are the forward and reverse reaction rate coefficients. The reaction rate coefficients vary with the length of the bond in accordance with [50]

$$\begin{aligned} k_f &= k_{f0} \exp\left[-\frac{\sigma_{ts}(x_b - \lambda)^2}{2k_b T} \right] \\ k_r &= k_{r0} \exp\left[\frac{(\sigma - \sigma_{ts})(x_b - \lambda)^2}{2k_b T} \right] \end{aligned} \tag{6.3}$$

where k_{f0} and k_{r0} are, respectively, the initial forward and reverse equilibrium reaction rates, σ_{ts} is the "transition state spring constant," k_b is the Boltzmann constant, and T is the absolute temperature. Although the values of these coefficients vary widely for different cell adhesion problems, typical ranges of variation are given in Table 6.2.

Table 6.2 Ranges of experimental values for parameters related to receptor–ligand binding (reproduced with permission from Chesnutt and Marshall [36])

Parameter	Range	Reference
ℓ, equilibrium bond length	5–50 nm	Bell et al. [47] Dembo et al. [50] Springer [46]
σ, spring constant	0.01–10 dyne/cm	Dembo et al. [50]
σ_{ts}, transition state spring constant	−5 to 5 dyne/cm	Dembo et al. [50]
N_{l0}, N_{r0}, initial ligand, receptor densities	10^9–10^{12} cm^{-2}	Dembo et al. [50]
k_{f0}, forward equilibrium reaction rate	10^{-12}–10^{-7} cm^2/s	Hammer and Apte [59]
k_{r0}, reverse equilibrium reaction rate	10^{-5}–10 s^{-1}	Bell [45] Ward and Hammer [60]

If the reaction rate coefficients are assumed to remain approximately constant during the binding process, the rate equation (6.2) has the form of a Ricotti equation for the number density $N_b(t)$. Taking as an initial condition a state in which there are no bonds at an initial time t_0, so that the two cells have just collided, an analytical solution for $N_b(t)$ is obtained as [36]

$$N_b(t) = \frac{2A \tanh\left[\frac{1}{2}(B^2 - 4AC)^{1/2}(t - t_0)\right]}{-B \tanh\left[\frac{1}{2}(B^2 - 4AC)^{1/2}(t - t_0)\right] + (B^2 - 4AC)^{1/2}} \tag{6.4}$$

where $A \equiv k_f N_{l0} N_{r0}$, $B \equiv -[k_f(N_{l0} + N_{r0}) + k_r]$, and $C \equiv k_f$. Since by definition B is negative and A and C are positive, the number of bonds given by (6.4) is always positive. The coefficients are restricted by the condition $B^2 - 4AC > 0$, which is satisfied for all cases provided that $N_{l0} = N_{r0}$. The value of $N_b(t)$ asymptotes at long time to equilibrium value $N_b(\infty)$ (Fig. 6.4), given by

$$N_b(\infty) = 2A/[-B + (B^2 - 4AC)^{1/2}] \tag{6.5}$$

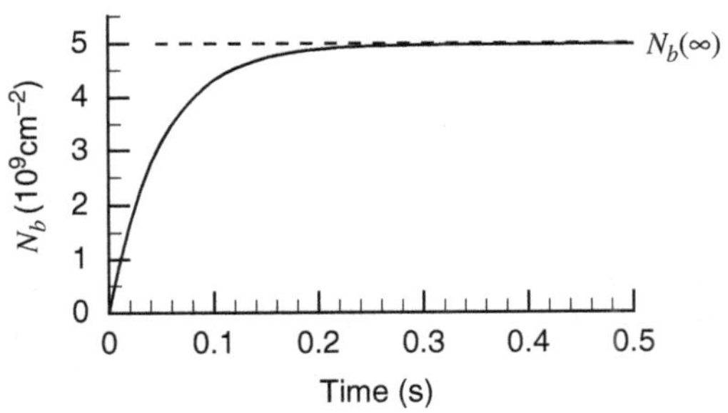

Fig. 6.4 Variation of the bond number density with time, as predicted from (6.4)

Based in part on results of microscale simulations [30, 31] as well as the many orders of magnitude difference in length scale between the cell diameter and the separation distance over which substantial adhesion occurs, we make the further assumption that the adhesive force occurs only within a flattened contact region on the cell surfaces, over which the cell separation distance is taken as constant, as determined from the minimum of the energy variation curve (e.g., Fig. 6.2). If this contact region is modeled as a circle of radius $a(t)$, then integration of the receptor–ligand binding energy over the contact region yields the net adhesion binding energy as [36]

$$U_S = -\pi\sigma(x_b - \lambda)^2 \int_0^a N_b(t, r)\, r\mathrm{d}r \tag{6.6}$$

The bond number density depends on radius r through the impact time $t_0(r)$, which is defined as the time at which the surfaces first approach to within the equilibrium separation distance at the given radius within the contact region. The time scale for bond formation T_B can be approximated from (6.4) as $T_B = 2/(B^2 - 4AC)^{1/2}$. When T_B is much smaller than the collision timescale, T_C, the additional approximation can be made that the impact time is uniform over the contact region, such that (6.6) reduces to

$$U_S = -\frac{\pi\sigma}{2}\,(x_b - \lambda)^2\, N_b(t)\, a^2(t) \tag{6.7}$$

Results (6.6) and (6.7) are equivalent to assuming a time-dependent effective surface energy density $\gamma(t)$ of

$$\gamma(t) = -\frac{1}{2}\sigma(x_b - \lambda)^2 \left[\int_0^1 N_b(t - t_0(s))\, s\, ds \right] \tag{6.8}$$

which for $T_B << T_C$ reduces to the equilibrium value

$$\gamma = -\frac{1}{4}\sigma(x_b - \lambda)^2 N_b(\infty) \tag{6.9}$$

There have been a variety of extensions of this simple kinetic model for ligand–receptor binding. Of particular interest is the observation by a number of investigators that the receptor proteins can move along the interface during the bonding process [51 – 53]. A model for this motion was developed by Tozeren et al. [53], who argue that receptor motion can be represented by adding a diffusive term to the right-hand side of (6.2). Such a model was used for microscale simulations of cell adhesion by Agresar [48], whereas other simulation approaches neglect protein mobility [54]. Experimental approaches for accurate measurement of receptor–ligand binding kinetics are described by Chesla et al. [39] and Tachev et al. [55]. The effects of competition between multiple ligand–receptor species are discussed by Zhu and Williams [56] and Coombs et al. [57]. An interesting review of receptor–ligand binding is given by Zhu [58].

6.4 Microscale Modeling: Deformable Blood Cells

As defined in the Introduction to this chapter, microscale models attempt to accurately resolve both the deformation and the flow field around each blood cell. Some aspects of the numerical techniques that apply to the computations of microscale blood cell dynamics are covered in Chapter 3 and Chapter 5 of this volume. Computational restrictions limit microscale computations to small clusters of cells, but within these limits, micromodels are very useful for examining details of the cell response to shear and collisions with other cells and surfaces. A wide variety of different computational methods have been used for microscale simulations, the most common of which are the lattice-Boltzmann method [61], boundary-element methods for low-Reynolds-number flows [28, 32, 62, 63], and various types of immersed boundary methods, in which the fluid is solved on a Cartesian grid with different densities and viscosities inside and outside of the cells.

Although the basic idea is similar, there are many different variations of this latter approach, including the volume-of-fluid method [64], an immersed finite-element method [65], the traditional immersed boundary method [27, 31], and the front-tracking method [30, 66]. A review of different variations of immersed boundary methods for biological transport problems is given by Shyy et al. [67]. In addition to variations in the fluid flow simulation method, the different microscale studies employ a variety of different models for the cell membrane, as well as for the constitutive behavior of the cytoplasmic fluid. Discussions of these issues are given by Lac et al. [63] and Marella and Udaykumar [68] in the context of computational models. More detailed reviews of the cell membrane mechanics and its relevance to blood flow are given by Chien [69] and Hochmuth [70]. The purpose of this section, however, is not to present a review of the methodologies used in these microscale studies, but rather to examine the primary results obtained by certain of these studies which most bear on the simplifications that might be used for purposes of mesoscale modeling.

For purposes of this review, the various microscale models are loosely divided into three categories: those dealing with single cells in a shear flow, those dealing with collision of two or more cells, and those dealing with adhesive cells. Two earlier papers in the first category, by Keller and Skalak [29] and Barthes-Biesel and Sgaier [26], are of particular interest. Keller and Skalak [29] examine analytically the problem of an ellipsoidal fluid capsule immersed in a shear flow, in which the ratio of fluid viscosity inside and outside of the capsule is varied. In agreement with experimental observations for red blood cells [71], the computations indicate that the viscosity ratio is primarily responsible for a transition between a tank-treading state, where the particle remains at a constant orientation and the membrane rotates around in the direction of shear flow vorticity, and a flip-flopping state, in which blood cells rotate in the shear field in a manner similar to rigid particles. Also, factors in this transition are the particle aspect ratio and sphericity index. The analysis was extended by Barthes-Biesel and Sgaier [26] by including a more realistic membrane model. Particularly, this latter study demonstrates that the membrane viscosity will have a significant effect on prediction of the transition from the tank-treading to the flip-flopping states. For the typical RBC shape and viscosity values, with cytoplasm viscosity about five times greater than the plasma viscosity, both the analysis and experiments indicate that the flip-flopping state will be observed for RBCs in a uniform shear flow, with little cell deformation.

Simulations of a deformable red blood cell in a shear flow using an immersed boundary method are reported by Eggleton and Popel [27] for cases where the plasma and cytoplasm viscosities are the same, which leads to tank-treading motion of the RBCs. Cases with properties closer to the physiologically observed values, leading to flip-flopping RBC motion, were examined for fully deformable cells by Pozrikidis [28] using a boundary-element method together with a reasonably sophisticated membrane mechanics model. Pozrikidis [28] shows that although the RBCs exhibit an oscillatory flip-flopping motion, they also deform during the motion from the biconcave shape typical of cells with no shear into an S-type shape, as illustrated in Fig. 6.5 using a time series of the cell cross section during one cycle of the flip-flopping oscillation.

The deformation of two deformable cells upon collision in a shear flow was examined using the immersed boundary method for the fluid flow coupled to a finite-element computation of the cell membrane deformation by Jadhav et al. [31]. The computational results, shown in Fig. 6.6, indicate that the cell membranes develop

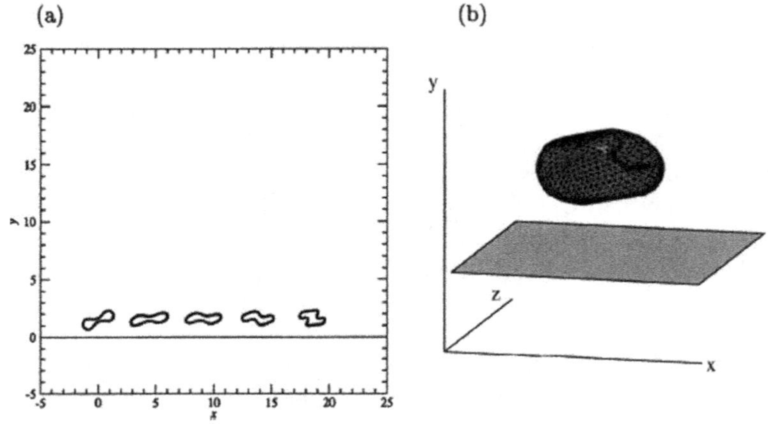

Fig. 6.5 (**a**) Time series of the cross section of an RBC in shear flow during one cycle of the flip-flopping motion and (**b**) shape of the cell in three dimensions at the final time (reproduced with permission from Pozrikidis [28])

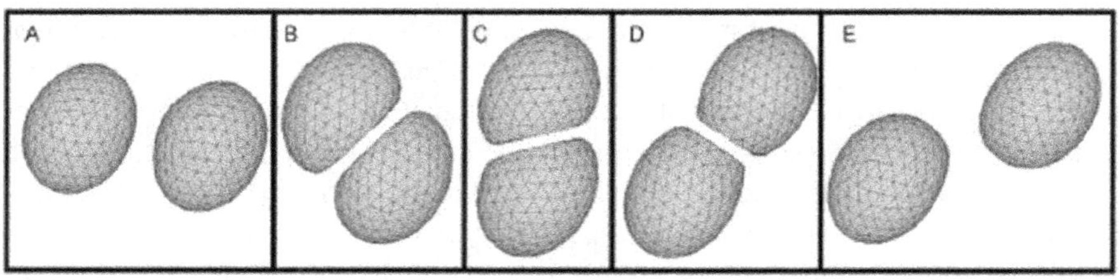

Fig. 6.6 Time series showing collision of two cells in a shear flow with a capillary number of Ca = 0.04 (reproduced with permission from Jadhav et al. [31])

a flattened region near the point of collision but do not deform substantially outside of this region. Within this "contact region," the cell surfaces are observed to be fairly flat at lower shear rates and to develop wrinkles for higher shear rates. These wrinkles serve to decrease the area available for receptor–ligand bonding of the cell surfaces. The behavior of the cell is associated with the capillary number $\mathrm{Ca} = \mu S d / 2E_\mathrm{m}$, where d is the cell diameter, μ is the viscosity of the external fluid, S is shear rate, and E_m is the membrane elastic modulus. The capillary number is a measure of the fluid viscous forces to the tensile force of the cell membrane. Jadhav et al. [31] report that cases with $\mathrm{Ca} \leq 0.004$ exhibit relatively modest deformations with approximately flat contact surfaces and that the radius of the contact region approaches the order of magnitude of the cell for Ca greater than about 0.01. Using RBC properties, a capillary number of 0.004 corresponds to a shear rate of 300 s^{-1}, above which it is doubtful that significant RBC aggregation would occur. We also note that these computations are performed with a viscosity ratio of unity, and one would expect that the contact region would be substantially smaller and the surface wrinkling less pronounced at larger viscosity ratios.

A number of recent papers have reported computational studies of collision of adhesive deformable cells in a shear flow [30, 61, 64, 65]. The paper by Liu and Liu [65] focuses on rouleau structures formed by RBCs and their evolution in shear flow. Computations with different values of the RBC membrane deformability are reported, for calculations with up to 10 cells. A typical result is given in Fig. 6.7, showing the deformation of a rouleau structure formed of four cells as a shear flow is imposed, leading to tilting of the structure on the direction of shear and a shift in the contact locations of the RBCs. Whereas in the initial configuration, prior to imposing the shear, the axes of the RBCs are all aligned parallel to the elongational direction of the rouleau structure, in the plot at long time the RBC axes are aligned nearly orthogonal to the elongational direction of the rouleau.

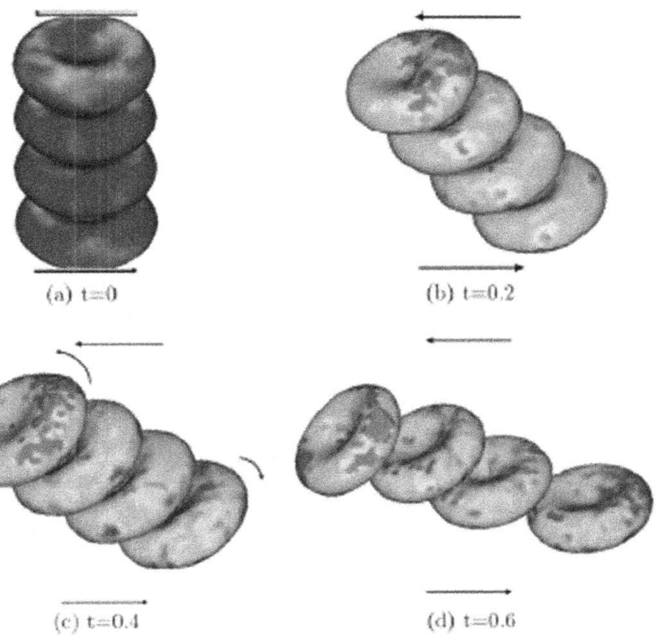

Fig. 6.7 Deformation of a four-cell rouleau under a shear flow with shear rate of 3.0 s^{-1}. The shading on the RBC surfaces indicates the shear stress magnitude (reproduced with permission from Liu and Liu [65])

6.5 Mesoscale Modeling Using the Discrete Element Method

The idea of explaining the dynamics of blood cells by use of either experimental or computational models based on tracking the motion of solid particles has a long history in blood flow literature. Of particular note among the early experimental work is that of Goldsmith (e.g., Karino and Goldsmith [72], Goldsmith and Karino [73]), who performed a long series of experiments comparing particle motion with

RBC and platelet motion in different flow fields, such as channel flow and vortex flows, with different hematocrit and channel and particle sizes. Such models generally neglect blood cell deformation, but as noted in Section 6.4, significant blood cell deformation is not usually observed in flows within the arterioles or larger blood vessels for physiologically observed viscosity values. Computational modeling for red cell motion based on use of noninteracting (i.e., dilute) dynamics of rigid particles is reported by numerous investigators, including the work of Hyun et al. [74], El-Kareh and Secomb [75], and Longest et al. [76] for cell motion in blood vessel bifurcations. Other investigators have used rigid particles to examine the motion of platelets in different blood flow systems, either experimentally (e.g., [77]) or via numerical computations with noninteracting particles [78, 79], including recent work by Krishnan et al. [25] that modeled shear-induced platelet activation during closing of a bileaflet artificial heart valve by examining the motion and stress acting on rigid particles suspended in the flow (Fig. 6.8).

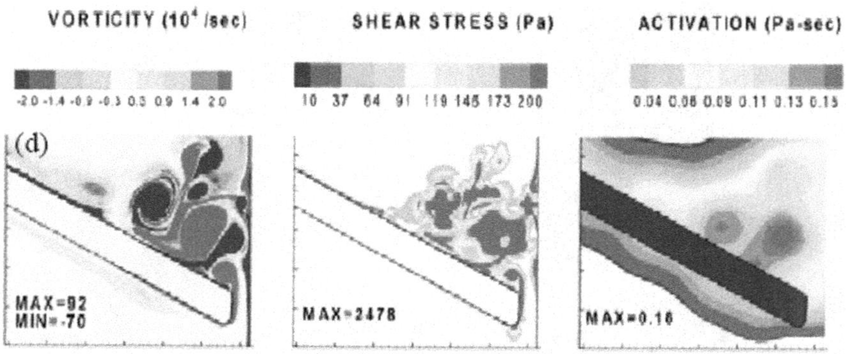

Fig. 6.8 Vorticity field, shear stress, and platelet activation level during closure of a bileaflet valve (reproduced with permission from Krishnan et al. [25])

Of course, as noted by Goldsmith and Karino [73], particle interactions play an important role in all blood flows, even at modest hematocrit levels. A number of authors have presented computations that include blood cell collisions using a relatively small number of rigid particles. For instance, a lattice-Boltzmann model using rigid particles to examine the enhancement caused by red blood cell collisions on leukocyte rolling along a wall is presented by Sun et al. [4] and Migliorini et al. [3] (Fig. 6.9). A level-set method was used by AlMomani et al. [80] to examine margination of platelets by red blood cells in a channel flow (Fig. 6.10). The study found that the red blood cells cause the platelets to tend to drift toward the wall even at modest hematocrit levels (above about 10%) and that the fluctuating shear levels induced by the RBCs have a substantial effect on the shear-induced activation parameter of the platelets in comparison to flow with no RBCs. This study used a simple method to account for homogeneous elongation of the particles, called the *pseudo-rigid body* model in the solid mechanics literature [81], but the computational results indicated that little particle deformation occurs (see also AlMomani

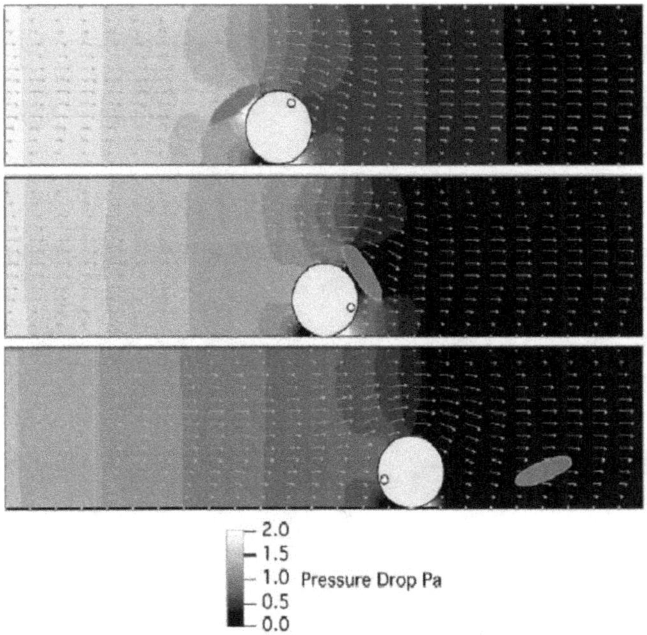

Fig. 6.9 Time series showing a simulation of the collision of a red blood cell with a leukocyte rolling along a wall (reproduced with permission from Migliorini et al. [3])

[82]). A computational model for simulation of large numbers of rigid adhesive particles of ellipsoidal shape was recently proposed by Chesnutt and Marshall [36, 37], which is described in more detail below.

In order to simplify the calculation of cell collisions and aggregation, it is common in discrete element models to make three approximations:

1. Colliding cells are assumed to approximately maintain their shape, with the exception of a flattened contact region of effective radius $a(t)$.
2. The receptor–ligand binding is assumed to occur only within the contact region.
3. The separation distance between two colliding cells within the contact region is assumed to be constant, set by a local equilibrium between the attractive and repulsive forces discussed in Section 6.3, such that motion of the cells toward or away from each other results only in a change of the contact region area.

For ease of presentation, a brief summary of the governing equations for two-dimensional discrete element calculations with ellipsoidal particles is presented below. The reader is referred to Chesnutt and Marshall [36] for the full three-dimensional theory.

The discrete element method follows the motion of individual rigid particles based on solution of the particle momentum and angular momentum equations

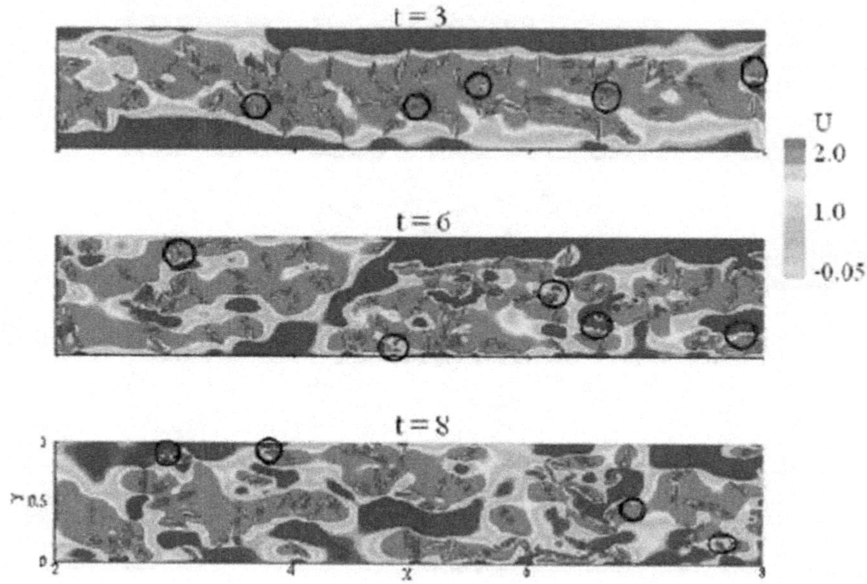

Fig. 6.10 Computation examining the effect of red blood cells (*ellipses*) on platelets (*circled*), including margination and effect on shear-induced activation. *Shades* represent contours of the streamwise velocity component (reproduced with permission from AlMomani et al. [80])

$$m\,\frac{\mathrm{d}\mathbf{v}}{\mathrm{d}t} = \mathbf{F}_\mathrm{F} + \sum_{j\neq i} \mathbf{F}_{\mathrm{A}j}, \quad I_\mathrm{z}\frac{\mathrm{d}\Omega_\mathrm{z}}{\mathrm{d}t} = M_{\mathrm{F,z}} + \sum_{j\neq i} M_{\mathrm{A}j,\mathrm{z}} \tag{6.10}$$

where m is the particle mass, $\mathbf{v}$ is the particle velocity, Ω_z is the particle rotation rate, and d/dt is the time derivative following a moving particle. The mass moment of inertia of a spheroidal particle with semiaxis lengths a along the particle symmetry axis and b in the direction orthogonal to the symmetry axis is $I_\mathrm{z} = \frac{1}{5}m\left(a^2 + b^2\right)$. The fluid-induced force and torque on the particle are given by $\mathbf{F}_\mathrm{F}$ and $M_{\mathrm{F,z}}$, and the force and torque due to collision and adhesion of particle i with another particle j are $\mathbf{F}_\mathrm{A}$ and $M_{\mathrm{A,z}}$. The sum is performed over all particles j in contact with particle i.

The fluid drag force on a spheroidal particle with aspect ratio $\beta \equiv a/b$ at low particle Reynolds number is given by Gallily and Cohen [83] as

$$\mathbf{F}_\mathrm{F} = \mu\hat{\mathbf{K}}\cdot(\mathbf{u}-\mathbf{v})f \tag{6.11}$$

where $\hat{\mathbf{K}}$ is the particle frame translation tensor, $\mathbf{u}$ is the undisturbed fluid velocity at the particle center, and μ is the fluid (plasma) viscosity. The friction factor f is used to account for the effects of particle crowding, such that $f = 1$ for an isolated particle. Di Felice [84] developed a correlation based on fluidized-bed experiments that expresses f in terms of the local particle concentration field. The particle frame translation tensor is the diagonal matrix [85]

$$\hat{\mathbf{K}} = 16\pi\, ab^2 \left(\frac{\mathbf{e}_{\hat{x}} \otimes \mathbf{e}_{\hat{x}}}{\chi_0 + a^2\alpha_0} + \frac{\mathbf{e}_{\hat{y}} \otimes \mathbf{e}_{\hat{y}}}{\chi_0 + b^2\beta_0} + \frac{\mathbf{e}_{\hat{z}} \otimes \mathbf{e}_{\hat{z}}}{\chi_0 + b^2\beta_0} \right) \tag{6.12}$$

where $\mathbf{e}_{\hat{x}}$, $\mathbf{e}_{\hat{y}}$, and $\mathbf{e}_{\hat{z}}$ are unit vectors in an orthogonal particle frame coordinate system advected with the particle, such that $\mathbf{e}_{\hat{x}}$ coincides with the particle symmetry axis and $\mathbf{e}_{\hat{z}}$ is orthogonal to the plane of motion, and $\otimes$ denotes the tensor product. In the simplified two-dimensional theory, the particles are assumed to move such that the symmetry axis remains oriented in the plane of motion, but in the more general three-dimensional theory, the angular momentum equation is written in the particle frame. The coefficients α_0 and β_0 can be written in terms of the aspect ratio β as

$$\alpha_0 = 2(1 - \beta_0) \quad \beta_0 = (-\beta^2 + \bar{\chi}_0/2)/(1 - \beta^2) \tag{6.13}$$

where for $\beta < 1$, as is typical of RBCs and platelets,

$$\bar{\chi}_0 = \frac{\chi_0}{b^2} = \frac{2\beta}{(1-\beta^2)^{1/2}} \left[\frac{\pi}{2} - \tan^{-1}\left(\frac{\beta}{(1-\beta^2)^{1/2}} \right) \right] \tag{6.14}$$

The particle torque induced by the fluid shear is [86]

$$M_{F,z} = \frac{16\pi\,\mu\, ab^2}{3\left(a^2\alpha_0 + b^2\beta_0\right)} \left[\left(a^2 - b^2\right) d_{\hat{y}\hat{x}} + \left(a^2 + b^2\right)\left(w_{\hat{y}\hat{x}} - \Omega_z\right) \right] \tag{6.15}$$

where the components of the rate of deformation and vorticity tensors in the particle frame are

$$d_{\hat{i}\hat{j}} = \frac{1}{2}\left(\frac{\partial u_{\hat{i}}}{\partial x_{\hat{j}}} + \frac{\partial u_{\hat{j}}}{\partial x_{\hat{i}}} \right) \quad w_{\hat{i}\hat{j}} = \frac{1}{2}\left(\frac{\partial u_{\hat{i}}}{\partial x_{\hat{j}}} - \frac{\partial u_{\hat{j}}}{\partial x_{\hat{i}}} \right) \tag{6.16}$$

It is necessary to also include added mass and pressure gradient forces (inertial buoyancy force) on the particles.

For colliding particles, it is necessary to introduce a rapid method both to detect particle collision and to determine the locations of the contact point once a collision has occurred. Several algorithms have been proposed in the granular flow literature for modeling collision of elliptical and ellipsoidal particles [87, 88]; however, certain methods are restricted to two-dimensional flows and others are known to exhibit stability problems. A robust and rapid method for detection of the overlap between two ellipsoidal shapes in three dimensions is given by the methods proposed by Alfano and Greer [89] and Chan [90]. Once a collision has been identified as having occurred, the location of the contact point on each ellipsoid is found using the *level surfaces* approach [91], in which a level surface is a set of similar surfaces extending both outward and inward from each ellipsoidal particle. The contact point of one ellipsoid in a colliding pair is found by examining its intersection with level surfaces of the other ellipsoid, and vice versa for the other ellipsoid of the pair. The collision detection algorithm requires solution of a fourth-order polynomial in 3D and a third-order polynomial in 2D, and the collision point identification algorithm requires solution of a sixth-order polynomial in 3D and a fourth-order polynomial in 2D, which are usually performed using the LAPACK subroutine package.

Collision and adhesion forces and torques acting on two colliding particles consist of normal and tangential components. A unit normal vector $\mathbf{n}$ at the contact point of a particle colliding with another particle is defined by the average of the unit normals at the contact points of the two colliding particles. The normal force $F_{\mathrm{n}}\mathbf{n}$ results from a combination of elastic repulsion of the particles and dissipation, as well as adhesive binding for cases with adhesion between the particles. The normal force produces a normal torque $F_{\mathrm{n}}\mathbf{r}_{iC} \times \mathbf{n}$, where $\mathbf{r}_{iC}$ is the vector that extends from the center of particle i to the contact point. For problems without adhesion the classical Hertz [92] collision formula $F_{\mathrm{ne}} = -K\delta_{\mathrm{N}}^{3/2}$ can be used to relate the elastic part of the normal force to the 3/2-power of the normal overlap δ_{N}, which is defined by the distance between the collision points on two colliding particles. For adhesive particles, it is noted by Chesnutt and Marshall [36] that the adhesive force due to ligand–receptor binding is mathematically analogous to van der Waals adhesion with a time-varying surface energy, so the well-known theory of Johnson et al. [93] for adhesive particles can be used. Most of the normal dissipative force during collision of two blood cells is due to the squeeze-film viscous fluid forces, which for particles of this size result in zero restitution coefficient at rebound [94]. Two particles may slide against each other producing a resistance that results in a sliding force $F_{\mathrm{s}}\mathbf{t}_{\mathrm{S}}$ and corresponding sliding torque $F_{\mathrm{s}}\mathbf{r}_{iC} \times \mathbf{t}_{\mathrm{S}}$, where $\mathbf{t}_{\mathrm{S}}$ is a unit vector in the direction of relative motion of the particle surfaces at the contact point projected onto the plane orthogonal to $\mathbf{n}$. A simple sliding resistance model is given by Cundall and Strack [95], which was extended to adhesive contacts by Thornton [96]. For three-dimensional flows, an associated torque $M_{\mathrm{t}}\mathbf{n}$ acts on the particle due to resistance from twisting. In the presence of adhesion, particles experience a strong rolling resistance due to the adhesive force [97 – 99], resulting in a rolling torque $M_{\mathrm{r}}\mathbf{t}_{\mathrm{R}} \times \mathbf{n}$, where $\mathbf{t}_{\mathrm{R}}$ is the direction of rolling velocity. If all of these effects are present, the net collision and adhesion force $\mathbf{F}_{\mathrm{A}}$ and torque $\mathbf{M}_{\mathrm{A}}$ acting on particle i due to collision with particle j are

$$\mathbf{F}_{\mathrm{A}j} = F_{\mathrm{n}}\mathbf{n} + F_{\mathrm{s}}\mathbf{t}_{\mathrm{S}} \quad \mathbf{M}_{\mathrm{A}j} = F_{\mathrm{n}}\,\mathbf{r}_{iC} \times \mathbf{n} + F_{\mathrm{s}}\,\mathbf{r}_{iC} \times \mathbf{t}_{\mathrm{S}} + M_{\mathrm{r}}\,\mathbf{t}_{\mathrm{R}} \times \mathbf{n} + M_{\mathrm{t}}\mathbf{n} \quad (6.17)$$

Comparison of discrete element computational results with experimental data and previous theoretical models is presented by Chesnutt and Marshall [36, 37] for aggregation of red blood cells in channel and shear flows. Both three- and two-dimensional cases are examined, using up to 13,720 particles for the three-dimensional calculations. As demonstrated in Figs. 6.11 and 6.12, the aggregate size distribution and the mean aggregate size are found to be in good agreement with experimental data for RBCs in shear and channel flows. Also, Chesnutt and Marshall [37] demonstrate that predicted viscosity values from the DEM computations are consistent with experimental data and theoretical models from previous investigators (e.g., [100, 101]) for blood viscosity under shear flow. Chesnutt and Marshall [36] also demonstrate that the discrete element method predictions yield

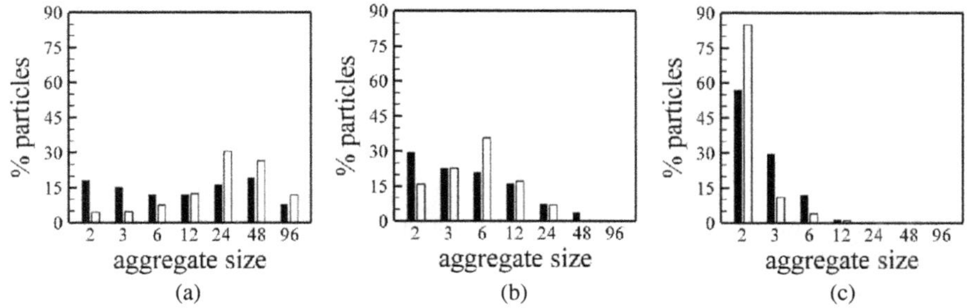

Fig. 6.11 Aggregate size distribution showing percent of aggregating particles contained in aggregates of the given size range at equilibrium under 3D channel flow, from DEM computations (*black bars*) and experimental data of Chen et al. [102] (*white bars*). Values of average shear rates include (**a**) 10.4 s^{-1}, (**b**) 41.7 s^{-1}, and (**c**) 166.7 s^{-1} (reproduced with permission from Chesnutt and Marshall [36])

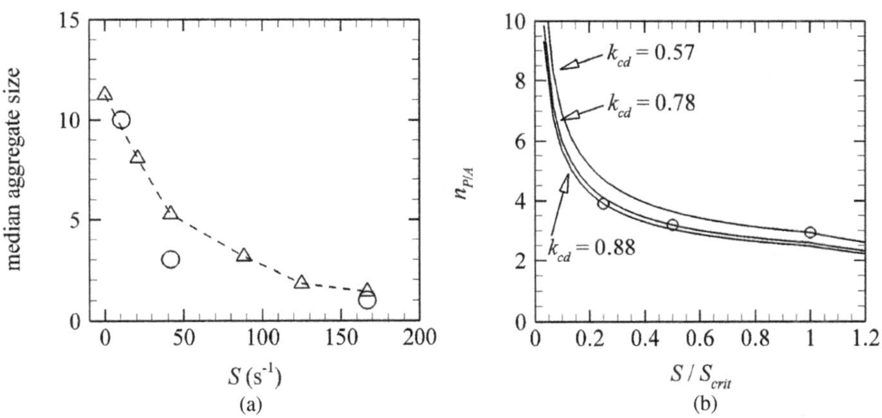

Fig. 6.12 Median aggregate size at equilibrium as a function of average shear rate for (**a**) DEM computations in 3D channel flow (*circles*) versus experimental data of Chen et al. [102] (*triangles—dashed line*) and (**b**) DEM computations in 3D shear flow (*circles*) versus theoretical predictions of Murata and Secomb [22] (*lines*) for different values of parameter k_{cd} (reproduced from Chesnutt and Marshall [37])

the expected margination of leukocytes and platelets in the presence of RBC adhesion.

6.6 Mesoscale Modeling Using Dissipative Particle Dynamics

One of the challenges with the discrete element method is the issue of solving for the fluid flow in between and surrounding the particles. As noted in the previous section, a wide variety of methods have been used to address this issue, including the level-set method and other types of immersed boundary methods, the lattice-Boltzmann method, and boundary-element methods. For flows with low Reynolds number, such as in arterioles and capillaries, the method of Stokesian dynamics [103], as well as the related force-coupling method [104], can be efficiently employed. Nevertheless, any of these approaches will severely restrict the number of blood cells that can be included in the DEM simulation. For certain problems at low hematocrit, it may even suffice to neglect the effect of the blood cells on the flow; however, this is not possible for the high hematocrit values present in most physiological blood flow problems. To further complicate the matter, blood plasma is not a Newtonian fluid, but rather contains about 7% protein macromolecules of various types. As observed in the experiments of Rand et al. [101] and Chien et al. [105], these blood proteins play a significant role in the observed shear-thinning behavior of blood, as well as other transport issues. Moreover, as discussed in Section 6.3, certain of these proteins play a critical role in the formation of blood clots and thrombi.

Although still in its early stages of development, the dissipative particle dynamics (DPD) method discussed in this section is a promising mesoscale approach for addressing both the challenge of simulating flow about the blood cells and incorporating the effects of protein microstructures on the flow. The DPD method was originally proposed by Hoogerbrugge and Koelman [106] as an extension of the molecular dynamics (MD) method. In MD, a "particle" represents a single atom, which is subject to both thermal oscillations and bonds to other atoms simulated using relatively sharp Lennard-Jones-type potentials. Both of these effects lead to the necessity of very small time scales for MD simulations, with a typical simulation covering a region with a length scale of about 1 nm and a time interval of about 1 ns. Problems that entail collective motion on much larger length and time scales, as is certainly the case for blood flow phenomena, are not addressable by the MD method, at least for the foreseeable future. In contrast to this, a single DPD "particle" represents a collection of many molecules, and the equations of motion and interaction of DPD particles are such that a much larger time scale can be used than for the MD model.

The equations governing interaction and motion of DPD particles were first put on firm footing in a paper by Español and Warren [107], in which the Fokker–Planck equation corresponding to the DPD particle interaction forces was derived, along with the fluctuation–dissipation theorem for the DPD method. In the basic form of the method, the velocity $\mathbf{v}_i$ of a DPD particle i is governed by Newton's second law under the action of three types of forces, given by

$$m\frac{\mathrm{d}\mathbf{v}}{\mathrm{d}t} = \sum_{j\neq i}\mathbf{F}_j^{\mathrm{C}} + \sum_{j\neq i}\mathbf{F}_j^{\mathrm{D}} + \sum_{j\neq i}\mathbf{F}_j^{\mathrm{R}} \tag{6.18}$$

where the sums are performed over all DPD particles within some critical distance r_{C} of the centroid of the given particle i. Each particle j within this critical distance imposes three forces on the particle i—a *conservative* force $\mathbf{F}_j^{\mathrm{C}}$, a *dissipative* force $\mathbf{F}_j^{\mathrm{D}}$, and a *random* force $\mathbf{F}_j^{\mathrm{R}}$. Galilean invariance requires that these forces depend only on the vector separating their centroids $\mathbf{r}_{ij} = \mathbf{r}_i - \mathbf{r}_j$ and the relative centroid velocity $\mathbf{v}_{ij} = \mathbf{v}_i - \mathbf{v}_j$, rather than the centroid positions and velocities independently.

Letting $\mathbf{n} \equiv (\mathbf{r}_i - \mathbf{r}_j)/r_{ij}$ be the unit vector pointing from the centroid $\mathbf{r}_i$ of particle i to the centroid $\mathbf{r}_j$ of particle j, where $r_{ij} = |\mathbf{r}_{ij}|$, expressions for the dissipative and random forces are proposed by Español and Warren [107] as

$$\mathbf{F}_j^{\mathrm{D}} = -\gamma\omega_{\mathrm{D}}(r_{ij})(\mathbf{n}\cdot\mathbf{v}_{ij})\mathbf{n}, \quad \mathbf{F}_j^{\mathrm{R}} = \sigma\omega_{\mathrm{R}}(r_{ij})\zeta_{ij}\mathbf{n} \tag{6.19}$$

Here, ζ_{ij} is a symmetric Gaussian white noise, and γ and σ are adjustable coefficients. The balance between the dissipative and random terms serves as the thermostat for the DPD simulation. Increase in the dissipation terms will tend to cool the system, such that if $\mathbf{F}^{\mathrm{R}}$ were set to zero all motion would eventually come to rest. Conversely, increase in the random terms tends to heat up the system, such that if $\mathbf{F}^{\mathrm{D}}$ were set to zero the particle motion would increase without bound. The fluctuation–dissipation theorem for DPD [107] requires that the coefficients of these two terms are related to the absolute temperature T such that

$$\omega_{\mathrm{R}}^2(r) = \omega_{\mathrm{D}}(r), \quad \sigma^2 = 2k_{\mathrm{B}}T\gamma \tag{6.20}$$

where k_{B} is Boltzmann's constant. Restriction (6.20) guarantees that the equilibrium Gibbs–Boltzmann distribution will be retained in the presence of nonzero random and dissipative forces.

The conservative force $\mathbf{F}^{\mathrm{C}}$ is typically selected to have the form of a soft potential, without the rapid variation as particles approach each other that is characteristic of the Lennard-Jones potential. A simple linear form for the potential has been motivated by averaging over the potential field of a group of atoms [108– 110] as

$$\mathbf{F}_j^{\mathrm{C}} = \begin{cases} a[1 - r_{ij}/r_{\mathrm{C}}]\,\mathbf{n} & \text{for } r_{ij} < r_{\mathrm{C}} \\ 0 & \text{for } r_{ij} > r_{\mathrm{C}} \end{cases} \tag{6.21}$$

where a is a parameter. Following Groot and Warren [109], the weighting function $\omega_{\mathrm{R}}(r)$ can similarly be taken to be a linear function, such that

$$\omega_{\mathrm{D}}(r) = \omega_{\mathrm{R}}^2(r) = \begin{cases} 1 - r/r_{\mathrm{C}} & \text{for } r < r_{\mathrm{C}} \\ 0 & \text{for } r > r_{\mathrm{C}} \end{cases} \tag{6.22}$$

The resulting equations have two adjustable parameters that can be used to distinguish between different types of materials—the repulsion parameter a and the noise parameter σ (or alternatively the friction factor γ). The repulsion parameter is related to the material compressibility and the noise parameter is related to the response rate of the system to temperature change. Guidance on selection of these two parameters is given by Groot and Warren [109]. A time step limitation for DPD is presented by Español and Warren [107] as $\Delta t/t_{\mathrm{C}} << 1$, where $t_{\mathrm{C}} = r_{\mathrm{C}}/v_{\mathrm{rms}}$ and $v_{\mathrm{rms}} = (3k_{\mathrm{B}}T/m)^{1/2}$ is the root-mean-square particle velocity. This limitation is equivalent to the requirement that two particles are not allowed to pass through each other during a time step without "seeing" each other.

While the primitive form of the DPD model discussed above contains only forces acting at the particle centers, extensions of the method have been developed that include shear forces based on a Voronoi tessellation of the fluid [111 – 113]. As discussed by Serrano and Español [112], the dissipative particle dynamics formulation can be viewed as a special case of the spherical particle hydrodynamics (SPH) method, originally developed by Lucy [114] for simulation of astrophysical flows and reviewed by Monaghan [115], and since applied to sprays and various other laboratory-scale flow fields.

A number of researchers have extended the DPD method to model polymeric solutions, where one type of DPD particle is used as a bead-spring-type representation of the polymer chains immersed in a second type of "fluid" DPD particle [116 – 119]. In its simplest form, this can be accomplished by adding to the different forces given in (6.18) a simple spring force between the DPD "beads" along the polymer chain, of the form

$$\mathbf{F}_j^{\mathrm{S}} = K\,\mathbf{r}_{ij} \tag{6.23}$$

where K is the spring constant. Other alternatives for the additional force between bead particles are discussed by Symeonidis et al. [117].

Application of the DPD method to colloidal fluids is discussed by Dzwinel and Yuen [120] (see also Boryczko et al. [121]) where again two distinct types of DPD particles are employed. One type of DPD particle represents the colloidal particles, which is immersed in a smaller second type of particle that discretizes the surrounding fluid. The fluid DPD particles are evolved using the fluid particle model of Español [111], so as to include interparticle shearing forces. The colloidal particles interact both with each other and with the fluid particles. In order to more accurately model the interaction of colloidal particles, the conservative force $\mathbf{F}^{\mathrm{C}}$ for these particles is modified so as to exhibit a soft attractive force at larger separation distances and a sharp repulsive force at small distances, in a manner similar to the MD model for colloidal particles.

The DPD model has been recently applied to blood flow problems. Dzwinel et al. [122] extended their colloidal fluid model to flow of red blood cells in narrow vessels. Each RBC is represented by a 3D mesh on which is placed interaction "solid particles," which interact with each other through conservative elastic forces and a dissipative force, but no random force. This crude discretization allows simple RBC deformation with a minimal number of DPD particles. A similar network of "solid particles" is used to represent the endothelial cells lining the vessel walls, but in this case the cells are not allowed to move. The RBCs are immersed in a sea of fluid particles, modeled using the fluid particle model of Español [111]. Example results from this method for RBC transport in channels of two different sizes are shown in Fig. 6.13.

The classical DPD model is used by Filipovic et al. [123] to model platelet transport and wall adhesion in blood flow, where the effect of RBCs on the platelet transport is neglected. The platelets are assumed to be spherical and rigid. Platelet interactions with each other and with the fluid particles are modeled using the same

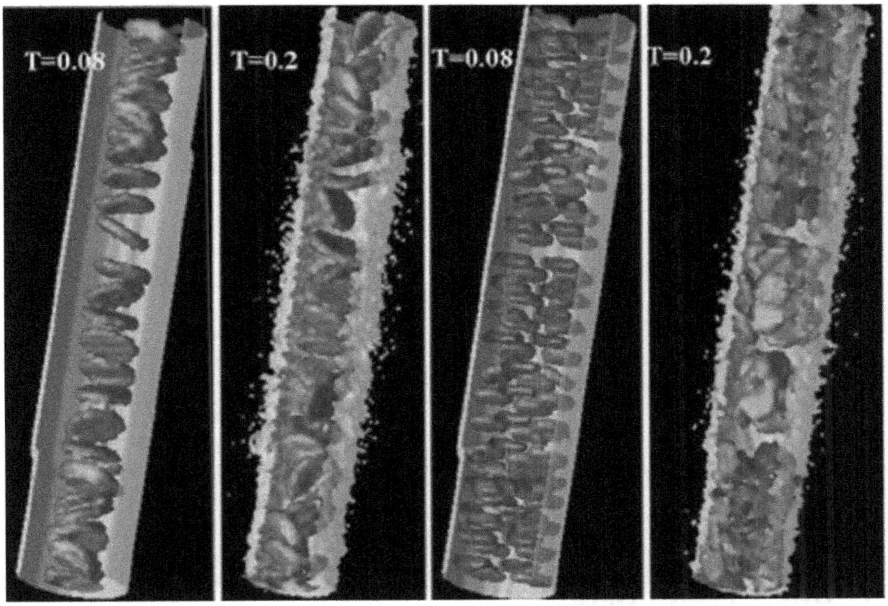

Fig. 6.13 Motion of RBCs in tubes of diameter (**a**) 10 μm and (**b**) 25 μm, at two different times for each case (reproduced with permission from Dzwinel et al. [122])

DPD particle force expression given by (6.18), (6.19), (6.20), (6.21) and (6.22), but with different parameter values. An additional attraction force is introduced to represent adhesion of platelets to vessel walls. Comparison of the model to experimental results for platelet deposition is reported by Filipovic et al. [124]. A model for the effect of red blood cells on aggregation of activated platelets is presented by Pivkin et al. [125], in which separate DPD particles representing RBCs and platelets are utilized. The Pivkin et al. [125] model does not model blood proteins explicitly, but rather accounts for the effects of blood proteins via a particle adhesion force. The model compares the rate of accumulation of platelets in a thrombus with and without the presence of red blood cells.

A DPD model of blood clotting was presented by Boryczko et al. [42], which extends the Dzwinel et al. [122] model by inclusion of additional bead-spring-type DPD particles representing fibrin monomers. The fibrin monomers are formed of two particle "beads" connected by a spring, which join with other fibrin monomers within a prescribed distance with a certain prescribed probability at each time step, thereby creating a fibrin chain. The RBC particles are attracted to the fibrin particles at long range and repelled at short range. Moreover, the fibrin chains can also break at each time step with another prescribed probability. Example results from this model for blood flow in a capillary tube are shown in Fig. 6.14, comparing the RBC positions both with and without fibrin monomers present.

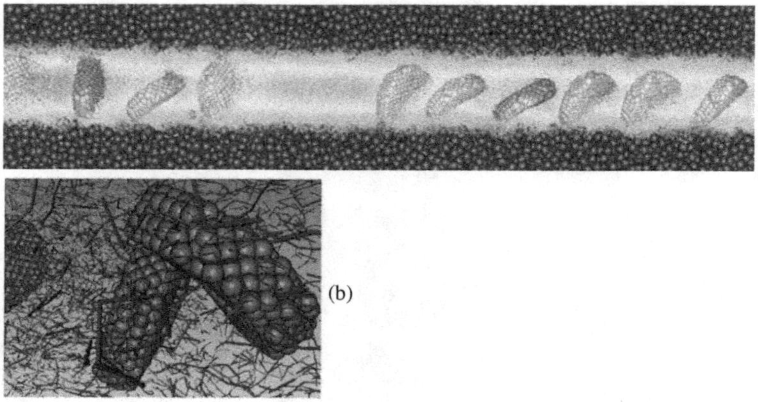

Fig. 6.14 Simulation of RBC motion in straight capillary channel (**a**) without fibrin and (**b**) a close-up view of RBCs with fibrin chains (reproduced with permission from Boryczko et al. [42])

6.7 Bridging the Scales

This chapter has reviewed a number of different methods that attempt to capture cooperative phenomena involving large numbers of blood cells, while still modeling at the cell level. In order to incorporate sufficient numbers of blood cells in the simulations, it is necessary to introduce a coarse representation of both the blood cells and the surrounding fluid flow, and their interactions, while still keeping true to the underlying physics and scaling involved in the real process. The shortcomings of such mesoscale models must be balanced against the physical requirements of the phenomena under investigation, so as to maximize the use of computational resources while still ensuring reasonable accuracy of the simulation results.

In incorporating these mesoscale models into a multiscale framework for blood flow simulation, a variety of strategies can be employed. The typical multiscale approach uses a hierarchical approach, in which simulations are performed first at fine scales and then at progressively larger scales. Information is extracted from the simulations at each scale and incorporated into models at larger scales. By way of example, we might perform a simulation of the electron wave field around an atom or small set of atoms using the quantum ab initio method to deduce the potential field surrounding each atom and the form of the bonding forces. These results could then be introduced into a molecular dynamics model, and the average of the potentials of a group of atoms could be averaged over space and a small time increment to obtain expressions for the soft-potential conservative force in the dissipative particle dynamics method. Similarly, the MD results for binding force of a section of a protein or between two different proteins could be used to develop an approximate model for protein bonding force in the DPD model, or of the ligand–receptor binding force in the DEM model. Microscale-level results for the deformation of blood cells under different fluid flows and under conditions of particle collision, as we have discussed in Section 6.3, can be used to justify approximations for particle collision and restricted modes of particle deformation in either DEM or DPD models.

Another multiscale modeling approach involves using different models in different regions of a flow. For instance, in a valve closure problem, as shown in Fig. 6.8, one might use a standard continuum CFD method for the blood flow outside of the high-shear region near the valve closure point and a mesoscale model such as DEM or DPD within this high-shear region. Domain-bridging approaches of this type have been introduced in solid mechanics problems, such as those involving local deformation of nanotubes, in which finite-element computation of the continuum equations is coupled to MD simulations within a small high-deformation region [126 – 128]. Domain-bridging methods of this type might also be useful for problems with stents or bifurcations, where cell-level representation of the blood flow dynamics is required in a small region of the flow field.

Much research remains to be done to validate and further explore both the DEM and DPD models for blood flow applications. Of particular importance is the problem of understanding the effects of spatial resolution on the solution accuracy. For instance, how many fluid particles in a DPD model are necessary to accurately approximate the flow around an RBC? Or, how small must the fibrin monomers be made to accurately approximate formation of a blood clot? Second is the issue of accurate modeling of the various adhesive processes between blood cells. Scaling estimates indicate adhesion length scales for processes such as ligand–receptor binding that are three orders of magnitude smaller than the typical cell size. Solution of the squeeze-film hydrodynamics within a gap of this small size, as well as the fluid dissipation as the cells approach each other to within this distance, introduces severe resolution issues for the DPD method, but also opens the door to appropriate simplified models such as those developed in the DEM models of cell aggregation. In all such mesoscale model development, a critical balancing of the model structure with the physical requirements of the cell interaction phenomena under physiologically realistic parameters and scaling is essential for future progress toward a reliable simulation methodology.

Acknowledgments The authors acknowledge financial support to J.K.W. Chesnutt from the Presidential Fellowship and IIHR–Hydroscience and Engineering at the University of Iowa and from the School of Engineering at the University of Vermont. Funding for J.S. Marshall was provided by the U.S. Department of Transportation (grant number DTOS59-06-G-00048) and by Vermont EPSCoR (grant number EPS 0701410).

References

1. Tsukada K, Sekizuka E, Oshio C, Minamitani H (2001) Direct measurement of erythrocyte deformability in diabetes mellitus with a transparent microchannel capillary model and high-speed video camera system. Microvasc Res 61:231–239
2. Melder RJ, Yuan J, Munn LL, Jain RK (2000) Erythrocytes enhance lymphocyte rolling and arrest in vivo. Microvasc Res 59:316–322
3. Migliorini C, Qian Y, Chen H, Brown EB, Jain RK, Munn LL (2002) Red blood cells augment leukocyte rolling in a virtual blood vessel. Biophys J 83:1834–1841

4. Sun C, Migliorini C, Munn LL (2003) Red blood cells initiate leukocyte rolling in postcapillary expansions: a lattice Boltzmann analysis. Biophys J 85:208–222
5. Reimers RC, Sutera SP, Joist JH (1984) Potentiation by red blood cells of shear-induced platelet aggregation: relative importance of chemical and physical mechanisms. Blood 64:1200–1206
6. Alkhamis TM, Beissinger RL, Chedian J (1987) Effect of red blood cells on platelet adhesion and aggregation in low-stress shear flow. Trans Am Soc Artif Intern Organs 33: 636–642
7. Joist JH, Bauman JE, Sutera SP (1998) Platelet adhesion and aggregation in pulsatile shear flow: effects of red blood cells. Thromb Res 92(6 Suppl 2):S47–S52
8. Pearson MJ, Lipowsky HH (2000) Influence of erythrocyte aggregation on leukocyte margination in postcapillary venules of rat mesentery. Am J Physiol 279:H1460–H1471
9. Cadroy Y, Hanson SR (1998) Effects of red blood cell concentration on hemostasis and thrombus formation in a primate model. Blood 75(11):2185–2193
10. Aarts PAMM, van den Broek SAT, Prins GW, Kuiken GDC, Sixma JJ, Heethaar RM (1988) Blood platelets are concentrated near the wall and red blood cells, in the center in flowing blood. Arterioscler 8:819–824
11. Nash GB, Watts T, Thornton C, Barigou M (2008) Red cell aggregation as a factor influencing margination and adhesion of leukocytes and platelets. Clin Hemorheol Microcirc 39:303–310
12. Goldsmith HL, Bell DN, Spain S, McIntosh FA (1999) Effect of red blood cells and their aggregates on platelets and white cells in flowing blood. Biorheol 36:461–468
13. Tilles AW, Eckstein EC (1987) The near-wall excess of platelet-sized particles in blood flow: its dependence on hematocrit and wall shear rate. Microvasc Res 33:211–223
14. Buchanan JR, Kleinstreuer C, Comer JK (2000) Rheological effects on pulsatile hemodynamics in a stenosed tube. Comput Fluids 29:695–724
15. Srivastava VP, Saxena M (1997) Suspension model for blood flow through stenotic arteries with a cell-free plasma layer. Math Biosci 139:79–102
16. Sharan M, Popel AS (2001) A two-phase model for flow of blood in narrow tubes with increased effective viscosity near the wall. Biorheology 38:415–428
17. Das B, Enden G, Popel AS (1997) Stratified multiphase model or blood flow in a venular bifurcation. Ann Biomed Eng 25:135–153
18. Bertoluzzo SM, Bollini A, Rasia M, Raynal A (1999) Kinetic model for erythrocyte aggregation. Blood Cells Mol Dis 25(22):339–349
19. Chen J, Huang Z (1996) Analytical model for effects of shear rate on rouleau size and blood viscosity. Biophys Chem 58:273–279
20. Kounov NB, Petrov VG (1999) Determination of erythrocyte aggregation. Math Biosci 157:345–356
21. Lim B, Bascom PAJ, Cobbold RSC (1998) Simulation of red blood cell aggregation in shear flow. Biorheology 34(6):423–441
22. Murata R, Secomb TW (1988) Effects of shear rate on rouleau formation in simple shear flow. Biorheology 25:113–122
23. Petrov VG, Edissonov I (1996) The role of aggregation kinetics in the sedimentation of erythrocytes. Biorheology 33(4):353–364
24. Huang PY, Hellums JD (1993) Aggregation and disaggregation kinetics of human blood platelets, Parts I and II: development and validation of a population balance method. Biophys J 65:334–353
25. Krishnan S, Udaykumar HS, Marshall JS, Chandran KB (2006) Dynamic study of platelet activation during mechanical heart valve operation. Ann Biomed Eng 34:1519–1534
26. Barthes-Biesel D, Sgaier H (1985) Role of membrane viscosity in the orientation and deformation of a spherical capsule suspended in shear flow. J Fluid Mech 160:119–135
27. Eggleton CD, Popel AS (1998) Large deformation of red blood cell ghosts in a simple shear flow. Phys Fluids 10(8):1834–1845

28. Pozrikidis C (2003) Numerical simulation of the flow-induced deformation of red blood cells. Ann Biomed Eng 31:1194–1205
29. Keller SR, Skalak R (1982) Motion of a tank-treading ellipsoidal particle in a shear flow. J Fluid Mech 120:27–47
30. Bagchi P, Johnson PC, Popel AS (2005) Computational fluid dynamic simulation of aggregation of deformable cells in a shear flow. J Biomech Eng 127:1070–1080
31. Jadhav S, Chan KY, Konstantopoulos K, Eggleton CD (2007) Shear modulation of intercellular contact area between two deformable cells colliding under flow. J Biomech 40:2891–2897
32. Freund JB (2007) Leukocyte margination in a model microvessel. Phys Fluids 19:023301-1–023301-13
33. Dulinska I, Targosz M, Strojny W, Lekka M, Czuba P, Balwierz W, Szymonski M (2006) Stiffness of normal and pathological erythrocytes studied by means of atomic force microscopy. J Biochem Biophys Methods 66:1–11
34. Chien S, Jan KM (1973) Ultrastructural basis of the mechanism of rouleaux formation. Microvasc Res 5:155–166
35. Neu B, Meiselman HJ (2002) Depletion-mediated red blood cell aggregation in polymer solutions. Biophys J 83:2482–2490
36. Chesnutt JKW, Marshall JS (2009) Blood cell transport and aggregation using discrete ellipsoidal particles. Comput Fluids 38:1782–1794
37. Chesnutt JKW, Marshall JS (2010) Structural analysis of red blood cell aggregates under shear flow. Ann Biomed Eng. 38(3):714–728
38. Bark N, Foldes-Papp Z, Rigler R (1999) The incipient stage in thrombin-induced fibrin polymerization detected by FCS at the single molecule level. Biochem Biophys Res Commun 260:35–41
39. Chesla SE, Selvaraj P, Zhu C (1998) Measuring two-dimensional receptor-ligand binding kinetics by micropipette. Biophys J 75:1553–1572
40. Bouchard BA, Butenas S, Mann KG, Tracy PB (2007) Interactions between platelets and the coagulation system. In: Michelson AD (ed) Platelets, 2nd edn. Academic/Elsevier, San Diego, CA, pp 377–402
41. Wootton DM, Ku DN (1999) Fluid mechanics of vascular systems, diseases, and thrombosis. Annu Rev Biomed Eng 1:299–329
42. Boryczko K, Dzwinel W, Yuen DA (2004) Modeling fibrin aggregation in blood flow with discrete-particles. Comput Methods Programs Biomed 75:181–194
43. Bongrand P, Bell GI (1984) Cell-cell adhesion: parameters and possible mechanisms. In: Perelson AS, DeLisi C, Wiegel FW (eds) Cell surface dynamics: concepts and models. Marcel Dekker, New York, NY, pp 459–493
44. Napper DH (1977) Steric stabilization. J Colloid Interface Sci 58:390–407
45. Bell GI (1978) Models for the specific adhesion of cells to cells. Sci 200:618–627
46. Springer TA (1990) Adhesion receptors of the immune system. Nature 346:425–434
47. Bell GI, Dembo M, Bongrand P (1984) Cell adhesion, competition between nonspecific repulsion and specific bonding. Biophys J 45:1051–1064
48. Agresar G (1996) A computational environment for the study of circulating cell mechanics and adhesion. Ph.D. dissertation, University of Michigan, Ann Arbor
49. Chang KC, Hammer DA (1999) The forward rate of binding of surface-tethered reactants: effect of relative motion between two surfaces. Biophys J 76:1280–1292
50. Dembo M, Torney DC, Saxman K, Hammer D (1988) The reaction-limited kinetics of membrane-to-surface adhesion and detachment. Proc R Soc Lond B 234:55–83
51. Torney DC, Dembo M, Bell GI (1986) Thermodynamics of cell adhesion II. Freely mobile repellers. Biophys J 49:501–507
52. Tozeren A (1990) Cell-cell, cell-substrate adhesion: theoretical and experimental considerations. J Biomech Eng 112:311–318
53. Tozeren A, Sung KLP, Chien S (1989) Theoretical and experimental studies on cross-bridge migration during cell disaggregation. Biophys J 55:479–487

54. N'Dri NA, Shyy W, Tran-Son-Tay R (2003) Computational modeling of cell adhesion and movement using a continuum-kinetics approach. Biophys J 85:2273–2286
55. Tachev KD, Angarska JK, Danov KD, Kralchevsky PA (2000) Erythrocyte attachment to substrates: determination of membrane tension and adhesion energy. Colloids Surf B: Biointerfaces 19:61–80
56. Zhu C, Williams TE (2000) Modeling concurrent binding of multiple molecular species in cell adhesion. Biophys J 79:1850–1857
57. Coombs D, Dembo M, Wofsy C, Goldstein B (2004) Equilibrium thermodynamics of cell-cell adhesion mediated by multiple ligand-receptor pairs. Biophys J 86:1408–1423
58. Zhu C (2000) Kinetics and mechanics of cell adhesion. J Biomech 33:23–33
59. Hammer DA, Apte SM (1992) Simulation of cell rolling and adhesion on surfaces in shear flow: general results and analysis of selectin-mediated neutrophil adhesion. Biophys J 63:35–57
60. Ward MD, Hammer DA (1993) A theoretical analysis of the effect of focal contact formation on cell-substrate attachment strength. Biophys J 64:936–956
61. Zhang J, Johnson PC, Popel AS (2008) Red blood cell aggregation and dissociation in shear flows simulated by lattice Boltzmann method. J Biomech 41:47–55
62. Lac E, Barthès-Biesel D (2005) Deformation of a capsule in simple shear flow: effect of membrane prestress. Phys Fluids 17:072105-1–072105-8
63. Lac E, Barthès-Biesel D, Pelekasis NA, Tsamopoulos J (2004) Spherical capsules in three-dimensional unbounded Stokes flows: effect of the membrane constitutive law and onset of buckling. J Fluid Mech 516:303–334
64. Khismatullin DB, Truskey GA (2005) Three-dimensional numerical simulation of receptor-mediated leukocyte adhesion to surfaces: effects of cell deformability and viscoelasticity. Phys Fluids 17:031505–031521
65. Liu Y, Liu WK (2006) Rheology of red blood cell aggregation by computer simulation. J Comput Phys 220:139–154
66. Agresar G, Linderman JJ, Tryggvason G, Powell KG (1998) An adaptive, Cartesian, front-tracking method for the motion, deformation and adhesion of circulating cells. J Comput Phys 143:346–380
67. Shyy W, Francois M, Udaykumar HS, N'Dri N, Tran-Son-Tay R (2001) Moving boundaries in micro-scale biofluid dynamics. Appl Mech Rev 54:405–453
68. Marella S, Udaykumar HS (2004) Computational analysis of the deformability of leukocytes modeled with viscous and elastic structural components. Phys Fluids 16(2):244–264
69. Chien S (1987) Red cell deformability and its relevance to blood flow. Annu Rev Physiol 49:177–192
70. Hochmuth RM (1987) Properties of red blood cells. In: Skalak R, Chien S (eds) Handbook of bioengineering. McGraw-Hill, New York, NY, pp 12.1–12.17
71. Goldsmith HL, Marrow J (1972) Flow behavior of erythrocytes. I. Rotation and deformation in dilute suspensions. Proc R Soc Lond 182: 351–384
72. Karino T, Goldsmith JL (1977) Flow behavior of blood cells and rigid spheres in an annular vortex. Philos Trans R Soc Lond B 279:413–445
73. Goldsmith HL, Karino T (1977) Microscopic considerations: the motion of individual particles. Ann N Y Acad Sci 283:241–255
74. Hyun S, Kleinstreuer C, Archie JP Jr (2001) Computational particle-hemodynamics analysis and geometric reconstruction after carotid endarterectomy. Comput Biol Med 31:365–384
75. El-Kareh AW, Secomb TW (2006) A model for red blood cell motion in bifurcating microvessels. Int J Multiphase Flow 26:1545–1564
76. Longest PW, Kleinstreuer C, Buchanan JR (2004) Efficient computation of micro-particle dynamics including wall effects. Comput Fluids 33:577–601
77. Cao J, Rittgers SE (1998) Particle motion within in vitro models of stenosed internal carotid and left anterior descending coronary arteries. Ann Biomed Eng 26:190–199

78. Longest PW, Kleinstreuer C (2003) Comparison of blood particle deposition models for non-parallel flow domains. J Biomech 36:421–430
79. Yeh C, Eckstein EC (1994) Transient lateral transport of platelet-sized particles in flowing blood suspensions. Biophys J 66:1706–1716
80. AlMomani T, Udaykumar HS, Marshall JS, Chandran KB (2008) Micro-scale dynamic simulation of erythrocyte-platelet interaction in blood flow. Ann Biomed Eng 36(6): 905–920
81. Cohen H, Mancaster RG (1988) The theory of pseudo-rigid bodies. Springer, New York, NY
82. AlMomani TD (2007) Micro-scale dynamic simulation of erythrocyte-platelet interaction. Ph.D. Dissertation, The University of Iowa, Iowa City, IA
83. Gallily I, Cohen AH (1979) On the orderly nature of the motion of nonspherical aerosol particles. J Colloid Interface Sci 68:338–356
84. Di Felice R (1994) The voidage function for fluid-particle interaction systems. Int J Multiph Flow 20:153–159
85. Happel J, Brenner H (1983) Low Reynolds number hydrodynamics. Martinus Nijhoff Publishers, The Hague
86. Jeffery GB (1922) The motion of ellipsoidal particles immersed in a viscous fluid. Proc R Soc Lond A 102:161–179
87. Rothenburg L, Bathurst RJ (1991) Numerical simulation of idealized granular assemblies with plane elliptical particles. Comput Geotech 11(4):315–329
88. Ting JM (1992) A robust algorithm for ellipse-based discrete element modeling for granular materials. Comput Geotech 13(3):175–186
89. Alfano S, Greer ML (2003) Determining if two solid ellipsoids intersect. J Guid Control Dyn 26:106–110
90. Chan K (2001) A simple mathematical approach for determining intersection of quadratic surfaces. In: Proceedings of the American Astronautical Society, AAS Paper 01–358, Jul–Aug
91. Schneider PJ, Eberly DH (2003) Geometric tools for computer graphics. Morgan Kaufmann, San Francisco, CA
92. Hertz H (1882) Über die Berührung fester elastische Körper. J Reine Angew Mathematik 92:156–171
93. Johnson KL, Kendall K, Roberts AD (1971) Surface energy and the contact of elastic solids. Proc R Soc Lond A 324:301–313
94. Joseph GG, Zenit R, Hunt ML, Rosenwinkel AM (2001) Particle-wall collisions in a viscous fluid. J Fluid Mech 433:329–346
95. Cundall PA, Strack ODL (1979) A discrete numerical model for granular assembles. Géotechnique 29(1):47–65
96. Thornton C (1991) Interparticle sliding in the presence of adhesion. J Phys D Appl Phys 24:1942–1946
97. Dominik C, Tielens AGGM (1995) Resistance to rolling in the adhesive contact of two elastic spheres. Philos Mag A 92:783–803
98. Marshall JS (2007) Particle aggregation and capture by walls in a particulate aerosol channel flow. J Aerosol Sci 38:333–351
99. Marshall JS (2009) Discrete-element modeling of particulate aerosol flows. J Comput Phys 228:1541–1561
100. Brooks DE, Goodwin JW, Seaman GVF (1970) Interactions among erythrocytes under shear. J Appl Physiol 28(2):172–177
101. Rand PW, Lacombe E, Hunt HE, Austin WH (1964) Viscosity of normal human blood under normothermic and hypothermic conditions. J Appl Physiol 19:117–122
102. Chen S, Barshtein G, Gavish B, Mahler Y, Yedgar S (1993) Monitoring of red blood cell aggregability in a flow-chamber by computerized image analysis. In: Proceedings of the 1st International 8th European Conference on Clinical Hemorheology, Vienna, 5–8 Jul
103. Brady JF, Bossis G (1988) Stokesian dynamics. Annu Rev Fluid Mech 20:111–157

104. Lomholt S, Maxey MR (2003) Force-coupling method for particulate two-phase flow: Stokes flow. J Comput Phys 184:381–405
105. Chien S, Usami S, Taylor HM, Lundberg JL, Gregersen MI (1966) Effects of hematocrit and plasma proteins on human blood rheology at low shear rates. J Appl Physiol 21:81–87
106. Hoogerbrugge PJ, Koelman JMVA (1992) Simulating microscopic hydrodynamic phenomena with dissipative particle dynamics. Europhys Lett 19(3):155–160
107. Español P, Warren P (1995) Statistical mechanics of dissipative particle dynamics. Europhys Lett 30(4):191–196
108. Forrest BM, Suter UW (1995) Accelerated equilibration of polymer melts by time-coarse-graining. J Chem Phys 102:7256–7266
109. Groot RD, Warren PB (1997) Dissipative particle dynamics: bridging the gap between atomistic and mesoscopic simulation. J Chem Phys 107(11):4423–4435
110. Flekkøy EG, Coveney PV (1999) From molecular dynamics to dissipative particle dynamics. Phys Rev Lett 83(9):1775–1778
111. Español P (1998) Fluid particle model. Phys Rev A 57(3):2930–2948
112. Serrano M, Español P (2001) Thermodynamically consistent mesoscopic fluid particle model. Phys Rev A 64:046115-1–046115-18
113. De Fabritiis G, Coveney PV, Flekkøy EG (2002) Multiscale dissipative particle dynamics. Philos Trans R Soc Lond A 360:317–331
114. Lucy LB (1977) A numerical testing of the fission hypothesis. Astron J 82:1013–1024
115. Monaghan JJ (1992) Smoothed particle hydrodynamics. Annu Rev Astron Astrophys 30:543–574
116. Akkermans RLC, Briels WJ (2000) Coarse-grained dynamics of one chain in a polymer melt. J Chem Phys 113(15):6409–6422
117. Symeonidis V, Karniadakis GE, Caswell B (2005) Dissipative particle dynamics simulations of polymer chains: scaling laws and shearing response compared to DNA experiments. Phys Rev Lett 95:076001-1–076001-4
118. Van Vliet RE, Hoefsloot HCJ, Hamersma PJ, Iedema PD (2000) Pressure-induced phase separation of polymer-solvent systems with dissipative particle dynamics. Macromol Theory Simul 9:698–702
119. Wijmans CM, Smit B, Groot RD (2001) Phase behavior of monomeric mixtures and polymer solutions with soft interactions potentials. J Chem Phys 114(17):7644–7654
120. Dzwinel W, Yuen DA (2002) Mesoscopic dispersion of colloidal agglomerate in a complex fluid modelled by a hybrid fluid-particle model. J Colloid Interface Sci 247:463–480
121. Boryczko K, Dzwinel W, Yuen DA (2002) Parallel implementation of the fluid particle model for simulating complex fluids in the mesoscale. Concurr Comput Pract Exp 14:137–161
122. Dzwinel W, Boryczko K, Yuen DA (2003) A discrete-particle model of blood dynamics in capillary vessels. J Colloid Interface Sci 258:163–173
123. Filipovic N, Kojic M, Tsuda A (2008) Modelling thrombosis using dissipative particle dynamics method. Philos Trans R Soc Lond A 366:3265–3279
124. Filipovic N, Ravnic D, Kojic M, Mentzer SJ, Haber S, Tsuda A (2008) Interactions of blood cell constituents: experimental investigation and computational modeling by discrete particle dynamics algorithm. Microvasc Res 75:279–284
125. Pivkin IV, Richardson PD, Karniadakis GE (2009) Effect of red blood cells on platelet aggregation. IEEE Eng Med Biol Mag 28(2):32–37
126. Belytschko T, Xiao SP (2003) Coupling methods for continuum model with molecular model. Int J Multiscale Comput Eng 1(1):115–126
127. Curtin WA, Miller RE (2003) Atomistic/continuum coupling in computational materials science. Model Simul Mater Sci Eng 11(3):R33–R68
128. Wagner GJ, Liu WK (2003) Coupling of atomic and continuum simulations using a bridging scale decomposition. J Comput Phys 190:249–274

Part III
Applications of Computational Simulations in the Cardiovascular and Pulmonary Systems

Chapter 7
Arterial Circulation and Disease Processes

Tim McGloughlin and Michael T. Walsh

Abstract Atherosclerosis is an arterial disease resulting in thickening of the arterial wall and occlusion of the vessels in advanced stages. In addition to hereditary and environmental factors, the effect of fluid-induced stresses on the arterial wall has also been implicated on the etiology of the disease due to the fact that the lesions are found in arterial curvature and branching sites with complex flow dynamics. In this chapter, the computational modeling of the fluid dynamics in the coronary arteries and the aorta is discussed in order to determine the relationship between wall shear stress and its temporal and spatial gradients with the disease progression. The importance of the use of three-dimensional geometry of the region of interest from imaged data, the effect of boundary conditions as well as the unsteady flow analysis on the results are discussed. The modeling of the flow dynamics in the abdominal aortic aneurysm (AAA) geometrical models, as well as models of vascular graft anastomotic regions is also discussed.

7.1 Introduction

Arteriosclerosis is the name given to the diseases of the vasculature that are characterized by the thickening of the vessel wall in large and small arteries. Arteriosclerosis is usually referred to as atherosclerosis when concerned with the thickening and hardening of the large vessels of the cardiovascular system. Thickening of the vessel wall begins with the intimal layer and progresses into the middle layer. The lesion that forms the disease is essentially an overgrowth of the smooth muscle cells within the medial layer. This overgrowth pushes the intimal layer into the lumen of the vessel, restricting blood perfusion to distal biological

T. McGloughlin (✉)
Department of Mechanical and Aeronautical Engineering, Materials and Surface Science Institute, Centre for Applied Biomedical Engineering Research, University of Limerick, National Technological Park, Castletroy, Limerick, Ireland
e-mail: tim.mcgloughlin@ul.ie

K.B. Chandran et al. (eds.), *Image-Based Computational Modeling of the Human Circulatory and Pulmonary Systems*, DOI 10.1007/978-1-4419-7350-4_7,

structures. The cells eventually degenerate to a stage when they produce an accumulation of lipids in their cytoplasm. Finally, these cells calcify to form the hardened lesion that permanently restricts blood supply.

Atherosclerosis is a slow, complex disease that typically starts in childhood and progresses with age [1]. Many theories exist regarding the initiation and development of atherosclerosis, such as the infection–inflammation hypothesis and the hypothesis of monoclonal growth [2]. However, the general consensus regarding atherogenesis is that it develops via a response to injury, initiated by circulating factors and modulated by the local anatomy and hemodynamics [3]. Atherosclerosis is an inflammatory process that involves the thickening of the vascular intima by a lipid-rich gruel (atheroma) and connective tissue (sclerosis), as is illustrated in Fig. 7.1. Deposition of low-density lipoprotein, platelets, cellular waste products, calcium, and other substances contributes to the progression of atherosclerosis.

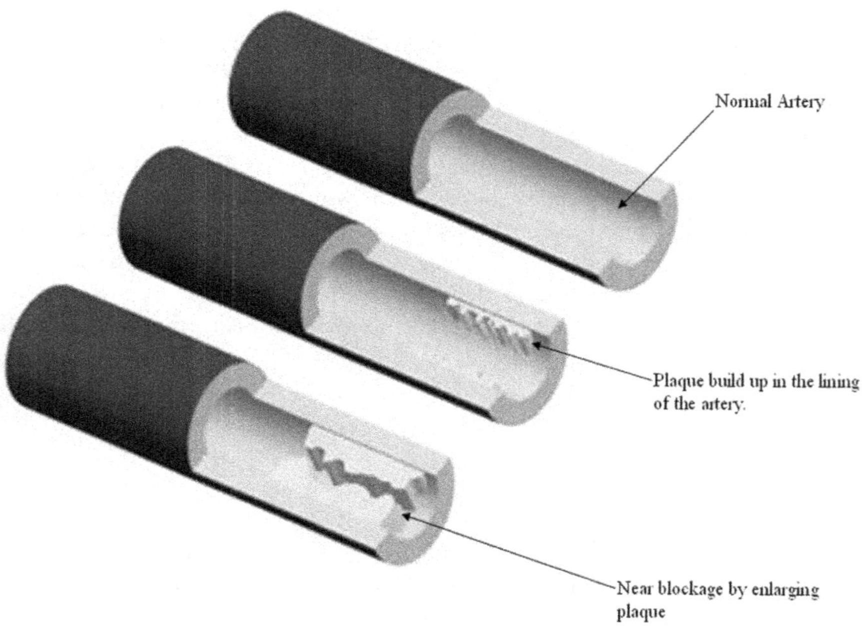

Fig. 7.1 A schematic diagram showing the progressive development of atherosclerotic plaque on the inner wall of an artery

If the progression of atherosclerosis goes unchecked, it can lead to a blockage of the lumen of the artery. Examples of how the influence of this disease varies with anatomical location demonstrate that if this blockage occurs in the coronary arteries, it can lead to a myocardial infarction. A stroke is the result of atherosclerotic plaque development and subsequent occlusion of the carotid artery. Plaque development in the femoral artery of the leg leads to intermittent claudication (pain when walking) and if untreated, gangrene can develop with resultant amputation of the leg. However, most deaths occur not from plaque occlusion of the artery but from ruptured plaques leading to the formation of thrombus on the surface [4]. This clot can break off and travel through the bloodstream until it is too large to pass through the arterial lumen, resulting in a sudden disruption of the supply of blood.

There are many processes involved in the formation of the disease, some of which are understood and some of which are not. Environmental and genetic factors, endothelial and smooth muscle cells, elements of blood, proteins, lipids, and connective tissue all have a role in the disease formation [5]. Different biochemical processes occur, from the activation of macrophages to the complex clotting cascade. These factors and processes are currently the focus of several studies, in an attempt to more fully understand and elucidate their contribution to disease formation.

7.2 Artery Wall Structure

The artery wall consists of three distinct layers: the tunica interna, the tunica media, and the tunica externa (Fig. 7.2). The tunica interna, also known as the tunica intima, is the inner-most layer of the artery wall. This layer consists of a lining of endothelial cells that rest on a layer of connective tissue. The outer boundary of the tunica

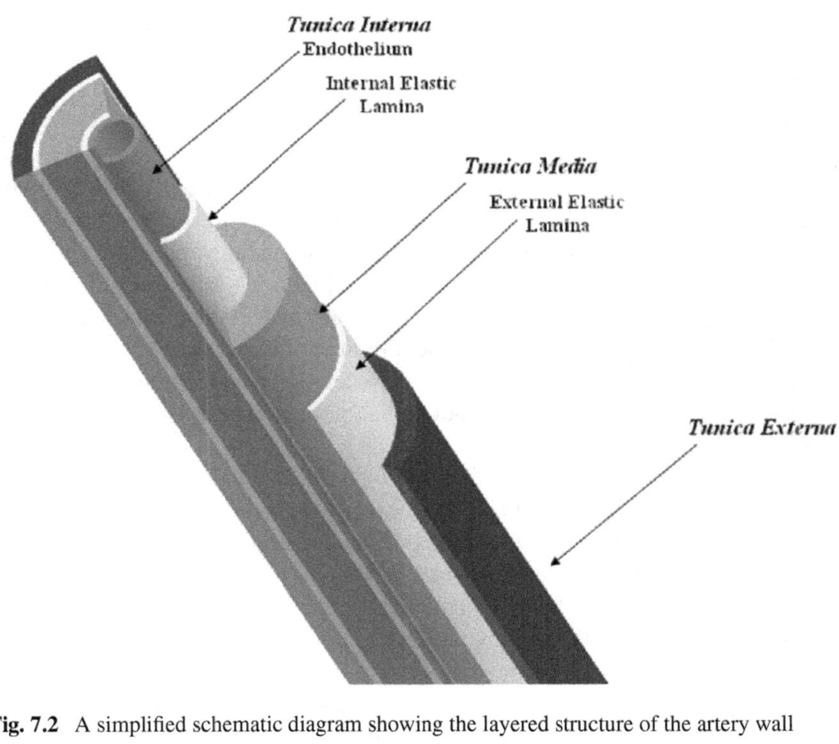

Fig. 7.2 A simplified schematic diagram showing the layered structure of the artery wall

interna is defined by a thick layer of elastic fibers known as the internal elastic lamina (IEL).

The tunica media, the middle layer, contains concentric sheets of smooth muscle tissue in a framework of loose connective tissue. It is the contraction of these sheets of smooth muscle tissue that allows the artery wall to contract and relax, thus changing lumen diameter. The tunica media is commonly the thickest layer of the small arteries. It is separated from the tunica externa by a thin band of elastic fibers known as the external elastic lamina (EEL). The tunica externa is the outer-most layer of the artery wall and forms a connective tissue sheath around the vessel. These connective tissue fibers blend into surrounding connective tissue fibers close to the artery, stabilizing and anchoring the blood vessel.

The composite layer design of the artery wall provides considerable strength to the vessel. The muscular and elastic components permit controlled alterations in diameter of the artery in response to changes in blood pressure and flow rate. Of the three layers present in the artery wall, the tunica intima is the most complex in function. This is due to the presence of the endothelial cells in this layer. The inner surface of the entire cardiovascular system is lined by a monolayer of endothelial cells (ECs), i.e., the endothelium. The integrity of the endothelium is very crucial to the arterial function. It regulates the mass transport between the blood stream and the arterial wall by providing a selective permeable barrier to the passage of macromolecules. It is also a perfect non-thrombogenic surface protecting the formation of thromboses in the arterial system.

7.3 Endothelium

The endothelium is a flat monolayer of cells that coat the vascular lumen throughout the body, from the largest arteries and veins down to the smallest capillaries. In a person with a body weight of 70 kg, the endothelium covers an area of approximately 700 m^2 and weighs almost 1.5 kg [6]. Long believed to be simply a non-thrombogenic barrier between the artery wall and flowing blood, it is only recently that the importance of the endothelium in vascular hemostasis has become evident. With the ability to secrete an array of bioactive substances that have varying effects on vessel tone, structure, and endothelial permeability, it is described as the largest organ in the body [7].

Intrinsic in the ability of the endothelium to elicit such wide and varying effects on the vessel wall is its location. Positioned directly between the media, containing smooth muscle cells, and the flowing blood, the endothelium is in prime position to regulate homeostasis. It is here that the endothelial cells have the ability to detect and react to changes in the local hemodynamic environment. Hemodynamic variations are transmitted to the endothelium by the alteration of the fluid shear stress exerted on the cells. Laminar shear stress promotes an anti-inflammatory, anti-oxidative, and anti-hypertrophic endothelial state, whereas non-laminar oscillatory flow promotes an oxidative state and an increase in endothelial permeability [8].

Two mechanisms exist through which the endothelium may transport blood-borne molecules across its surface. The first is receptor-mediated endocytosis, which supplies nutrients for the cell's own metabolic needs and accounts for only 5% of transendothelial transport. The other mode of transport is transendothelial transport, which accounts for the remaining 95% of transport. Paracellular transport (transport through the clefts between adjacent endothelial cells) is restricted to molecules smaller than 4 nm in diameter and is the pathway for water. Larger molecules with diameters up to 30 nm, such as low-density lipoprotein, are transported through vesicles that are located on the surface of the cell [9].

There is a very slow turnover of ECs in uninjured vessels, which release an array of factors responsible for the maintenance of vascular tone, the structure of the vessel wall, the regulation of vasomotion, and the conservation of a thrombo-resistant intimal surface [10]. Endothelial cells can perceive and respond to mechanical, chemical, and humoral changes in the vasculature. They act as a mediator and regulator of inflammatory and immunological responses by producing growth-promoting and growth-inhibiting molecules and by modulating the composition of the extracellular matrix (ECM) [11].

Physiologically, fully differentiated smooth muscle cells (SMCs) are typified by an abundance of contractile proteins, principally smooth muscle actin and myosin and little endoplasmic reticulum, and they exhibit a very slow turnover with low rates of proliferation and death [3]. SMCs and the ECM provide vascular dimension and tone, while SMC also synthesize and secrete elements of the ECM. These secreted elements have a number of functions, which include mechanical support, cell adhesion, and motility to proliferation. In response to injury, SMCs are capable of rapid division, migration, and proliferation regulated by growth factors [3]. Once activated, SMC lose their differentiated state, exhibit a synthetic phenotype and acquire abundant endoplasmic reticulum, and initiate synthesis and secretion of ECM throughout the vasculature. This requires a change in the regulation of active genes, referred to as phenotypic modulation, and is a precondition for SMC migration and proliferation. It can also result in the altered regulation of genes for several ECM molecules [3].

7.4 Mechanical Forces on the Arterial Wall

The arterial wall experiences three main hemodynamic forces: fluid shear stress, circumferential strain, and hydrostatic pressure. The wall shear stress is a product of the frictional force due to the velocity gradient at the wall and the blood viscosity. It varies spatially due to the geometry of the artery and temporally due to the pulsatility of the blood flow. As the blood pulses through the artery, it increases the diameter of the lumen, resulting in a compressive strain being produced in the arterial wall. The hydrostatic pressure results in a compressive stress that acts normal to the vessel wall.

While lifestyle and genetics result in varied levels of disease development from one individual to the next, the focal accumulation of plaques in similar regions of the circulatory system in each person affected by this disease cannot be accounted for by these systemic factors. The tortuous geometries of the arterial network produce disturbed blood flow patterns that induce complex and varied hemodynamic forces on the arterial wall. Interestingly, it is only in the last 40 years that the link between the hemodynamic environment and atherosclerosis has been investigated. The first studies investigating blood flow patterns in relation to atherosclerotic plaque development were conducted in 1966 [12]. Following steady-flow analysis, they hypothesized that boundary layer separation, which occurs in the same regions as the lesions, could result in the deposition of blood-borne platelets and cholesterol. Developing the knowledge in this field, Fry [13] demonstrated that shear stress greater that 38 Pa can damage the endothelial monolayer that lines the arterial lumen. However, a subsequent study by Friedman [14] examining pulsatile flow through a symmetrical branch indicated that these high values of shear are unlikely to be reached in the circulatory system. In 1971, Caro [15] conducted steady-flow simulations in a cast of a human aorta and demonstrated that lesions were found preferentially in regions believed to experience low wall shear stress.

7.5 Wall Shear Stress

Wall shear stress is an important determinant of endothelial cell function. It induces the release of vasoactive compounds affecting blood vessel diameter. Several authors have found that if the flow rate and thus mean wall shear stress is forced to change from its physiological state by a shunt or by an arterial ligation-positioned downstream, arterial diameter adapts and mean wall shear rate is restored toward its baseline value. If viscosity, and thus mean wall shear stress, is increased, mean wall shear stress is reduced toward its baseline level, also by an increase in arterial diameter. Optimum value for mean wall shear stress should be around 1.5 Pa.

7.6 Mechanisms of Disease Formation

For many years, it was thought that atherosclerotic plaques developed on passive arterial walls, but it is now understood that the endothelial layer lining the inside of arteries responds to these hemodynamic forces and plays a very active role in the formation of the disease [16–19]. Following extensive investigations comparing the hemodynamic environment with the locations in the circulatory system where plaques develop, wall shear stress has been hypothesized to be the hemodynamic factor with the strongest role in the initiation of this disease [20, 21].

Significant evidence exists to support the theory that hemodynamics has an important role to play in atherosclerosis. Studies have shown that changing the flow patterns of blood can affect the wall shear stress gradients and magnitudes, which in turn may lead to atherosclerosis and intimal hyperplasia, or thickening of the artery wall [5, 22–30]. Abnormal blood flow patterns are initiated by a change in the geometry of the vessel through which the blood flows. Within bifurcations and curved tubes, blood flow patterns may exhibit flow separation, recirculation, and flow stagnation with associated secondary velocities. Arteries that do not induce such flow patterns tend to be spared from the disease. As a result, non-uniform hemodynamics has been implicated as a major contributor to the disease initiation and development [31].

7.7 Flow in Small Vessels Hemodynamic Modelling of Coronary Flows

Hemodynamics in healthy vessels can be used to investigate the influence of the wall shear stress on the initiation of atherosclerotic plaques. A set of healthy arteries was used to develop three-dimensional models of coronary vessels as shown in Fig. 7.3.

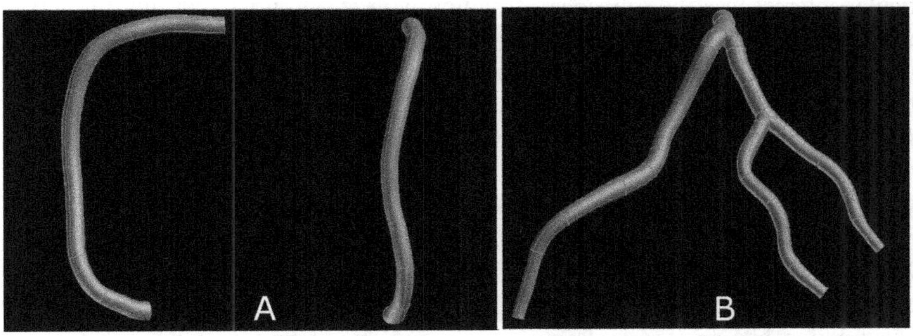

Fig. 7.3 Three-dimensional model of the *right* (**a**) and *left* (**b**) coronary arteries. The right coronary artery model consists of the main trunk as all minor branching is neglected. The left coronary model consists of the left main, left circumflex and left anterior descending arteries

These arteries were selected in consultation with a cardiologist to ensure that no significant plaques were present. During the three-dimensional construction of the models of the right and left coronary arteries, a number of assumptions needed to be made to reflect minor differences from physiological geometries. These assumptions included the following:

- A varying circular cross section in the models. The internal lumen can be determined by intravascular ultrasound (IVUS) but this method is not routinely used clinically for diagnostic purposes and is quite invasive. Thus, in the models presented here an alternative approach was adapted. The models shown here incorporated the changes in arterial radius along the length of the vessel and represent healthy vessels, which results in the assumption of a circular cross-sectional area being acceptable [32].

- The wall of the vessel modeled as rigid. This is an acceptable assumption for the coronaries, as they are less elastic than the peripheral arteries [33, 34].
- All branching in the right coronary artery (RCA) and in the minor branches of the left coronary artery (LCA) (diameter less than 25% of primary vessel) can be neglected. It has been determined that these minor branches measure less that 25% of the diameter of the primary vessel and therefore do not greatly influence the flow fields [35, 36].

7.8 The Influence of Wall Motion

During the beating of the heart, the movement of the heart walls results in vessel motion during the pulsatile cycle. The influence of this wall motion on the flow behavior within the vessels attached to the heart walls has been the focus of many studies. Zeng [37] conducted an investigation of the influence of vessel motion on flow patterns and associated wall shear stresses (WSSs). Their study aimed to determine if such wall motion could influence atherosclerotic development. Of particular interest was the influence of motion with steady-flow and pulsatile inlets. The investigation considered only the simpler right coronary artery and there were also significant geometrical differences between the model shown in Fig. 7.3a and the RCA described by Zeng. Notably, however, the peak time-averaged wall shear stresses for the pulsatile flow were of similar magnitude (approximately 2 Pa). It was also evident from the Zeng study that the influence of the motion of the vessel on WSS was negligible. More recent studies by Ramaswamy [38, 39] did reveal an influence on wall shear stress distribution, time-averaged wall shear stresses, and oscillatory shear index (OSI) due to motion of the arterial vessels. These investigations were, however, conducted on a single case in which the subject under examination had an atherosclerotic lesion, and this feature coupled with the singularity of the data makes it difficult to arrive at meaningful conclusions. A detailed computational modeling study by Theodorakakos [40] further investigated the influence of cardiac motion on flow development. Their study concluded that myocardial motion had only a minimal effect on flow distribution, as they found only minor differences in WSS when the motion of the vessel is included in the analysis. Thus, while there is some evidence indicating that the motion of the coronary arteries does influence the WSS patterns in the vessels, geometry and pulsatility appear to play a more dominant role in the development of atherosclerotic plaques. Therefore, a static average model incorporating the range of curvature present throughout the pulse cycle is a reasonable approximation when modeling the coronary arteries. This approach was used in the results presented in this chapter.

7.9 Boundary Conditions for Coronary Flows

Boundary conditions have a significant influence on the results obtained in flow modeling and thus it is necessary to ensure that the inputted boundary conditions are as close to the physiological environment as possible.

7.10 Velocity

A number of published velocity profiles for each artery were examined and the common characteristics of the profiles were determined. These profiles, which can be seen in Figs. 7.4 and 7.5, were used as a realistic representation of velocities for each artery [41–46].

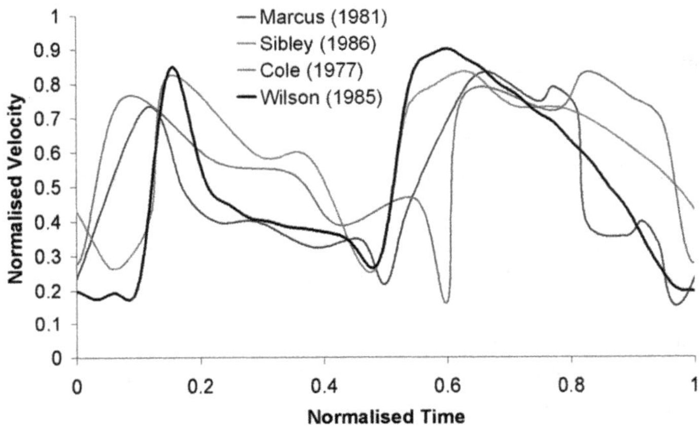

Fig. 7.4 Velocity profiles for the RCA [41–44]

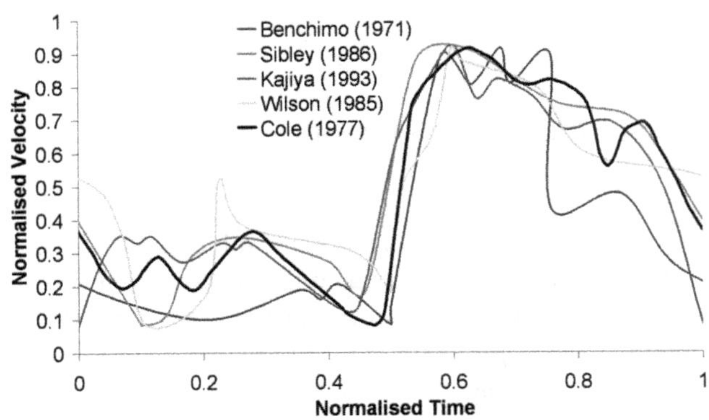

Fig. 7.5 Velocity profiles for the LCA [42, 43, 45, 46]

It can be seen from these profiles that the common characteristic of the RCA velocity profile is two peak pulses in flow, while the majority of LCA flow occurs during diastole. The profile recorded by Cole [43] was selected as the input for the LCA, while that published by Wilson [44] was chosen for the RCA models, as these are representative of the 'average' velocity profiles present in these vessels. These profiles were then scaled to provide a time-averaged Reynolds number of 200 at the inlet, which compares well with the values used in published work. The Womersley parameter was 2.76 for the RCA and 3.84 for the LCA. A fully developed velocity profile was used as the input boundary condition for both the steady and pulsatile right and left coronary models presented. The flow entering the coronary arteries is not parabolic due to the branch angles between these arteries and the aorta and the complex nature of aortic flow. Myers [36] studied the effect of three different inlet velocity profiles: a flat inlet, fully developed Poiseuille flow, and Dean-type inlet. They determined that the WSS distributions for the blunt and parabolic inlet profiles were very similar, differing by less than 10% of the inlet Poiseuille magnitude. An appropriate inlet length for both the RCA and LCA was determined. An outlet length of five diameters was added to the 3D geometries to eliminate outlet effects.

7.11 Outlet Boundary Conditions for Coronary Flows

Since published data for pressures at the distal portions of the coronary arteries or the distribution of flow between the branches of the LCA are not available, the outlets of the models were all set to a pressure of zero. This assumption is unlikely to greatly affect the flow profiles of the RCA, as the Reynolds number used for inlet boundary condition compares well to published work. Flow in the LCA would be influenced if the distribution of flow between the left anterior descending (LAD) and the left circumflex artery (LCX) is not a function of the areas of the daughter branches. However, ventricle contraction has the greatest influence on left coronary artery peripheral resistance and it is likely that it affects all left coronary artery (LCA) vessels similarly.

7.12 Numerical Model Development

Computational modeling is a very useful tool for examining fluid flow through complex geometries for which an exact theoretical solution does not exist. The influence of various boundary conditions can be efficiently analyzed without the need to redesign the 3D model. However, numerical analysis does have its limitations and is only as accurate as the mesh and computational parameters allow. It is critical that the influence of the mesh and other modeling variables is determined and minimized as far as is computationally possible. For the results presented here, the study involved the use of a steady-flow analysis and was performed on an extremely high-density mesh of the RCA, which was used as a base marker by which to compare the lower element meshes. This lead to the selection of a four-layer boundary layer mesh with 900 elements per cross-sectional area for the 3D models of the RCA and LCA. This boundary layer mesh was assessed for pulsatile flow models by comparing results with a high-density mesh, the results of which demonstrated that the WSS values remained within 2% of the high-density mesh. Independence tests were then performed for time step, pulse period, and residual convergence criterion. For all studies a low percentage difference in WSS was noted between the higher parameters and optimized modeling boundary conditions. The meshes and solvers used in this study have been used previously to model flow in the aorta and femoral bypass grafts as described in [47–49], and these computational models have been validated by comparison with laser Doppler anemometry measurements and previously published research [50–52], which indicates that the mesh and solvers used in this study are appropriate for hemodynamic modeling. Good agreement between Womersley's theoretical solutions and computational results was achieved which added to the confidence in the numerically generated grid selected.

7.13 Coronary Flow Analysis

Realistic coronary geometries were reconstructed from biplane cineangiograms obtained from the Midwestern Regional Hospital, Limerick. These geometries were used to perform computational flow analyses in order to determine the hemodynamic profiles present in these arteries and develop a deeper understanding of the role these forces play in atherosclerotic plaque development. Steady and pulsatile studies were conducted using FLUENT 6.0 as both provide information regarding the complex flow fields present in these arteries. Steady-flow studies demonstrate the influence of the geometrical features on the fluid mechanics of the blood, while the pulsatile analysis displays the temporal variations in hemodynamics introduced by the pumping of the heart.

7.14 Steady Flow in the Right Coronary Artery

Steady-flow computational analyses of the right coronary artery were conducted using the realistic geometry and the computational mesh presented earlier. The geometry of the RCA consists of a primary plane of curvature, which will be referred to as the medial plane, that contains two main bends as the artery arches around the heart. There is also secondary, out of plane, curvature that has an influence over the flow fields. The Reynolds number was set to 200 and the flow entering the artery was fully developed. The curvature of the artery varied greatly along the length and this curvature skewed the flow toward the outer wall and induced secondary, Dean-like vortices in the flow fields. Because of the significant level of curvature, with Dean's numbers reaching 130, secondary flow fields are induced with flow in the center of the artery moving toward the outer wall along the diameter of the vessel and returning to the inner wall along the circumference of the tube. A series of slices through the vessel at both these regions were examined to analyze the flow patterns present (Fig. 7.6).

The parabolic flow is skewed shortly after entering the RCA and outer wall shear magnitudes 3.3 times those of the inner wall are observed with the average outer WSS value 2.3 times that of the inner wall. When examining the overall wall shear stress contours (Fig. 7.7) in this steady-flow model of the RCA, it can be observed that much of the shear stress experienced by the wall lies within the range predicted

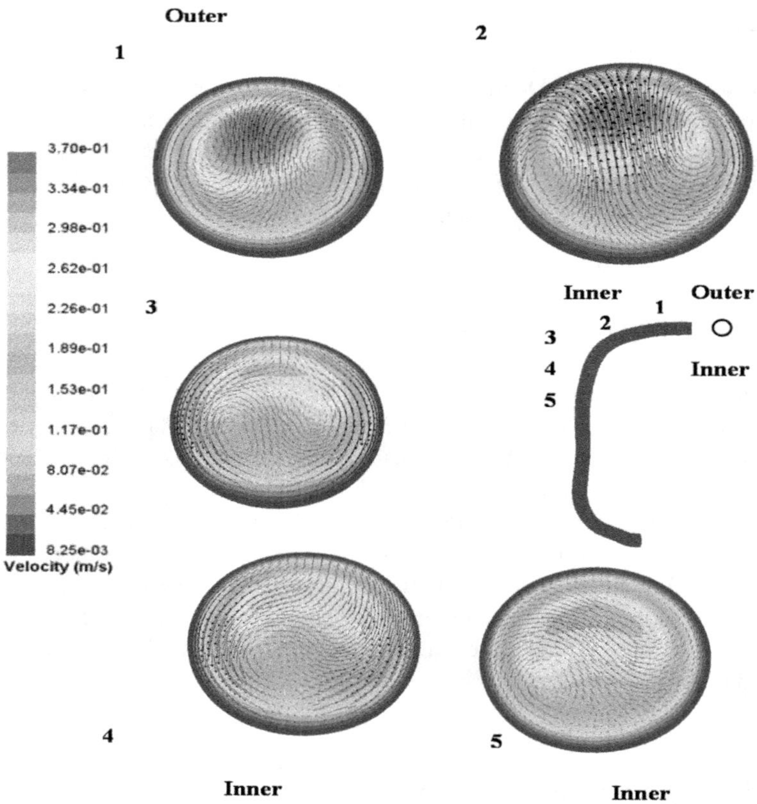

Fig. 7.6 Primary contours and secondary vectors of velocity in the RCA using the inlet profile from [44]

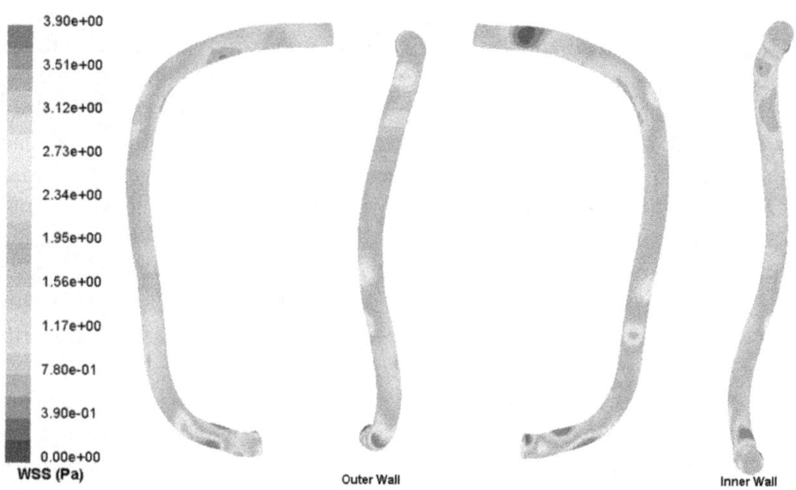

Fig. 7.7 The wall shear stress profiles in the RCA showing the variation in shear stress caused by the complex geometry. The inlet velocity profile from [43]

by the Poiseuille analysis of 1.45 ± 0.5 Pa. This variation in shear stress caused by the complex geometry of the RCA is thought to cause the localization of atherosclerotic plaques. Not only is there a wide variation in shear magnitudes present in the RCA but also there are sizeable differences in shear within a small region of the arterial wall due to the skewing of flow.

7.15 Pulsatile Flow in the Right Coronary Artery

While the steady-flow analyses can provide useful information regarding the influence of the arterial geometry on the hemodynamics in the vessel, it is not physiologically accurate to assume steady-flow boundary conditions, as the pumping of the heart results in significant temporal variations in the inlet velocity. As shown earlier the inlet velocity profile was obtained from a publication by Wilson [44]. This profile can be described by a Fourier series waveform with 13 harmonics and it was scaled to have a time-averaged Reynolds number of 200, as with the steady-flow studies. The period of the pulse was 0.8 s, which is equivalent to 75 beats/min, representing in vivo resting conditions. Therefore, the Womersley parameter was 2.76. Examining the velocity and acceleration of this waveform, a number of key time points during the pulse cycle can be observed, such as the maximum magnitude of inlet velocity, peak acceleration, and peak deceleration.

Pulsatile flow produces more flattened profiles because the layers of fluid close to the wall move slower than those in the center of the vessel, due to the influence of viscosity. The higher the Womersley parameter, the more pronounced the effect. Therefore, because of their lower momentum, these laminae can reverse quicker when the flow accelerates and decelerates. During the acceleration phase, a flattened profile is observed without much skewing of flow. When the flow reaches its peak and starts to decelerate, this flattened profile across the tube develops into a more skewed profile in the curved sections of the artery and a more parabolic profile in the straight section. The highest level of skewing is observed during the deceleration period. When the change in velocity occurs more slowly, the flow field tends toward its steady-state condition. The wall shear stresses, which were determined from the pulsatile model of the RCA, are shown in Fig. 7.8.

7.16 Steady Flow in the Left Coronary Artery

The left coronary artery consists of the left main coronary artery, the left circumflex artery, and the left anterior descending artery, which between them create the two main bifurcations which dominate the hemodynamics of this artery (Fig. 7.9). The diameter of the artery reduces along the length of the artery, resulting in a taper effect at the distal end of the left circumflex and the left anterior descending. There is also significant curvature along the length of this artery, although not at the same

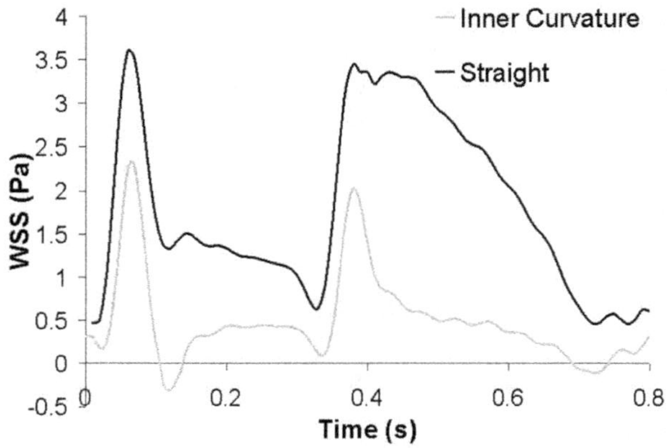

Fig. 7.8 Wall shear stress profiles selected from the RCA displaying the temporal variation at the two spatial locations indicated (Used with permission from ASME [54])

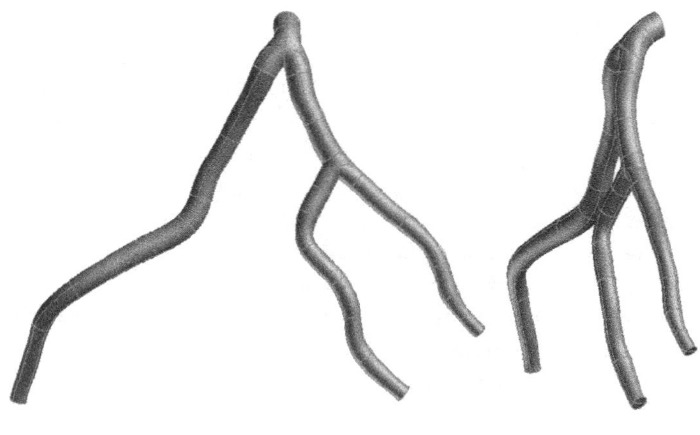

Fig. 7.9 The geometry of the left coronary artery of the left coronary arterial model containing the left main, the left circumflex, and the left anterior descending arteries

magnitude as that of the right coronary artery. The first bifurcation occurs at a location where there is also significant curvature, which results in complex flow profiles at this junction. Steady-flow analyses were conducted in the vessel using the same meshing scheme as that of the right coronary artery. The Reynolds number was 200 and the flow entering the artery was fully developed.

At the two bifurcations, the blood is skewed toward the flow divider, resulting in very low velocities along the outer wall of the bifurcation (see Fig. 7.10).

The first bifurcation, between the left anterior descending and the left circumflex, is not symmetrical and also contains a significant level of curvature that results in the flow being heavily skewed toward the left anterior descending. This results in a recirculation region occurring in the left circumflex artery at the inner wall of the curvature. As a consequence of this, a stagnation point is created and the velocity gradient close to the wall is extremely low. The bifurcation further along the artery

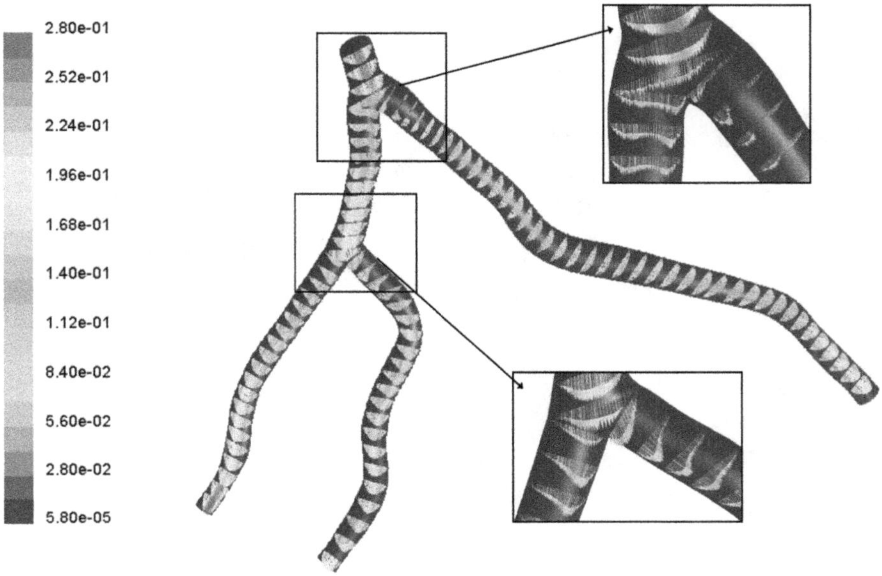

Fig. 7.10 Velocity vectors along the length of the LCA. At the two bifurcations, the blood is skewed toward the flow divider resulting in very low velocities along the outer wall of the bifurcation. The first bifurcation, between the left anterior descending and the left circumflex, is not symmetrical and also contains a significant level of curvature that results in the flow being heavily skewed toward the left anterior descending. This results in a recirculation region occurring in the left circumflex artery at the inner wall of the curvature and a stagnation point is created

in the left anterior descending occurs in a straighter section and is more symmetrical, with a predominantly even distribution of flow entering each daughter branch. Again the flow is skewed toward the flow divider with far lower velocities along the outer wall of the bifurcation. There is no recirculation region in this bifurcation for the steady-flow analysis. The same degree of sharp curvature that was present in the RCA is not observed in the LCA. However, the artery is by no means straight and the flow is skewed by the curving of the artery as it reaches around the heart. In one segment of the left anterior descending artery, counter-rotating Dean-like vortices develop as the flow skews around a relatively sharp bend. The primary flow at this section is directed toward the outer wall of curvature (Fig. 7.11).

High velocities occur at both flow dividers and the outer wall of curvature, resulting in higher shear stresses at these regions. The outer wall of the bifurcation experiences extremely low wall shear stresses due to the skewing of the flow. In the recirculation region in the first bifurcation, there is a region where the wall shear stress is reversed in the opposite direction to the primary flow. This could be physiologically relevant for atherosclerotic plaque development. The highest spatial gradients in wall shear stress in the left coronary artery occur at the bifurcations. There is an extremely large variation in the shear experienced by the outer wall of the bifurcation and the flow divider, with WSS at the outer wall of the left circumflex

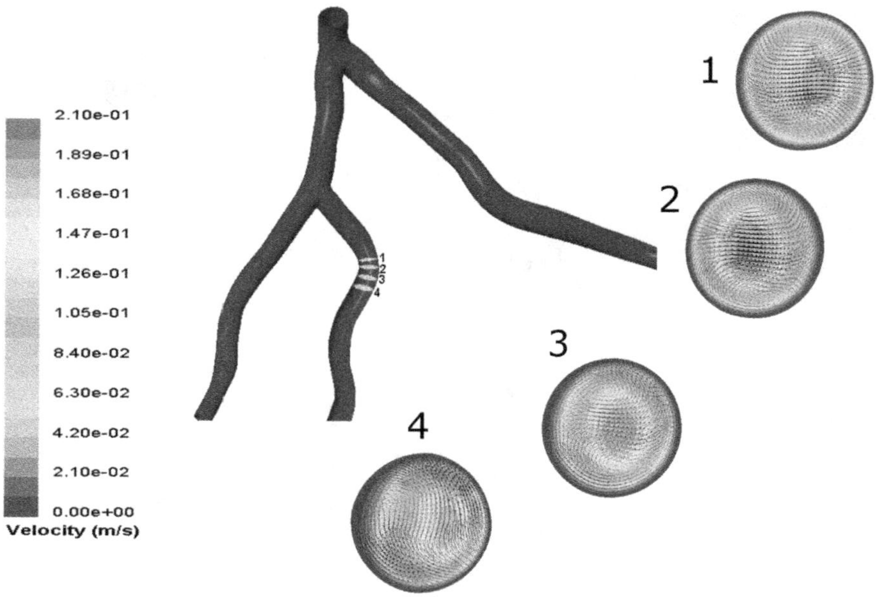

Fig. 7.11 Flow profiles at a region of high curvature in the LCA. The development of counter-rotating Dean-like vortices as the flow skews around a relatively sharp bend is displayed in slices 1–4

almost 15 times smaller than that at the flow divider. However, the disparity in WSS does normalize and the spatial gradient reduces within a relatively short distance from the bifurcation.

7.17 Pulsatile Flow in the Left Coronary Artery

As with the RCA, pulsatile flow analyses were performed to provide a more physiological view of the hemodynamics present in the left coronary artery during the pulse cycle. The inlet velocity profile was obtained from a published work [43] and was scaled to provide a time-averaged Reynolds number of 200 and a pulse period of 0.8 s. This resulted in a Womersley parameter of 3.84. The LCA velocity inlet profile was reduced to a trigonometric approximation described by a Fourier series waveform with 13 harmonics. Once more, key time points during the pulse period such as the maximum velocity, acceleration, and deceleration were examined; the flow at maximum acceleration is presented and was examined in greater detail.

The velocity profiles along the length of the LCA are strikingly different during the acceleration phases when compared with the rest of the pulse cycle. During this period the velocity profile across the diameter of the vessel is largely flat, with predominantly positive flow. Very little recirculation is observed when the flow is accelerating, even in the first bifurcation where a strong recirculation region is

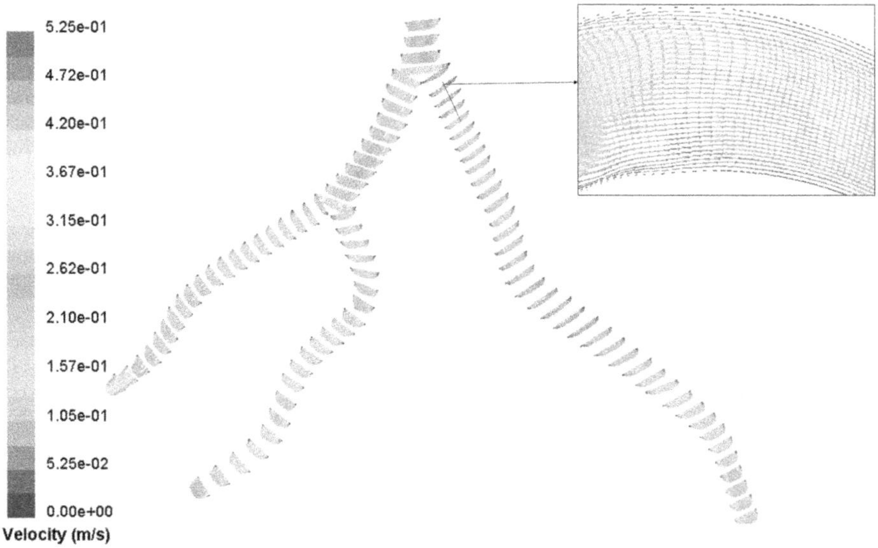

Fig. 7.12 Velocity profiles during acceleration phases in the LCA. During this period the velocity profile across the diameter of the vessel is largely flat, with predominantly positive flow. Very little recirculation is observed when the flow is accelerating, even in the first bifurcation where a strong recirculation region is established for the rest of the pulse cycle

established for the rest of the pulse cycle. There is, however, a minimum period over which acceleration needs to occur for these flattened profiles to dominate the flow. The velocity profile for one of those time points is shown in Fig. 7.12.

At a later phase in the cycle, reverse flow is seen at the wall along the majority of the length of the vessel, which could be highly significant for atherosclerotic plaque development. The same velocity profile is seen along much of the artery, consisting of a parabolic profile in the center of the tube, reducing to zero out toward the wall and continuing to decrease into reverse flow next to the wall. The recirculation region at the entrance to the left circumflex contains strong secondary flow profiles. For much of the rest of the pulse cycle, the flow is skewed toward the flow divider of the bifurcations and the outer walls of curvature. Parabolic flow enters the left main coronary artery but is immediately skewed upon entering the bifurcation between the left circumflex and the left anterior descending. Along the length of one of the branches of the LAD, the flow is skewed left and right as the artery curves, resulting in low velocities and wall shear stresses along the inner wall of curvature.

There is a huge variation in WSS in the LCA during one pulse due to the geometrical features of the LCA and the temporal velocity input. As seen from the velocity profiles, the regions of high shear are at the apex of the bifurcations and the outer walls of curvature (Fig. 7.13). As the geometry of LCA changes significantly along the length, there are large spatial gradients in shear. The WSS magnitudes in the LCA are lower than the RCA, with only very minor regions experiencing shear values greater than 3 Pa at the peak velocity of the pulse cycle. Extremely low WSS

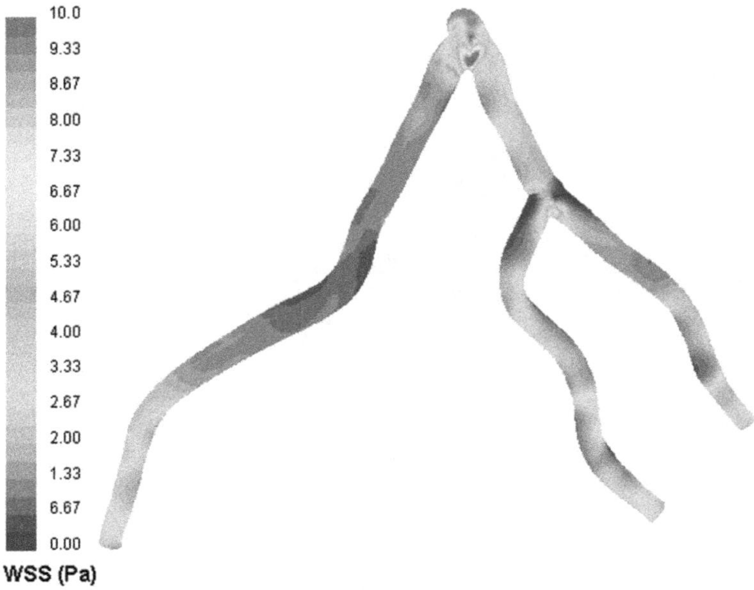

Fig. 7.13 WSS contours along the LCA during the peak velocity input. The regions of high shear at the apex of the bifurcations and the outer walls of curvature are highlighted. As the geometry of the LCA changes significantly along its length, there are large spatial gradients in shear and the magnitudes of the shear stresses in the LCA are lower than those in the RCA

magnitudes are present for the entire duration of the pulse cycle at some locations in the LCA, particularly at the outer wall of the bifurcations where the flow alters from minor forward flow during the acceleration periods to reverse and recirculating flow when the velocity inputs are decelerating.

7.18 Discussion

The geometries of the right and left coronary arteries are relatively complex, thus producing varied hemodynamic profiles along the length of the vessels. FLUENT 6.0 was used to perform computational analyses on both the RCA and the LCA using the developed geometries and the mesh and boundary conditions presented above. The predominant features of the flow mechanics in these vessels are the skewing of flow toward the outer wall of curvature and the flow divider of the bifurcations. These findings allowed WSS profiles from different locations in the arteries that represented the physiological types of shear profile that could participate in the promotion or the prevention of atherosclerotic plaque development to be identified. Two of these profiles were from regions that indicate a high predisposition to lesion development [34, 53], the inner wall of curvature, and the outer wall of a bifurcation. In vitro EC studies [54] showed how these shear profiles could be applied in an in vitro test facility to assess upregulation in adhesion molecule expression and monocyte adhesion. The profile from a "healthy" region where the shear stress magnitudes are close to the mean physiological values for the arterial network (1.5 Pa) was used to provide a useful comparison of cell adhesion molecule (CAM) expression between these types of shear profiles.

7.19 Flow in Large Vessels – Hemodynamic Modeling of Aortic Flows

Expanding our discussion to large vessels, a similar approach is adapted, and to demonstrate this approach, three example cases are shown which demonstrate many of the features of flow in healthy and diseased aortic vessels. The initial model shown was extracted from a magnetic resonance image (MRI) scan as shown in Fig. 7.14 [55] and was used to generate a model of a healthy subject. Although the resolution is low, the aorta (circular white region) is clearly evident in the image. The scans were each examined in order to reveal that no obvious disease was present and that the aorta appeared to be functioning as normal.

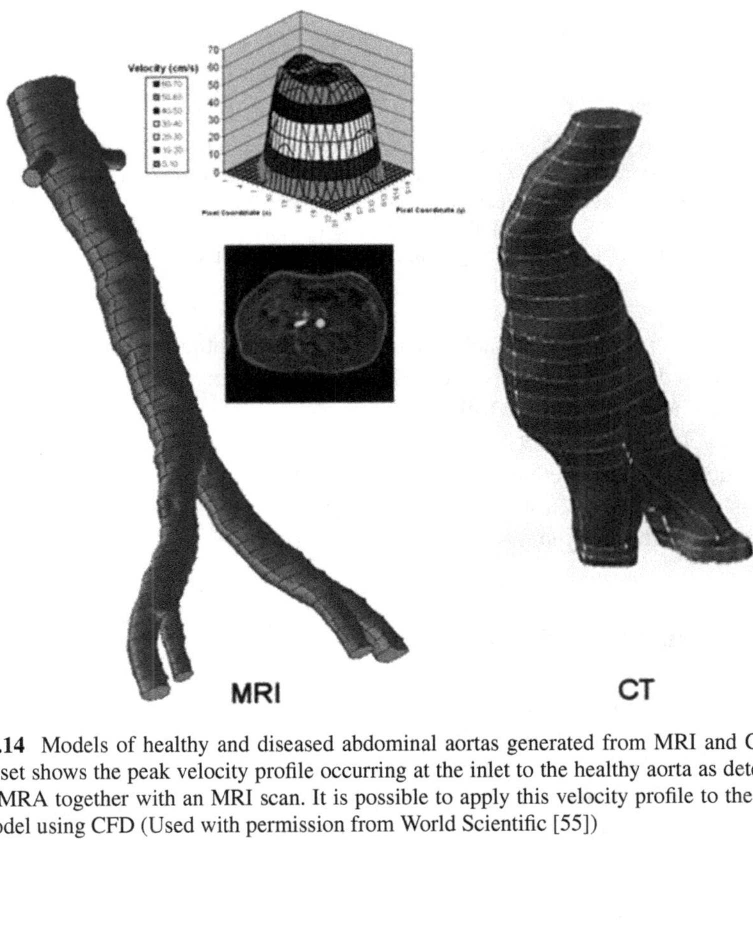

Fig. 7.14 Models of healthy and diseased abdominal aortas generated from MRI and CT data. The inset shows the peak velocity profile occurring at the inlet to the healthy aorta as determined using MRA together with an MRI scan. It is possible to apply this velocity profile to the inlet of the model using CFD (Used with permission from World Scientific [55])

Figure 7.14 also shows the inlet velocity profile at peak velocity and illustrates the nature of the skewed velocity profile. This skewed profile in the descending aorta is due to the flow regime setup during the motion around the aortic arch.

Models 1 and 2 are based on the realistic human aorta as shown in Fig. 7.15a, b. Model 3 is an idealized abdominal aortic aneurysm (AAA) model which was based on average values as obtained from the EUROSTAR Data Registry Center (2001) and is shown in Fig. 7.15c. The realistic models have an out–of-plane curvature, non-homogenous size, and quasi-elliptical cross sections. The branches have been omitted for the first model and included in the second model. This will show what effect that overall curvature alone has on the flow patterns.

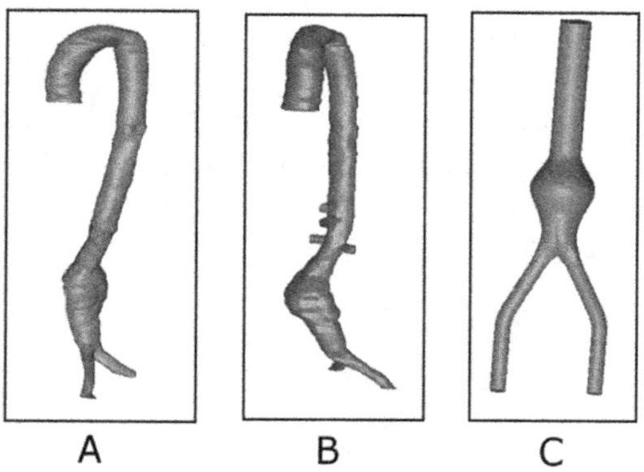

Fig. 7.15 Reconstructed models of realistic AAAs without branching (**a**), with branching (**b**), and a model of an idealized AAA (**c**). Realistic models have out-of-plane curvatures, non-homogenous sizes, and quasi-elliptical cross sections. Branching arteries were omitted from model (**a**) and included in model (**b**). This allowed the effect overall curvature on flow patterns to be assessed. Model (**c**) was based on average values as obtained from the EUROSTAR Data Registry Center [45]

In order to ensure stability and accuracy of the computational approach to the flow analysis of these complex models, a range a numerical tests needed to be conducted. These included the following:

Pulse cycle independence; for larger diameter arteries, time-periodic independence is normally achieved at 4–5 cycles. Validation was achieved using analytical solutions to the Navier–Stokes equations, which are available for straight circular cylinders for both steady and transient flows.

7.20 Boundary Conditions

Infrarenal blood flow characteristics are extremely complex and are affected by the four major outflow vessels just below the diaphragm; these are the celiac, the superior mesenteric, and the left/right renal arteries [56]. During resting conditions, these four vessels that branch off the aorta receive approximately two-thirds of the total descending thoracic aorta blood flow. During exercise conditions there is an increase in the cardiac output with most of the blood supplying the lower limbs. As a result of this increase of blood supply to the legs, the percentage of thoracic aorta blood flow exiting the four vessels at the diaphragm decreases from 66% to about 20% [56–61]. Table 7.1 shows the percentage of blood flow through the major vessels in the human aorta for resting, medium exercise, and exercise.

Table 7.1 Proportion of blood flow through the major arteries of the aorta [60, 61]

Artery	Rest (%)	Medium exercise (%)	Exercise (%)
Celiac	21.7	9.7	4.4
Superior mesenteric (SMA)	14.7	6.9	3.5
Left renal	14.7	7.6	4.4
Right renal	14.7	7.6	4.4
Inferior mesenteric (IMA)	4.8	3.9	0.007
Left iliac	14.7	32.2	40.5
Right iliac	14.7	32.2	40.5

The medium exercise as given in Table 7.1 is taken as the average blood flow through each vessel for the resting and exercising conditions. Cheng [57, 58] conducted numerous MRI scans on a range of subjects varying in age from 20 to 70 years. For resting conditions they found on average 3 and 1 l/min of blood flow at the supraceliac and infrarenal locations, respectively, and for exercising conditions, 7.6 and 5.9 l/min, respectively. For medium exercise conditions, this would correspond to 5.3 and 3.5 l/min for supraceliac and infrarenal locations, respectively. For our simulations, flow though the inferior mesenteric artery (IMA) was neglected. This is a valid assumption for AAA blood flow modeling, as due to the presence of thrombus in the sac and the insertion of a stent graft across the aneurysm sac, flow is prevented from traveling through this artery.

7.21 Steady-Flow Boundary Conditions

For the purpose of a steady-flow simulation, a medium exercise condition was selected. This is the average blood flow for resting and exercising conditions as given in Table 7.1. A flat inlet profile was applied to the aortic inlet of both the realistic AAA models (see Fig. 7.22). The assumption of a flat velocity profile at the aortic inlet has been confirmed by various in vivo measurements using hot film anemometry on different animal models distal to the aortic valve [62–65]. The velocity input for the realistic AAA without the four branches was 3.5 l/min, while the input for the realistic AAA model with the branches was 5.3 l/min.

To assess the influence of boundary conditions, three different inlet velocity conditions (see Fig. 7.22) and three geometries (see Fig. 7.15) were examined.

7.22 Steady Flow – Realistic Model

Figure 7.16 shows the overall contour vector plot of the two realistic models (model 1 without the branches and model 2 with the branches) before the AAA sac. Both models had the same flow rate at the outlet, which resulted in a higher input velocity for model 2 which included the four branches. Both models showed skewing of the flow toward the inner wall at the entrance of the aortic arch and midway along the aortic arch. Also there is a recirculation region at the inner wall at the exit of the arch with the fluid skewed toward the inner wall. This flow phenomenon was demonstrated experimentally in a number of studies [63–69].

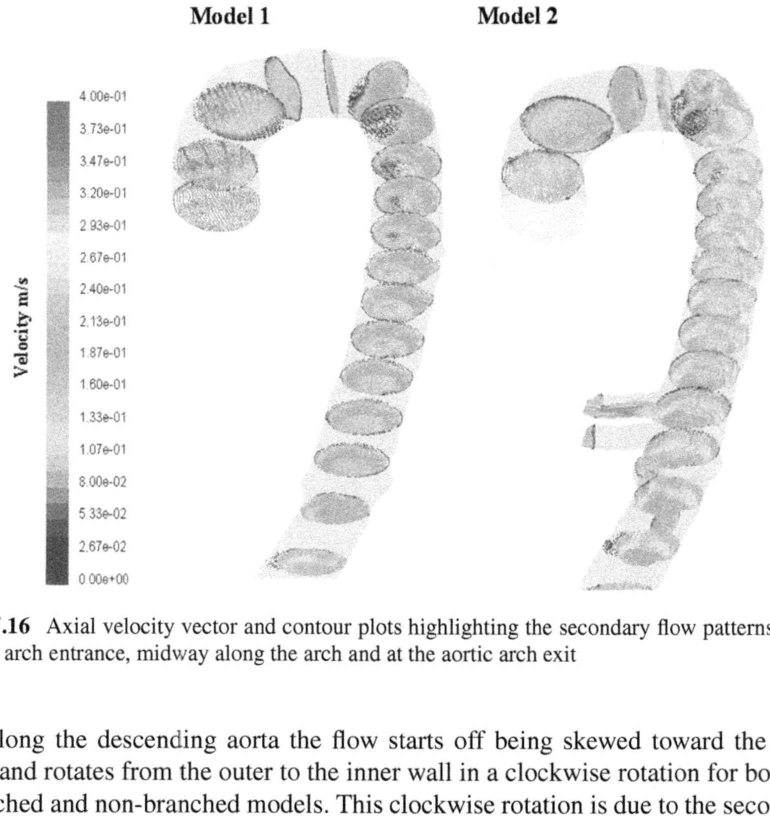

Fig. 7.16 Axial velocity vector and contour plots highlighting the secondary flow patterns at the aortic arch entrance, midway along the arch and at the aortic arch exit

Along the descending aorta the flow starts off being skewed toward the outer wall and rotates from the outer to the inner wall in a clockwise rotation for both the branched and non-branched models. This clockwise rotation is due to the secondary flows in the descending aorta. Model 2 shows the greatest skewing and a larger boundary layer from the inner to the outer wall. In vivo studies have shown the existence of this clockwise rotation in the descending aorta during systole [67]. These features are shown in Figs. 7.17 and 7.18.

The influence the four branches have on the flow patterns are shown in Fig. 7.19. As can be seen there is a greater skewing of the blood in the form of a moon contour with greater secondary flows for model 2. Further downstream before the AAA sac, the flow is still skewed toward the inner wall with an increased boundary layer.

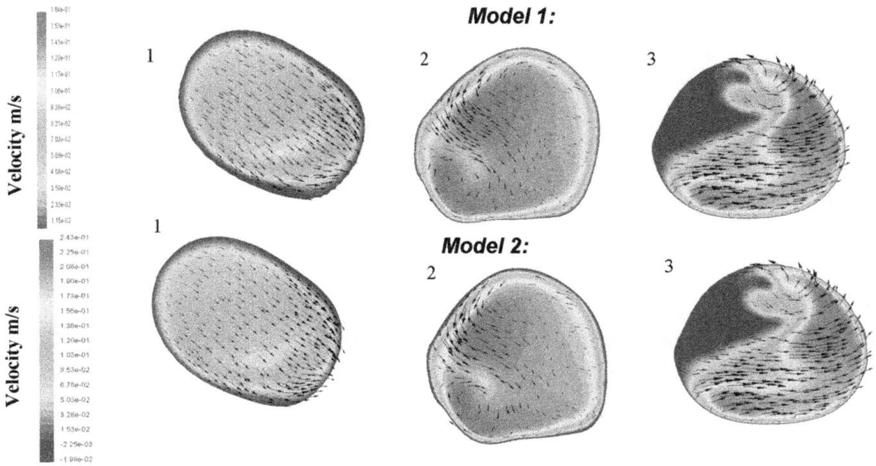

Fig. 7.17 Axial velocity vector and contour plots along the aortic arch including the secondary flow vector directions. For slice locations, see Fig. 7.16

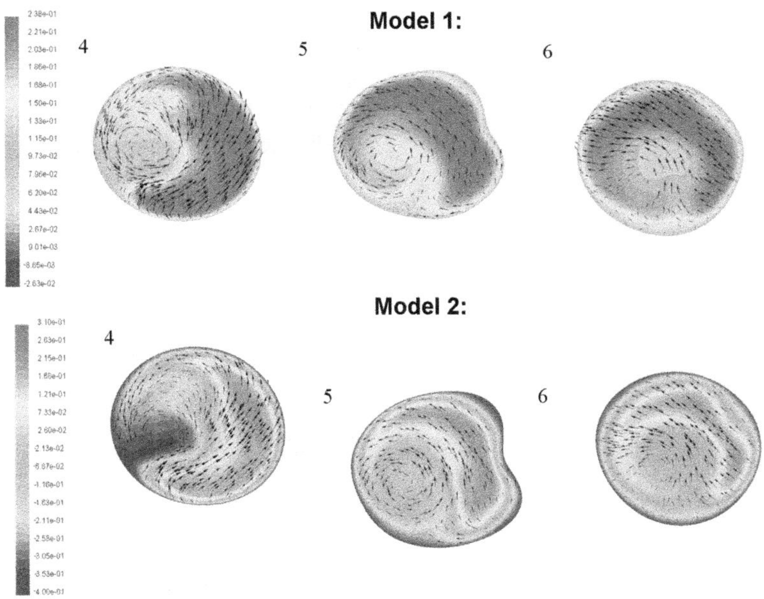

Fig. 7.18 Axial velocity vector and contour plots along the descending aorta above the celiac artery including the secondary flow vector directions. For slice locations, see Fig. 7.16

Without the inclusion of the branches, the blood flow tries to become more parabolic as can be seen from the contour plot (Fig. 7.19).

Based on these variations on the flow pattern, it was decided for subsequent simulations that the branches are required when modeling the effects of blood flow

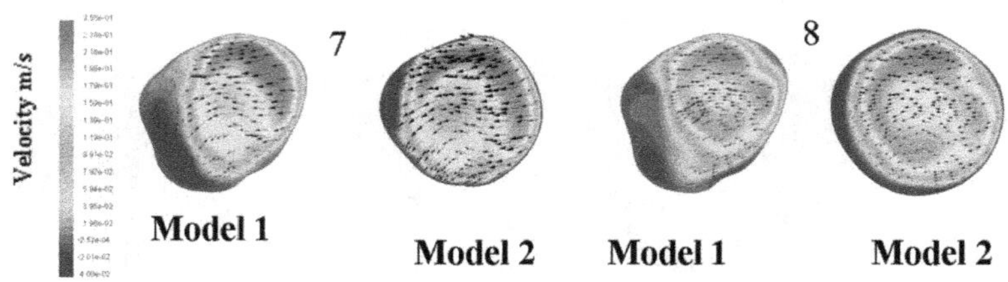

Fig. 7.19 Axial velocity vector and contour plots along the abdominal aorta including the secondary flow vector directions. For slice locations, see Fig. 7.16

through the abdominal aorta. The inclusion of the branches creates a greater skewness and secondary velocity on the flow patterns. In vivo MRI velocity contour slices have also shown skewness of the blood flow patterns at the supraceliac and infrarenal locations [58]. Figure 7.20 shows the velocity profiles through the AAA sac with the branches included. The flow entering the AAA sac is skewed toward the outer wall with secondary flows. The flow moves from the outer wall toward the center of the AAA sac, thus creating a low velocity and recirculating region in the AAA sac. This recirculating region is mainly due to the expansion of the geometry. The low-velocity region continues distal of the AAA sac toward the bifurcation and the secondary flows are reducing. At the bifurcated junction the velocity profile skews toward both iliac legs with a significant increase in secondary flows.

Figure 7.21 shows the velocity profiles in both the left leg (LL) and the right leg (RL). Both iliac legs have a complex flow pattern due to the out-of-plane curvature.

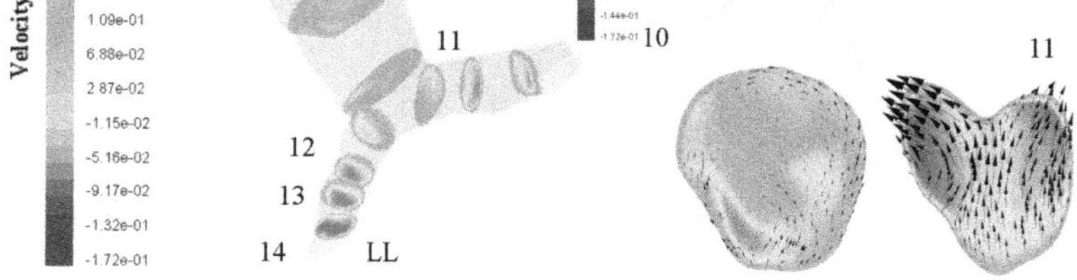

Fig. 7.20 Velocity vector and contour plots through a realistic abdominal aortic aneurysm. The flow entering the AAA sac is skewed toward the outer wall with secondary flows (see slice 8). The flow moves from the outer wall toward the center of the AAA sac, thus creating a low velocity and recirculating region in the AAA sac (see slice 9). This recirculating region is mainly due to the expansion of the geometry. The low-velocity region continues distal of the AAA sac toward the bifurcation and the secondary flows are reducing (see slice 10). At the bifurcated junction the velocity profile skews toward both iliac legs with a significant increase in secondary flows (see slice 11)

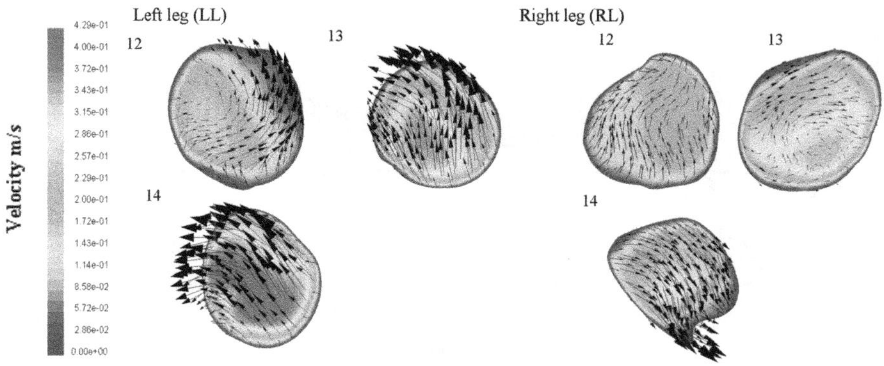

Fig. 7.21 Velocity vector and contour plots through both iliac legs in a model of a diseased aorta. The velocity profiles in both the *left* leg (LL) and the *right* leg (RL) can be seen (see Fig. 7.20 for location references). Both iliac legs have complex flow patterns due to the out-of-plane curvature

The flow through the left leg has a greater secondary velocity profile. The flow is skewed toward the outer wall for both legs with a boundary layer from the inner wall to the outer wall.

7.23 Influence of Steady Input Boundary Conditions

To analyze the effects of applying different boundary conditions, three different inlet velocity profiles (flat, parabolic, and realistic) were inputted as shown in Fig. 7.22. Figure 7.23 shows a line plot of the velocities across the center point. The influence of boundary conditions on the flow patterns was analyzed on the idealized AAA model as shown in Fig. 7.24.

As can be seen from Fig. 7.24, the flat inlet profile is becoming more parabolic at the center of the AAA sac, similar to the parabolic input. Along the proximal

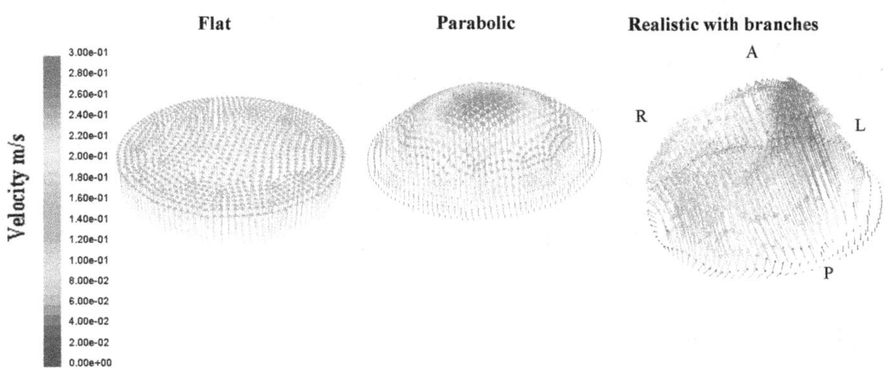

Fig. 7.22 Velocity vector plots of a flat, parabolic, and realistic inlet velocity profile. On the realistic profile, anterior (*A*), posterior (*P*), left (*L*), and right (*R*) directions are indicated

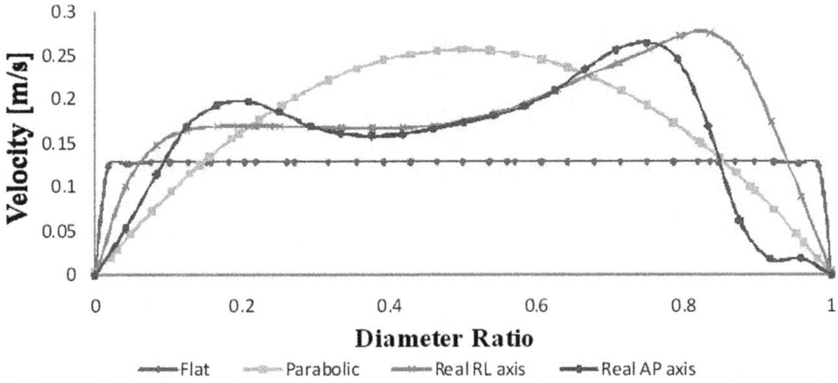

Fig. 7.23 Two-dimensional velocity profiles for the inlet boundary conditions shown in Fig. 7.22. The realistic profile in Fig. 7.22 is represented by 2D profiles in the anterior–posterior (AP) direction and the right–left (RL) direction

end of the idealized AAA model, the secondary flows for the branch input have reduced significantly and the secondary motion is recreated again at the bifurcated junction as shown in Fig. 7.19. The branch input created a complex skewing profile throughout the idealized model, which creates a much greater recirculation region in the AAA sac. The flat and parabolic inputs show very similar flat velocity profiles at the bifurcation point with skewness toward the inner wall and then toward the outer wall after the bend in both iliac legs in a half moon shape. The flow is skewed toward the outer wall after the bifurcation for the realistic input, similar to the realistic AAA of Fig. 7.20.

7.24 Pulsatile Flow in a Bifurcation

Pulsatile flow through a bifurcation is also needed as a validation for the CFD. Using the pulsatile flow circuit and the idealized AAA silicone model as described in [70], a typical physiological pulse as found in the abdominal aorta was created. A developed flow input was applied and laser Doppler anemometry (LDA) was used to provide experimental validation of the computational model. Good agreement between theory and experiments was achieved, thus giving confidence in the numerically generated grid. Perhaps the key parameter of the flow analysis of the aorta relates to the geometry being examined and this is presented in the next section. Previous publications [71–73] also show aspects of this work.

7.25 Geometrical Effects

The main influencing factors for the velocity profiles through all of the models are the geometrical differences that are associated with each model. The change

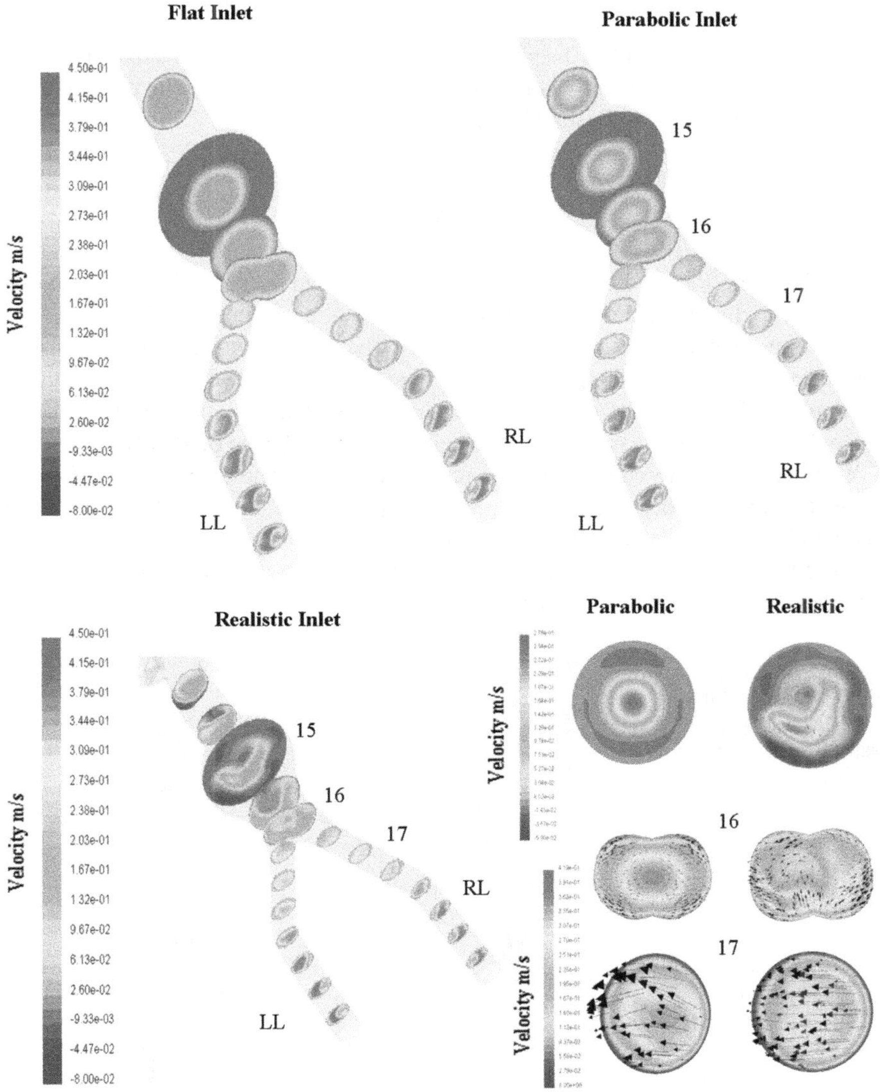

Fig. 7.24 Velocity contour plots in an idealized abdominal aortic aneurysm model with a flat, parabolic, and realistic inlet velocity profile (see Fig. 7.22). The left iliac (LL) and right iliac (RL) arteries are indicated. Also highlighted are the differences in velocity vector and contour plots between the parabolic and realistic boundary condition models at the locations of the aneurysm (15), the bifurcation (16), and the right iliac (17)

in radius of curvature along the axis is the major difference between the realistic models and the idealized models. It is known from the literature that curvature effects throughout the arterial system induce Dean forces that increase in magnitude with increasing Dean number [36, 62, 69–76]. Dean number (De) is the ratio of the effective centrifugal inertial forces to the viscous forces. With increasing De, the effects of the centrifugal forces become stronger and increase the secondary velocities. Dean number is an analytical solution to fully developed flow in curved tubes and is given by a non-dimensional parameter

$$\mathrm{De} = \mathrm{Re}\sqrt{\left(\frac{a}{R}\right)}$$

where Re is the Reynolds number, R is the radius of curvature of the arch, and a is the radius of the lumen along the line of the radius of curvature [77, 78].

Dean number (De) can be used to assess the variation due to the radius of curvature effects along the axis of all the models. For each model the radius of curvature varies along the axis. Dean number can be used to analyze the different geometrical differences associated with each model and hence predict which model will create the greatest skewness, recirculation, and secondary velocities. Dean's number is inversely proportional to the radius of curvature and directly proportional to the radius of the lumen and Reynolds number. As can be seen from Fig. 7.25, the major difference between the idealized AAA and the realistic AAA is the locus of the centerline through all the centroids.

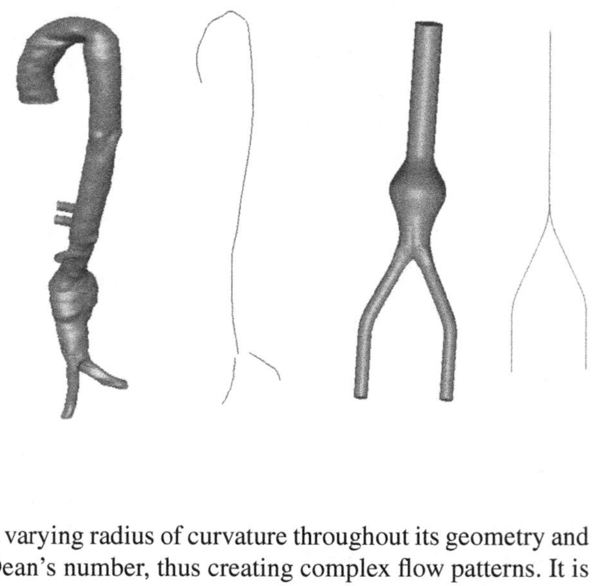

Fig. 7.25 Plots of the *centerline* curvature of realistic and idealized AAA models

The realistic model has a varying radius of curvature throughout its geometry and this will lead to a varying Dean's number, thus creating complex flow patterns. It is possible to predict complex flow patterns such as increased skewness, recirculation, and secondary flows by analyzing the radius of curvature and thus Dean's number along an arterial section. For example, along the aortic arch in Fig. 7.26 there is a low radius of curvature, thus resulting in a high Dean's number. This high Dean's number creates secondary flows and recirculation of the flow and this was found for both the steady and pulsatile cases.

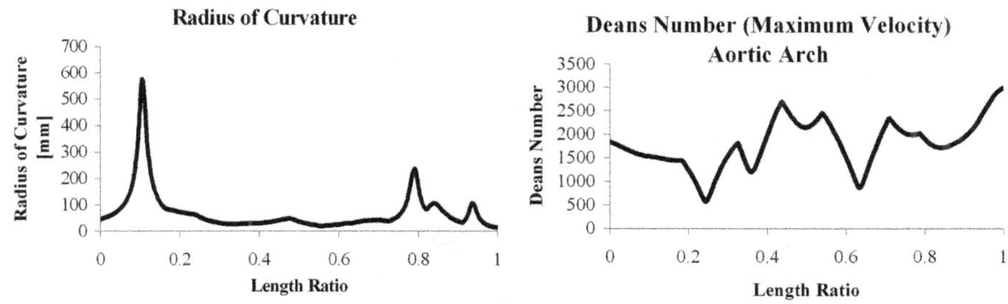

Fig. 7.26 Plots of the radius of curvature and Dean's number along the aortic arch of the realistic model shown in Fig. 7.25

7.26 Geometrical Differences Associated with the Realistic and Idealized AAA Models

Figure 7.27 shows the radius of curvature and Dean's number along the descending aorta. There is a high Dean's number at the entrance that suddenly drops due to an increasing radius of curvature, which reaches a peak around the location of the renal arteries. The radius of curvature then lowers again, thus increasing Dean's number and the secondary velocities distal to the renal arteries. The four branches around this location add to this secondary flow pattern and create an irregular skewing effect on the flow entering the AAA sac.

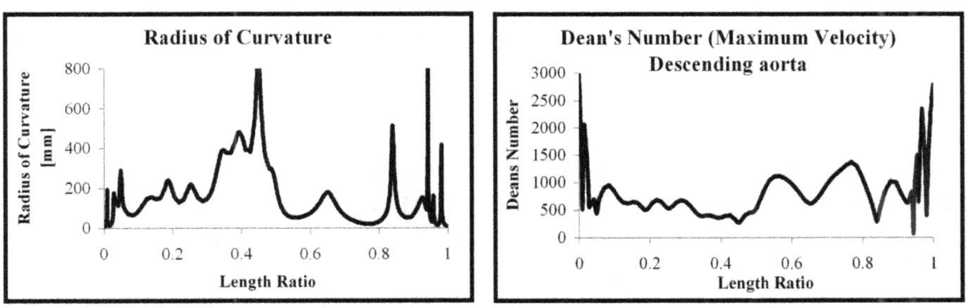

Fig. 7.27 Plots of the radius of curvature and Dean number along the descending aorta of the realistic model shown in Fig. 7.25

The idealized model has an infinite radius of curvature (see Fig. 7.25) along the AAA sac except at the bifurcation point where secondary flows are reintroduced again. This infinite radius of curvature results in a zero Dean's number, which damps out the inputted secondary velocities. The greatest secondary flows are generated when the radius of curvature is the lowest. Around the aneurysmal sac there is a low WSS due to the recirculation flow patterns. The highest WSS is found at the bifurcation and throughout the iliac legs for all models (Fig. 7.28).

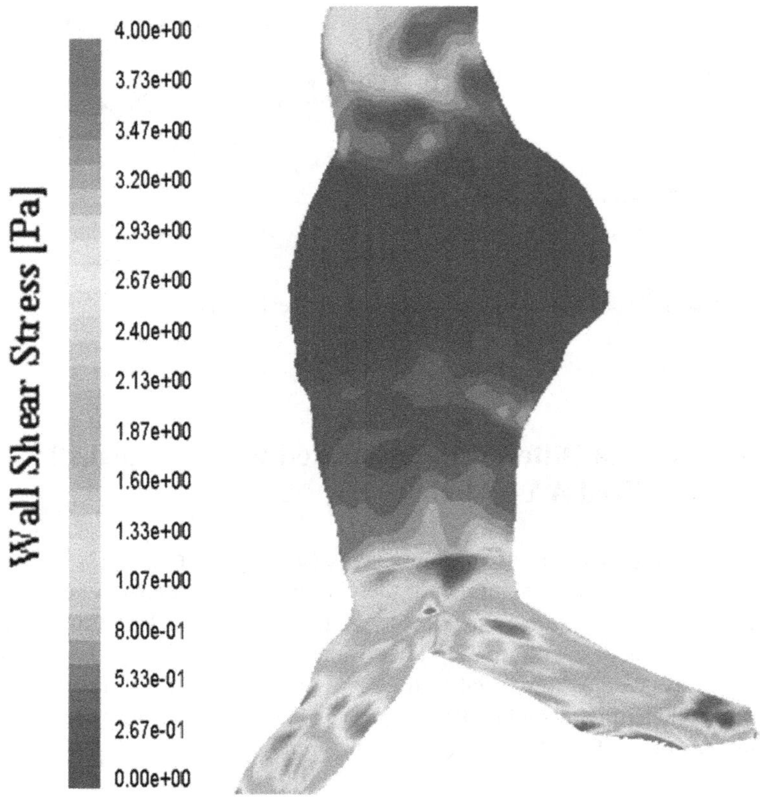

Fig. 7.28 Wall shear stress distribution around the aneurysm sac of the realistic model shown in Fig. 7.25 at the maximum velocity of the pulse

The correct input boundary conditions must be implemented for the hemodynamic modeling of blood flow through idealized or realistic AAA and stent grafts. The exclusion of the ascending aorta, aortic arch, and descending aorta has been a major assumption for numerical models found in the literature. The complexity of the flow patterns, degree of skewness, and secondary flows is influenced by the inclusion of the full human aorta and the four branches. Assuming developed flow before the AAA or stent graft models neglects the influence of the inputted secondary flows. The realistic input boundary condition creates a moon-like skewing profile, which results in a greater recirculation region in the AAA sac as opposed to inputting a developed flow profile. Recirculation zones formed in an aneurysm cavity create conditions that promote thrombus formation and the viability of rupture. Platelets with an elevated shear history are readily deposited to the wall in areas of low wall shear stress. Due to this recirculation, the aneurysmal wall becomes lined with a mural thrombus, an aggregate of blood cells, platelets, blood protein, and cellular debris. The thrombus grows as material is deposited from the flowing blood and protrudes into the lumen. Particles of the mural thrombus may break off and be carried downstream; these emboli can be lodged in smaller blood vessels and interrupt the flow of blood. The growing mural thrombus may also occlude blood flow through the aneurysm [79, 80]. The out-of-plane curvature influences the local characteristics of the velocity profiles. For the realistic geometries there is a varying radius of curvature and hence varying Dean's number. The radius of curvature and Dean's number can predict the location along the model in which the greatest secondary flows will occur. Therefore, these two parameters alone are needed when predicting atherosclerosis and thrombus formation. Arterial plaque is known to localize along the inner curvature of bends and near sites of bifurcations and is predominant in the abdominal aorta in the infrarenal and iliac regions [56, 57, 59, 60, 81, 82]. Atherosclerotic plaque thickening occurs with low magnitudes of near-wall velocity and rapid changes in direction [60]. Secondary flows have been shown to activate platelets [83]. It is therefore possible to predict from MRI or CT scans which location along the aorta and iliac arteries will be more susceptible to atherosclerosis lesions based on the radius of curvature. The idealized models have an infinite radius of curvature and thus zero Dean's number. This reduces or eliminates inputted secondary flows.

The previously presented results have shown the complex nature of the flow in healthy blood vessels. Certain arterial geometrical setups may predispose the subject to the development of arterial disease. Such arterial diseases require surgical intervention to restore blood flow to distal biological structures. The treatment of the coronary artery, peripheral, and aortic disease is varied, and some approaches are presented in the subsequent sections.

7.27 Treatment of Arterial Disease

There are several different surgical methods available to treat atherosclerosis. Balloon angioplasty, stents, endarterectomy, and grafts are all widely used [84, 85] and research into drug treatments, gene therapies, and vaccinations is ongoing. Using balloon angioplasty, stenting or endarterectomy results in the restoration of blood supply without changing the route of the flow. The use of grafts, however, involves the re-routing of flow from the blood vessel in the graft and then back into the blood vessel. This re-routing of the flow leads to a further mechanism of disease formation in the arterial system and this is discussed further in the next section.

Vascular grafting procedures are important and necessary for maintaining healthy blood supply to distal biological structures. Surgical grafting procedures fall into two categories, bypass grafting and aneurysm grafting. Vascular aneurysm grafting involves the implantation of a graft to exclude the aneurysm from blood pressure, eliminating the possibility of aneurysm rupture. Vascular bypass grafting restores hemodynamic flow beyond the diseased arterial section. Both these grafting approaches are described in the next sections using examples of peripheral bypass grafting and abdominal aortic aneurysm grafting.

7.27.1 Vascular Aneurysm Grafting

Traditional surgical repair of abdominal aortic aneurysms involves opening the chest or the abdomen, gaining temporary vascular control of the aorta and below the lesion, opening the aneurysmal sac, and suturing a prosthetic graft to the healthy aorta within the aneurysm itself (see Fig. 7.29a). The sac of the aneurysm is then sutured over the graft, thus excluding the Dacron or PTFE graft from direct contact with bowel and other viscera.

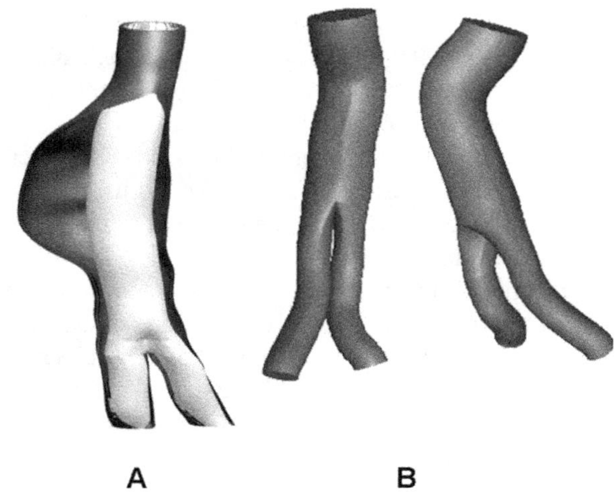

Fig. 7.29 Illustration of a graft/stent graft in an abdominal aortic aneurysms (**a**). The graft/stent graft geometry is highlighted from two views (**b**)

The flow through the graft is not re-routed as per a bypass graft. It does however follow a slightly different path with the bifurcation point appearing more proximal in the graft than in the host artery. The resulting flow patterns through the graft are presented in Fig. 7.30. The degree of secondary flows is evident. The flow

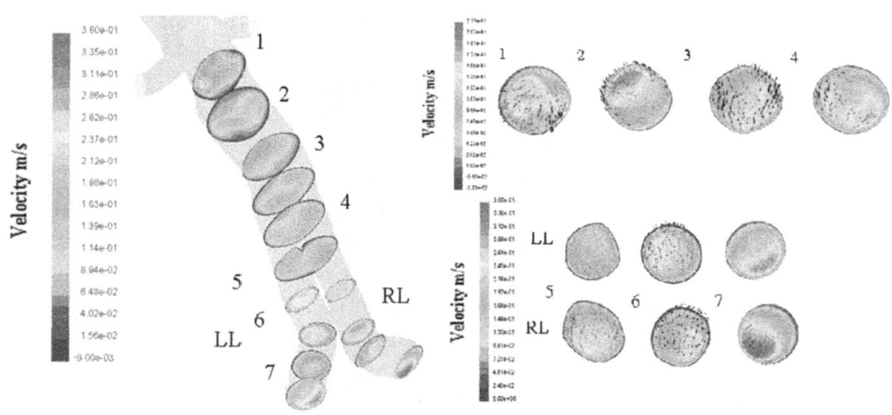

Fig. 7.30 Velocity vector and contour plots present in the graft/stent graft shown in Fig. 7.29. Slices 1–4 highlight the flow patterns in the graft/stent graft before the bifurcation. The flow patterns at specific axial locations in the left iliac (LL) and the right iliac (RL) are also shown

patterns through the graft are highly skewed with a slight recirculation region developing in the proximal end. Skewing and secondary flows exist at the entrance to the bifurcation.

The outcome of standard surgical abdominal aortic aneurysm (AAA) repair has proven to be excellent, with mortality rates in the range of 3–5%. However, standard AAA repair is not perfect, and the quality of life after this repair is impaired by postoperative pain, sexual dysfunction, and a lengthy hospital stay resulting in high health costs. All these negative effects are related to the large incision and extensive tissue dissection and have very little to do with the functioning of the graft. This is not the case with vascular bypass grafting where the re-routing of the blood leads another type of arterial disease, intimal hyperplasia, to form at the junction where the flow rejoins the host vessel.

7.27.2 Vascular Bypass Grafting

Vascular bypass grafting, when used to treat peripheral arterial disease, has moderate long-term patency rates [86]. Restenosis, when it occurs, has a predilection for the distal anastomosis, forming predominantly at the heel and toe of the anastomosis and on the artery bed, opposite the anastomosis [47] (see Fig. 7.31). Most commonly the graft is attached with a side-to-end junction proximal to the blockage and an end-to-side junction distal to the junction (see Fig. 7.31).

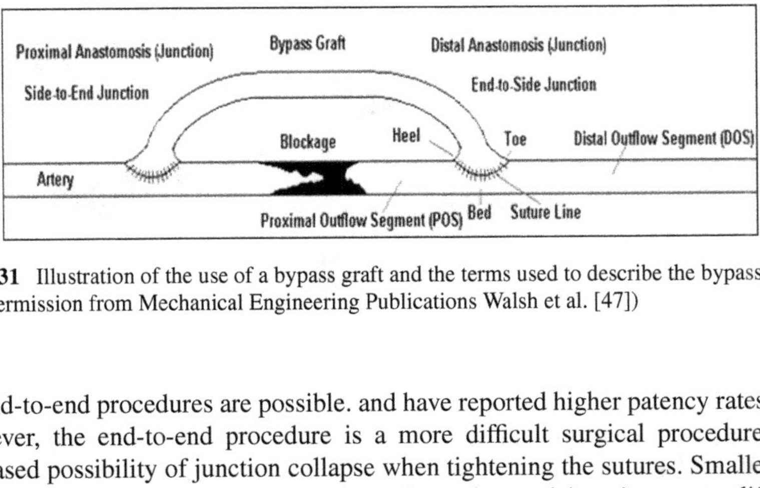

Fig. 7.31 Illustration of the use of a bypass graft and the terms used to describe the bypass (Used with permission from Mechanical Engineering Publications Walsh et al. [47])

End-to-end procedures are possible. and have reported higher patency rates [87]. However, the end-to-end procedure is a more difficult surgical procedure with increased possibility of junction collapse when tightening the sutures. Smaller vessels, such as those in the lower limb, make end-to-end junctions very difficult. End-to-end junctions also do not allow for the possibility of outflow in the proximal outflow segment of the distal junction (see Fig. 7.31). As a result, end-to-side anastomoses are the preferred surgical procedure in the lower limb.

The surgical procedure used to create an end-to-side anastomosis is entirely controlled by the surgeon. The patency rate of vascular bypass grafts varies remarkably from one surgical group to another, from vein grafts to synthetic grafts, and from one anatomic location to another location [88]. The success of vascular bypass surgery relies on the personal experience of the surgeons and the surgical techniques employed. It is possible for a surgeon to unknowingly create a graft/artery junction geometry, which could result in detrimental hemodynamic flow patterns in the junction that may lead to early restenosis. Therefore, it remains a challenge to determine the ability of surgeons to modify these flow patterns and in doing so to avoid graft/artery junction geometries that develop hemodynamic flow patterns which promote restenosis.

Either autologous vein or a synthetic material can be used for the bypass graft. In approximately one-third of patients with occlusive peripheral disorders, autologous vein grafts are either unacceptable or unavailable for bypass [89]. Therefore synthetic materials are often used in bypass surgery. Various surgical techniques have been developed to increase the patency rates of the end-to-side anastomoses. A number of graft/artery junction geometries exist in the present array of bypass surgical techniques. For example, a standard end-to-side anastomosis using both autologous and synthetic materials can vary in graft/artery diameter ratio and graft/artery angle [90]. Also, several surgical techniques exist that use vein cuffs or patches to alter the junction geometry, attempting to improve patency rates of bypass procedures (see Fig. 7.32).

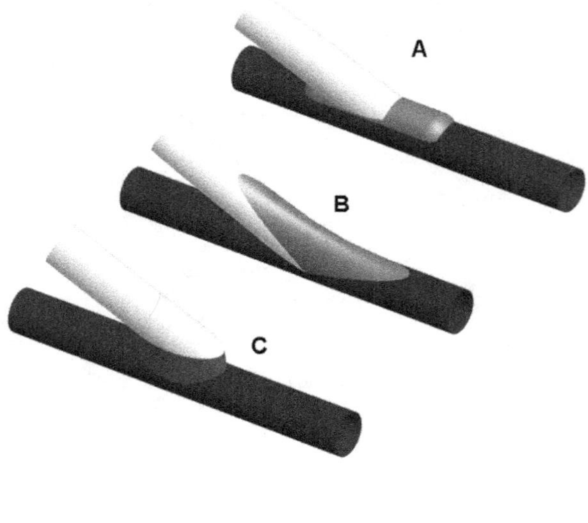

Fig. 7.32 Illustrations of surgical techniques using vein patches or cuffs. (**a**) The Linton patch, (**b**) the Taylor patch, and (**c**) the Miller cuff

The surgical techniques illustrated in Fig. 7.32 are used with synthetic graft material. Miller cuff geometries have significantly different flow patterns than the standard end-to-side anastomoses. Noori [91] found that the Miller cuff, with its increased internal junction area, shows lower bed stagnation point (BSP) movement and has a persistent wash out of the junction area with a large recirculation region found in the cavity. However, the patency rates associated with these procedures are moderate. The Taylor patch adopts a different approach. The Taylor patch geometry attempts to reduce the flow disturbances in the junction area by reducing the graft/artery junction angle. The separation region at the toe is also reduced. However, the reduced graft/artery angle increases the BSP movement [92]. The Linton patch attempts to achieve the same hemodynamic patterns as the Taylor patch procedure. All of these techniques involve harvesting autologous vein and using it in a manner illustrated in Fig. 7.32. Each technique is said to have an added advantage over conventional bypass grafting with synthetic grafts in that the use of autologous vein helps buffer compliance mismatch between the elastic host artery and the rigid graft when synthetic grafts are used. Each technique is also thought to improve the flow patterns in the junction area, and thus improve the patency rates of the bypass procedure. It is evident that these surgical approaches can be represented by two trends, using a patch geometry or a cuff geometry. Reported patency rates for the Taylor Patch approach include 54% at 5 years [93], 35% at 3 years [94], and 45% at 3 years [95]. Reported patency rates for the Miller cuff approach include 35% after 3 years [96] and 38% at 3 years [97]. This suggests that the results for each of these techniques are similar to those for conventional bypass grafts, 43% at 3 years [95].

The role of disease formation on the bed of the junction in the blocking of bypass grafts is highlighted when one considers that the patency rates of these procedures that use vein patches and cuffs are only moderately better than the conventional bypass surgery. This implies that compliance mismatch does not contribute as significantly as bed disease formation to the blocking of distal graft/artery junctions.

Three methods of disease formation exist in the graft–artery junction: (a) disease formation due to the abnormal flow patterns in the junction area, (b) disease formation due to the natural healing response of the artery to surgical injury, and (c) disease formation due to the material mismatch present at the anastomosis.

At the junction, disease forms at the heel, toe, and bed of the junction (see Fig. 7.31). At the heel and toe, all three of the above-mentioned disease-promoting factors influence the formation of the disease, namely surgical injury, material mismatch, and abnormal wall shear stresses (WSSs) around the junction due to the flow separation and recirculation regions. The suture line intimal thickening represents vascular healing and the greater prominence of intimal thickening around the suture line in prosthetic grafts may be related to compliance mismatch [98]. However, disease formation on the bed of the junction is entirely due to the abnormal flow patterns created as the flow exits the graft and enters the artery, impinging on the bed of the junction [99]. Such flow behavior is abnormal in the human circulatory system. Thus, hemodynamic flow patterns in distal end-to-side anastomoses are widely implicated in the initiation of the disease formation process [90, 91, 100, 101]. These abnormal hemodynamics are presented for an idealized end-to-side junction using pathlines in Fig. 7.33 and velocity vectors in Fig. 7.34. Also, presented in Fig. 7.34 are the corresponding wall shear stress distribution and Sherwood number distributions for the mean velocity in the decelerating phase of a resting femoral pulse. The wall shear stress distribution represents the abnormal frictional forces imposed on endothelial cells on the surface of the bed of the junction. In a healthy ideal case, the wall shear stress distribution would have a constant value of approximately 1 Pa for the corresponding time point in the pulse. The Sherwood number represents the

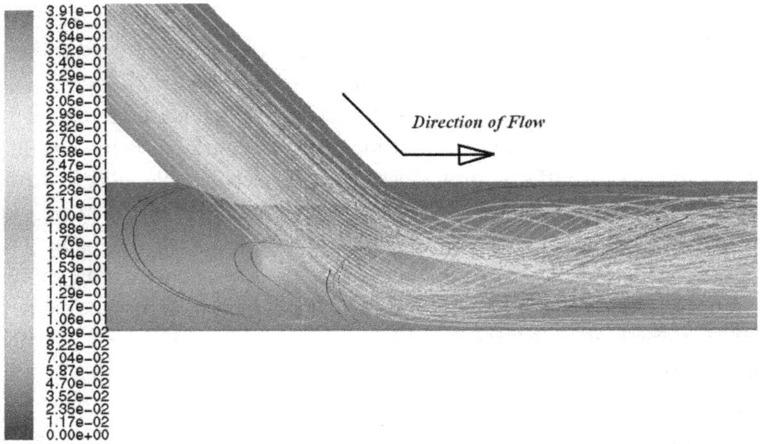

Fig. 7.33 Flow pathlines in an idealized femoral artery end-to-side bypass geometry at the maximum velocity of a typical femoral pulse [46]. The *pathlines* illustrate the complex nature of the flow in the junction area

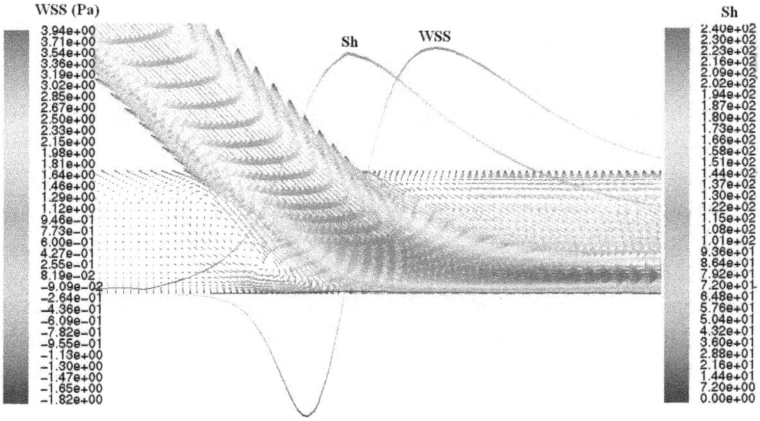

Fig. 7.34 Velocity vectors in an idealized femoral artery end-to-side bypass at the maximum velocity of a typical femoral pulse [46]. Also shown are the associated wall shear stress distribution and Sherwood number distribution on the bed of the junction

mass transport in the junction area and the abnormal Sherwood number (Sh) distribution shown in Fig. 7.34 results in abnormal mass transport to the endothelial cells. This further emphasizes the role of abnormal hemodynamics in the restenosis of the distal junction of bypass grafts.

Significantly, the precise mechanisms for disease formation and proliferation at graft/artery junctions are still unknown. The flow patterns created by end-to-side distal anastomoses exert abnormal wall shear stress distributions on the endothelial cells on the bed of the junction [102, 103]. The role of these wall shear stresses in intimal thickening has been the subject of many hypotheses. Of all of these, the theories that are based on the change in wall shear stresses that occur around the distal anastomosis are most common. This is because almost all endothelial damage is caused by wall shear stresses and the atherosclerotic lesions are well correlated with the spatial variation of wall shear stress [88].

Several hypotheses for disease formation which implicate WSSs have been put forward: a high wall shear stress theory, a low wall shear stress theory, a wall shear stress gradient theory, and an oscillating wall shear stress theory which has been widely used [13, 15, 104]. All of these theories can be supported by research findings and it is probable that each aspect has a role to play in the restenosis process. This is a result of the complex nature of the flow patterns associated with end-to-side anastomoses. These patterns have been analyzed previously [47, 105, 106], and all of the above wall shear stress theories can be supported by their analyses. Flow impinging on the arterial bed creates a stagnation point. A stagnation point is always associated with a low wall shear stress region as by definition the wall shear stress at the stagnation point is zero. However, with flow impinging on the arterial bed, a region of high wall shear stress is also created. Considering the low and high wall shear stress regions together, though they are at different locations on the bed of the junction, a significant wall shear stress gradient is clearly present. Since the flow is pulsatile in nature, the flow impinging on the arterial bed produces an oscillating stagnation point. Thus, the very nature of the flow in an end-to-side anastomosis causes all four of the above-stated wall shear stress theories to be present. As such, one theory cannot be assumed to be the initiating factor at this stage and all should be considered as possible initiating factors in the disease formation process.

7.28 Future Trends in Vascular and Cardiovascular Disease Modeling

Arterial hemodynamics has a major role to play in the development of atherosclerotic and aneurysmal disease in the vasculature. Flow in the coronary vessels has been shown to be highly complex in nature and there are indications that there may be regions within these vessels that are more prone to disease formation than are others. A range of shear stresses exist in the coronary arteries, with very different magnitudes of time-averaged and peak shears evident when comparing WSS profiles from straight sections of the right coronary artery with those at the outer wall of the LAD bifurcation. Therefore, these computational studies provide a wide span of WSS profiles that can be used for in vitro studies to provide an indication of endothelial response to the different shears present in vivo. Further refinement of the model could include coupling the aorta to the coronary branches and this could provide further insights.

In aortic flows, realistic models have an out-of-plane curvature, non-homogenous size, and quasi-elliptical cross sections, and geometrical effects greatly influence the hemodynamic characteristics. Thus hemodynamic factors are a key factor in the localization of vascular diseases, stent graft thrombosis, micro-embolism, and graft occlusions in both untreated and treated subjects. In the context of analysis of aortic flows, the input boundary conditions and geometrical effects have been shown to be influential factors in the creation of secondary flows and skewness of the flow. Exclusion of the recirculation effects at the aortic arch exit seems to underestimate the skewing and secondary flow patterns generated at the infrarenal location.

It may be that certain subjects have geometrical regimes which predispose them to disease formation. With the continuing improvements being made in diagnostic imaging such as ultrasound, CT scanning, and MRI, and the ongoing enhancement in computational modeling software capability, the approaches described here may lead to a greater role for flow mechanists in the development of diagnostic and treatment regimes for cardiovascular disease. These approaches could include the development of treatment devices and systems that incorporate flow regimes which are less prone to atherogenesis.

Acknowledgments The authors would like to acknowledge the following former and current members of the CABER team for their contribution to the research described above: Dr. Thomas O'Brien, Dr. Liam Morris, Dr. Siobhan O'Callaghan, Dr. Paul Devereux, Dr. Lucy O'Keeffe, and Dr. Grainne Carroll, and special thanks to William Denny and Stephen Broderick for assistance on preparation of graphics.

References

1. McGill HC Jr, McMahan CA, Herderick EE, Malcom GT, Tracy RE, Strong JP (2000) Origins of atherosclerosis in childhood and adolescence. Am J Clin Nutr 72:1307S–1315S
2. Kádár A, Glasz T (2001) Development of atherosclerosis and plaque biology. Cardiovasc Surg 9:109–121
3. Newby AC (2000) An overview of the vascular response to injury: a tribute to the late Russell Ross. Toxicol Lett 112–113:519–529
4. Martini F (2001) Fundamentals of anatomy and physiology, 5th edn. Prentice Hall, Upper Saddle River, NJ
5. Nerem RM (1992) Vascular fluid mechanics, the arterial wall, and atherosclerosis. J Biomech Eng 114:274–280
6. Luscher TF, Barton M (1997) Biology of the endothelium. Clin Cardiol 8(Suppl II):II3–II10
7. Endemann DH, Schiffrin EL (2004) Endothelial dysfunction. J Am Soc Nephrol 15: 1983–1992
8. Cunningham KS, Gotlieb AI (2005) The role of shear stress in the pathogenesis of atherosclerosis. Lab Invest 85:9–23
9. Tedgui A (1996) Endothelial permeability under physiological and pathological conditions. Prostaglandins Leukot Essent Fatty Acids 54:27–29
10. Wu S, Shung KK (1996) Cyclic variation of Doppler power from whole blood under pulsatile flow. Ultrasound Med Biol 22(7):883–894
11. Davies PF, Tripathi SC (1993) Mechanical stress mechanisms and the cell. Circ Res 72: 239–245
12. Fox J, Hugh AE (1966) Localization of atheroma: a theory based on boundary layer separation. Br Heart J 26:388–399
13. Fry DL (1973) Responses of the arterial wall to certain physical factors. In: Porter R, Knight J (eds) Atherosclerosis: initiating factors. Associated Scientific Publisher, Amsterdam, pp 93–125
14. Friedman M, O'Brien V, Ehrlich LW (1975) Calculations of pulsatile flow through a branch. Circ Res 36:277–285

15. Caro C, Fitz-Gerald JM, Schroter RC (1971) Atheroma and arterial wall shear. Observation, correlation and proposal of a shear dependent mass transfer mechanism for atherogenesis. Proc R Soc Lond B 177(46):109–159
16. Malek A, Izumo S (1995) Control of endothelial cell gene expression by flow. J Biomech 28(12):1515–1528
17. Papadaki M, Eskin SG (1997) Effects of fluid shear stress on gene regulation of vascular cells. Biotechnol Prog 13:209–221
18. Traub O, Berk BC (1997) Laminar shear stress; mechanisms by which endothelial cells transduce an atheroprotective force. Arterioscler Thromb Vasc Biol 18:677–685
19. Garcia-Cardena G, Comander J, Anderson KR, Blackman BR, Gimbonre MA Jr (2001) Biomechanical activation of vascular endothelium as a determinant of its functional phenotype. Proc Natl Acad Sci USA 98(8):4478–4485
20. Dewey CFJ (1984) Effects of fluid flow on living vascular cells. J Biomech Eng 106: 31–35
21. Nerem R, Alexander RW, Chapell DC, Medford RM, Varner SE, Taylor R (1998) The study of the influence of flow on vascular endothelial biology. Am J Med Sci 316(3):169–175
22. Ballyk PD, Walsh C, Butany J, Ojha M (1998) Compliance mismatch may promote graft–artery intimal hyperplasia by altering suture-line stresses. J Biomech 31:229–237
23. Deplano V, Souffi M (1999) Experimental and numerical study of pulsatile flows through stenosis: wall shear stress analysis. J Biomech 32:1081–1090
24. Friedman MH, Bargeron CB, Duncan DD, Hutchins GM, Mark FF (1992) Effects of arterial compliance and non-Newtonian rheology on correlations between intimal thickness and wall shear. J Biomech Eng 114:317–320
25. Hayashi K, Yanai Y, Naiki T (1996) A 3D LDA study of the relation between WSS and intimal hyperplasia in a human aortic bifurcation. J Biomech Eng 118:273–279
26. Hofer M, Rappitsch G, Perktold K, Trubel W, Schima H (1996) Numerical study of wall mechanics and fluid dynamics in end-to-side anastomoses and correlation to intimal hyperplasia. J Biomech 29(10):1297–1308
27. Keynton RS, Rittgers SE, Shu MCS (1991) The effect of angle and flow-rate upon hemodynamics in distal vascular graft anastomoses: an in vitro model study. J Biomech Eng 113:458–463
28. Moore JA, Steinman DA, Prakash S, Johnston KW, Ethier CR (1999) A numerical study of blood flow patterns in anatomically realistic and simplified end-to-side anastomoses. J Biomech Eng 121:265–272
29. da Silva AF, Carpenter T, How TV, Harris PL (1997) Stable vortices within vein cuffs inhibit anastomotic myointimal hyperplasia. Eur J Vasc Endovasc Surg 14:157–163
30. Stanton AV (1999) Hemodynamics, wall mechanics and atheroma: a clinicians perspective. Proc Inst Mech Eng 213:385–390
31. Tarbell JM (2003) Mass transport in the arteries and the localisation of atherosclerosis. Ann Rev Biomed Eng 5:79–118
32. Berthier B, Bouzerar R, Legallais C (2002) Blood flow patterns in an anatomically realistic coronary vessel: influence of three different reconstruction methods. J Biomech 35: 1347–1356
33. Asakura T, Karino T (1990) Flow patterns and spatial distribution of atherosclerotic lesions in human coronary arteries. Circ Res 66:1045–1066
34. Perktold K, Rappitsch G (1995) Computer simulation of local blood flow and vessel mechanics in a compliant carotid artery bifurcation model. J Biomech 28:845–856
35. Kirpalani A, Park H, Butany J, Johnston KW, Ojha M (1999) Velocity and wall shear stress patterns in the human right coronary artery. J Biomech Eng 121:370–375
36. Myers J, Moore JA, Ojha M, Johnston KW, Ethier CR (2001) Factors influencing blood flow patterns in the human right coronary artery. Ann Biomed Eng 29:109–120
37. Zeng D, Ding Z, Freidman MH, Ethier CR (2003) Effects of cardiac motion on right coronary artery hemodynamics. Ann Biomed Eng 31:420–429

38. Ramaswamy SD, Vigmostad SC, Wahle A, Lai YG, Olszewski ME, Braddy KC, Brennan TMH, Rossen JD, Sonka M, Chandran KB (2004) Fluid dynamic analysis in a human left anterior descending coronary artery with arterial motion. Ann Biomed Eng 32(12): 1628–1641
39. Ramaswamy SD, Vigmostad SC, Wahle A, Lai YG, Olszewski ME, Braddy KC, Brennan TMH, Rossen JD, Sonka M, Chandran KB (2006) Comparison of left anterior descending coronary artery hemodynamics before and after angioplasty. J Biomech Eng 128:40–48
40. Theodorakakos A, Gavaises M, Andriotis A, Zifan A, Liatsis P, Pantos I, Efstathopoulos EP, Katritsis D (2008) Simulation of cardiac motion on non-Newtonian, pulsating flow development in the human left anterior descending coronary artery. Phys Med Biol 53: 4875–4892
41. Marcus M, Wright C, Doty D, Eastham C, Laughlin D, Krumm P, Fastenow C, Brody M (1981) Measurements of coronary velocity and reactive hyperemia in the human coronary circulation of humans. Circ Res 49:877–891
42. Sibley D, Millar HD, Hartley CJ, Whitlow PL (1986) Subselective measurement of coronary blood flow velocity using a steerable Doppler catheter. J Am Coll Cardiol 8:1332–1340
43. Cole J, Hartley CJ (1977) The pulsed Doppler coronary artery catheter. Circulation 56(1):18–25
44. Wilson R, Laughlin DE, Ackell PH, Chilian WM, Holida MD, Hartley CJ, Armstrong ML, Marcus ML, White CW (1985) Transluminal, subselective measurement of coronary artery blood flow velocity and vasodilator reserve in man. Circulation 72(1):82–92
45. Benchimol A, Segall HF, Gartlan JL, Ariz P (1971) New method to measure phasic coronary blood velocity in man. Am Heart J 81(1):91–101
46. Kajiya F, Matsuoka S, Ogasawara Y, Hiramatsu O, Kanazawa S, Wada Y, Tadaoka S, Tsujioka K, Fujiwara T, Zamir M (1993) Velocity profiles and phasic flow patterns in the non-stenotic human left anterior descending coronary artery during cardiac surgery. Cardiovasc Res 27:845–850
47. Walsh MT, Kavanagh EG, O'Brien T, Grace PA, McGloughlin T (2003) On the existence of an optimum end-to-side junctional geometry in peripheral bypass surgery – a computer generated study. Eur J Vasc Endovasc Surg 26(6):649–656
48. O'Brien T, Walsh M, McGloughlin T (2005) On reducing abnormal hemodynamics in the femoral end-to-side anastomosis: the influence of mechanical factors. Ann Biomed Eng 33(3):310–322
49. Morris L, Delassus P, Walsh M, McGloughlin T (2004) A mathematical model to predict the in vivo pulsatile drag forces acting on bifurcated stent grafts used in endovascular treatment of abdominal aortic aneurysms (AAA). J Biomech 37(7):1087–1095
50. Walsh M, McGloughlin T, Liepsch DW, O'Brien T, Morris L, Ansari AR (2003) On using experimentally estimated wall shear stresses to validate numerically predicted results. Proc Inst Mech Eng H J Eng Med 217(H2):77–90
51. Morris L, Delassus P, Grace P, Wallis F, Walsh M, McGloughlin T (2006) Effects of flat, parabolic and realistic steady flow inlet profiles on idealized and realistic stent graft fits through abdominal aortic aneurysms (AAA). Med Eng Phys 28(1):19–26
52. O'Brien T, Morris L, Walsh M, McGloughlin T (2005) That hemodynamics and not material mismatch is of primary concern in bypass graft failure: an experimental argument. J Biomech Eng Trans ASME 127(5):881–886
53. Fox B, Seed WA (1981) Location of early atheroma in the human coronary arteries. J Biomech Eng 3:208–212
54. O'Keeffe LM, Muir G, Piterina AV, McGloughlin T (2009) Vascular cell adhesion molecule – 1 expression in endothelial cells exposed to physiological coronary wall shear stresses. J Biomech Eng 131(8):081003–081009
55. O'Brien TP, Walsh MT, Morris L, Grace PA, Kavanagh EG, McGloughlin TM (2007) Numerical and experimental techniques for the study of biomechanics in the arterial system. In: Leondes CT (ed) Biomechanical systems technology, vol 4. World Scientific Hackensack, NJ, USA

56. Moore JE Jr, Ku DH, Zarins CK, Glagov S (1992) Pulsatile flow visualization in the abdominal aorta under differing physiologic conditions: implications for increased susceptibility to atherosclerosis. J Biomech Eng 114:391–353
57. Cheng CP, Parker D, Taylor CA (2002) Quantification of wall shear stress in large blood vessels using Lagrangian interpolation functions with cine phase-contrast magnetic resonance imaging. Ann Biomed Eng 30:1020–1032
58. Cheng CP, Herfkens RJ, Taylor CA (2003) Abdominal aortic hemodynamic conditions in healthy subjects aged 50–70 at rest and during lower limb exercise: in vivo quantification using MRI. Atherosclerosis 168(2):323–331
59. Pedersen EM, Sung HW, Yoganathan AP (1994) Influence of abdominal aortic curvature and resting versus exercise conditions on velocity fields in the normal abdominal aortic bifurcation. J Biomech Eng 166:347–353
60. Moore JE Jr, Ku DH (1994a) Pulsatile velocity measurements in a model of the human abdominal aorta under simulated exercising and postprandial conditions. J Biomech Eng 116:107–111
61. Moore JE Jr, Ku DH (1994b) Pulsatile velocity measurements in a model of the human abdominal aorta under resting conditions. J Biomech Eng 116:337–346
62. Fung YC (1997) Biomechanics circulation, 2nd edn. Springer, New York, NY
63. Nerem RM, Rumberger M, Gross DR, Hamlin RL, Geiger GL (1974) Hot-film anemometry velocity measurements of arterial blood flow in horses. Circ Res 10:301–313
64. Falsetti H, Kiser KM, Francis GP, Belmore ER (1972) Sequential velocity development in the ascending and descending aorta of the dog. Circ Res 21:328–338
65. Seed WA, Wood NB (1971) Velocity patterns in the aorta. Cardiovasc Res 5:319–330
66. Tortoli P, Bambi G, Guidi F, Muchada R (2002) Toward a better quantitative measurement of aortic flow. Ultrasound Med Biol 28(2):249–257
67. Chandran KB (1993) Flow dynamics in the human aorta. J Biomech Eng 115:611–616
68. Chandran KB, Yearwood TL (1981) Experimental study of physiological pulsatile flow in a curved tube. J Fluid Mech 111:59–85
69. Yearwood TL, Chandran KB (1984) Physiological pulsatile flow experiments in a model of the human aortic arch. J Biomech 15(9):683–704
70. O'Brien T, Morris L, O'Donnell M, Walsh M, McGloughlin T (2006) Injection-moulded models of major and minor arteries: the variability of model wall thickness owing to casting technique. Proc Inst Mech Eng H J Eng Med 219(H5):381–386
71. O'Brien T, Walsh M, Kavanagh EG, Finn SP, Grace PA, McGloughlin T (2006) The surgical feasibility of a novel vascular graft developed for the treatment of peripheral arterial disease using in vitro experimental design techniques. Ann Vasc Surg 21(5):611–617
72. O′Brien T, Walsh M, McGloughlin T (2006) Altering end-to-side anastomosis junction hemodynamics: the effects of flow-splitting. Med Eng Phys 28(7):727–733
73. Molony D, Callanan A, Morris L, Walsh M, McGloughlin T (2008) Geometrical enhancements for abdominal aortic aneurysm grafts and stent-grafts. J Endovasc Ther 15(5): 518–529
74. Papaharilaou Y, Doorly DJ, Sherwin SJ (2002) The influence of out-of-plane geometry on pulsatile flow within a distal end-to-side anastomosis. J Biomech 35:1225–1239
75. Pedersen EM, Sung HW, Yoganathan AP (1994) Influence of abdominal aortic curvature and resting versus exercise conditions on velocity fields in the normal abdominal aortic bifurcation. J Biomech Eng 166:347–353
76. Naruse T, Tanishita K (1996) Large curvature effect on pulsatile entrance flow in a curved tube: model experiment simulating blood flow in an aortic arch. J Biomech Eng 118: 180–186
77. Dean WR (1927) Note on the motion of fluid in a curved pipe. Philos Mag 20(7):208–223
78. Dean WR (1928) The streamline motion of fluid in a curved pipe. Philos Mag 30(7):673–693
79. Bluestein B, Niu L, Schoephoerster RT, Dewanjee MK (1996) Steady flow in an aneurysm model: correlation between fluid dynamics and blood platelet deposition. J Biomech Eng 118:280–286

80. Egelhoff CJ, Budwig RS, Elger DF, Khraishi TA, Johansen KH (1999) Model studies of the flow in abdominal aortic aneurysms during resting and exercise conditions. J Biomech 32:1319–1329
81. Shipkowitz T, Rodgers VGJ, Frazin JF, Chandran KB (2000) Numerical study on the effects of secondary flow in the human aorta on local shear stresses in abdominal branches. J Biomech 33:717–728
82. Taylor CA, Hughes TJR, Zarins CK (1999) Effect of exercise on hemodynamic conditions in the abdominal aorta. J Vasc Surg 26(6):1070–1089
83. Walsh PW, Chin-Quee S, Moore JE Jr (2003) Med Eng Phys 25:299–307
84. Melliere D, Cron J, Allaire E, Desgranges P, Becquemin JP (1999) Indications and benefits of simultaneous endoluminal balloon angioplasty and open surgery during elective lower limb revascularisation. Cardiovasc Surg 7(2):242–246
85. Gallard RB, Magee TR, Berridge DC, Hopkinson GB, Lewis MH, Parvin SD (1999) Variation in vascular activity in the United Kingdom: an analysis of five regions. Cardiovasc Surg 7(7):694–698
86. Streinchenberger R, Barjoud H, Adeleine P, Larese A, Nemoz C, Chatelard P, Nedey C, Sabben F, Ganichot F, Jurus C (2000) Venous allografts preserved at 4°C for infrainguinal bypass : long-term results from 170 procedures. Ann Vasc Surg 14(6):553–560
87. Madiba TE et al (1997) Choosing the proximal anastomosis in aortobifemoral bypass. Br J Surg 84(10):1416–1418
88. Lei M, Kleinstreuer C, Archie JP (1997) Hemodynamic simulations and computer-aided designs of graft–artery junctions. J Biomech Eng 119:343–348
89. Strandness DE, Didisheim P Jr, Clowes AW, Watson JT (1987) Vascular diseases current research and applications. Grune & Stratton, Orlando
90. Giddens DP, Zarins CK, Glagov S (1993) The role of fluid mechanics in localisation and detection of atherosclerosis. J Biomech Eng 115:588–594
91. Noori N, Scherera R, Perktoldb K, Czernya M, Karnerb G, Trubela M, Polterauera P, Schimaa H (1999) Blood flow in distal end-to-side anastomoses with PTFE and a venous patch: results of an in vitro flow visualisation study. Eur J Endovasc Surg 18:191–200
92. Fei DY, Thomas JD, Rittgers SE (1994) The effect of angle and flow rate upon hemodynamics in distal vascular graft anastomoses: a numerical model. J Biomech Eng 116:331–336
93. Taylor RS, Loh A, McFarland RJ, Cox M, Chester JF (1992) Improved technique for PTFE bypass grafting: long-term results using anastomotic vein patches. Br J Surg 79:348–354
94. Eagleton MJ, Ouriel K, Shortell C, Green RM (1999) Femoral–infrapopliteal bypass with prosthetic grafts. Surgery 26(4):759–764
95. Fichelle JM, Marzelle J, Colacchio G, Gigou F, Cormier F, Cormier JM (1995) Infrapopliteal polytetrafluoroethylene and composite bypass: factors influencing patency. Ann Vasc Surg 9(2):187–196
96. Chesire NJ, Wolfe JH (1992) How to select the treatment of choice of critical leg ischemia. Ann Chir Gynaecol 81(2):141–152
97. Kreienberg PB, Darling C, Chang BB, Paty PSK, Lloyd WE, Shah DM (2000) Adjunctive techniques to improve patency of distal prosthetic bypass grafts: polytetrafluoroethylene with remote arteriovenous fistulae versus vein cuffs. J Vasc Surg 31(4):696–701
98. Bassiouny HS, White S, Glagov S, Choi E, Giddens DP, Zarins CK (1992) Anastomotic intimal hyperplasia: mechanical injury of flow induced. J Vasc Surg 15:708–717
99. Staalsen NH, Ulrich M, Winther J, Pedersen EM, How T, Nygaard H (1995) The anastomosis angle does change the flow fields at vascular end-to-side anastomoses in-vivo. J Vasc Surg 21(3):460–471
100. Kleinstreuer C, Lei M (1995) Hemodynamics of a femoral graft artery connector mitigating restenosis. Adv Bioeng BED 31:171–172
101. Friedman MH (1983) Arterial geometry affects hemodynamics. Atherosclerosis 46:225–231
102. Xiao Y, Truskey GA (1997) Effect of flow recirculation upon endothelial cell height and shape. Bioeng Conf BED 35:537–538

103. White SS, Zarins CK, Giddens DP (1993) Hemodynamics patterns in two models of end-to-side vascular graft anastomoses: effects of pulsatility, flow division, Re. No. and hood length. J Biomech Eng 15:104–111
104. Kleinstreuer C, Lei M, Archie JP (1996) Flow input waveform effects on the temporal and spatial wall shear stress gradients in a femoral graft–artery connector. J Biomech Eng 118:506–510
105. Ojha M, Ethier CR, Johnston KW, Cobbold RS (1990) Steady and pulsatile flow field in an end-to-side arterial anastomosis model. J Vasc Surg 12(6):747–753
106. Ojha M, Cobbold RS, Johnston KW (1994) Influence of angle on wall shear stress distribution for an end-to-side anastomosis. J Vasc Surg 19(6):1067–1073

Chapter 8
Biomechanical Modeling of Aneurysms

Madhavan L. Raghavan and David A. Vorp

Abstract Aneurysms are abnormal dilations in the arterial wall and predominantly occur in the aorta and in the cerebral vasculature. The aortic aneurysm is predominantly found in the infrarenal abdominal region and the cerebral aneurysms generally occur in or near the circle of Willis. There is considerable interest in understanding the mechanism underlying aneurysm growth, diagnosing their severity or propensity to rupture, and developing endovascular implants to treat the same. Biomechanical simulations are being employed to improve our understanding of factors that trigger aneurysms and mechanisms for growth/rupture of the lesions. In this chapter, studies on reconstruction of the three-dimensional geometry of the aneurysms, material modeling of the arterial wall and aneurysm components, and biomechanical analyses toward prediction of potential rupture are discussed.

8.1 Introduction

8.1.1 Incidence and Epidemiology

Aneurysms are abnormal dilations in blood vessel walls. If left untreated, some continue to dilate and eventually rupture, causing hemorrhage and likely death. Aneurysms can occur in any blood vessel in the body, although they tend to predominantly occur in the aorta and the cerebral vasculature. Estimates of the incidence of aneurysms in the general population have been difficult to ascertain owing to the generally non-symptomatic nature of these lesions. Estimates range from 2 to 10% of the population [1–3].

Aortic aneurysms occur predominantly in the infrarenal abdominal aorta. Here, the normally 2-cm-diameter abdominal aorta can dilate in a fusiform fashion to as

M.L. Raghavan (✉)
Department of Biomedical Engineering, 1136 Seamans Center, College of Engineering, The University of Iowa, Iowa City, IA 52242-1527, USA
e-mail: ml-raghavan@uiowa.edu

K.B. Chandran et al. (eds.), *Image-Based Computational Modeling of the Human Circulatory and Pulmonary Systems*, DOI 10.1007/978-1-4419-7350-4_8,

large as 5–15 cm in diameter over a course of years if they do not rupture (Fig. 8.1). Rupture of an abdominal aortic aneurysm (AAA) can occur at any size, although larger aneurysms are more likely to rupture than smaller ones [4]. If the AAA is identified prior to being symptomatic, physicians typically repair it if it has grown to a size of 5.0–5.5 cm or larger. Treatments usually involve an open surgical resection or minimally invasive placement of an endovascular graft. Because the risk of mortality and morbidity for treatment in some patients can be high, an assessment of rupture risk would help better judge whether intervention is necessary. While AAA size remains a good indicator of rupture risk [4], smaller AAAs do rupture in some patients and larger AAAs remain unruptured in many. Therefore, size alone is not a fully suitable predictor.

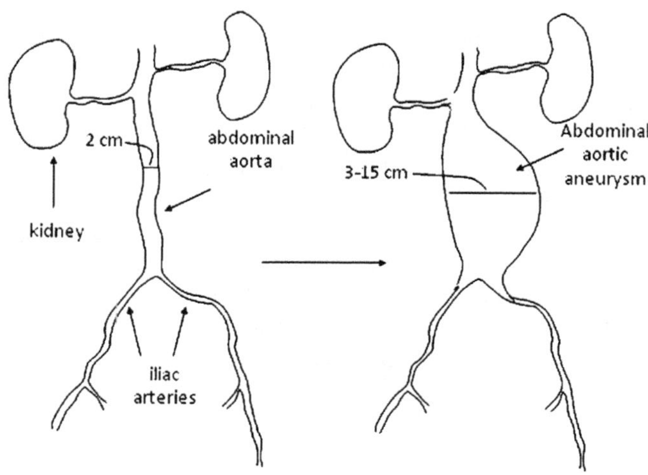

Fig. 8.1 Schematic illustration of an infrarenal abdominal aortic aneurysm. A 2-cm-diameter abdominal aorta may, in a period of years, bulge to a maximum dimension of 4–15 cm

Cerebral aneurysms occur predominantly in or near the circle of Willis—a circular configuration of arteries in the base of the brain (Fig. 8.2). Unlike aortic aneurysms, these are usually saccular outpouchings from vessel bifurcations or branches (bifurcating aneurysms) or from arterial segments (sidewall aneurysms). Cerebral aneurysm diameters are usually anywhere from a few millimeters (small aneurysms) to a few centimeters (giant aneurysms). The treatment modalities involve microsurgical clipping or endovascular coiling. Rupture of these lesions is rarer than in aortic aneurysms, with rough estimates ranging from 0.1 to 1% per year [2, 5]. Here too, assessment of rupture risk can aid in the clinical management of patients given that a subset of the patients present with high treatment risk.

8.1.2 Role for Biomechanical Modeling and Simulation

There is considerable interest in understanding the mechanisms underlying aneurysm growth, diagnosing their severity or propensity to rupture, and developing

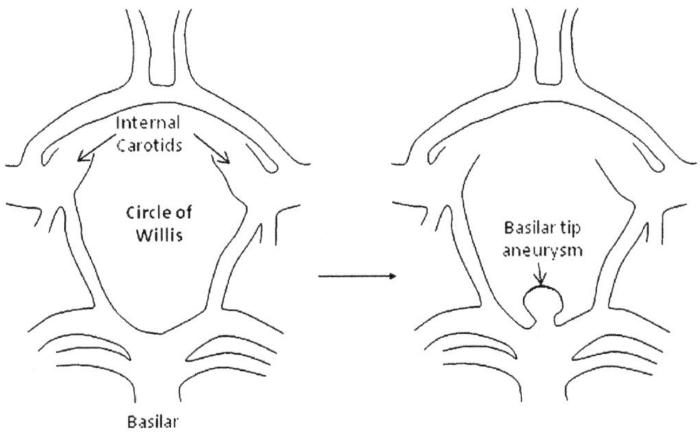

Fig. 8.2 Schematic illustration of a cerebral aneurysm. Note the aneurysm at the basilar tip location in the *right* compared to a normal circle of Willis network on the *left*

devices and implants to treat them. For effective clinical management of aneurysm patients, an understanding of factors that trigger aneurysms and the mechanisms of growth/rupture can be quite valuable. Further, a platform for rapid assessment of implant design performance can aid in the choice of patients who get these implants and in optimizing the device designs. Indeed, in all these aspects, biomechanical modeling and simulation can play an important role. With regard to aneurysm pathogenesis, biomechanical growth models, if validated effectively, may be interrogated to test postulates on injury mechanisms, triggers that aggravate growth rates, or mechanisms of failure [6–8]. Such simulations may result in specific hypotheses that can then be tested in animal models of aneurysms or epidemiological/longitudinal outcomes data in the patient population. With regard to diagnosis, biomechanical simulations of stress/strain in the aortic wall may be compared to stochastic information on tissue failure properties to help identify aneurysms with impending rupture [9, 10]. In the design of implants, biomechanical models may be used to test the performance of planned design improvements [11, 12]. This chapter will introduce readers to the current state of the literature on biomechanical modeling of aortic and cerebral aneurysms; specifically, it will address developments in computational simulation of hemodynamics and mechanical stress distributions, modeling of aneurysm growth, and integrative biomechanical tools to assess rupture risk.

8.2 Geometric Modeling of Aneurysms

The first aspect of a computational model for a solid structure is an accurate estimation and reconstruction of the three-dimensional geometry. For computational studies that seek to understand general biomechanical aspects of aneurysm disease, such as parametric analyses on the effects of shape on wall stress distribution or testing hypotheses on mechanobiology in these lesions, hypothetical geometries may suffice. But for studies that seek to delve into patient-specific aspects, such as those that attempt to predict rupture risk of individual aneurysms, imaging data become the key source in the development of geometric models. The reconstruction of patient-specific 3D models of these lesions can also help guide the development of population average generic models.

Aneurysms present with different sizes and shapes, though in general they are either fusiform or saccular. Naturally, therefore, reliable characterization of geometrical variations at a population level and on a patient-specific basis can aid our understanding of the natural history of these lesions. The detection of the presence of an aneurysm is almost always made using diagnostic imaging—the most common modalities are ultrasound (for AAA) and digital subtraction angiograms (for cerebral aneurysms). Because traditional applications of both these modalities result in two-dimensional information, they lend themselves to reasonable estimations of size but poor approximations of shape. Subsequent to diagnosis, most patients with aneurysms inevitably receive some form of three-dimensional scans such as computed tomography (CT), magnetic resonance imaging (MRI), or rotational angiography (RA) (in the case of cerebral aneurysms), which lend themselves to effective characterization of both size and shape with a significant level of accuracy and completeness.

8.2.1 Abdominal Aortic Aneurysms

Of the common imaging modalities, transabdominal ultrasound and intravascular ultrasound (IVUS) do not lend themselves to three-dimensional reconstruction, essential for accurate modeling. In the former, the resolution is usually insufficient (Fig. 8.3). For IVUS, the inability to maintain a controlled orientation of the sensing tip of the catheter inside the large aneurysm sac is a major limitation?a problem not present when imaging atherosclerotic lesions, as the lumen area is small enough to maintain the catheter-sensing tip coincident with the vessel axis. Thus, MRI (Fig. 8.4) and CT (Fig. 8.5) are the best alternatives. Most patients who are diagnosed with abdominal aortic aneurysms and are deemed fit for treatment are scanned by CT or MRI with or without contrast in order to plan for treatment, be it endovascular or open resection. Even most patients who are followed with a "watchful waiting" strategy tend to get one of these scans. The scan data thus obtained provide ample information for characterization of their geometries, be it for a population-based or an individual-based study.

The key aspects of an AAA geometric model are the dilated aortic wall, any associated intraluminal thrombus (ILT), localized regions of calcifications, and, where possible, the regional variations in aortic wall thickness. Raghavan et al. and Thubrikar et al. reconstructed 3D geometric models of the dilated aortic wall of patients from CT data by manually segmenting the cross sections, converting into surface triangulations and smoothing shape artifacts using a Laplacian approach [13–15]. These simple and straightforward methods possibly resulted in limited

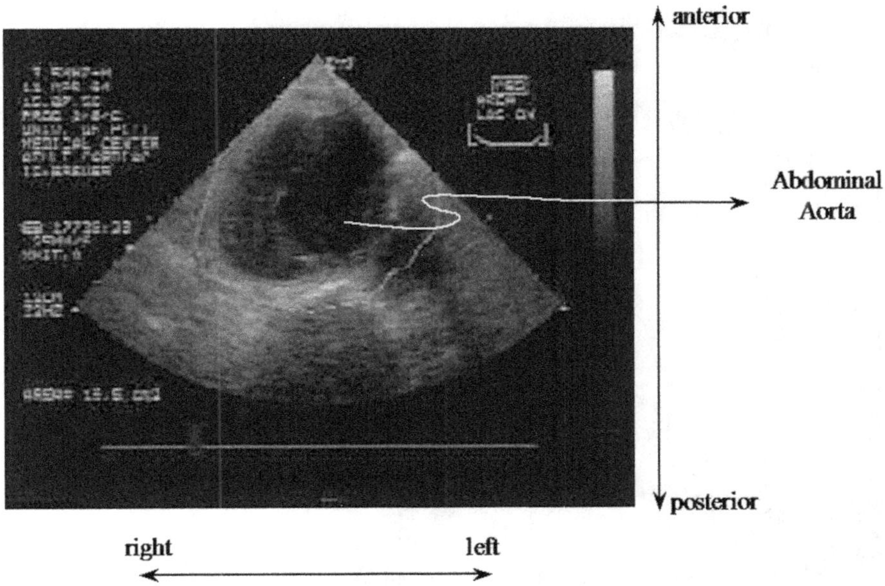

Fig. 8.3 Transabdominal ultrasound of human AAA. The imaging modality may be well suited for diagnosing the presence (or the absence) of an AAA, but perhaps of insufficient resolution for extracting its 3D geometry accurately

accuracy and increased user sensitivity, however. Subsequently, Wang et al. [16] reported on the reconstruction of both the AAA wall and the ILT from CT scans of patients (Fig. 8.6).

Responding to demands from physicians performing pre-surgical planning for accurate 3D models of AAA, Medical Media Systems (now M2S Inc., Lebanon, NH) used proprietary semi-automated methods to reconstruct the AAA wall, ILT, and regions of calcifications on a patient-specific basis (Fig. 8.7). These models were used by physicians for ascertaining the dimensions of endovascular grafts to be implanted in AAA patients and by investigators for computational stress analysis [17–19]. de Bruijne and colleagues [20] employed a parametric approach to 3D reconstruction where a deformable template was used to fit to the scan data. This approach allows for controlled and consistent smoothing of the AAA surface and lends itself to a quantified inference of the geometry itself. Recently, Martufi et al. [21] reported on the reconstruction of AAA, ILT, and estimates of regional variations in aortic wall thickness. Development of 3D geometric models of AAA aside, there is also interest in making inferences on morphology from such models. Sacks and colleagues [15] performed a surface curvature analysis of AAA and reported the existence of substantial axial asymmetry, and included adjacent elliptical and hyperbolic regions. Based on a more direct dimensional analysis, Fillinger and colleagues [22] reported that when matched for age, gender, and diameter, ruptured AAAs tend to be less tortuous, yet have greater cross-sectional diameter asymmetry.

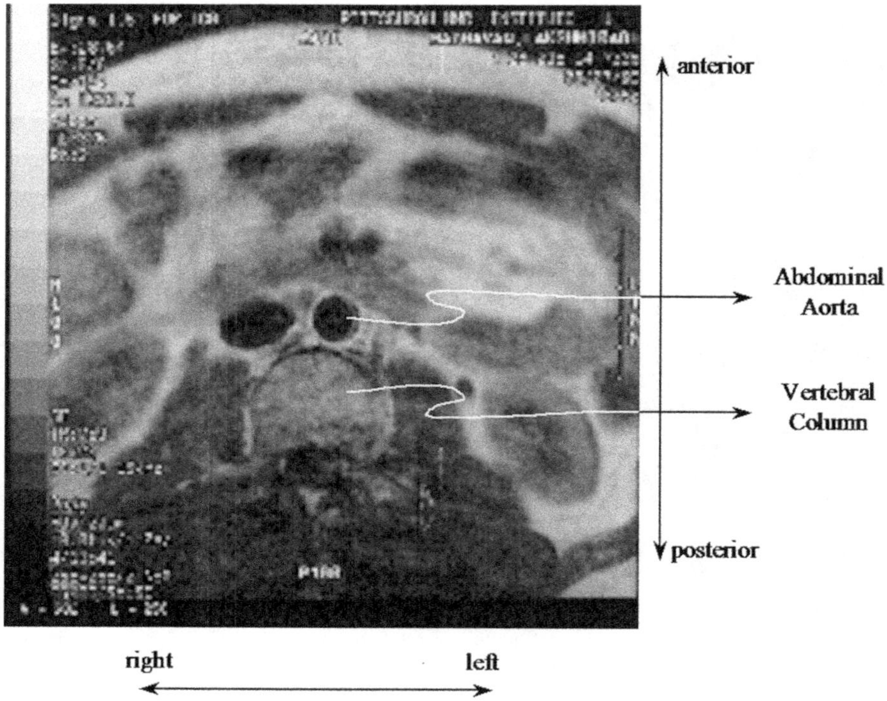

Fig. 8.4 MRI of normal infrarenal abdominal aorta

8.2.2 Cerebral Aneurysms

Biplane angiogram is the traditional imaging modality for diagnosis and surgical planning of cerebral aneurysms, but it is not sufficient for complete characterization of the 3D geometry of these lesions. However, contrast-enhanced CT and MR (CTA and MRA) and rotational angiography are becoming routine in the clinical management of patients.

As with abdominal aortic aneurysms, the two approaches to 3D reconstruction were the semi-automated segmentation-based approach and the use of deformable models. Steinman and colleagues [23] reported on a semi-automatic 3D reconstruction of a giant posterior communicating artery aneurysm using a 3D discrete dynamic contour segmentation approach. Subsequently, Cebral and colleagues [24] performed 3D reconstruction of cerebral aneurysms using two approaches—a segmentation approach using level sets and a segmentation approach using deformable models. In the former, a rough segmentation of the aneurysm is performed using the fast marching method, which is then smoothed using a level set algorithm. In the latter, a rough isosurface extracted using a region growing segmentation is used to initialize a deformable model template which may then be deformed such that

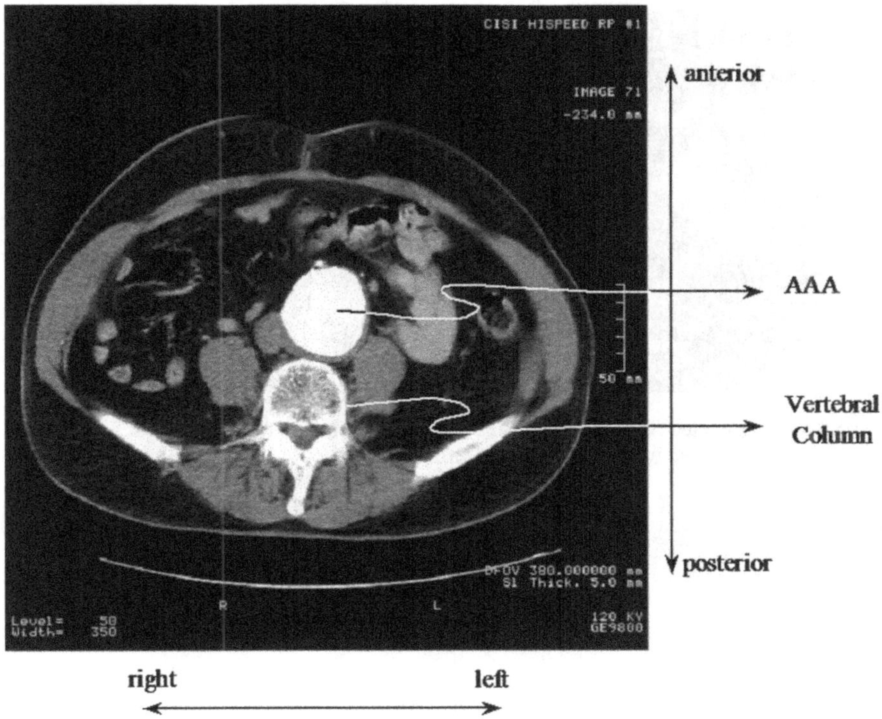

Fig. 8.5 CTA of human AAA. The aneurysm is contrast enhanced in this scan

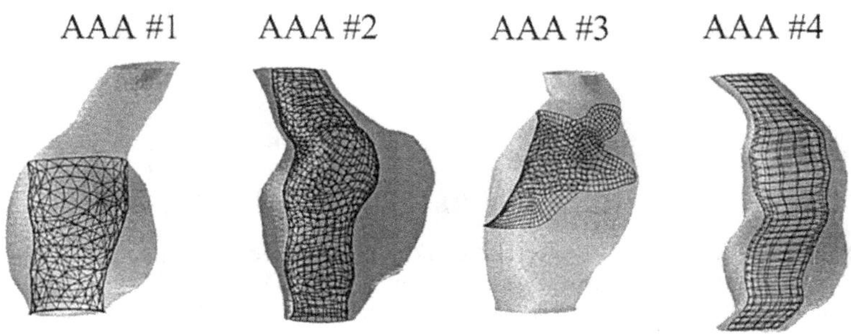

Fig. 8.6 Reconstructed model of patient-specific AAA with ILT by Wang et al. The ILT is shown inside the AAA wall sac

an optimized surface is obtained that is a compromise between smoothing constraints and closeness of fit to the rough segmented model. Based on subjective visualization, they showed that the two approaches result in similar reconstructions. Ma and colleagues [25] reported on a semi-automated segmentation followed by a curvature-assisted non-shrinking smoothing approach to geometric reconstruction (Fig. 8.8).

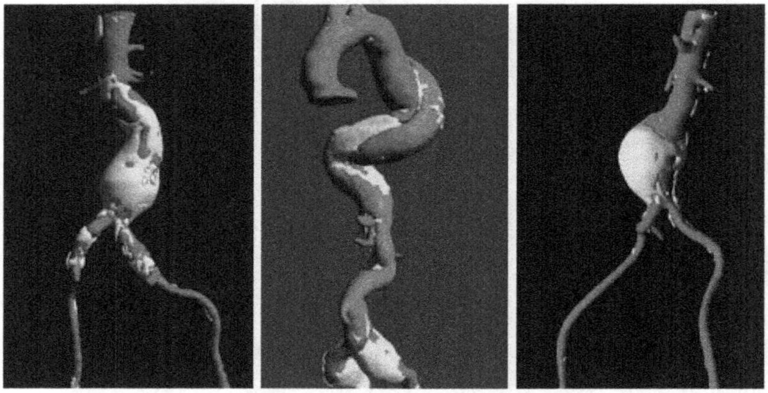

Fig. 8.7 Reconstructed AAA model by M2S Inc., a commercial service. *Shades* of gray indicate regions of wall with and without intraluminal thrombus

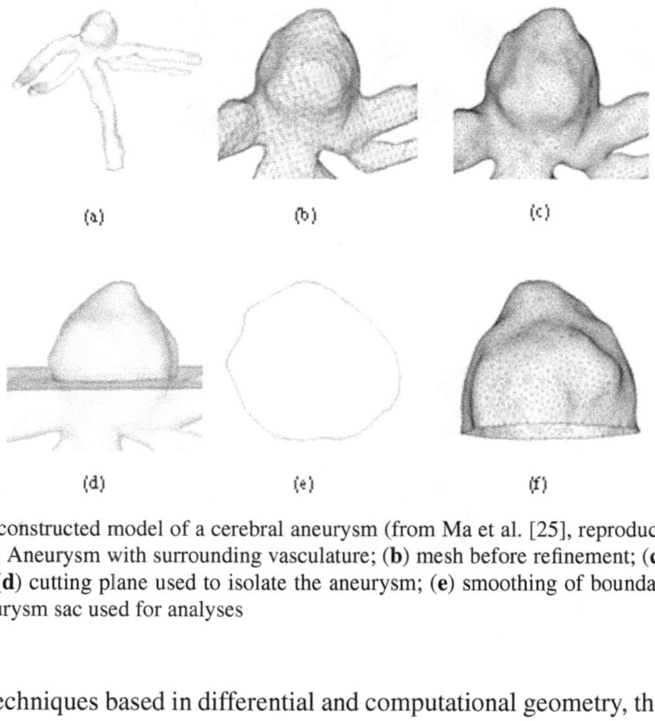

Fig. 8.8 Reconstructed model of a cerebral aneurysm (from Ma et al. [25], reproduced with permission). (**a**) Aneurysm with surrounding vasculature; (**b**) mesh before refinement; (**c**) mesh after refinement; (**d**) cutting plane used to isolate the aneurysm; (**e**) smoothing of boundary curve; (**f**) isolated aneurysm sac used for analyses

Using techniques based in differential and computational geometry, they isolated the aneurysm from contiguous vasculature using a cutting plane and estimated 3D measures of surface size and shape that may be used to infer the global geometric characteristics of cerebral aneurysms [25, 26]. Banatwala and colleagues reported on a novel parameterization of the aneurysm geometry using Legendre polynomials such that the entire aneurysm may be approximated using a minimal number of parameters which may then permit quantified inferences of size and shape. Such parameterized surfaces have the distinct advantage of permitting quick, reliable, and relatively noise-free estimates of surface curvature and pressure-induced wall tension but are limited in their ability to conform to all the surface features in a patient-specific aneurysm. In a remarkable series of reports on geometric modeling of cerebral aneurysms, Frangi and colleagues developed and assessed various approaches using both 3DRA and CTA data [27–29]. For segmentation of 3DRA data, they used a deformable surface model and for CTA, a region-based implicit surface model. 3D moment invariants were estimated and used to infer shape of these lesions [27]. They further assessed the methods used to isolate the aneurysm and reported on the sensitivity of shape measures to the method of aneurysm isolation (Fig. 8.9).

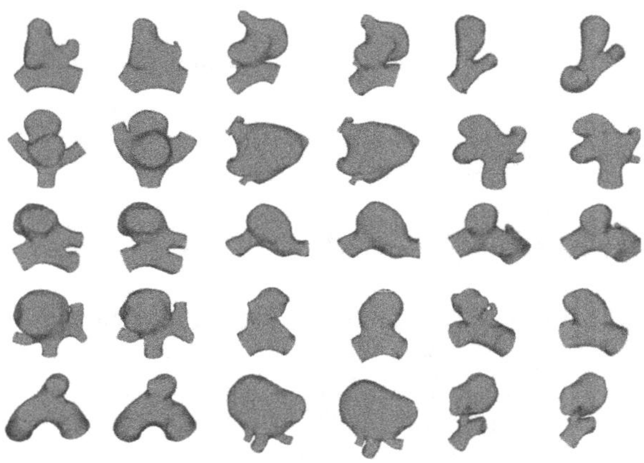

Fig. 8.9 Three-dimensional aneurysm models reported in Millan et al. Fifteen aneurysms are shown in three pairs of columns. Each pair shows the same aneurysm reconstructed by using two different segmentation models; *left*: based on deformable surface models and *right*: based on region-based implicit deformable surfaces

8.2.3 Summary

Development of geometric models of AAA and cerebral aneurysms is of great interest to investigators. Current methods continue to be semi-automated, requiring user interactions. While this may be necessary to avoid common image-processing pitfalls, it also constrains this technology from being used in large population studies, which will be necessary to test hypotheses on aneurysm pathophysiology with reliability. Future developments will hopefully make inroads in automating this process while also allowing for effective quantification so that practical inferences on patient-specific aneurysm geometry may be made.

8.3 Material Modeling of Aneurysms

Material models form the foundation on which the computational simulation of mechanical behavior of aneurysms rests. Not surprisingly then, there is a body of literature on material modeling of aneurysm tissue. Studies on aneurysm mechanical properties may be broadly divided into those that seek to make descriptive interpretations in order to understand disease pathogenesis and those that seek to develop predictive constitutive models as a step toward computational simulation. In the former, there is an emphasis on relating structure to function and on failure properties. These studies generally show that aneurysms have lower failure strength than do normal arterial tissues [10, 30–37] and may help form a basis for interpreting stress estimations from computational modeling. This section will instead focus on constitutive modeling efforts on aortic and cerebral aneurysms as they are directly relevant to biomechanical modeling. To develop constitutive models, two approaches may be taken—testing of excised lesions under controlled conditions or reverse estimation from dynamic in vivo motion. In practice, however, the latter is quite difficult to accomplish and therefore, all relevant reports in literature use the former approach of testing excised lesions in a lab setting.

8.3.1 Abdominal Aortic Aneurysms

There are multiple solid entities in an AAA—the aneurismal aortic tissue, the intraluminal thrombus, and calcified regions being the most prominent constituents. Constitutive models for abdominal aortic aneurysm tissue and thrombus have been reported by mainly testing specimens harvested during surgical resection of unruptured AAA. It is generally recognized that aneurysm tissue undergoes finite strains prior to failure, requiring the use of finite-elastic constitutive models. Based on uniaxial extension testing of flat strips of tissue excised from the anterior midsection of about 60 patients, Raghavan and Vorp reported an isotropic hyperelastic material model described by a second-order polynomial strain energy density function [38] with population mean parameter values (Equation (8.1)).

$$W = a(I_c - 3) + b(I_c - 3)^2 \tag{8.1}$$

where W is the strain energy density function, I_c is the first invariant of the right Cauchy Green stretch tensor, and a and b are its material parameters.

They noted that specimens oriented longitudinally had different average parameters compared to those oriented circumferentially, although without statistical difference, suggesting that an anisotropic model might be more appropriate. Subsequently, Vande Geest et al. [39] tested square specimens excised during surgical resection by biaxial extension from 26 AAA patients and reported on an orthotropic four-parameter exponential strain energy function with mean parameter values (Equation (8.2)):

$$W = b_0 \left(e^{\frac{1}{2} b_1 E_{\theta\theta}^2} + e^{\frac{1}{0} b_Z E_{LL}^2} + e^{b_E E_{\theta\theta} E_{LL}} - 3 \right) \tag{8.2}$$

where, $E_{\theta\theta}$ and E_{LL} are the Green strain components in the circumferential and axial orientations and $b_i (i = 0 - 3)$ are the material parameter values.

Similarly, Wang et al. [40] tested strips of intraluminal thrombus at various locations within the aneurismal sac from 50 patients and reported on an isotropic, hyperelastic, two-parameter strain energy density function with population mean parameter values (Equation (8.3)):

$$W = c_1(II_B - 3) + c_2(II_B - 3)^2 \tag{8.3}$$

Vande Geest et al. [39] then explored the biaxial behavior of ILT and found that the material does indeed behave as an isotropic material (as concluded by the earlier Wang et al. study [40]), but they noted significant errors when predicting the biaxial constitutive response using Equation (8.3).

Notably, all the above material models capture solely the passive elastic behavior of the respective structures. The challenges in specimen procurement and experimental techniques have generally limited the development of more sophisticated models that capture other aspects of behavior such as viscoelasticity or growth/remodeling characteristics. In recent years, there have been some efforts toward developing structurally based models that incorporate the individual contributions of elastin and collagen and thereby accommodate mechanisms of growth. But these remain postulated models that have not been evaluated fully with experimental/empirical data for reliable parameter estimations. These growth models are discussed under Section 8.7.

8.3.2 Cerebral Aneurysms

In cerebral aneurysms, presence of intraluminal thrombus is relatively less common and therefore reports have been confined to arterial wall tissue assessments. As with AAA, there have been studies aimed at gaining a descriptive understanding of the tissue [37]. Toward development of constitutive modeling, a series of reports by one group remain the only known effort. Hsu et al. [41, 42] developed a multiaxial experimental system and studied the pressure-inflation behavior of two cerebral aneurysms harvested as a whole from cadavers. Using their sophisticated test system, they were able to capture multiaxial deformations at multiple locations under varying pressures. Despite the rather small study population, these studies remain the most comprehensive characterization of elastic behavior of these lesions. They proposed the non-linear theory of elastic membranes as an appropriate theoretical framework for saccular aneurysms. Later work by Seshaiyer et al. [43] found that the Fung-type anisotropic pseudostrain hyperelasticity model described the experimental results well, for which they reported estimated material parameter values:

$$W = c(e^Q - 1);\ Q = c_1E_{11}^2 + c_2E_{22}^2 + 2c_3E_{11}E_{22} \tag{8.4}$$

where E_{ii} are the in-plane membrane normal strain components and c and $c_i (i = 1-3)$ are the material parameter values.

One major challenge in the development of constitutive models for cerebral aneurysms is the scarce availability of specimens because the common clinical practice of surgical clipping does not easily lend itself to safe harvesting of large enough tissue samples for testing. This limitation continues to inhibit effective and reliable development of constitutive models for cerebral aneurysms with the work by Humphrey and colleagues being the only reported effort in this important area.

8.4 Computational Simulations of Intra-aneurysmal Hemodynamics

Alteration in hemodynamics of blood flow is believed to influence vascular wall remodeling. With the morphological changes effected by development of aneurysms, there is considerable interest in characterizing the altered hemodynamics and understanding what role if any that these alterations play in the remodeling of aneurismal tissue and consequently growth and rupture risk. Hemodynamics studies discussed in this section are generally those that do not account for aneurismal wall compliance. Those that do are discussed under Section 8.6.

8.4.1 Abdominal Aortic Aneurysm

Computational fluid dynamic simulations of blood flow within the aortic aneurysm sac have been reported widely. Peattie and colleagues [44–47] reported on a series of studies evaluating AAA hemodynamics using idealized or realistic AAA geometries, assuming a rigid aneurismal wall and a Newtonian model for blood under steady or pulsatile conditions. Finol and colleagues [48–50] also reported on similar studies. They generally found vortex formations in large AAA sacs, relatively minor increase in blood pressure within the sac, and a significant increase in intra-aneurismal turbulence proportional to size and asymmetry. Boundary conditions used were usually idealized velocity profiles at the inlet, resulting in a time-averaged flow rate of about 2–2.5 l/min and constant pressure outlet boundary conditions. Dalman and colleagues [51] reported on pulsatile blood flow simulations in patient-specific AAA under patient-specific inlet velocity profiles measured from phase-contrast MRI and (estimated) impedance boundary conditions at the iliac vessel outlets, allowing for a more realistic flow separation at the iliac bifurcation and consequently, more realistic flow characteristics. Many of these and other studies suggest that the significant thrombus deposition inside the sac is likely a result of hemodynamic issues such as flow stagnation. AAAs tend to have significant amounts of thrombus depositions. Hans and colleagues [52] estimated that on average, half the volume of AAA is filled with ILT. In this context, it is worth noting that except in rare cases, most portions of the aneurysm wall are unlikely to experience direct contact with flowing blood. Therefore, traditional hemodynamic measures such as peak wall shear stress or other derivatives of this measure such as oscillatory shear index require further evaluation and improvement.

8.4.2 Cerebral Aneurysms

Most intracranial aneurysms do not contain significant amounts of intrasac thrombus covering the wall and therefore, flow-induced vascular remodeling is likely more relevant compared to that for AAA. Computational hemodynamic simulations of intracranial aneurysms have mostly assumed the Newtonian model for blood, a rigid wall, velocity boundary condition at the inlet, and constant pressure at the outlet. Ujjie used a 2D axisymmetric model for CFD analysis and reported that the height-to-neck ratio can significantly affect flow-induced wall shear stress. Steinman and colleagues [23] reported on a computational simulation of blood flow in a giant intracranial aneurysm. They found high-speed flow entering the aneurysm at the proximal and distal ends of the neck, promoting the formation of both persistent and transient vortices within the aneurysm sac. This produced dynamic patterns of elevated and oscillatory wall shear stresses distal to the neck and along the sidewalls of the aneurysm (Fig. 8.10).

These findings were consistently noted in other independent studies as well [53–55]. Cebral et al. [53] simulated the blood flow in 62 patient-specific 3D cerebral aneurysm models assuming Newtonian blood, idealized velocity profile inlet, and constant pressure outlet. Based on the simulated flow into and out of the aneurysm,

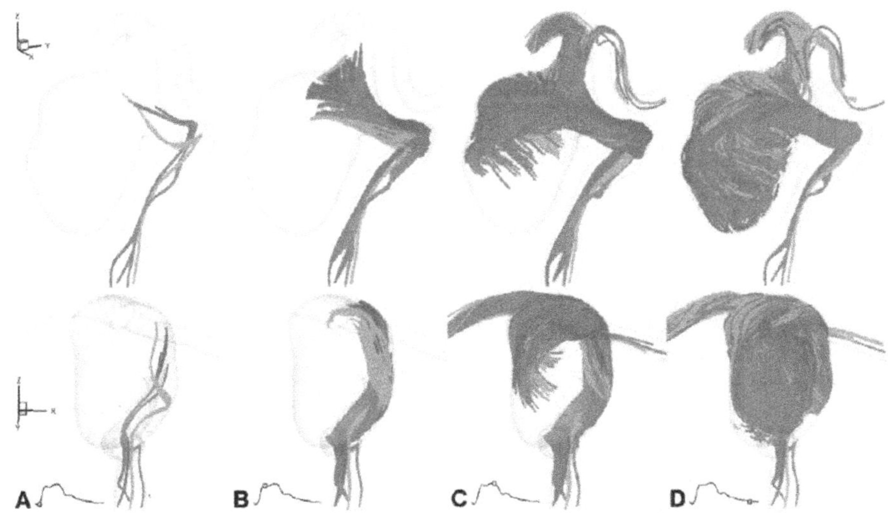

Fig. 8.10 Virtual slipstreams, shown at selected times in lateral (*top row*) and anteroposterior (*bottom row*) oblique views, provide an overview of the cerebral aneurysm hemodynamics (from Steinman et al. with permission)

they characterized aneurysm flow into four major types: type I—unchanging direction of inflow jet with a single associated vortex; type II—unchanging direction of inflow jet with multiple associated vortices but no change in the number of vortices during the cardiac cycle; type III—changing direction of inflow jet with creation of a single vortex; and type IV—changing direction of the inflow jet with creation or destruction of multiple vortices (Fig. 8.11). A total of 44% of their study population had type I flows, with the rest roughly divided among the other three types. They also noted that the aneurysms with small impingement sizes (of the main jet entering the aneurysm sac and impacting the dome region) were 6.3 times more likely to have experienced rupture compared with aneurysms with relatively large impingement zones. Shojima et al. [56] performed 3D CFD simulations to study the bloodstream impacting force and the local pressure elevation at the aneurysm. They concluded that the pressure elevation at the area of flow impact and at the aneurysm constituted only 1–2% of the peak intravascular pressure. Using hypothetical 3D geometric models of aneurysms, Utter and Rossmann [57] assessed the relationship between morphometric variables and flow characteristics. They reported that the ellipticity of an aneurysm was observed to be strongly correlated with wall shear stress at the aneurysm fundus, while its non-sphericity, volume, and degree of undulation [26] were more weakly correlated. Challenges in accurately characterizing cerebral aneurysm blood flow include determining the flow domain (for example, should the entire circle of Willis or just the aneurysm with contiguous branch vessels be included?) and accounting for vessel wall compliance, the outlet boundary conditions (impedances for various branch vessels), and the external pressure that may act on the outer wall of the arteries and some aneurysms.

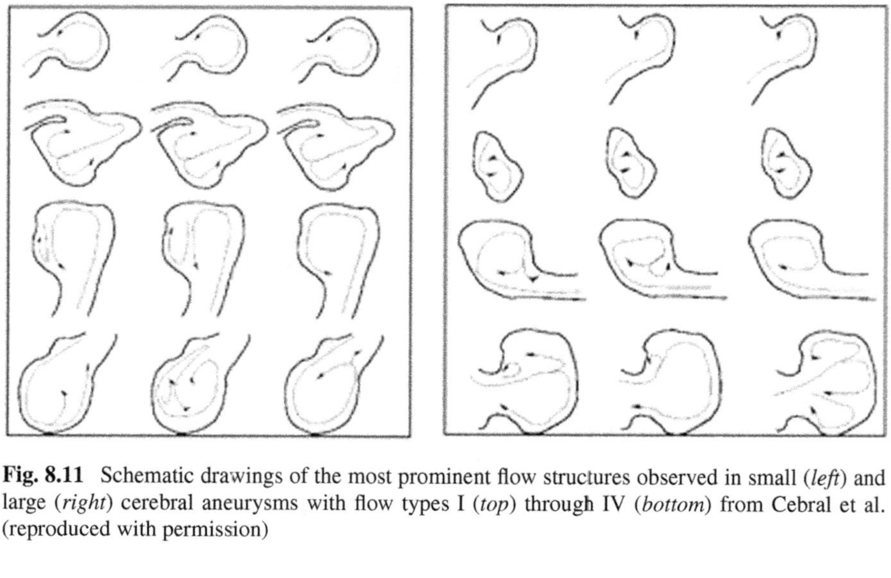

Fig. 8.11 Schematic drawings of the most prominent flow structures observed in small (*left*) and large (*right*) cerebral aneurysms with flow types I (*top*) through IV (*bottom*) from Cebral et al. (reproduced with permission)

8.4.3 Challenges in Aneurismal Hemodynamic Simulations

The scope and accuracy of simulations of blood flow in aneurysms is constrained to some extent by lack of sufficient information, theoretically sound formulations at low spatial scales, and high computational cost. There is no consensus on the appropriate constitutive model to use for whole blood in these simulations. Common models include the Newtonian model, the Casson's model, and the Carreau model. While there is little doubt about the shear-thinning behavior of blood, whether the inclusion of this behavior in the fluid model with the associated uncertainties (multiple functional forms—Casson's versus Carreau, perturbations in estimated parameters due to experimental complexity, etc.) will improve accuracy of simulations remains a point of debate. Nevertheless, some have assessed what effect, if any, the inclusion of shear-thinning behavior has on the simulation results and interpretation. Perktold and colleagues [58] found little "essential" difference in simulations of a saccular aneurysm, while both Khanafer et al. [59] and Valencia et al. [60] reported for aortic and cerebral aneurysms, respectively, that the shear stress increases while the flow becomes more stable due to inclusion of the shear-thinning behavior. Two challenges in reconciling these contrasting findings are the lack of consensus on the indices of flow that are relevant to aneurismal growth and rupture and in ascertaining how much difference in a particular hemodynamic index is clinically "significant". Blood modeling aside, our understanding of boundary conditions at the inlet (e.g., velocity profiles) and outlet (e.g., impedance or pressure) of the aneurismal flow domain remains marginal, especially regarding how these boundary conditions vary with exercise or other activities [51]. Some of the best available information may still only facilitate continuum-level simulations, which may fail to elucidate the precise mechanisms—platelet adhesion, for example—inherent in thrombus formation within the aneurysm sac. This may have important consequences to aneurysm growth and rupture. For that, multi-scale simulations of the kind described in earlier chapters may be indispensable, but these are so far lacking for aneurismal flows. And finally, the issue of computational cost continues to restrain our ability to be more ambitious in modeling. Recent advances in grid and/or cloud computing may need to be leveraged in order to reliably address these emerging challenges.

8.5 Computational Estimations of Aneurysmal Wall Stress and Strain

Aneurysm rupture is essentially a mechanical failure of the tissue and therefore pressure-induced stress has been suggested as an indicator of rupture risk. There have been several studies aimed at estimating the stress induced in the aneurysm wall. Computational methods usually employ the finite-element method for estimations.

8.5.1 Abdominal Aortic Aneurysm

The earliest studies involved the use of idealized geometric and material modeling for estimating stress in generalized models. Stringfellow et al. [61] performed an FE stress analysis of axisymmetric aneurysms using a linearized elasticity model and reported the relationship between peak wall stress and AAA geometry. Vorp and colleagues [62] developed hypothetical 3D models of AAA with varying size and asymmetry and showed that not only size but also shape can be an important factor in wall stress. Raghavan and colleagues [13] used patient-specific 3D models of AAA reconstructed from CT data, modeled the tissue behavior using Equation (8.1), assumed that the in vivo geometry is stress free, and ignored the role of thrombus and a uniform wall thickness to estimate the pressure-induced AAA wall stress. Subsequently, Wang et al. [16] included the role of ILT (modeled as a hyperelastic material using Equation (8.3)) in the computational model and reported that ILT plays a protective role in the AAA, effectively reducing the pressure-induced stress (Fig. 8.12). The material model for ILT did not account for its porosity whose effects remain poorly understood [63]. Fillinger and colleagues performed a prospective longitudinal cohort assessment of 103 AAA patients, all of whom presented unruptured in the clinic (mean follow-up time 14±2 months) using an automated stress analysis and post-processing algorithm. They found that initial peak wall tension (determined using the geometry reconstructed from the very first CT scan before being placed on follow-up) differentiated the AAAs that remained unruptured under observation (N=42) and those that had elective surgery (N=39) from those that went on to rupture (N=22). AAA diameter was also different between the groups, but peak wall stress differentiated the AAA that went on to rupture substantially better (3% difference in diameter between the groups versus 38% difference in peak wall stress). A receiver operating characteristic curve that allows for a comparison of the predictive indices confirmed that peak wall tension as calculated using finite element analysis is a significantly improved predictor of rupture over diameter or diameter × pressure (which is an equivalent of tension calculated simplistically using Laplace's law).

Doyle et al. [64] compared various modeling techniques including linear versus non-linear AAA tissue models and with versus without ILT, and reported that the inclusion of ILT had the most impact on computed stresses. Vande Geest et al. [65] incorporated an anisotropic model (Equation (8.2)) and reported on differences in stress estimate compared to an isotropic model. Speelman and colleagues [66] included the role of calcification, whose properties were characterized in a small population experimental study, and noted that calcifications could increase peak stress by up to 22% and that the properties, size, and shape of the calcification were factors in this increase. On calcification, the ability to characterize their mechanical property and the nature of their attachment to the aortic wall remains poorly understood. In a study of AAA harvested as a whole from autopsies, Raghavan et al. [31] anecdotally noted that the load required to tear apart rectangular tissue specimens with calcifications was not significantly lower than that for similar-sized tissue specimens without calcification. This anecdotal observation is not consistent

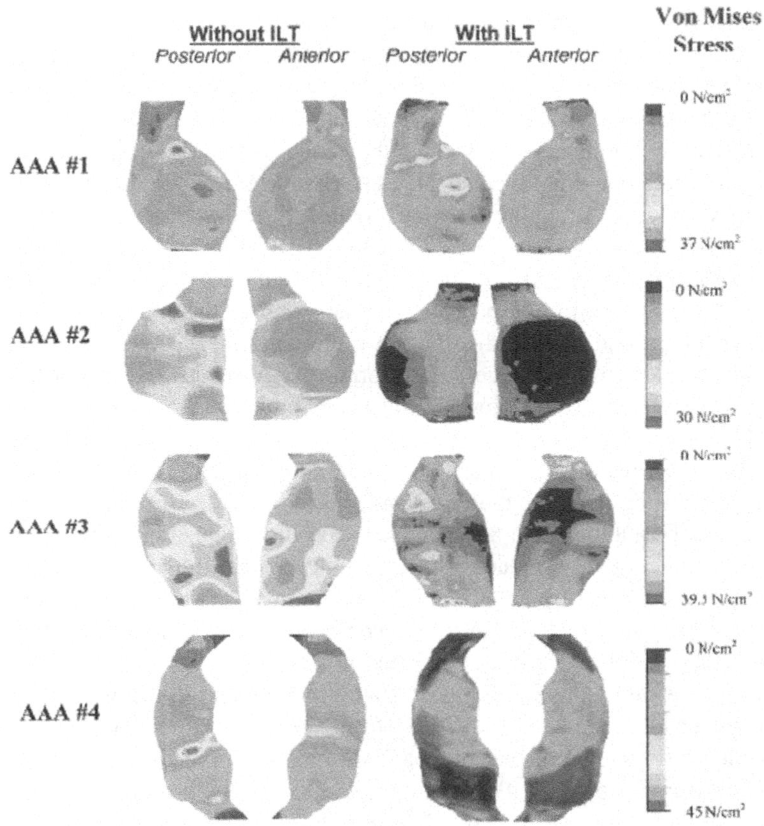

Fig. 8.12 Comparison of 3D wall stress distribution between AAA models with and without ILT (from Wang et al. with permission). The report emphasized the role of ILT in protecting the aneurysm wall pressure induced from stress

with Speelman's finding [66] because if calcifications were to increase the induced stress, the calcified specimens are likely to have failed at a lower load. Raghavan et al. [31] also noted that calcified tissues were also invariably thicker, suggesting that mere alteration of local elastic properties may not be sufficient to model a calcification. Additional studies are needed to understand the precise nature of the mechanics of calcification in AAA walls.

One common limitation of all of the above studies is the assumption that the geometry reconstructed from in vivo imaging is stress free, as well as the use of the forward approach to stress estimation. Rather, the AAA is in a pre-stressed state (at mean aortic pressure) during imaging. Raghavan et al. [67] suggested that the error from this assumption is minimal, but it is likely exacerbated in hypertensive patients, where the deviation between the true stress-free state and the in vivo state could be expected to be greater. Lu et al. [68] reported on a novel inverse elastostatics method for stress analysis of AAA where the stress-free assumption is not made. The problem is correctly posed as a pre-deformed surface existing under a pre-existing pressure for which the stress field may be directly solved. Further, because the problem is properly posed, stress calculations correctly become independent of material properties, as this is essentially an equilibrium problem where both the deformed geometry and the boundary conditions at this deformed geometry are known. The method is schematically shown in Fig. 8.13. In the traditional forward stress analysis, the equilibrium equation is written in referential form, relative to the undeformed state (Div $\boldsymbol{P} = 0$). The conceptual uniqueness of this approach is that the equilibrium equation is written in spatial form relative to the deformed state (div $\sigma = 0$). The equilibrium problem is interpreted as the solution for the zero-tension geometry and tension distribution in the current geometry such that "div $\boldsymbol{\sigma} = 0$" holds. Since there is a one-to-one correspondence between the positions of a point in the undeformed and deformed states, the motion of the body can be thought inversely. Further, since for elastic material the stress depends only on deformation gradient, one can parameterize the Cauchy stress ($\boldsymbol{\sigma}$) in terms of the inverse deformation gradient. In this case, the Eulerian form (div $\boldsymbol{\sigma} = 0$) is naturally chosen, as the differential equation is defined relative to the current position, which is known. The equilibrium equation reduces to a second-order non-linear differential equation for the undeformed shape. The equation can be solved using techniques that apply to the forward problem. Once the undeformed shape is found, the stress distribution in the deformed stated is obtained in standard manner. A patient-specific AAA was assessed by this inverse elastostatics method and shown to result in peak stress estimates that are 2–15% different from traditional estimates (Fig. 8.14). de Putter et al. [69] reported the use of a backward increment method toward a similar objective and also showed that stress estimates can vary by a small, but significant, amount. Caution is however warranted in considering the impact of these more accurate methods for most patient-specific AAA, because while they are likely to alter the estimates, they may alter them consistently for most patients so that the interpretations made may not necessarily change in a substantive manner. It is likely that

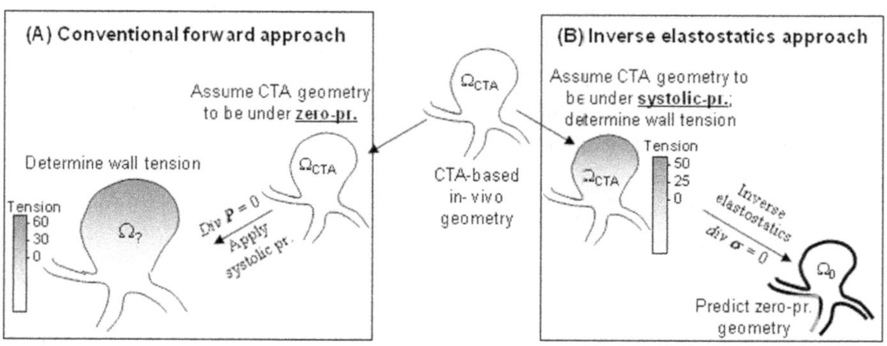

Fig. 8.13 Schematic illustration of the inverse elastostatics method. In this method, the problem is posed as a pre-deformed surface existing under a pre-existing pressure for which the stress field may be directly solved

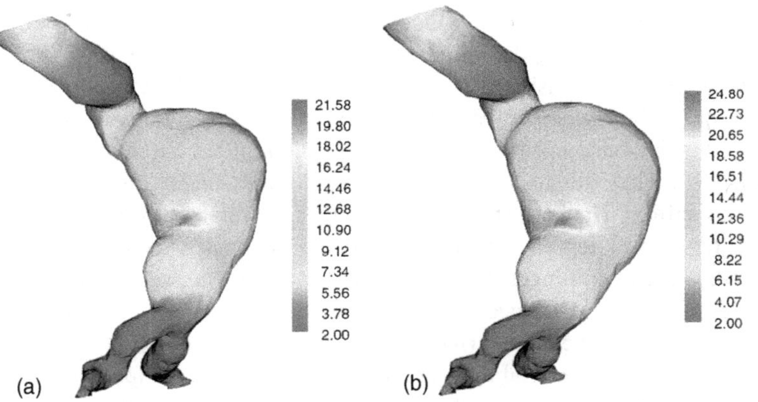

Fig. 8.14 von Mises stress at 100 mmHg arterial pressure in a reconstructed AAA model. (**a**) Inverse analysis; (**b**) forward analysis using the in vivo configuration as referenced from Lu et al.

this method will have a real impact in patients with chronic hypertension, but this too remains to be verified in a large enough patient population.

In the stress analysis of AAA, there has been considerable progress over the last decade or two. Some challenges that remain are accurately characterizing the role of thrombus by appropriate material model and computational methods, acquiring information on the regional variations in wall thickness, and understanding how localized malformations such as calcifications and blisters (tiny outpouchings in the AAA wall) influence the wall stresses. But perhaps the greatest impact will come from an improved understanding of the intra-aneurismal and inter-aneurismal failure properties of these lesions. Thubrikar et al. [33] and Raghavan et al. [31] reported that there are significant regional variations in failure strength of the tissue. In a series of reports on the topic [10, 35, 70–73], Vorp and colleagues leveraged experimental data on tissues harvested from a large population of patients to develop a stochastic model for the prediction of failure strength in AAA based on demographic and morphometric factors. If predicted accurately, these strength estimates may be used to develop a rupture potential index (RPI)—the stress–strength ratio. With the development of more sophisticated imaging modalities, better experimental techniques, and improved computational methods, AAA biomechanics appears to hold some promise. Longitudinal cohort studies with a large study population are also essential to reliably verify the clinical utility of this promising technology.

8.5.2 Cerebral Aneurysms

In cerebral aneurysms, computational estimation of wall stress has not received as much attention as in AAA or as in hemodynamic simulations. Hademenos et al. [74, 75] employed Laplace law to document wall stress in axisymmetric models of different sizes and shapes. Humphrey and colleagues [9, 43, 76–78] utilized experimentally derived material model (Equation (8.4)) and idealized geometries to computationally (FEM) determine the effect of shape and size of these lesions on wall stress resultant (tension). In one report on the elastodynamics of aneurysms, this group studied whether cerebral aneurysms are dynamically unstable (i.e., whether they resonate in response to pulsatile blood flow) and concluded that they are not [79]. They suggested that estimate of stress under quasi-static pressure may itself suffice. They also noted that contact constraints may elevate the stresses around the contact region with implications to patients who present with symptoms indicative of impingement of the aneurysm on a nerve ending in the brain [80]. Ma et al. [81] performed finite element stress analysis on 28 patient-specific models of cerebral aneurysms (reconstructed from CTA data) and an anisotropic material model with material fiber orientations along the principal curvature directions. They found that peak wall stresses varied regionally and between patients in the range of 0.3–1.1 MPa (Fig. 8.15). They also reported that the strain energy stored is minimized when the material fiber directions lie along the principal curvature directions. One limitation in the reported methodology was the assumption that the CTA reconstructed shape was stress free, similar to most studies on AAA. Here too, Lu and colleagues [82] utilized the inverse elastostatics approach to eliminate this assumption improving accuracy, although its impact remains to be verified in the patient population.

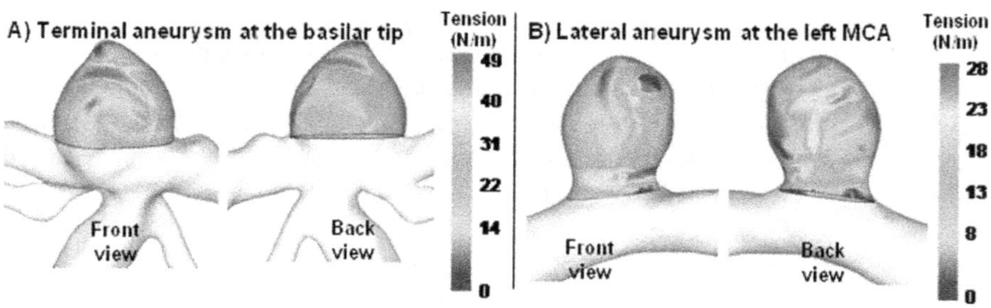

Fig. 8.15 Stress distribution in reconstructed model of a terminal (**a**) and lateral (**b**) aneurysm from a study reported by Ma et al. The stress component shown is along the presumed collagen fiber direction (maximum principal curvature direction)

Challenges in the computational estimations of pressure-induced wall stress in cerebral aneurysms are similar to those in AAA—lack of reliable information on wall thickness, material behavior, failure properties, and empirical outcome data from prospective large population longitudinal cohorts.

8.6 Fluid–Structure Interaction Studies

In most studies, the simulations of fluid-level events are decoupled from the structure-level events presumably for computational ease. However, the blood flow is likely affected by vessel wall motion or compliance and the boundary conditions on the wall are in turn affected by blood flow. It is conceivable that modeling this coupled fluid–structure interaction could reveal additional insights into the mechanics of aneurysms.

Using an arbitrary Lagrangian–Eulerian (ALE) approach, Di Martino and colleagues [83] studied the fluid–structure interaction (FSI) in a patient-specific AAA employing the commercial code FIDAP (Fluent Inc., now ANSYS Inc., Pittsburgh, PA). Blood was modeled as a Navier–Stokes fluid under pulsatile flow and the AAA wall and thrombus as isotropic linear-elastic materials. Li and Kleinstreuer [84] also reported on a similar approach, but on AAA with an implanted stent graft. Scotti and colleagues [85] employed finite strain material models for an FSI analysis and reported that the inclusion of AAA wall compliance affects wall shear stress estimates significantly. Bluestein and colleagues [86, 87] incorporated anisotropic finite strain models and the role of thrombus to report on a fully coupled FSI simulation of AAA. Physiologically estimated pressures [88] were applied at the inlet and outlet of the AAA flow domain. The AAA was assumed to initially be in a zero stress state. The flow domain was pressurized from 0 Pa to 110 kPa with zero flow for 1 s before the waveforms were applied at the inlet and the outlet. After establishing that, periodicity was achieved by the second cycle and each computation was continued for four complete flow cycles using the waveform and a time step of 0.004 s. The Newton iteration scheme was used for the sparse matrix solver, with 0.001 relative tolerance for the degrees of freedom. They found that while the locations of high and low stresses are consistent for both isotropic and anisotropic material models, the differences between them become pronounced at large values of strain.

On cerebral aneurysms, Torii and colleagues [89] reported on FSI simulations comparing the use of linear versus finite elastic material models. The maximum displacement computed with their finite elastic model was 36% smaller compared to the linearly elastic material model, but the displacement patterns such as the site of local maxima are not sensitive to the wall models. The blood near the apex of an aneurysm was mostly stagnant, causing very low wall shear stress. They also noted that relatively high flow velocities due to the interaction between the blood flow and the aneurismal wall were independent of the wall model. Ahmed et al. [90] also performed FSI simulation of a wide-necked cerebral aneurysm assuming non-Newtonian blood (Carreau model) and a linear-elastic wall using the ANSYS-CFX commercial solver (ANSYS, Inc., Pittsburgh, PA). They used Womersley velocity profile at the inlet and documented the effect of higher pressure and changing modulus on the wall stress. As expected, increased pressure and lowered modulus increased stress and deformation.

8.7 Framework for Biomechanical Modeling of Growth and Remodeling

Computational simulations of aneurysms described in earlier sections provide the investigator insights into the state of the aneurysm at one snapshot in time. However, aneurysms are clearly evolving structures with ongoing growth and remodeling.

While studies on flow and wall stress may provide some insights into the current state of the aneurysm, they fail to provide any information about how it got there or where it will go from there. Exploring precisely such questions is essential toward understanding the pathogenesis of this deadly disease and developing long-term strategies for treatment or observation, as well as for management of the patients in the clinic, irrespective of whether the aneurysms are repaired or merely watched. As with any biological entity, the processes are clearly biological and complex, but the underlying mechanisms that trigger or guide such processes may have some basis in mechanobiology. In this context, biomechanical modeling can be a very valuable tool. If a reasonably reliable mechanobiological model for aneurysm growth were to be developed, it may permit quick and easy testing of hypotheses on the underlying biomechanical mechanisms during aneurysm growth.

Aortic and cerebral aneurysms are thought to have very different etiology and pathological processes underlying their developments. Whereas in AAA there is significant elastin degradation, there is still some elastin content. In cerebral aneurysms, there is virtually no elastin, with the connective tissue in the wall predominated by collagen. The growth processes are also quite different as reflected in studies on the outcome of patients with unruptured aneurysms on untreated follow-up. Of 103 unruptured, untreated AAA patients followed by Fillinger et al. [17], about 21% over an average follow-up period of 1 year had confirmed or symptomatic rupture. This is strongly contrasted with the International Study of Unruptured Intracranial Aneurysms (ISUIA) where of 1,692 patients followed over an average of 5 years, only 3% experienced rupture [2]. On growth rates, whereas AAAs tend to grow rather steadily [3], most cerebral aneurysms rarely do [2, 5]. In cerebral aneurysm, many lesions are thought to stabilize and seldom rupture [9, 76], while in AAA, most lesions are expected to continue to grow (even if at different rates) and eventually rupture if left untreated [17]. Ironically in this context, the theoretical framework used may be applicable (with some appropriate modifications to specifics) to modeling the growth of both these lesions.

Watton and Hill [7] proposed a mathematical model to account for the evolution of the abdominal aortic aneurysm. The artery was modeled as a two-layered, cylindrical membrane using non-linear elasticity and a physiologically realistic constitutive model. They simulated changes in the microstructure of the arterial wall that led to the development of AAA by assuming that collagen remodels to compensate for the loss of elastin. The underlying premise is that when elastin loss causes increases in collagen strain, collagen fibers will seek to remodel so as to return to a homeostatic strain range. As the aneurysm develops in size, the naturally occurring process of fiber turnover acts to restore the strain in the fibers. They introduced recruitment variables, which define the factors by which the tissue must be stretched with respect to the undeformed configuration, for the collagen to begin to bear load, thereby simulating the gross mechanical effects of fiber turnover in altered configurations. With remodeling of the material constituents, the new equilibrium displacement field for the arterial wall was estimated. The wall is treated as a membrane and so a variational equation over a 2D domain governed the ensuing deformation and evolution of the aneurysm. The resulting equations were solved using FEM. Realistic rates of dilation for physiologically realistic geometric, material, and remodeling parameters were estimated. Dilation rates were seen to increase in hypertensive conditions. In a follow-up study, Watton and Hill [6] studied the influence of the model's parameters on growth rates of the AAA. Their models suggested an exponential growth rate for AAA. Volokh and Vorp postulated an alternative approach to modeling AAA growth—a coupled mathematical model of growth and failure of AAA. The failure portion of the model was based on the constitutive theory of softening hyperelasticity where the classical hyperelastic law is enhanced with a new constant indicating the maximum energy that an infinitesimal material volume can accumulate without failure. The new constant controls material failure and it can be interpreted as the average energy of molecular bonds from the microstructural standpoint. The coupled theory was used to simulate growth and rupture of an idealized spherical AAA. They noted that qualitatively, the results were consistent with observations, but a quantitative calibration of the model is required.

In cerebral aneurysms, Humphrey [9] postulated a mechanism for the growth, remodeling, and rupture of cerebral aneurysms (Fig. 8.16). An initial insult (generally unknown) may cause a local weakening of the wall and thus a mild dilatation. This raises the local stress field above a homeostatic range, thus setting into motion

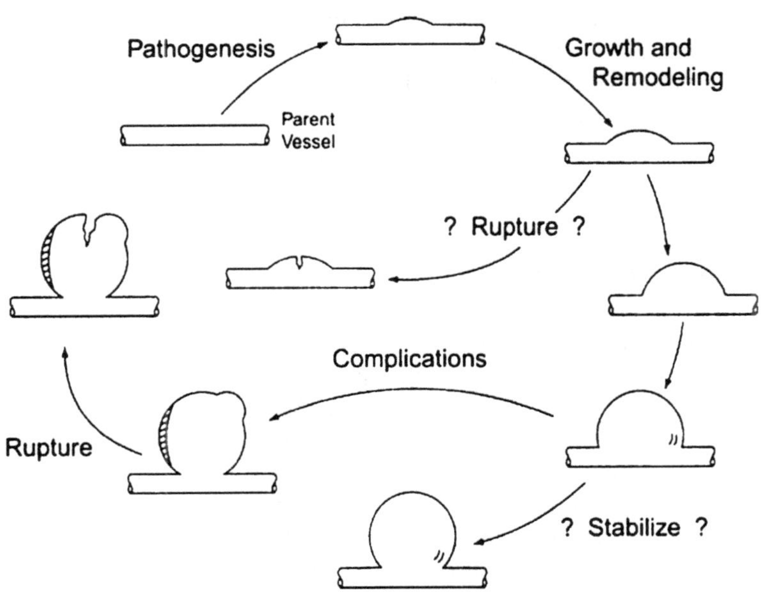

Fig. 8.16 Postulated mechanism for growth, remodeling, and rupture of cerebral aneurysms by Humphrey. An initial insult may cause local weakening of the wall and thus a mild dilatation. This raises the local stress field, setting into motion a growth and remodeling process that may result in a larger but stable lesion (if the degradation and the deposition of collagen are well balanced) or one that progresses to rupture (if degradation of collagen exceeds deposition). Even a larger stable lesion may over time progress to rupture if the degradation/deposition balance is adversely impacted over time

a growth and remodeling process that attempts to reduce the stresses toward the homeostatic range. If the requisite degradation and deposition of collagen are well balanced, this could produce a larger, but stable lesion. If the degradation/deposition balance is perturbed beyond some threshold, this could lead to rupture. Even in the lesions that do stabilize, because of sustained hemodynamic loading and possibly additional complications (insults), there may be a need for maintaining a sustained degradation/deposition balance or else rupture may occur. Postulated as it may be, it provided a reasonable framework for development of a theoretical framework. In a series of reports, Humphrey and Taylor forcefully make a case for fluid–structure growth modeling of aneurysms and emphasize that there is a pressing need to embed rational growth and remodeling equations within a robust FSI model [93]. Baek et al. proposed that growth and remodeling are stress mediated [94]. Similar to the strain-based approach of Watton and Hill, perturbations in stress beyond the homeostatic range are postulated to trigger remodeling. A 2D constrained mixture model for growth and remodeling of an ellipsoidally shaped saccular aneurysm was proposed and enlargement and changes in material symmetry in the aneurismal wall were numerically simulated. Results suggested that ellipsoidal aneurysms tend toward spherical shapes. On the remodeling mechanism, they noted that a balance between radial expansion and wall thickening plays a critical role in determining the stability of an enlarging lesion.

In summary, computational models are unlikely to be infallible, but infallibility is a lofty standard. Even if these models can help identify the potentially promising hypotheses, this may help direct biological and clinical investigators toward designing experiments or patient outcome data collection strategies for a more reliable verification.

8.8 Future Directions

Computational biomechanical modeling of aneurysms has come a long way. This is as well, given the mechanical issues inherent to the growth, rupture, and treatment. The underlying objective for much of the biomechanical modeling is to help distinguish aneurysms with high rupture risk compared to those without. The immediate need of the physician managing the aneurysm patient for predicting rupture risk is the primary driver. Naturally, therefore, acquiring information from and developing biomechanical measures (be they an index of peak wall stress or shear stress or a shape/size factor) of a single snapshot in time for a given aneurysm may itself be a productive and clinically translational endeavor. Studies aimed at estimating normal and shear stresses in the AAA wall fall into this category. A second objective is to understand the growth and remodeling mechanisms in the disease process—an effort grounded in basic science with perhaps long-term implications for prevention, early diagnosis, and/or treatment. Currently in early stages of theoretical development, these studies leverage advancements in theoretical mechanics and mechanobiology to construct a framework where scientists may filter and refine hypotheses on disease pathogenesis before undertaking experimental and/or clinical studies. But perhaps the greatest impact from biomechanical modeling may come from studies with a third objective—the use of biomechanical modeling for aiding the treatment of aneurysms. Here, computational models may be used for the design of implants [12, 91], the selection of patients who are best suited for a particular treatment strategy or implant design, the prediction of outcome following a particular method of treatment [92], and testing the efficacy of novel treatment strategies. This chapter did not discuss efforts toward this third objective, but they are worth noting.

Despite progress in this field, many challenges remain. In the modeling of aneurysms, future efforts that are likely to have a substantial impact are obtaining additional measured information, using it for computational simulations and verification of these methods in prospective longitudinal cohorts. New measured information could be (1) dynamic 3D data (from gated MR or CT) for identification of regional properties; (2) higher resolution imaging that permits the extraction of regionally varying wall thickness; and (3) independent measures of wall caliber on a patient-specific basis—be it via stochastic models, a better understanding of the relationship between localized malformations (e.g., calcification or blister) and mechanical properties, or from blood samples (e.g., MMP9 has been proposed to be correlated with collagen content and hence tissue properties), if these are shown to be of value. Another effort that is indispensable is the collection of empirical data through prospective longitudinal cohort studies. Here, patients may be followed over long periods of time (years, not months) with time series imaging data and other information (e.g., plasma MMP9 levels, presence/absence of hypertension) over that period. Such empirical information is essential for reliable assessment of the true clinical impact of the computational simulations of wall stress and in the validation/parameter estimation for growth/remodeling models.

References

1. Bengtsson H, Sonesson B, Bergqvist D (1996) Incidence and prevalence of abdominal aortic aneurysms, estimated by necropsy studies and population screening by ultrasound. Ann NY Acad Sci 800:1–24
2. Wiebers DO et al (2003) Unruptured intracranial aneurysms: natural history, clinical outcome, and risks of surgical and endovascular treatment. Lancet 362(9378):103–110
3. Keen RR, Dobrin PB (2000) Development of aneurysms. In: Medical intelligence unit 17. Landes Bioscience, Georgetown, TX, p 235 (NetLibrary Inc.)
4. Darling RC et al (1977) Autopsy study of unoperated abdominal aortic aneurysms. The case for early resection. Circulation 56(3 Suppl):II161–II164
5. Weir B (2002) Unruptured intracranial aneurysms: a review. J Neurosurg 96(1):3–42
6. Watton PN, Hill NA (2009) Evolving mechanical properties of a model of abdominal aortic aneurysm. Biomech Model Mechanobiol 8(1):25–42
7. Watton PN, Hill NA, Heil M (2004) A mathematical model for the growth of the abdominal aortic aneurysm. Biomech Model Mechanobiol 3(2):98–113
8. Watton PN, Ventikos Y, Holzapfel GA (2009) Modelling the growth and stabilization of cerebral aneurysms. Math Med Biol 26:133–164
9. Humphrey JD (2002) Cardiovascular solid mechanics: cells, tissues, and organs. Springer, New York, NY, p xvi, 757

10. Vorp DA, Vande Geest JP (2005) Biomechanical determinants of abdominal aortic aneurysm rupture. Arterioscler Thromb Vasc Biol 25(8):1558–1566
11. Kleinstreuer C et al (2008) Computational mechanics of Nitinol stent grafts. J Biomech 41(11):2370–2378
12. Kleinstreuer C, Li Z, Farber MA (2007) Fluid–structure interaction analyses of stented abdominal aortic aneurysms. Ann Rev Biomed Eng 9:169–204
13. Raghavan ML et al (2000) Wall stress distribution on three-dimensionally reconstructed models of human abdominal aortic aneurysm. J Vasc Surg 31(4):760–769
14. Thubrikar MJ, Al-Soudi J, Robicsek F (2001) Wall stress studies of abdominal aortic aneurysm in a clinical model. Ann Vasc Surg 15(3):355–366
15. Sacks MS et al (1999) In vivo three-dimensional surface geometry of abdominal aortic aneurysms. Ann Biomed Eng 27(4):469–479
16. Wang DH et al (2002) Effect of intraluminal thrombus on wall stress in patient-specific models of abdominal aortic aneurysm. J Vasc Surg 36(3):1–7
17. Fillinger MF et al (2003) Prediction of rupture risk in abdominal aortic aneurysm during observation: wall stress versus diameter. J Vasc Surg 37(4):724–732
18. Fillinger MF et al (2002) In vivo analysis of mechanical wall stress and abdominal aortic aneurysm rupture risk. J Vasc Surg 36(3):589–597
19. Raghavan ML et al (2005) Automated methodology for determination of stress distribution in human abdominal aortic aneurysm. J Biomech Eng 127(5):868–871
20. de Bruijne M, van Ginneken B, Viergever MA, Niessen WJ (2003) Three-dimensional point distribution models for tubular objects. Department of Information and Computing Sciences, Utrecht University, Utrecht
21. Martufi G, Di Martino ES, Amon CH, Muluk SC, Finol EA (2009) Three-dimensional geometrical characterization of abdominal aortic aneurysms: image-based wall thickness distribution. J Biomech Eng 131(6):061015
22. Fillinger MF et al (2004) Anatomic characteristics of ruptured abdominal aortic aneurysm on conventional CT scans: implications for rupture risk. J Vasc Surg 39(6):1243–1252
23. Steinman DA et al (2003) Image-based computational simulation of flow dynamics in a giant intracranial aneurysm. AJNR Am J Neuroradiol 24(4):559–566
24. Cebral JR et al (2005) Efficient pipeline for image-based patient-specific analysis of cerebral aneurysm hemodynamics: technique and sensitivity. IEEE Trans Med Imaging 24(4): 457–467
25. Ma B, Harbaugh RE, Raghavan ML (2004) Three-dimensional geometrical characterization of cerebral aneurysms. Ann Biomed Eng 32(2):264–273
26. Raghavan ML, Ma B, Harbaugh RE (2005) Quantified aneurysm shape and rupture risk. J Neurosurg 102(2):355–362
27. Millan RD et al (2007) Morphological characterization of intracranial aneurysms using 3-D moment invariants. IEEE Trans Med Imaging 26(9):1270–1282
28. Radaelli AG et al (2008) Reproducibility of haemodynamical simulations in a subject-specific stented aneurysm model—a report on the Virtual Intracranial Stenting Challenge 2007. J Biomech 41(10):2069–2081
29. Zhang C et al (2009) Morphodynamic analysis of cerebral aneurysm pulsation from time-resolved rotational angiography. IEEE Trans Med Imaging 28(7):1105–1116
30. He CM, Roach MR (1994) The composition and mechanical properties of abdominal aortic aneurysms. J Vasc Surg 20(1):6–13
31. Raghavan ML et al (2006) Regional distribution of wall thickness and failure properties of human abdominal aortic aneurysm. J Biomech 39(16):3010–3016
32. Raghavan ML, Webster MW, Vorp DA (1996) Ex vivo biomechanical behavior of abdominal aortic aneurysm: assessment using a new mathematical model. Ann Biomed Eng 24(5): 573–582
33. Thubrikar MJ et al (2001) Mechanical properties of abdominal aortic aneurysm wall. J Med Eng Technol 25(4):133–142

34. Vallabhaneni SR et al (2004) Heterogeneity of tensile strength and matrix metalloproteinase activity in the wall of abdominal aortic aneurysms. J Endovasc Ther 11(4):494–502
35. Vande Geest JP et al (2006) A biomechanics-based rupture potential index for abdominal aortic aneurysm risk assessment: demonstrative application. Ann N Y Acad Sci 1085:11–21
36. Vorp DA et al (1996) Wall strength and stiffness of aneurysmal and nonaneurysmal abdominal aorta. Ann N Y Acad Sci 800:274–276
37. Toth M et al (1998) Sterically inhomogenous viscoelastic behavior of human saccular cerebral aneurysms. J Vasc Res 35(5):345–355
38. Raghavan ML, Vorp DA (2000) Toward a biomechanical tool to evaluate rupture potential of abdominal aortic aneurysm: identification of a finite strain constitutive model and evaluation of its applicability. J Biomech 33(4):475–482
39. Vande Geest JP, Sacks MS, Vorp DA (2006) The effects of aneurysm on the biaxial mechanical behavior of human abdominal aorta. J Biomech 39(7):1324–1334
40. Wang DH et al (2001) Mechanical properties and microstructure of intraluminal thrombus from abdominal aortic aneurysm. J Biomech Eng 123(6):536–539
41. Hsu FPK et al (1995) A triplane video-based experimental system for studying axisymmetrically inflated biomembranes. IEEE Trans Biomed Eng 42(5):442–450
42. Hsu FPK et al (1994) Identification of response functions from axisymmetrical membrane inflation tests – implications for biomechanics. Int J Solids Struct 31(24):3375–3386
43. Seshaiyer P, Humphrey JD (2003) A sub-domain inverse finite element characterization of hyperelastic membranes including soft tissues. J Biomech Eng Trans Asme 125(3): 363–371
44. Peattie RA et al (1996) Steady flow in models of abdominal aortic aneurysms. Part II: Wall stresses and their implication for in vivo thrombosis and rupture. J Ultrasound Med 15(10):689–696
45. Peattie RA, Riehle TJ, Bluth EI (2004) Pulsatile flow in fusiform models of abdominal aortic aneurysms: flow fields, velocity patterns and flow-induced wall stresses. J Biomech Eng 126(4):438–446
46. Peattie RA et al (1996) Steady flow in models of abdominal aortic aneurysms. Part I: Investigation of the velocity patterns. J Ultrasound Med 15(10):679–688
47. Peattie RA et al (1994) Development of turbulence in steady flow through models of abdominal aortic aneurysms. J Ultrasound Med 13(6):467–472
48. Finol EA, Amon CH (2001) Blood flow in abdominal aortic aneurysms: pulsatile flow hemodynamics. J Biomech Eng 123(5):474–484
49. Finol EA, Amon CH (2002) Flow-induced wall shear stress in abdominal aortic aneurysms: Part I?steady flow hemodynamics. Comput Methods Biomech Biomed Eng 5(4):309–318
50. Finol EA, Keyhani K, Amon CH (2003) The effect of asymmetry in abdominal aortic aneurysms under physiologically realistic pulsatile flow conditions. J Biomech Eng 125(2):207–217
51. Dalman RL et al (2006) AAA disease: mechanism, stratification, and treatment. Ann N Y Acad Sci 1085:92–109
52. Hans SS et al (2005) Size and location of thrombus in intact and ruptured abdominal aortic aneurysms. J Vasc Surg 41(4):584–588
53. Cebral JR et al (2005) Characterization of cerebral aneurysms for assessing risk of rupture by using patient-specific computational hemodynamics models. AJNR Am J Neuroradiol 26(10):2550–2559
54. Cebral JR, Hendrickson S, Putman CM (2009) Hemodynamics in a lethal basilar artery aneurysm just before its rupture. AJNR Am J Neuroradiol 30(1):95–98
55. Ma B (2004) Modeling the geometry, hemodynamics and tissue mechanics of cerebral aneurysms (Thesis advisor: Raghavan ML). In: Biomedical Engineering. University of Iowa, Iowa City, IA, p 155
56. Shojima M et al (2005) Role of the bloodstream impacting force and the local pressure elevation in the rupture of cerebral aneurysms. Stroke 36(9):1933–1938

57. Utter B, Rossmann JS (2007) Numerical simulation of saccular aneurysm hemodynamics: influence of morphology on rupture risk. J Biomech 40(12):2716–2722
58. Perktold K, Peter R, Resch M (1989) Pulsatile non-Newtonian blood flow simulation through a bifurcation with an aneurysm. Biorheology 26(6):1011–1030
59. Khanafer KM et al (2006) Modeling pulsatile flow in aortic aneurysms: effect of non-Newtonian properties of blood. Biorheology 43(5):661–679
60. Valencia AA et al (2006) Blood flow dynamics in saccular aneurysm models of the basilar artery. J Biomech Eng 128(4):516–526
61. Stringfellow MM, Lawrence PF, Stringfellow RG (1987) The influence of aorta-aneurysm geometry upon stress in the aneurysm wall. J Surg Res 42(4):425–433
62. Vorp DA, Raghavan ML, Webster MW (1998) Mechanical wall stress in abdominal aortic aneurysm: influence of diameter and asymmetry. J Vasc Surg 27(4):632–639
63. Di Martino ES, Vorp DA (2003) Effect of variation in intraluminal thrombus constitutive properties on abdominal aortic aneurysm wall stress. Ann Biomed Eng 31(7):804–809
64. Doyle BJ, Callanan A, McGloughlin TM (2007) A comparison of modelling techniques for computing wall stress in abdominal aortic aneurysms. Biomed Eng Online 6:38
65. Vande Geest JP et al (2008) The effects of anisotropy on the stress analyses of patient-specific abdominal aortic aneurysms. Ann Biomed Eng 36(6):921–932
66. Speelman L et al (2007) Effects of wall calcifications in patient-specific wall stress analyses of abdominal aortic aneurysms. J Biomech Eng 129(1):105–109
67. Raghavan ML, Ma B, Fillinger MF (2006) Non-invasive determination of zero-pressure geometry of arterial aneurysms. Ann Biomed Eng 34(9):1414–1419
68. Lu J et al (2005) Prediction of aneurysm stress based on deformed geometry using inverse finite element formulation. In: 2005 Summer Bioengineering Conference, Vail, CO
69. de Putter S et al (2007) Patient-specific initial wall stress in abdominal aortic aneurysms with a backward incremental method. J Biomech 40(5):1081–1090
70. Vande Geest JP et al (2006) Gender-related differences in the tensile strength of abdominal aortic aneurysm. Ann N Y Acad Sci 1085:400–402
71. Vande Geest JP, Sacks MS, Vorp DA (2004) Age dependency of the biaxial biomechanical behavior of human abdominal aorta. J Biomech Eng 126(6):815–822
72. Vande Geest JP et al (2006) Towards a noninvasive method for determination of patient-specific wall strength distribution in abdominal aortic aneurysms. Ann Biomed Eng 34(7):1098–1106
73. Vorp DA (2007) Biomechanics of abdominal aortic aneurysm. J Biomech 40(9):1887–1902
74. Hademenos GJ et al (1994) A nonlinear mathematical model for the development and rupture of intracranial saccular aneurysms. Neurol Res 16(5):376–384
75. Hademenos GJ et al (1998) Anatomical and morphological factors correlating with rupture of intracranial aneurysms in patients referred for endovascular treatment. Neuroradiology 40(11):755–760
76. Humphrey JD (2002) Intracranial aneurysms. In: Humphrey JD (ed) Cardiovascular solid mechanics: cells, tissues, and organs. Springer, New York, NY, pp 386–428
77. Humphrey JD, Canham PB (2000) Structure, mechanical properties, and mechanics of intracranial saccular aneurysms. J Elasticity 61(1–3):49–81
78. Kyriacou SK, Humphrey JD (1996) Influence of size, shape and properties on the mechanics of axisymmetric saccular aneurysms. J Biomech 29(8):1015–1022
79. Shah AD, Humphrey JD (1999) Finite strain elastodynamics of intracranial saccular aneurysms. J Biomech 32(6):593–599
80. Seshaiyer P, Humphrey JD (2001) On the potentially protective role of contact constraints on saccular aneurysms. J Biomech 34(5):607–612
81. Ma B et al (2007) Nonlinear anisotropic stress analysis of anatomically realistic cerebral aneurysms. J Biomech Eng 129(1):88–96
82. Lu J, Zhou X, Raghavan ML (2008) Inverse method of stress analysis for cerebral aneurysms. Biomech Model Mechanobiol 7(6):477–486

83. Di Martino ES et al (2001) Fluid–structure interaction within realistic three-dimensional models of the aneurysmatic aorta as a guidance to assess the risk of rupture of the aneurysm. Med Eng Phys 23(9):647–655
84. Li Z, Kleinstreuer C (2005) Blood flow and structure interactions in a stented abdominal aortic aneurysm model. Med Eng Phys 27(5):369–382
85. Scotti CM, Finol EA (2007) Compliant biomechanics of abdominal aortic aneurysms: a fluid–structure interaction study. Comput Struct 85(11–14):1097–1113
86. Bluestein D et al (2009) Intraluminal thrombus and risk of rupture in patient specific abdominal aortic aneurysm – FSI modelling. Comput Methods Biomech Biomed Eng 12(1):73–81
87. Rissland P et al (2009) Abdominal aortic aneurysm risk of rupture: patient-specific FSI simulations using anisotropic model. J Biomech Eng 131(3):031001
88. Olufsen MS et al (2000) Numerical simulation and experimental validation of blood flow in arteries with structured-tree outflow conditions. Ann Biomed Eng 28(11):1281–1299
89. Torii R et al (2008) Fluid–structure interaction modeling of a patient-specific cerebral aneurysm: influence of structural modeling. Comput Mech 43(1):151–159
90. Ahmed S et al (2007) Fluid structure interaction modelling of a patient specific cerebral aneurysm: effect of hypertension and modulus of elasticity. In: Proceedings of the 16th Australasian fluid mechanics conference (AFMC). School of Engineering, The University of Queensland, Gold Coast, Queensland
91. Morris L et al (2004) A mathematical model to predict the in vivo pulsatile drag forces acting on bifurcated stent grafts used in endovascular treatment of abdominal aortic aneurysms (AAA). J Biomech 37(7):1087–1095
92. Di Martino E et al (2004) Wall stresses before and after endovascular repair of abdominal aortic aneurysms. In: International mechanical engineering congress and exposition, Anaheim, CA
93. Humphrey JD and Taylor CA, "Intracranial and Abdominal Aortic Aneurysms: Similarities, Differences, and Need for a New Class of Computational Models", Annn Rev Biomed Eng 10: 221–246
94. Baek S, Rajagopal KR, Humphrey JD, "A theoretical model of enlarging intracranial fusiform aneurysms", J Biomech Eng. 2006 Feb;128(1):142–149

Chapter 9
Advances in Computational Simulations for Interventional Treatments and Surgical Planning

Diane A. de Zélicourt, Brooke N. Steele, and Ajit P. Yoganathan

Abstract Computational analyses of blood flow through the vascular system have the potential to improve medical care by identifying and quantifying hemodynamics relevant to the protection from and initiation and progression of vascular disease. In addition to identifying relevant mechanics, computational analyses coupled with advanced simulation environments could also be used to predict the hemodynamics associated with alternate anatomic or hemodynamic scenarios and to predict the overall performance of each scenario. Such simulations can be used by clinicians to design optimized interventional treatments and by engineers to design optimal vascular devices. In this chapter, the application of computational simulations for the prediction of vulnerable atherosclerotic plaques, improved understanding of the mechanics of balloon angioplasty, design of endovascular stents, and patient-specific surgical planning is discussed.

9.1 Introduction

Mechanical forces are known to influence vascular disease, such as the localization of atherosclerotic plaques in regions with low or oscillating wall shear stress [1] and arterial remodeling in regions of high wall stress [2]. In addition to local mechanics, other pathologic forces, such as hypertension due to focal changes in geometry, are known to cause vascular complications such as arterial remodeling and organ damage [3]. Because these mechanical forces often cannot be measured directly or diagnosed precisely, computational analyses of blood flow through the vascular system have the potential to improve medical care by identifying and quantifying hemodynamics relevant to the protection from or initiation and progression

A.P. Yoganathan (✉)
Wallace H. Coulter School of Biomedical Engineering, Georgia Institute of Technology and Emory University, 313 Ferst Drive, Atlanta, GA 30332-0535, USA
e-mail: ajit.yoganathan@bme.gatech.edu

K.B. Chandran et al. (eds.), *Image-Based Computational Modeling of the Human Circulatory and Pulmonary Systems*, DOI 10.1007/978-1-4419-7350-4_9,

of vascular disease. In addition to identifying relevant mechanics, computational analyses coupled with advanced simulation environments could be used to predict the hemodynamics associated with alternate anatomic or hemodynamic scenarios and to predict the overall performance of each scenario. Such simulation environments may then be used by clinicians to design optimized interventional treatments, and by engineers to design optimal vascular devices [4, 5].

In the past decade, advances in clinical imaging, robust numerical schemes, and improved access to computational resources have facilitated the development of computational analyses to complement clinical in vivo data. This chapter will review recent efforts in the development of combined application for planning, design, and optimization of cardiovascular treatments or devices. The exact methodologies implemented for each application are dependent on the motivation for analyzing the region of interest. Once clinical concerns are identified, clinicians and researchers work together to identify the relevant parameters and determine criterion for their optimization. Analyses may be targeted toward idealized models, such as in device design where a variety of subjects must be considered, or toward patient-specific models where optimized treatment strategies are desired. The selection of the mathematical model depends on the clinical parameters of interest. High-resolution models are required to investigate local properties such as wall stress, while lower resolution models are used to investigate global properties such as systemic flow and pressure distribution. The complexity of the model increases with the number of parameters that are used and the accuracy of the model is challenged by the need to estimate model parameters. Of particular concern is the ability to incorporate the physiologic responses to mechanical loading that would influence the optimization process or analysis validation.

This chapter includes a review of two classes of computational analysis: solid modeling of arteries and fluid modeling of blood. For the former category, we support our discussion by taking the example of endovascular treatments. The goal of these endovascular interventions is to increase the arterial lumen size to improve circulation, but treatments demonstrate varying levels of efficacy and negative remodeling consequence. Solid modeling of arteries and devices thus seeks to compare different treatment options, investigate the impact of device design on the arterial wall, and identify the most appropriate treatment strategy for a given clinical scenario. Similar approaches have been undertaken for the optimization of procedures and devices based on hemodynamic parameters. This is addressed in the second part of this chapter, focused on the application of blood flow simulations toward the planning and optimization of cardiovascular surgical procedures. The goal of surgical planning is to identify the procedure that promotes the most favorable hemodynamic outcome. Due to the size and complexity of the vascular system, both one-dimensional (1D) and three-dimensional (3D) CFD methods are used to predict the postoperative hemodynamics and reviewed here. Finally, the third section provides a global prospective on the current status and future work required for the planning and optimization of cardiovascular devices and surgeries.

9.2 Analysis for Endovascular Treatment and Device Design

9.2.1 Introduction

One consequence of atherosclerosis is the focal accumulation of atheromatous lesions that obstruct and reduce blood flow. Treatment options depend on the location and severity of the disease and include bypass grafting, endovascular balloon angioplasty, and stent placement. When applicable, endovascular techniques (i.e., balloon angioplasty and stenting) are preferred due to the benefits of reduced trauma, recovery time, mortality, and morbidity. In balloon angioplasty, a catheter with a deflated balloon on its tip is advanced to the lesion site where the balloon is inflated to expand the luminal area by deforming the atheroma and vessel wall. In certain cases, a stent is deployed with the balloon to serve as a scaffold to maintain distortion of the atheroma and prop the vessel open.

Each treatment option alters the mechanical transduction pathways of the vessel wall, which will promote remodeling response of the vessel, the plaque, or both. Rigid, calcified plaque shields vessels from fluid shear and pulsatile stress and strain, which will lead to wall remodeling and further plaque growth. Stents increase the complexity of these interactions by adding load to portions of the walls while shielding other portions. Stents are also subjected to cyclical mechanical loading that may lead to fatigue or structural failure. While it is known that device wall interactions are important, it is not possible to experimentally measure these interactions. Computational analysis techniques including finite element modeling (FEM) can be used to compute the mechanical consequence of each treatment option. Below we will look at how computational modeling was used to improve the understanding of plaque vulnerability, angioplasty and stenting procedures.

9.2.2 Identification of Vulnerable Plaque

The first decision a physician must make is whether an intervention is necessary. This decision is straightforward when a significant arterial blockage is identified, but less so when lesions are not large enough to diminish flow. In the second case, the physician must determine the stability of a plaque to assess if it is likely to become dislodged and cause a heart attack or stroke [6]. Because not all plaques are vulnerable, detailed 3D analyses of known stable and unstable plaques are needed to identify the difference in mechanical loading that may lead to rupture. In efforts to address this clinical problem, Ohayon et al. [7] used computational modeling to evaluate the residual stress strain state of vulnerable plaque caps ex vivo in order to reconstruct the residual stress strain state of these plaques in vivo. In this study, Ohayon et al. were able to demonstrate that plaque rupture is a consequence of both external loading and intraplaque stresses and strains. In another study, Ohayon et al. [8] used intravascular ultrasound (IVUS) imaging to reconstruct nonruptured in vivo plaque geometries and analyzed the impact of morphological plaque features that might be used to predict plaque stability. Further research into the detection and identification of unstable lesions could lead to the development of patient-specific treatment strategies for the prevention of acute thrombotic events.

9.2.3 Endovascular Balloon Angioplasty

Percutaneous transluminal coronary angioplasty (PTCA) to increase vessel lumen diameter was first reported by Dotter and Judkins in 1964 [9] and implemented by Gruentzig in 1977. Initially, its success was attributed to the compression and redistribution of the atheromatous plaque material. However, because atherosclerotic plaques have been found to be relatively incompressible, it is now accepted that it is the nonrecoverable deformation of the arterial wall that results in lumen diameter gains. As a consequence, the deformed arterial walls undergo remodeling to repair the vessel tissues. In some cases, this remodeling is exaggerated, resulting in restenosis or decreased vessel lumen diameter. In order to minimize the risk and maximize the benefits of angioplasty, computational models have been used to improve the understanding of the underlying mechanical processes associated with balloon angioplasty.

The most challenging aspect of modeling the diseased vessel is the specification of material properties. While it is possible to quantify plaque constituents and material properties ex vivo [10, 11], this information cannot yet be determined in vivo. Therefore, patient-specific modeling is not yet possible. Instead, researchers rely on ex vivo data collected from excised tissues. The majority of researchers use simplified, elastic, isotropic models for vessel and plaque [12]. A notable exception and most complete model of a diseased vessel to date is that developed by Holzapfel et al. [11], which consists of eight distinct layers exhibiting anisotropic, inelastic, highly nonlinear mechanical characteristics. Holzapfel used this detailed diseased vessel description to model the mechanical procedure of a balloon angioplasty. This study identifies the mechanical stresses that could lead to restenosis and plaque rupture for this particular specimen. To understand the consequence of using a simplified model, Holzapfel repeated the computation using three common simplification techniques: neglecting axial prestretch, assuming plane strain, and assuming isotropy. The simplification leads to maximum stress deviations of up to 600%. This study highlights the significance of lesion-specific modeling and the potential error with idealized models. Therefore, before modeling can be used to address patient-specific planning, new diagnostic techniques are required to noninvasively quantify lesion-specific constitutive models.

9.2.4 Endovascular Stents

The use of an endovascular stent was first suggested by Dotter and Judkins in 1964 [9] as an improvement to angioplasty. Sigwart placed the first intercoronary stent into a human in 1987 [13]. Clinical trials conducted in the 1990 s showed that stents in coronary arteries offered an improvement over angioplasty alone, including a 10% reduction in 6-month restenosis rates [14–16]. While these improvements lead to the routine clinical implantations of stents [17], the use of endovascular implants introduces negative vascular response that suggests that implants may not always be the best course of action.

The vascular response to endovascular implants includes four phases: thrombus formation, inflammation, proliferation, and arterial remodeling [2]. These phases can be related to the design of the stent in that vascular response is directly related to the position and penetration of stent struts into the vessel wall. Inflammation and proliferation responses are focused on the portions of the internal elastic membrane that are stretched by the stent struts. Vascular wall remodeling occurs in response to mechanotransduction of fluid shear and (pulsatile) blood pressure, both of which are hindered by the presence of a rigid stent. Kastrati et al. [18] correlated restenosis rates with implanted stent models and the results suggest that stent design is a contributing risk factor to restenosis. Computational modeling has been employed to investigate the mechanical interactions between the stent and arterial wall. Below is a brief review of computational models that seek to identify optimal stent designs that will maintain an effective lumen while minimizing the negative vascular response mechanisms.

Lally et al. [19] investigated the difference in the mechanical loading caused by two different commercial stents, the S7 (Medtronic AVE, Minnesota, USA) and the NIR (Boston Scientific, Massachusetts, USA), which have been clinically shown to have different incidence of restenosis. Compared to the S7, the NIR stent has larger cells and thicker struts. The material properties of the vessel and plaque were modeled using hyperelastic, isotropic constitutive equations fit to experimental data for femoral artery and calcified plaque. The analyses utilized the commercial finite element software MSC Marc/Mentat (Santa Ana, California). In the analysis, the lumen of the artery was pressurized to simulate deployment, and the stents were added to support the vessel when deployment pressure was removed. Results showed that the NIR model with thicker struts yielded more dimensional (radial) stability, which introduced higher principle stress and tissue prolapse between the struts. When the vascular response to implants is considered, it is clear that (1) the higher stress concentrations will elicit more thrombus formation, (2) prolapse coupled with high degree of dimensional stability will stretch the tissue, encouraging inflammatory response, and (3) the rigidity of the implant will shield the vessel wall from pulsatile motion, leading to enhanced remodeling. Therefore, it was found that the higher incidence of restenosis in the NIR [18] could be explained by the mechanical design of the stent.

Stents are made from a variety of metals, including stainless steel and Nitinol. The stainless-steel stents are collapsed over a balloon and later expanded by inflating the balloon. Nitinol stents are compressed inside a sheath and self-expand when the sheath is removed. Migliavacca et al. [20] compared the difference in mechanical loading of a stainless-steel and Nitinol stent of the same geometry, based on the Palmaz–Schatz design [21]. The artery and plaque were modeled as hyperelastic isotropic using literature data to fit the parameters. In addition to the material property, Migliavacca et al. also parametrically modified the metal-to-artery ratio of the stent, degree of stenosis, and the stiffness of the plaque. Analyses were performed using ABAQUS (Hibbit Karlsson & Sorenses, Inc, Pawtucket, Rhode Island, USA) by applying pressure to the lumen of the artery construct to achieve a lumen diameter of 4 mm and inserting the expanded stents. After expansion and insertion of the stent, the pressure was removed and the artery walls were allowed to recoil against the deployed stent. The wall stresses imparted by the stent in the different cases were compared. Due to the inherent flexibility of Nitinol, the Nitinol stent was found to maintain a smaller lumen diameter and impart approximately 50% less Von Mises stresses in the arterial wall than the stainless-steel counterpart. These findings suggest that softer materials, such as Nitinol, diminish wall stress concentration, tissue prolapse, and pulsatile shielding that promote thrombus formation, inflammatory response, and enhanced remodeling, respectively. These findings are supported by Sabeti et al. [22], who found a significantly reduced risk of restenosis with Nitinol stents compared to stainless steel.

Holzapfel et al. [10] developed a methodology to optimize a patient-specific stent design. A unique feature of this study is the use of a realistic arterial model with a complex, nonsymmetric plaque [11]. Using three different commercial stents as templates: (a) the Multi-Link-Tetra™ stent (Guidant), (b) the NIROYAL™ Elite stent (Boston-Scientific), and (c) the InFlow™-Gold-Flex stent (InFlow Dynamics), Holzapfel et al. investigated the consequence of modifying parameterized strut geometry and overall stent dimensions. All stents were modeled as having the material properties of stainless steel. This analysis utilized a custom solver to apply a deforming pressure load to the struts of the stents. Results show that stent geometry impacts the mechanical loading. The stent based upon the Multi-Link-Tetra™ design was shown to minimize injurious stress distributions, while the stent based upon the InFlow™-Gold-Flex design was shown to have the highest stress distributions of the three. These findings are supported by Kastrati et al. [18] with a 20% restenosis rate for the Multi-Link and 38 and 50% restenosis rates for InFlow Steel and InFlow Gold stents, respectively.

In an effort to identify an optimum stent geometry, Bedoya et al. [23] examined the mechanical consequence of modifying the strut spacing and sinusoidal shape (amplitude and radial curvature) of a generic stent cell. Analysis was performed using MSC Patran/Marc (Santa Ana, California). The artery was modeled as a symmetric cylinder with a homogeneous, nonlinear hyperelastic constitutive model with parameters fit to experimental data. In the analysis, the artery was tethered and pressurized before the stent was included in the model. Pressure was returned to diastole and the stent-induced loading was evaluated. Results show that a larger cell design minimized the stent contact with the artery, which in turn minimized the areas of stress concentration. The larger cell spacing resulted in more radial compression of the stent. These results are somewhat contradictory when compared with Lally et al. [19], where the stent with smaller cell spacing was found to minimize wall stress. The difference between the two studies can be found in the individual strut geometry. In Bedoya et al. [23], the individual strut dimensions did not differ with cell sizing, so that a stent with smaller cells was also more rigid and had a greater metal-to-artery ratio. In Lally et al. [19], the S7 had smaller cells but was also more compliant, allowing radial compression of the stent. Therefore, both studies identify radial compliance as a key factor to minimizing stress. Because different tissue models, strut dimensions, and stent material properties were used, different conclusions were drawn as to the optimal cell size. Timmins et al. [24, 25] extended the work of Bedoya et al. [23] by using an optimization algorithm to further refine the parameters (strut spacing, strut amplitude, and radius of curvature). They highlight selection criterion when using stents in tapered vessels: A stent with higher radial rigidity will provide more luminal gain with greater wall shear stress than a stent with less radial rigidity [25]. They also suggest that stent design could be optimized for the type of lesion, using designs that maximize luminal gains with stiff lesions, and designs that minimize wall shear stress for softer lesions [24].

In summary, all studies indicate that vessel wall stress is directly proportional to the final radius of the stented lumen. The best design results were indicated by a flexible, small-celled stent that more evenly distributes the minimal load required to maintain patency. Stent design is continually improving, as evident with the development of drug-eluting stents [26], compliance matching stents [27], and hybrid dynamic stents that have biodegradable components [28]. The analyses reviewed herein focused on comparing the relative stress distribution between stents, and as such idealized vessel properties were used by most of the researchers. Holzapfel et al. [11] highlight the complexity of the material properties of diseased vascular tissue, which may ultimately play a part in optimal stent design.

9.3 Patient-Specific Surgical Planning

Surgical planning remains a broad and vague term that has been used to describe a wide variety of applications, ranging from pure imaging to solid and fluid mechanics modeling. As a general rule, surgical planning describes any approach helping in the planning and decision-making process. The first occurrences of "surgical planning" in literature refer to the progress in clinical imaging, 3D anatomy reconstruction, and visualization. Vannier and Marsh report the advantage of 3D reconstruction from CT images for the evaluation and planning of craniofacial surgeries as early as 1984 [29].

With rapid advances in high-speed computation, the task of assembling and visualizing clinical data has been greatly facilitated, creating new opportunities for real-time, interactive computer applications during surgical procedures. During the past two decades, medical imaging methods have grown from their initial use as instantaneous snapshots of human anatomy to applied computer vision and graphic techniques for planning and analyzing surgical procedures [30]. Image guidance is now routinely applied in neurosurgery [31], spine surgery, abdominal surgery, catheterization [32], and orthopedic surgery [33–35]. Detailed preoperative images of the lesion and surrounding organs are registered onto the patient's anatomy at the time of intervention and serve as virtual maps to guide the surgeons' gestures, allowing for minimally invasive interventions. Multidisciplinary modules, combining clinical imaging with computer-aided design, are also gaining interest for the design and surgical placement of bone or dental implants [36].

Turning toward the future, more advanced surgical planning approaches are under development that incorporate virtual reality environments coupled with mechanical and/or fluid modeling to not only reproduce the surgeon's gesture but also to predict the resulting patient's outcome. Of particular interest for our discussion are the work conducted by Taylor et al. and Yoganathan and colleagues. Taylor et al. have developed an integrated framework for the automated optimization of cardiovascular geometries [37–40], such as the optimization of aorto-iliac bypass graft design. Yoganathan and colleagues, on the other hand, have focused their attention on patient-specific repairs of congenital heart defects [41, 42] and the incorporation of virtual reality environments [41, 42].

To provide support to our discussion, we will take the example of the clinical management of patients born with single-ventricle heart defects. These patients present a challenging scenario to clinicians with complex cardiac lesions and numerous complications on the one hand, and multiple treatment options on the other hand. The complexity of the anatomy and pathology confront surgeons and cardiologists with complex decision-making processes, for which they often have to resort either to their intuition or to previous experience gained on sufficiently similar cases. Such a setting is the typical example of a clinical scenario that could benefit from the recent advances in virtual surgery simulations and outcome optimizations.

9.3.1 Single-Ventricle Heart Defects: Review of the Clinical Problem

Single-ventricle heart defect (SVHD): The incidence of children born with a SVHD, in which there is one effective pumping chamber, is about 2 per 1,000 births. In patients with such an anatomy, oxygenated and deoxygenated blood mix in the single ventricle (Fig. 9.1b).

Without surgical intervention this set of lesions is generally lethal, resulting in 95% mortality within the first month of life. The concept of a total right ventricular bypass, first introduced by Fontan and Baudet in 1971 [43], is a palliative surgical procedure aimed at separating the systemic and pulmonary circulations, thus eliminating venous blood mixing (Fig. 9.1c). The currently accepted palliation strategy culminates in the total cavopulmonary connection (TCPC), which is generally performed in three stages, progressively separating the systemic and pulmonary circulations. In the first stage, performed when pulmonary vascular resistance is relatively high, the systemic and pulmonary circulations are connected in parallel, often utilizing a shunt between the systemic and pulmonary arteries to provide pulmonary blood flow. This shunt is removed in the second stage, and the superior vena cava (SVC) is connected to pulmonary arteries (PAs), resulting in a superior

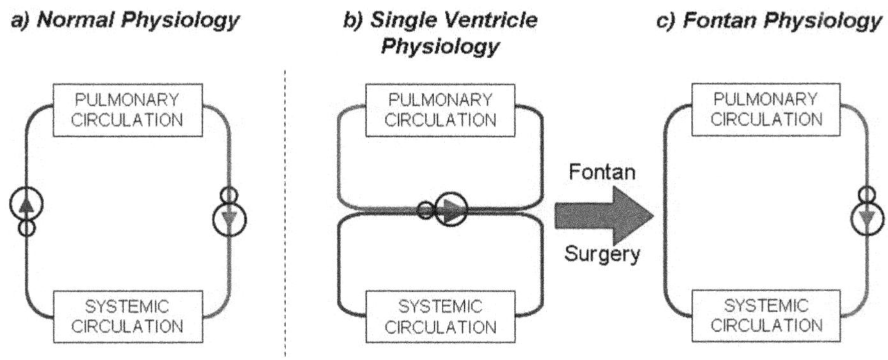

Fig. 9.1 Cartoon depicting the (**a**) normal, (**b**) single ventricle, and (**c**) Fontan physiology

cavopulmonary connection and partial right ventricular bypass. The TCPC is completed in the third stage with the connection of the inferior vena cava (IVC) to the superior cavopulmonary connection.

Long-term outcomes and problems: The modifications to the original Fontan procedure, as well as improved management and care, have steadily improved surgical outcomes, reducing postoperative mortality to the level of many biventricular congenital heart disease repairs [44, 45]. However, Fontan patients are still susceptible to numerous, long-term complications [46, 47]. The fact is that the Fontan procedure leaves the single ventricle to pump blood through both the systemic and the pulmonary vascular beds (Fig. 9.1c). As a result, the single ventricle experiences a lower preload and subsequent increase in venous pressure due to the lack of pressure step-up typically provided by a right ventricle.

Critical parameters in TCPC design: A few of the late Fontan complications, such as limited exercise performance, liver dysfunction, and pulmonary arteriovenous malformations (PAVMs), may be directly impacted by poor TCPC design and hemodynamics. Clinical studies have shown that, unlike healthy individuals, Fontan patients only respond to exercise by increasing their heart rate, while the stroke volume, which is typically limited by the preload, remains approximately constant [48, 49]. On the other hand, liver dysfunction has been correlated with the high central venous pressures observed in Fontan patients [3, 50]. While seemingly independent, both of these observations stem from the fact that, in the absence of a right pumping chamber, the pressure difference between the IVC and the left atrium is the only force left to drive blood through the lungs. Therefore, the higher the vascular resistance downstream of the liver, the higher the pressure difference required to achieve a given cardiac output. The absence of a pulmonary ventricle amplifies any increase in resistance, making the TCPC geometry and the pulmonary vascular resistance (PVR) critical to long-term outcome. Vasodilating agents and respiration exercises have been utilized successfully to lower PVR. However, in Fontan patients all blood flow has to first travel through the TCPC, such that if the effective TCPC resistance is high, improvements in PVR alone may be insufficient to improve preload and cardiac output, especially under the increased demands of exercise or other high output states (e.g., illness, pregnancy). Even when cardiac output is not measurably affected, lowering TCPC resistance might allow for reduced medication and small but important decreases in central venous pressure such that long-term outcome and quality of life are improved. Finally, it has been hypothesized that hepatic flow plays a major role in both lung development and the prevention of PAVMs [51–54]. For a given PVR, the geometry of the TCPC, in particular of the IVC anastomosis and its orientation relative to the SVC, will dictate hepatic flow distribution to the lungs.

From the above clinical follow-up studies, which demonstrate the morbidity and decreased functional status of Fontan patients, it is clear that improvements are needed. Several palliative options have been discussed in the literature [53, 55, 56], but clearly the wide variety of patient anatomies makes it difficult to design a general procedure that fits all patients. In parallel, the complexity of in vivo anatomies poses significant challenges to identify the best-suited surgical option for a given patient, i.e., the one that will best distribute hepatic flow to the lungs and offer the lowest vascular resistance, for example. This setting is the typical example of a clinical scenario that could benefit from the recent advances in virtual surgery simulations and outcome optimizations. This area has also been explored by a number of research groups and as such provides an ideal support to our discussion.

9.3.2 Comparing Global Outcome and Cardiovascular Function

Lumped-parameter models have been widely employed to investigate the function of the whole cardiovascular system. This approach exploits an analogy between fluid flow and electric circuits where pressure and flow rate would be analogous to voltage and current, and where organs are modeled by their global hemodynamic status rather than based on anatomic configuration. The behavior of a vascular tree is represented by a combination of resistors, inductors, and capacitors to represent the viscous and inertial properties of blood and the elastic wall [57–59]. The contraction of the heart is modeled using time-varying capacitors, etc.

From an optimization standpoint, lumped parameters are useful in that they provide global research directions. Going back to single-ventricle patients, for example, lumped-parameter models have been used to investigate the true impact of the TCPC geometry on the global Fontan circulation [60]. Using a lumped-parameter model developed by Lucas et al. [61], Sundareswaran et al. [60] modeled the TCPC as a pure resistor. In these simulations, patient specificity was accounted for by first matching global parameters such as cardiac output and cuff pressures and then deriving each patient's equivalent TCPC resistance from 3D CFD simulations. Results showed that (1) TCPC resistance for patients without medication was about 50% of the pulmonary vascular resistance, (2) cardiac output in Fontan patients strongly depended on vascular resistance, and (3) subjects with a lower TCPC resistance may increase their cardiac output more effectively than those with high-resistance TCPCs. This simple calculation demonstrated how higher TCPC power loss, even on the order of mW, may translate into a significantly decreased exercise capacity for single ventricle patients. Such a global approach allows for the comparison of widely different treatment approaches such as (1) the prescription of vasodilators to reduce pulmonary vascular resistance, (2) a reoperation for the optimization of the TCPC and subsequent TCPC resistance reduction, and (3) the implementation of a vascular assist device on the systemic venous or arterial side.

Although lumped-parameter models are conceptually straightforward, the parameters in the model do not necessarily have physiologic significance in that they cannot be directly measured. As such geometrical changes, among others, cannot directly be accounted for. Instead, model parameters such as resistance, impedance, or compliance must be fit to pressure and flow data. In the study by Sundareswaran et al., for example, intermediate 3D CFD simulations were conducted to characterize the patient-specific TCPC resistances instead of directly inputting the patient-specific anatomies.

9.3.3 Comparing Performances of Different Design Variations

If the comparison of widely different approaches may be better addressed with global lumped-parameter models, the comparison of the two specific surgical designs requires a detailed analysis of the underlying hemodynamics. Modeling the impact of a change in vessel geometry (as a consequence of a surgical intervention, for example) is the ideal application setting for computational fluid dynamics (CFD) which offer a high degree of freedom regarding the geometry to be simulated, the possibility to be automated, and a representation of the flow field to a level of details that may not be accessed experimentally. A detailed review of all available methods to achieve this purpose is provided in Chapter 3 of this volume. Coupled with 3D reconstruction methods for clinical images, lumped parameters, 1D or 3D CFD models constitute a useful tool set that enables fast explorative and morphological studies of patient-specific anatomies and multiple different surgical options.

The pillars of the patient-specific planning process include (1) image acquisition, (2) volume rendering of medical images, (3) shape alteration, anatomical editing, (4) image-based numerical modeling, and (5) exchange of 3D geometrical information between the patient's data, the surgeon, CFD analysis, and numerical optimization. Such a framework is illustrated in Fig. 9.2 as applied to the TCPC.

Different methodologies have been developed to address all of the above. We will address each point in order in the following sections, falling back to single-ventricle examples for clarity.

9.3.3.1 Patient Data Acquisition

Optimization of a surgical design requires first a proper representation of the patient's anatomy and then functional input parameters, such as pressure and/or velocity information, to reproduce the patient's hemodynamics. While direct pressure measurements may only be performed using invasive pressure catheters

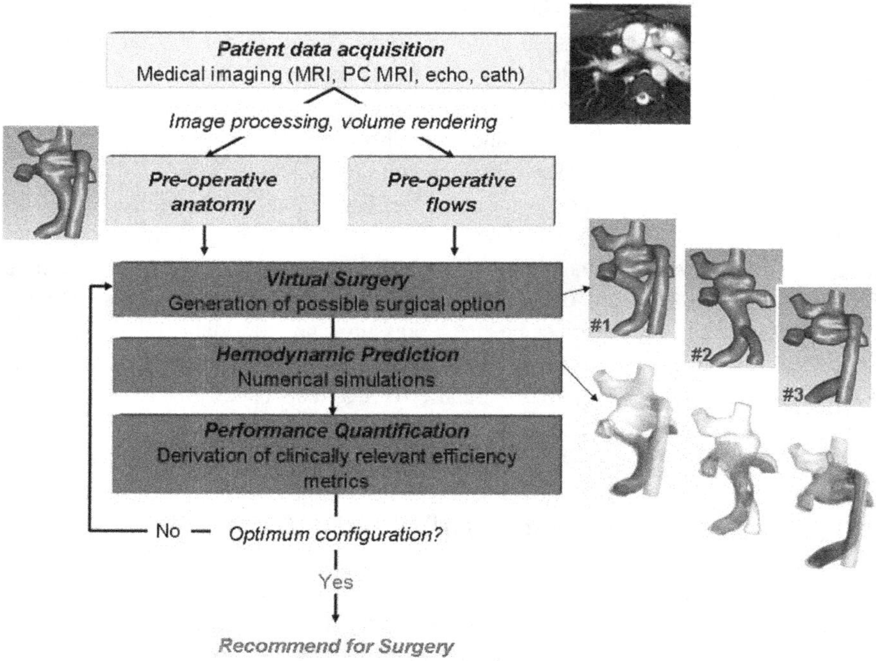

Fig. 9.2 Surgical planning overview flow chart

inserted into the vessel of interest, quite a few methods have been developed to quantify geometry and flows in vivo, with a trend toward noninvasive dynamic 3D approaches. A thorough review of these imaging modalities and associated 3D reconstructions is provided in Chapters 1 and 2. Briefly, the most widely used methods for clinical diagnosis or research purposes include ultrasound Doppler echocardiography, X-ray computed tomography (CT), and magnetic resonance imaging (MRI). Ultrasound echocardiography has the highest temporal resolution of all three modalities, which may be of interest especially for cardiovascular structures in the vicinity of the heart or highly pulsatile artery flows. However, echocardiographic imaging cannot be acquired through air or bones, which imposes a stringent limitation on the anatomic structures that may be accessed and imaged. CT and MRI scanning are far more versatile in that regard. CT scanning is typically used to look at anatomical features. Its advantages include very fast acquisition times, low cost, high spatial resolution, and the ability to image bone, soft tissue, and blood vessels all at the same time. Another significant advantage of CT imaging in the interventional and surgical fields is that, unlike MRI, it may be performed even in patients who have an implanted medical device of any kind. However, the radiation dose involved in CT scanning precludes it from being used routinely for research purposes, particularly on young subjects. On the opposite, MRI scans are harmless to the patient, allowing for frequent acquisitions if needed. Furthermore, MRI has the unique ability to characterize both geometry and 3D velocity fields, which makes it the radiographic procedure of choice for a wide range of cardiovascular modeling applications. Acquisitions may be synchronized with the cardiac (ECG-gated) and/or respiratory cycles for dynamic geometry and flow characterizations across the cardiac cycle.

9.3.3.2 Anatomy Reconstruction and Surrounding Organs Representation

With the recent progress in image interpolation and 3D rendering methods, detailed 3D geometries can be reconstructed from stacks of 2D images acquired with any of the above imaging modalities. Fast minimum or maximum intensity projection methods are now typically included in MRI and CT equipment packages. Even without using 3D volumetric acquisition sequences, these projection methods allow for the interactive generation of a projection image in any arbitrary viewing angle, drastically increasing the diagnosis tools available to the clinicians. The creation of 3D geometries for numerical modeling applications is more stringent than rendering for sole visual examination in that the final geometries should be watertight and sufficiently smooth. All methods thus typically involve segmentation of the vessels of interest in discrete 2D slices, and fitting of a smooth 3D surface.

Yoganathan et al. [62, 63] developed a three-step approach to reconstruct arbitrarily shaped cardiovascular anatomies from CMR data (Fig. 9.3). 1)Interpolation:

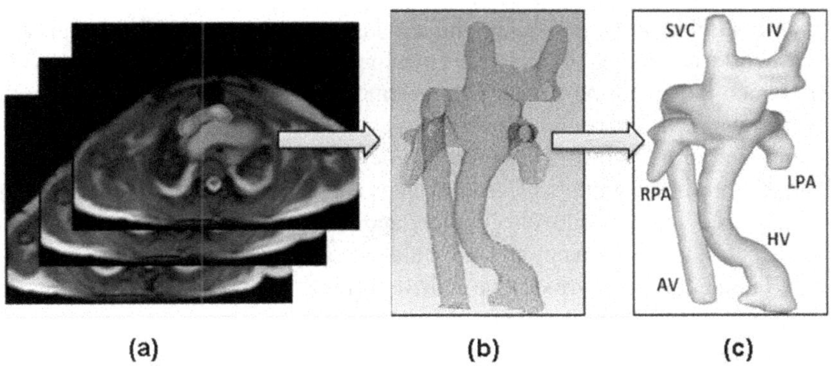

Fig. 9.3 Steps for the anatomic reconstruction of the TCPC from the stack of axial MRI. (**a**) The TCPC is first segmented in each slice. (**b**) The segmented contours are then imported into Geomagic Studio 9.0 (Geomagic Inc., NC, USA). (**c**) A smooth surface is fitted to the points to obtain a 3D anatomic reconstruction of the TCPC

First, an adaptive control grid interpolation is applied to the raw CMR data set to achieve isotropic voxel size and facilitate a smoother TCPC segmentation and 3D reconstruction [62, 63]. 2)Segmentation. After interpolation, the vessels of interest are segmented out of the enhanced MRI data set using a semiautomated bouncing ball algorithm [62]. 3) 3D Reconstruction: All segmented 2D contours are imported into Raindrop Geomagic software (Research Triangle Park, North Carolina) as a uniformly sampled set of points. The 3D geometry is finally obtained by fitting B-spline patches lofted together into a single surface.

Still focusing on cardiovascular applications, Taylor et al. developed a fast algorithm for the interactive segmentation of sufficiently smooth vessels [64]. In that approach, the user first roughly draws the vessel axis. This starting pathline is then used to generate a set of 2D projection images normal to the vessel path. Luminal contours are segmented in each of these normal planes using a level set method. Finally, all contours are lofted together with nonuniform rational B-spline surfaces. Geometries with branching vessels are rebuilt by reconstructing each branch independently and merging the reconstructed 3D volume with Boolean and smoothing operations.

Numbers of other approaches are available in literature or as commercial software packages. Variations between different approaches come from the segmentation technique used, surface fitting method, and degree of smoothing applied at each step. The quality and accuracy of the final surface is for the most part dictated by the quality of the 2D segmentation masks. While this may be an easy task for structures with high contrasts, it may actually be a critical consideration for lower contrast applications, such as cardiovascular applications. Furthermore, while some smoothing is necessary for the stability of the numerical flow solvers as well as to remove the unphysiological, jagged aspect of a stack of 2D images, a potential danger is the oversimplification of the underlying anatomy. Deviation between the reconstructed 3D surface and original images is thus a critical parameter for the evaluation of the reconstruction accuracy.

From a surgical planning standpoint the ideal scenario would be to apply the reconstruction algorithms to the entire anatomy, thus providing a virtual reality within which the operation should be performed. For the TCPC, for example, this would involve reconstruction of the heart, great vessels, lungs, spine, and ribs, with realistic material properties. However, geometric interpolation and reconstruction techniques for more comprehensive anatomies are associated with an inherent cost, which may become prohibitive to clinical application when applied to all cardiovascular, pulmonary, skeleton, and muscular structures surrounding the region of interest. In practice, only key structures are thus reconstructed. For TCPC surgeries, for example, Yoganathan et al. only consider the TCPC, heart, and great vessels. Combination of full 3D volumetric reconstruction and maximum image projection methods may allow the user to visualize all surrounding structures without actually reconstructing them (Fig. 9.4).

9.3.3.3 Modeling the Intervention

A major milestone for the planning of surgical procedures is the ability to modify the preoperative anatomy and model the different postoperative configurations. Studies limited to only a few cases have typically employed commercially available CAD tools to manually modify geometries [65, 66]. Such processes are cumbersome and depend on the skill of the operator, especially when applied to complex anatomies. Two main approaches have thus been derived to achieve anatomy modification: (1) advanced anatomy editing tools, with a special focus on easing the human–computer interaction and (2) parametric representation and modification of the anatomy.

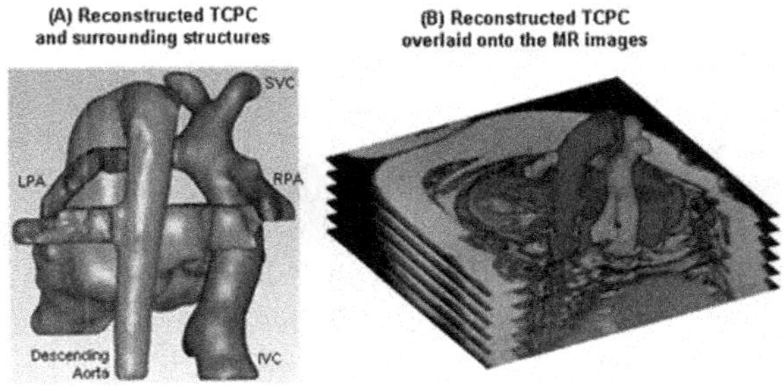

Fig. 9.4 Representation of the vessels to be operated and the surrounding anatomical constraints. (**a**) The TCPC lumen (*in red*) is reconstructed using the framework depicted in Fig. 9.3. The same procedure is applied to the heart and great vessels (*in blue*) to represent the spatial constraints imposed by the surrounding organs. (**b**) In order to avoid reconstructing the entire anatomy, the reconstructed vascular structures are registered onto the original CMR dataset, providing visual landmarks to the surgeon. Nomenclature of the main TCPC vessels: IVC, inferior vena cava; SVC, superior vena cava; RPA, right pulmonary artery; LPA, left pulmonary artery

The first category of tools seeks to closely reproduce the surgical gesture. These may be related to virtual sculpture and make use of the progress in geometrical handling, image rendering, and human–computer interactions. Pekkan and Rossignac [41] report a preliminary virtual surgery framework for cardiovascular surgeries that integrates 3D shape editing concepts and human–shape interaction technologies [41]. This framework is exemplified in Fig. 9.5 for the case of a bypass TCPC connection.

Similar frameworks have been developed for other applications. O'Leary et al. report an interface with sensitive feedback to mimic bone surgery [67]. These environments are user-intensive by design. Their realism and relevance will depend not

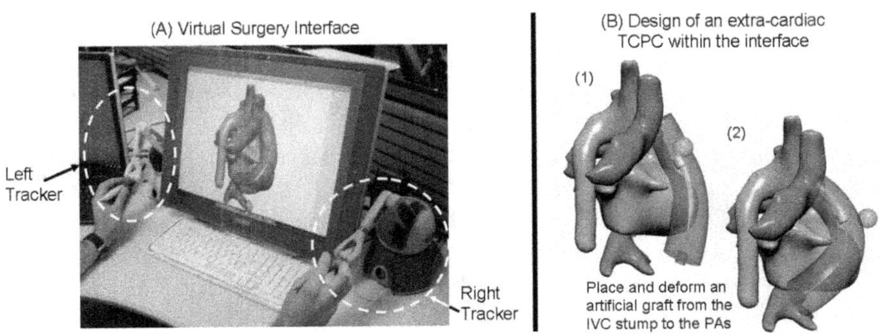

Fig. 9.5 Example of a virtual surgery environment. Patient TCPC (*shown in red*) and surrounding anatomical structures, such as the heart and great vessels (*shown in blue*), have been reconstructed from MRI and loaded into the virtual surgical interface. Using two haptic trackers, one in each hand, the user/surgeon may directly interact with the geometry, here placing and deforming an artificial graft to complete the TCPC. (**a**) Virtual surgery interface. (**b**) Design of an extracardiac TCPC within the interface

only on the extent to which the surrounding structures are represented in the interface, but also on the realism with which the surgeon may perform a procedure. Such virtual reality surgical simulators are gaining interest as training and evaluation tools for surgical trainees [68, 69].

The second category of tools includes work such as the one conducted by Mardsen et al., which does not reproduce the surgery per se but is directly focused toward automated shape optimization [38]. In this study, Mardsen et al. seek the optimal end-to-side anastomoses design to be used in bypass surgery. This study uses an idealized vessel representation, where the geometry of the end-to-side anastomosis is characterized by three different parameters: the diameter of the vessel, the diameter of the graft, and the connection angle. Surgical feasibility is accounted for by setting bounds to the parameter values. Parametric variation of the graft angle and diameter and automatic generation of the corresponding 3D geometry allow for the systematic investigation of the different surgical options.

9.3.3.4 Fast Performance Estimation and Optimization Using 1D FEA Modeling

Due to the computational expense associated with 3D FEA, researchers consider using more efficient analysis when possible. As mentioned earlier, lumped-parameter models are used to represent large vascular circuits, but require experimental data to assign component parameters. Because these parameters do not have physiologic meaning, it is not possible to represent a geometric modification or known hemodynamic variation using electric analogs alone. Therefore, the next simplified method available is the 1D FEA. Although 1D FEA cannot be used for detailed flow field evaluation, it provides very efficient estimation of global parameters such as flow rates, pressures, and derived metrics. The 1D equations are derived from Navier–Stokes using simplifying assumptions, including the assumption of an axisymmetric velocity profile across the vessel diameter, and that axially dominated flow has no in-plane velocity components [70]. In regions of flow separation, such as downstream of stenoses or near branches, minor loss coefficients must be incorporated into the 1D analysis to accurately model the pressure losses [71, 72].

This approach has been applied among others by Wan et al. to rank different treatment options for aorto-iliac occlusive disease [73]. In this study, 1D results were compared to 3D CFD [74], demonstrating a good match between the two methods but more importantly yielding the same ranking for tested treatment options. In addition, these studies underscored the computational speed of 1D FEA modeling when compared to 3D CFD. For the aorto-iliac simulations, for example, 3D CFD required days of preparation and mesh generation and approximately 25,000 cpu min of solution time, whereas the 1D solutions required about 1 h of model preparation and less than 5 cpu min of solution time [73].

More challenging to address with 1D modeling approaches are the connections with multiple inlets, featuring flow collisions and mixing. Efforts in that direction have been made by Steele et al., modeling the loss coefficients in a complex model

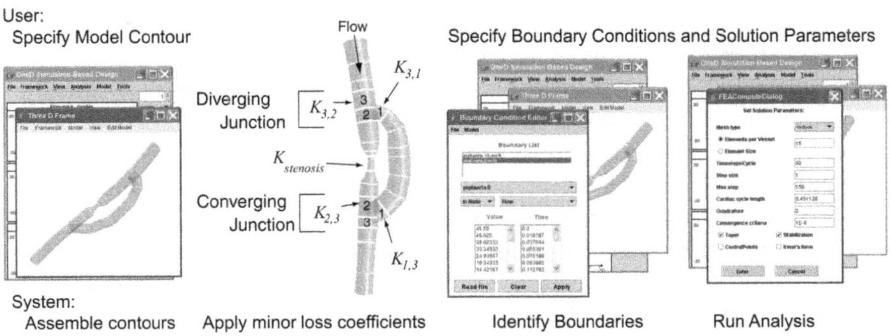

Fig. 9.6 Example of the 1D FEA environment used by Steele et al. in their study of a thoraco–thoraco bypass around a stenosis. After validation of the minor loss coefficients, 1D FEA modeling allows for the fast assessment of global hemodynamic parameters, such as pressure and flow distribution along the vessel axes

of a thoraco–thoraco bypass around a stenosis [39]. In that configuration (shown in Fig. 9.6), a bypass graft provided an alternate conduit around a stenosed region.

While the vast majority of the blood flow went through the low-resistance bypass graft, some remnant flow went through the stenosis, resulting in a diverging junction with one inlet and two outlets, and a converging junction with two inlets and one outlet. Both of these configurations were modeled as modified T-junctions. Flow distribution results were compared to 3D CFD as well as in vivo PC MRI measurements. All three methods were in close agreement, demonstrating that 1D models could adequately describe blood flow distribution in a configuration including stenosis and three-way junctions or bifurcations. TCPC geometries often contain four-way junctions, for which minor loss models have not been validated. Nevertheless, it is feasible to imagine that such minor loss coefficients may be developed, and as such, it may be possible to use 1D analyses to narrow the parameter estimation range for 3D simulations. Despite these advantages, 1D FEA cannot describe nonsymmetric properties such as wall shear stress, or the relative distribution of hepatic flow to the lungs. Applications requiring such parameter estimations require 3D CFD simulations.

9.3.3.5 Full Postoperative Hemodynamics Characterization and Optimization Using 3D CFD

3D CFD has been widely used and applied to numerous research areas including flow through large airways and lungs [75], blood flow through the carotid artery [76], coronary arteries [77], detailed aortic arch models [78], aneurysms [79], anastomosis and graft designs [37], heart development [80], and heart motion [81]. Knowledge of the whole flow field is required for the assessment of parameters such as fluid shear stresses, vorticity, turbulence, and local flow unsteadiness.

In the Fontan area, 3D CFD studies have explored the flow structure of intra-atrial tunnels [82] and extracardiac conduits [83]. The latter design was shown to have superior hemodynamics [84], which was in agreement with previous in vivo observations [85]. Parametric studies have focused on the design of the IVC anastomosis site [86], the influence of varying caval flow ratios on dissipation, flow structures, and shear stress [87], and the effect of pulmonary afterload [88]. The geometry of the TCPC has been modeled with increasing accuracy, from angular parametric models based on average anatomical measurements [82] to realistic models directly reconstructed from patient MRI data [86, 88–91].

Pioneering attempts toward patient-specific TCPC optimization using 3D CFD were performed by Yoganathan et al. for two commonly encountered scenarios among the Fontan patients (Fig. 9.7): (i) severe LPA stenosis due to aortic arch reconstruction [66] and (ii) large IVC-to-LSVC offset in dual SVC cases [65].

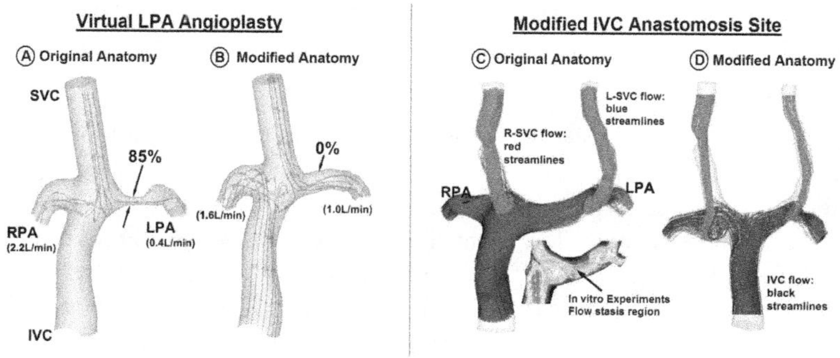

Fig. 9.7 (**a** to **b**): Virtual angioplasty of an LPA stenosis, resulting in a 61% improvement in left lung perfusion and 50% reduction in hemodynamic power loss as compared to the modified anatomy. (**c** to **d**): No flow-stasis at the PA branch in between the two SVCs after relocating the IVC

For the first case, a virtual LPA angioplasty was performed, while in the second case, the IVC was shifted toward the center of the connection, hoping for a better perfusion of the segment bounded by the two SVCs. Modified and original anatomies were run in CFD and their performances compared. For both cases, the virtual modifications brought in significant improvements in lung perfusion, cardiac output, and cardiac energy loss.

Soerensen and colleagues report the potential benefits of a bifurcated graft (in IVC, SVC, or both) to control the flow distribution to the lungs while concurrently avoiding blood mixing at the center of the connection [92]. Such a configuration is illustrated in Fig. 9.8, with an idealized IVC and bicaval bifurcations (dubbed as Optiflo by the authors). Although only investigated from a theoretical standpoint, this design showed promising performances, with energy dissipation levels 50% lower than in any other idealized configuration. The potential benefits of such bifurcated grafts over traditional nonbifurcated ones were further demonstrated in a patient-specific scenario by Mardsen et al. [93]. Similar to Soerensen et al., the authors report that bifurcating the IVC

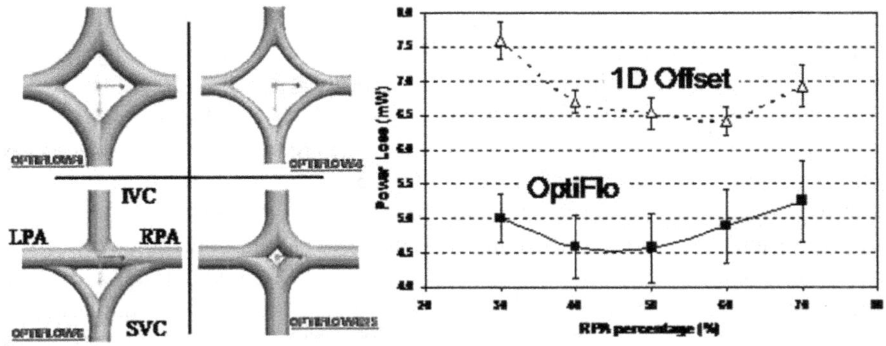

Fig. 9.8 Selected Optiflo® models [92]. These optimized configurations, featuring a bifurcated vena cava, are 41% superior to the best-known idealized model and can be potentially constructed from partial grafts allowing growth. In Optiflo-6 (*bottom left*), only the IVC split is provided, which may be more practical for surgical construction

yielded improved flow distribution to the lungs and significantly decreased energy dissipation.

The first implementation of a virtual surgical planning framework in a clinical setting is reported by Sundareswaran et al. [42], who applied to identify the best reoperation strategy for the case of a failing Fontan patient with severe pulmonary artery malformations and an arterial oxygen saturation of only 68%. This patient had a complex heterotaxy syndrome with an interrupted IVC and azygous continuation, resulting in a cardiovascular anatomy and TCPC geometry far more complex than a typical Fontan scenario. The patient underwent an MRI evaluation for 3D anatomy and flow reconstruction. Using the virtual surgery platform described in Section 9.3.3.3, multiple reoperation strategies were designed and an IVC to azygous vein conduit was retained as the best candidate to restore an equal distribution of hepatic nutrients to the lungs for that patient. The surgery was performed using this approach, and 5 months after surgery, oxygen saturations were up from 68 to 94%, demonstrating closure of the PAVMs.

9.3.3.6 Automated Optimization Methods Using 3D CFD

The case report of Sundareswaran et al. is not a real optimization per se, as only a limited number of configurations were manually designed and explored. However, only a few studies report a fully automated optimization framework for blood flow applications. Antaki et al. depict a gradient-based approach for the optimization of blood pump components under steady flow conditions [94]. Quarteroni et al. used gradient-based methods to optimize arterial bypass grafts in a simplified 2D setting [95]. The fact is that beyond the problem of an adequate parameterization of the geometry of interest, a major challenge to be overcome for the automated optimization of cardiovascular structures is the computational cost associated with the resolution of the 3D fluid structures for complex, time-dependent cardiovascular flows. The estimation of the cost function or cost function gradient may become extremely expensive as the problem complexity increases. An interesting piece of work in that regard is the integrated framework proposed by Taylor and colleagues including (1) 3D anatomical reconstruction, (2) a 3D CFD coupled with lower order models for realistic boundary conditions, and (3) a derivative-free approach for the efficient optimization of expensive functions [23]. Their derivative-free optimization uses a surrogate management framework together with mesh adaptive direct search methods. Briefly, the cost function is evaluated at a limited number of locations within the search space. A surrogate cost function is fitted onto these discrete values and used for a preliminary optimization round. Once a local minima, x', for the surrogate function is identified, the parameter space is interrogated in k evenly spaced locations around x', and the process repeated until convergence is achieved. This framework is exemplified on simple pipe geometries. Although optimization results may seem surprising from a fluid dynamic point of view, this framework is still remarkable as the first of its kind.

9.4 Including Surrounding Organs

Most of the studies mentioned in this section consider the TCPC as a stand-alone structure. While it is possible to visualize the interaction of the modeled region within a volumetric data set of the anatomy, analyses and optimizations to date disregard constraints imposed by the surrounding organs. Alterations to the anatomy will preferentially alter hemodynamics. To date, the response of surrounding organs to these hemodynamic changes has not been considered. We will address both geometric limitations and hemodynamic consequence of optimized TCPC geometries in the following sections.

9.4.1 Geometric Constraints of the Modified Configuration

Although flow modeling may suggest an optimal TCPC configuration, individual anatomy may prohibit that option in practice and the surgeon may have to compromise. Examples of limiting geometric factors include dextrocardia, abnormal pulmonary venous connections, bilateral SVC, and/or interruption of the IVC. In their recent publications, Yoganathan et al. [41, 42] account for these constraints by reconstructing the major surrounding structures (namely, the heart and great vessels) and incorporating them into the virtual surgery tool (Fig. 9.5). Anatomical constraints are thus accounted for at the design stage itself, allowing the user/surgeon to visualize the TCPC in its context and directly assess whether an option may seem feasible or not. More advanced processes could impose material properties to allow slight tissue deformations and still avoid situations where a vessel would intersect another. Within an automated optimization framework, such as the one proposed by Marsden et al. [38], the spatial constraints imposed by the surrounding organs will have to be translated into feasibility constraints imposed on the different parameters that govern the TCPC geometry. These may then be accounted for either by restricting the parameter space to be searched or by associating the nonfeasible combinations of parameters with high values for the cost function. Although simple in concept this implementation may be cumbersome in practice, especially when the geometry of the surrounding organs becomes more complex.

9.4.2 Adapting the Boundary Conditions to the Modified Configuration

A difficulty that is faced when modifying the design of a cardiovascular component is to not only predict changes in local hemodynamics, but also to predict the impact that these changes will have on the rest of the cardiovascular system. As mentioned above (Section 9.3.3), the TCPC circulation exposes both the upstream venous system and the downstream pulmonary vasculature to physiologically abnormal pressures. Normally, the valves in the right ventricle separate the venous circulation from the pulsatile driving force generated by the right ventricle to pump blood through the lung. The pulsatile nature of pulmonary arterial flow ensures low diastolic pressures (8 mmHg) and low pulmonary capillary pressure (7 mmHg). In the absence of this driving force, venous and pulmonary pressures become elevated and pulmonary flow loses its pulsatility, resulting in suboptimal performance of all organs. The goal of TCPC optimization is to mitigate this negative impact of a uni-ventricular anatomy, and to do so, boundary conditions must accurately represent the response of the surrounding vascular system.

9.4.2.1 Inlet and Outlet Boundary Conditions

The most straightforward way to specify boundary conditions is to use measured data. As described above (Section 9.3.3.1), the most accessible data are from non-invasive imaging techniques that can capture geometry as well as blood flow and velocity in the great vessels. These measured flow data are ideal for modeling the preoperative state, but pose some problems in modeling postoperative anatomies. The lumped-parameter studies of Pekkan et al. [96] and that of Sundareswaran and colleagues [60] provide a good illustration of how in vivo preoperative flow rates may be an inaccurate representation of postoperative TCPC flow rates. Indeed, geometrical alteration to the TCPC will most likely alter the associated power losses, which Pekkan et al.89] showed to directly reflect on cardiac output. This increase/decrease in cardiac output will translate in an equivalent increase/decrease in inflow to the TCPC structure, which will in turn impact the local hemodynamics and power losses. Outflow specifications of postoperative anatomies present an even greater challenge. The goal of a TCPC design is often to modify the distribution of blood flow between the lungs. In this case, preoperative outlet flows would not only be invalid, but the specification of a flow at the outlets would confound the design by forcing the solution to model the desired flow split. Ideally, the cardiovascular system would be represented as a closed loop allowing feedback mechanisms between outlets and inlets. Lumped-parameter models are the best choice for modeling a large system, offering efficiency by using a relatively small number of parameters to model a large, complex system. Lumped parameters have been developed for normal and diseased subjects, including Fontan patients [61, 97, 98]. However, these models are limited by the need to use known or desired hemodynamics to fit the model parameters and it is not clear how these parameters would be selected to represent postoperative hemodynamics in individual patients.

As an alternative to lumped-parameter boundary conditions, 1D FEA may be used to simulate surrounding structures based on individual patient geometry. Vignon-Clementel et al. used a combination of 1D FEA and lumped-parameter models to simulate surrounding structures and apply the appropriate boundary conditions on their 3D CFD model [40]. 1D FEA must also have boundary conditions, so impedance derived from lumped-parameter models or fractal-like trees have been employed [99–101]. Clipp et al. [99, 102] describe how the 1D FEA coupled with an impedance boundary condition based on parameterized fractal-like trees can be used to model the pulmonary vasculature based on a combination of patient-specific and organ-specific data (see Fig. 9.9).

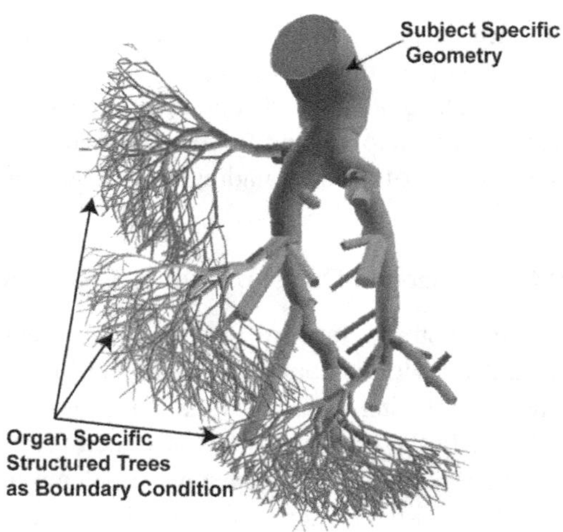

Fig. 9.9 Virtual pulmonary model in which parameterized structured trees are used to compute an impedance boundary condition for every outlet (only three are shown). Structured tree geometries can be modified to represent dynamic changes associated with respiration

This virtual pulmonary model includes the resistance vessels that may be constricted or dilated to represent response to changing inlet hemodynamics [99, 100, 102], thus providing a basis for a dynamic boundary condition. Dynamic impedance boundary conditions have been presented by Vignon-Clementel et al. [40], with time-varying lumped-parameter models, and by Clipp and Steele [99] using the time-varying geometry model. Lumped-parameter boundary conditions require experimental data to specify parameter values, whereas geometry can be estimated from vascular anatomy, making the geometry-based method easier to specify in the absence of hemodynamic data. To improve the usability of the geometry-based dynamic model, it may be possible to use the geometry-based impedance to specify lumped-parameter values.

9.4.2.2 Material Properties

Other boundary condition problems involve accounting for wall motion in cases where this motion is sufficient to alter the ranking of the different treatment options. Characterization of the vessel wall properties constitutes a research area by itself and is typically performed in vitro using axial or biaxial testing devices. However, the inclusion of these models in the computations is still sparse. An additional difficulty that is faced when modeling diseased vessels is the nonhomogeneity of the vessel wall properties. In the Fontan area, Orlando et al. looked into the impact of wall compliance on TCPC hemodynamics [103]. Using an FSI formulation, they modeled the TCPC vessel walls as uniformly elastic. Although interesting from a technical point of view, the clinical relevance of such an approach is questionable as arteries, veins, and artificial grafts feature drastically different wall properties and the numerous suture lines running through the connection locally constrain wall deformation.

A nice review of all the different factors that come into play when modeling vessel–blood interactions is provided by Vigmond et al. [104]. Their paper looks into the effect of bundle branch blockage on heart contractility and associated blood motion. Although applied to an extreme case of wall motion, this paper underscores the tight relationship that exists between blood flow patterns and vascular dynamics, mediated by mechanical sensors at the cell level. A fully coupled model involving all temporal and length scales from the chemical processes to the global 3D hemodynamics would be prohibitively expensive, but as lumped-parameter models were developed to model the bulk hemodynamic behavior of certain organs, there is a need for system-level models of the wall biology to be able to truly predict wall motion in surgical planning scenarios.

9.5 Future Direction for Biomedical Simulations

Simulation-based optimization techniques have long been used in traditional engineering such as automotive or aerospatial design. Their application to clinical problems is an emerging field which holds great promises for both medical device design and patient treatment planning. The development of comprehensive models could cut down the cost and time required for the design and testing of new medical devices by reducing the number of prototypes evaluated during in vitro and in vivo test phases. For the clinical community, these models could allow physicians to test out multiple treatment options and decide on the optimal approach for each patient. For cardiovascular diseases in particular, the knowledge of the effect of hemodynamics on vascular adaptation and cardiac function, and the determination of the optimal hemodynamic condition for an individual patient, would allow for the establishment of optimal surgical procedure and clinical care on a patient-specific basis, ultimately resulting in improved patient outcome and reduced clinical costs.

The examples provided in this chapter served as a basis to review the current state of the art in cardiovascular modeling and predictive simulations. Methods developed to compare the multiple options available to treat vascular stenosis and aneurysms involve detailed mechanical models of healthy and diseased vessel wall, simulations of different treatment options (dilating balloon, stents, bypass grafts), predictions of their mechanical impact on the vessel structure, simulations of the associated hemodynamics, and computation of the resulting wall stresses. Similarly, simulation-based virtual surgery environments provide means to reproduce a surgical procedure and predict the associated hemodynamics. Automated optimization frameworks are currently underway that will determine the optimal surgical design for an individual patient with minimal surgeon/user interaction.

However, what is also important to note from this chapter is that, while predictive simulations and virtual environment are gaining interest and popularity, they should still be taken with a pinch of salt. Past and ongoing efforts have established robust fluid dynamic solvers and fluid–structure interaction algorithms, models for blood rheology, inflow/outflow boundary conditions, vessel wall mechanics, and cell mechanics. But each of these models carries an inherent set of simplifications and assumptions, and their coupling into a single united framework is still limited. Some of the critical points to consider in regard to the patient-specific anatomic models include the assessment of the spatial and temporal accuracy of clinical imaging modalities to capture in vivo anatomies and flows, their suitability to serve as CFD boundary conditions, and an accuracy evaluation of the segmentation and reconstruction methods used to extract anatomic and flow data from these in vivo images. Simulations then require the critical evaluation of each modeling assumption, weighing their computational cost against the accuracy gain. Typical examples include the validity with which static models represent the dynamic cardiovascular environment, or whether blood may be modeled as a Newtonian fluid instead of using detailed rheological models. The creation of virtual posttreatment configurations is even more challenging in that models should not only reproduce the geometrical deformation imparted to a vessel by a device or surgery, but also the impact of the procedure on the entire physiology. This relates to the boundary condition issue raised in Section 9.4.2. Different TCPC designs will yield different resistances, and in turn different cardiac outputs. Similarly, the design of a TCPC will directly impact blood flow distribution to the lungs, such that preoperative inflow/outflow conditions may not be representative of the postoperative physiologic state. The extent of the vasculature and degree of details to which it should be modeled rise as a natural extension of these boundary condition questions. Taylor et al., Steele et al., and Wan et al. typically extend their simulation domains to the capillary beds downstream of the region of interest, imposing the resulting vascular resistances and impedance as outflow boundary conditions [40, 101, 105]. Sundareswaran et al. sought to model the adaptation of the ventricular output power to different TCPC resistances by means of lumped-parameter modeling [60]. Similar couplings have been established between fluid flow solvers and vascular wall mechanics. Vigmond and Peskin coupled fluidics and myocardial electromechanics to predict the effect of bundle blockage on ventricular dynamics [104]. However, most of these couplings are still unidirectional, and closed-loop fully-coupled multiscale simulations still represent a significant challenge. A final consideration is the simulation time scale. Most simulations to date report hemodynamic results over a few cardiac cycles, while cardiovascular adaptation or remodeling processes typically occur over several months or years. Evolutionary models are thus needed for the prediction of long-term outcome based on hemodynamic changes, which will also require a shift in the temporal length scales used in the simulations. A first step in that direction was taken by Figueroa et al., who implemented a global model for the response and remodeling of an arterial wall to mechanical stimuli and applied it to the formation of a fusiform aneurism in basilar arteries [106]. Similar models will have to be developed for all aspects of the cardiovascular systems, which will require a tight interaction with the biological and clinical community for their establishment and validation.

Beyond modeling challenges, the time required to conduct a given treatment optimization is a major issue for the deployment of numerical platforms in the clinical community. Clearly, for such platforms to represent an appealing and economically viable option, simulations should run in short time frames compromising between the competing demands for accuracy and efficiency. Improvement in hardware speed as well as the optimization and parallelization of the numerical flow solvers are many available options to reduce computation times. Of particular interest is the work by Steele et al., who developed an Internet-based approach to run simulations in parallel [39]. Generalization of this approach using multiple interconnected computers on internal or external networks could allow for massively parallel simulations even in clinical settings where large computer clusters may not be available.

While it is apparent that the application of numerical simulations to predict clinical outcome and rank treatment options still has quite a few challenges to face, we would like to conclude this chapter on an optimistic note. The authors intimately believe that simulation-based treatment planning will enhance clinical care and patient outcome. The set of tools described herein provide a means to explore human physiology to a level of temporal or spatial details not achievable in vivo. The development of mechanistic and evolutionary models and their validation on large patient populations will further increase our understanding of different disease mechanisms. Finally, predictive simulations present a unique setting to evaluate and compare a wide range of treatment options, which is simply not possible without them.

Acknowledgments The development of the surgical planning interface for Fontan patients developed by Yoganathan and colleagues was in majority supported by the NIH/NHLBI grant HL67622. This work was conducted in close collaboration between the authors and Dr. Jarek Rossignac, from the College of Computing at the Georgia Institute of Technology, for the development of the interactive surgical planning environment, and Dr. Fotis Sotiropoulos, from the University of Minnesota, for the establishment of a robust numerical flow solver for hemodynamic applications. Finally, none of these developments would have been possible without a close interaction with our clinical collaborators, namely, Dr. Mark Fogel from the Children's Hospital of Philadelphia,

Dr. Shiva Sharma from Pediatric Cardiology Services in Atlanta, Dr. Kirk Kanter from the Children's Healthcare of Atlanta, and finally Dr. Pedro del Nido from the Children's Hospital of Boston.

References

1. Chatzizisis YS, Coskun AU, Jonas M, Edelman ER, Feldman CL, Stone PH (2007) Role of endothelial shear stress in the natural history of coronary atherosclerosis and vascular remodeling: molecular, cellular, and vascular behavior. J Am Coll Cardiol 49(25):2379–2393
2. Edelman ER, Rogers C (1998) Pathobiologic responses to stenting. Am J Cardiol 81(7 A): 4–6
3. Procelewska M, Kolcz J, Januszewska K, Mroczek T, Malec E (2007) Coagulation abnormalities and liver function after hemi-Fontan and Fontan procedures—the importance of hemodynamics in the early postoperative period. Eur J Cardiothorac Surg 31(5):866–872
4. Basdogan C, De S, Kim J, Muniyandi M, Kim H, Srinivasan MA (2004) Haptics in minimally invasive surgical simulation and training. IEEE Comput Graph Appl 24(2):56–64
5. Soler L, Marescaux J (2008) Patient-specific surgical simulation. World J Surg 32(2): 208–212
6. Fuster V, Moreno PR, Fayad ZA, Corti R, Badimon JJ (2005) Atherothrombosis and high-risk plaque Part I: evolving concepts. J Am Coll Cardiol 46(6):937–954
7. Ohayon J, Dubreuil O, Tracqui P, Le Floc'h S, Rioufol G, Chalabreysse L, Thivolet F, Pettigrew RI, Finet G (2007) Influence of residual stress/strain on the biomechanical stability of vulnerable coronary plaques: potential impact for evaluating the risk of plaque rupture. Am J Physiol Heart Circ Physiol 293(3):H1987
8. Ohayon J, Finet G, Gharib AM, Herzka DA, Tracqui P, Heroux J, Rioufol G, Kotys MS, Elagha A, Pettigrew RI (2008) Necrotic core thickness and positive arterial remodeling index: emergent biomechanical factors for evaluating the risk of plaque rupture. Am J Physiol Heart Circ Physiol 295(2):H717–27
9. Dotter CT, Judkins MP (1964) Transluminal treatment of arteriosclerotic obstruction description of a new technic and a preliminary report of its application. Circulation 30(5):654–670
10. Holzapfel GA, Stadler M, Gasser TC, Press I, Ogden RW (2005) Towards a computational methodology for optimizing angioplasty treatments with stenting. In: Anonymous (ed) Mechanics of biological tissue. Springer, Heidelberg, pp 207–220
11. Holzapfel GA, Stadler M, Schulze-Bauer CAJ (2002) A layer-specific three-dimensional model for the simulation of balloon angioplasty using magnetic resonance imaging and mechanical testing. Ann Biomed Eng 30(6):753–767
12. Oh S, Kleinberger M, McElhaney JH (1994) Finite-element analysis of balloon angioplasty. Med Biol Eng Comput 32:108–114
13. Sigwart U, Puel J, Mirkovitch V, Joffre F, Kappenberger L (1987) Intravascular stents to prevent occlusion and restenosis after transluminal angioplasty. N Engl J Med 316(12): 701–706
14. Serruys PW, de Jaegere P, Kiemeneij F, Macaya C, Rutsch W, Heyndrickx G, Emanuelsson H, Marco J, Legrand V, Materne P (1994) A comparison of balloon-expandable-stent implantation with balloon angioplasty in patients with coronary artery disease. N Engl J Med 331(8):489–495
15. Fischman DL, Leon MB, Baim DS, Schatz RA, Savage MP, Penn I, Detre K, Veltri L, Ricci D, Nobuyoshi M (1994) A randomized comparison of coronary-stent placement and balloon angioplasty in the treatment of coronary artery disease. N Engl J Med 331(8):496–501
16. Versaci F, Gaspardone A, Tomai F, Crea F, Chiariello L, Gioffre PA (1997) A comparison of coronary-artery stenting with angioplasty for isolated stenosis of the proximal left anterior descending coronary artery. N Engl J Med 336(12):817–822

17. Dangas G, Fuster V (1996) Management of restenosis after coronary intervention. Am Heart J 132(2):428
18. Kastrati A, Mehilli J, Dirschinger J, Pache J, Ulm K, Schuhlen H, Seyfarth M, Schmitt C, Blasini R, Neumann FJ (2001) Restenosis after coronary placement of various stent types. Am J Cardiol 87(1):34–39
19. Lally C, Dolan F, Prendergast PJ (2005) Cardiovascular stent design and vessel stresses: a finite element analysis. J Biomech 38(8):1574–1581
20. Migliavacca F, Petrini L, Massarotti P, Schievano S, Auricchio F, Dubini G (2004) Stainless and shape memory alloy coronary stents: a computational study on the interaction with the vascular wall. Biomech Model Mechanobiol 2(4):205–217
21. Migliavacca F, Petrini L, Colombo M, Auricchio F, Pietrabissa R (2002) Mechanical behavior of coronary stents investigated through the finite element method. J Biomech 35(6):803–811
22. Sabeti S, Schillinger M, Amighi J, Sherif C, Mlekusch W, Ahmadi R, Minar E (2004) Primary patency of femoropopliteal arteries treated with nitinol versus stainless steel self-expanding stents: propensity score-adjusted analysis. Radiology 232(2):516–521
23. Bedoya J, Meyer CA, Timmins LH, Moreno MR, Moore JE Jr (2006) Effects of stent design parameters on normal artery wall mechanics. J Biomech Eng 128:757
24. Timmins LH, Moreno MR, Meyer CA, Criscione JC, Rachev A, Moore JE (2007) Stented artery biomechanics and device design optimization. Med Biol Eng Comput 45(5):505–513
25. Timmins LH, Meyer CA, Moreno MR, Moore JE (2008) Mechanical modeling of stents deployed in tapered arteries. Ann Biomed Eng 36(12):2042–2050
26. Sousa JE, Serruys PW, Costa MA (2003) New frontiers in cardiology drug-eluting stents: part I. Circulation 107(17):2274–2279
27. Berry JL, Manoach E, Mekkaoui C, Rolland PH, Moore JE, Rachev A (2002) Hemodynamics and wall mechanics of a compliance matching stent: in vitro and in vivo analysis. J Vasc Interv Radiol 13(1):97–105
28. Moreno MR, Bedoya J, Meyer C, Moore Jr., J.E., Holzapfel GA, Ogden RW (2005) Computational modeling of stented arteries: considerations for evolving stent designs. In: Anonymous (ed) Mechanics of biological tissue. Springer, Berlin, Heidelberg, pp 1283–1288
29. Vannier MW, Marsh JL, Warren JO (1984) Three dimensional CT reconstruction images for craniofacial surgical planning and evaluation. Radiology 150(1):179–184
30. Shahidi R, Tombropoulos R, Grzeszczuk RP (1998) Clinical applications of three-dimensional rendering of medical data sets. Proc IEEE 86(3):555–568
31. Foroglou N, Zamani A, Black P (2009) Intra-operative MRI (iop-MR) for brain tumour surgery. Br J Neurosurg 23(1):14–22
32. Razavi R, Hill DL, Keevil SF, Miquel ME, Muthurangu V, Hegde S, Rhode K, Barnett M, van Vaals J, Hawkes DJ, Baker E (2003) Cardiac catheterisation guided by MRI in children and adults with congenital heart disease. Lancet 362(9399):1877–1882
33. Liebergall M, Ben-David D, Weil Y, Peyser A, Mosheiff R (2006) Computerized navigation for the internal fixation of femoral neck fractures. J Bone Joint Surg Am 88(8):1748–1754
34. Dutton AQ, Yeo SJ, Yang KY, Lo NN, Chia KU, Chong HC (2008) Computer-assisted minimally invasive total knee arthroplasty compared with standard total knee arthroplasty. A prospective, randomized study. J Bone Joint Surg Am 90(1):2–9
35. Gray ML, Pearle AD (2009) Advanced imaging and computer-assisted surgery of the knee and hip. Introduction. J Bone Joint Surg Am 91(Suppl 1):1–2
36. Xiaojun C, Ming Y, Yanping L, Yiqun W, Chengtao W (2009) Image guided oral implantology and its application in the placement of zygoma implants. Comput Methods Programs Biomed 93(2):162–173
37. Ku JP, Draney MT, Arko FR, Lee WA, Chan FP, Pelc NJ, Zarins CK, Taylor CA (2002) In vivo validation of numerical prediction of blood flow in arterial bypass grafts. Ann Biomed Eng 30(6):743–752

38. Marsden AL, Feinstein JA, Taylor CA (2008) A computational framework for derivative-free optimization of cardiovascular geometries. Comput Methods Appl Mech Eng 197(21–24):1890–1905
39. Steele BN, Draney MT, Ku JP, Taylor CA (2003) Internet-based system for simulation-based medical planning for cardiovascular disease. IEEE Trans Inf Technol Biomed 7(2):123–129
40. Vignon-Clementel IE, Alberto Figueroa C, Jansen KE, Taylor CA (2006) Outflow boundary conditions for three-dimensional finite element modeling of blood flow and pressure in arteries. Comput Methods Appl Mech Eng 195(29–32):3776–3796
41. Pekkan K, Whited B, Kanter K, Sharma S, de Zelicourt D, Sundareswaran K, Frakes D, Rossignac J, Yoganathan AP (2008) Patient-specific surgical planning and hemodynamic computational fluid dynamics optimization through free-form haptic anatomy editing tool (SURGEM). Med Biol Eng Comput 46(11):1139–1152
42. Sundareswaran KS, de Zelicourt D, Sharma S, Kanter KR, Spray TL, Rossignac J, Sotiropoulos F, Fogel MA, Yoganathan AP (August 2009) Correction of pulmonary arteriovenous malformation using image-based surgical planning. JACC Cardiovasc Imaging 2(8):1024–1030
43. Fontan F, Baudet E (1971) Surgical repair of tricuspid atresia. Thorax 26:240–248
44. Gaynor JW, Bridges ND, Cohen MI, Mahle WT, Decampli WM, Steven JM, Nicolson SC, Spray TL (2002) Predictors of outcome after the Fontan operation: is hypoplastic left heart syndrome still a risk factor? J Thorac Cardiovasc Surg 123(2):237–245
45. Mair DD, Puga FJ, Danielson GK (2001) The Fontan procedure for tricuspid atresia: early and late results of a 25-year experience with 216 patients. J Am Coll Cardiol 37(3):933–939
46. Driscoll DJ, Feldt RH, Mottram CD, Puga FJ, Schaff HV, Danielson GK (1987) Cardiorespiratory response to exercise after definitive repair of univentricular atrioventricular connection. Int J Cardiol 17(1):73–81
47. Durongpisitkul K, Driscoll DJ, Mahoney DW, Wollan PC, Mottram CD, Puga FJ, Danielson GK (1997) Cardiorespiratory response to exercise after modified Fontan operation: determinants of performance. J Am Coll Cardiol 29(4):785–790
48. Senzaki H, Masutani S, Kobayashi J, Kobayashi T, Sasaki N, Asano H, Kyo S, Yokote Y, Ishizawa A (2002) Ventricular afterload and ventricular work in fontan circulation: comparison with normal two-ventricle circulation and single-ventricle circulation with Blalock-Taussig shunts. Circulation 105(24):2885–2892
49. Senzaki H, Masutani S, Ishido H, Taketazu M, Kobayashi T, Sasaki N, Asano H, Katogi T, Kyo S, Yokote Y (2006) Cardiac rest and reserve function in patients with Fontan circulation. J Am Coll Cardiol 47(12):2528–2535
50. Kiesewetter CH, Sheron N, Vettukattill JJ, Hacking N, Stedman B, Millward-Sadler H, Haw M, Cope R, Salmon AP, Sivaprakasam MC, Kendall T, Keeton BR, Iredale JP, Veldtman GR (2007) Hepatic changes in the failing Fontan circulation. Heart 93(5):579–584
51. Duncan BW, Desai S (2003) Pulmonary arteriovenous malformations after cavopulmonary anastomosis. Ann Thorac Surg 76(5):1759–1766
52. Pandurangi UM, Shah MJ, Murali R, Cherian KM (1999) Rapid onset of pulmonary arteriovenous malformations after cavopulmonary anastomosis. Ann Thorac Surg 68(1):237–239
53. Pike NA, Vricella LA, Feinstein JA, Black MD, Reitz BA (2004) Regression of severe pulmonary arteriovenous malformations after Fontan revision and "hepatic factor" rerouting. Ann Thorac Surg 78(2):697–699
54. Shinohara T, Yokoyama T (2001) Pulmonary arteriovenous malformation in patients with total cavopulmonary shunt: what role does lack of hepatic venous blood flow to the lungs play? Pediatr Cardiol 22(4):343–346
55. AboulHosn J, Danon S, Levi D, Castellon Y, Child J, Moore J (2007) Regression of pulmonary arteriovenous malformations after transcatheter reconnection of the pulmonary arteries in patients with unidirectional Fontan. Congenit Heart Dis 2(3):179–184
56. Stamm C, Friehs I, Duebener LF, Zurakowski D, Mayer JE, Jr., Jonas RA, del Nido PJ (2002) Improving results of the modified Fontan operation in patients with heterotaxy syndrome. Ann Thorac Surg 74(6):1967–1977; discussion 1978

57. De Pater L, van den Berg J (1964) An electrical analogue of the entire human circulatory system. Med Electron Biol Eng 2:161–166
58. Westerhof N, Bosman F, De Vries CJ, Noordergraaf A (1969) Analog studies of the human systemic arterial tree. J Biomech 2(2):121–143
59. Westerhof N, Noordergraaf A (1970) Errors in the measurement of hydraulic input impedance. J Biomech 3(3):351–356
60. Sundareswaran KS, Pekkan K, Dasi LP, Whitehead K, Sharma S, Kanter KR, Fogel MA, Yoganathan AP (2008) The total cavopulmonary connection resistance: a significant impact on single ventricle hemodynamics at rest and exercise. Am J Physiol Heart Circ Physiol 295(6):H2427–H2435
61. Lucas C, Ketner ME, Steele BN, Mill M, Sheridan B, Lucas W, Pekkan K, Yoganathan AP (2006) Importance of respiration and graft compliance in Fontan circulations: experimental and computational studies. Journal of Biomechanics 39(Suppl 1):S207
62. Frakes DH, Smith MJ, Parks J, Sharma S, Fogel SM, Yoganathan AP (2005) New techniques for the reconstruction of complex vascular anatomies from MRI images. J Cardiovasc Magn Reson 7(2):425–432
63. Frakes DH, Conrad CP, Healy TM, Monaco JW, Fogel M, Sharma S, Smith MJ, Yoganathan AP (2003) Application of an adaptive control grid interpolation technique to morphological vascular reconstruction. IEEE Trans Biomed Eng 50(2):197–206
64. Wang K, Dutton R, Taylor C (1999) Improving geometric model construction for blood flow modeling. IEEE Eng Med Biol Mag 18(6):33–39
65. de Zelicourt DA, Pekkan K, Parks J, Kanter K, Fogel M, Yoganathan AP (2006) Flow study of an extracardiac connection with persistent left superior vena cava. J Thorac Cardiovasc Surg 131(4):785–791
66. Pekkan K, Kitajima HD, de Zelicourt D, Forbess JM, Parks WJ, Fogel MA, Sharma S, Kanter KR, Frakes D, Yoganathan AP (2005) Total cavopulmonary connection flow with functional left pulmonary artery stenosis: angioplasty and fenestration in vitro. Circulation 112(21):3264–3271
67. O'Leary SJ, Hutchins MA, Stevenson DR, Gunn C, Krumpholz A, Kennedy G, Tykocinski M, Dahm M, Pyman B (2008) Validation of a networked virtual reality simulation of temporal bone surgery. The Laryngoscope 118(6):1040–1046
68. Rosen JM, Long SA, McGrath DM, Greer SE (2009) Simulation in plastic surgery training and education: the path forward. Plast Reconstr Surg 123(2):729–738; discussion 739–740
69. Hart R, Karthigasu K (2007) The benefits of virtual reality simulator training for laparoscopic surgery. Curr Opin Obstet Gynecol 19(4):297–302
70. Hughes TJR, Lubliner J (1973) On the one-dimensional theory of blood flow in the larger vessels. Math Biosci 18(1–2):161–170
71. Gardel A (1957) Les pertes de charge dans les écoulements au travers de branchments en té (Pressure drops in flows through T-shaped pipe-fittings). Bull Tech Suisse Romande 10: 143–148
72. Gardel A (1957) Les pertes de charge dans les écoulements au traverse de branchments en té (Pressure drops in flows through T-shaped pipe-fittings). Bulletin Technique de la Suisse Romande 9:123–130
73. Wan J, Steele BN, Spicer SA, Strohband S, Feijoo GR, Hughes TJR, Taylor CA (2002) A one-dimensional finite element method for simulation-based medical planning for cardiovascular disease. Comput Methods Biomech Biomed Eng 5(3):195–206
74. Taylor CA, Draney MT, Ku JP, Parker D, Steele BN, Wang K, Zarins CK (1999) Predictive medicine: computational techniques in therapeutic decision-making. Comput Aided Surg 4(5):231–247
75. Nowak N, Kakade PP, Annapragada AV (2003) Computational fluid dynamics simulation of airflow and aerosol deposition in human lungs. Ann Biomed Eng 31(4):374–390
76. Cebral JR, Yim PJ, Lohner R, Soto O, Choyke PL (2002) Blood flow modeling in carotid arteries with computational fluid dynamics and MR imaging. Acad Radiol 9(11):1286–1299

77. Johnston BM, Johnston PR, Corney S, Kilpatrick D (2004) Non-Newtonian blood flow in human right coronary arteries: steady state simulations. J Biomech 37(5):709–720
78. Shahcheraghi N, Dwyer HA, Cheer AY, Barakat AI, Rutaganira T (2002) Unsteady and three-dimensional simulation of blood flow in the human aortic arch. J Biomech Eng 124(4):378–387
79. Chatziprodromou I, Butty V, Makhijani V, Poulikakos D, Ventikos Y Pulsatile (2003) Blood flow in anatomically accurate vessels with multiple aneurysms: a medical intervention planning application of computational haemodynamics. Flow, Turbulence Combustion 71(1):333–346
80. DeGroff CG, Thornburg BL, Pentecost JO, Thornburg KL, Gharib M, Sahn DJ, Baptista A (2003) Flow in the early embryonic human heart: a numerical study. Pediatr Cardiol 24(4):375–380
81. Saber NR, Gosman AD, Wood NB, Kilner PJ, Charrier CL, Firmin DN (2001) Computational flow modeling of the left ventricle based on in vivo MRI data: initial experience. Ann Biomed Eng 29(4):275–283
82. de Leval MR, Dubini G, Migliavacca F, Jalali H, Camporini G, Redington A, Pietrabissa R (1996) Use of computational fluid dynamics in the design of surgical procedures: application to the study of competitive flows in cavopulmonary connections. J Thorac Cardiovasc Surg 111(3):502–513
83. Migliavacca F, de Leval MR, Dubini G, Pietrabissa R, Fumero R (1999) Computational fluid dynamic simulations of cavopulmonary connections with an extracardiac lateral conduit. Med Eng Phys 21(3):187–193
84. Hsia TY, Migliavacca F, Pittaccio S, Radaelli A, Dubini G, Pennati G, de Leval M (2004) Computational fluid dynamic study of flow optimization in realistic models of the total cavopulmonary connections. J Surg Res 116(2):305–313
85. Lardo AC, Webber SA, Friehs I, del Nido PJ, Cape EG (1999) Fluid dynamic comparison of intra-atrial and extracardiac total cavopulmonary connections. J Thorac Cardiovasc Surg 117(4):697–704
86. Migliavacca F, Dubini G, Bove EL, de Leval MR (2003) Computational fluid dynamics simulations in realistic 3-D geometries of the total cavopulmonary anastomosis: the influence of the inferior caval anastomosis. J Biomech Eng 125(6):805–813
87. Khunatorn Y, Shandas R, DeGroff C, Mahalingam S (2003) Comparison of in vitro velocity measurements in a scaled total cavopulmonary connection with computational predictions. Ann Biomed Eng 31(7):810–822
88. Guadagni G, Bove EL, Migliavacca F, Dubini G (2001) Effects of pulmonary afterload on the hemodynamics after the hemi-Fontan procedure. Med Eng Phys 23(5):293–298
89. Pekkan K, Dasi LP, de Zelicourt D, Sundareswaran KS, Fogel MA, Kanter KR, Yoganathan AP (2009) Hemodynamic performance of stage-2 univentricular reconstruction: Glenn vs. hemi-Fontan templates. Ann Biomed Eng 37(1):50–63
90. Pekkan K, de Zelicourt D, Ge L, Sotiropoulos F, Frakes D, Fogel MA, Yoganathan AP (2005) Physics-driven CFD modeling of complex anatomical cardiovascular flows—a TCPC case study. Ann Biomed Eng 33(3):284–300
91. Wang C, Pekkan K, de Zelicourt D, Horner M, Parihar A, Kulkarni A, Yoganathan AP (2007) Progress in the CFD modeling of flow instabilities in anatomical total cavopulmonary connections. Ann Biomed Eng 35(11):1840–1856
92. Soerensen DD, Pekkan K, de Zélicourt D, Sharma S, Kanter K, Fogel M, Yoganathan AP (2007) Introduction of a new optimized total cavopulmonary connection. Ann Thorac Surg 83(6):2182–2190
93. Marsden AL, Bernstein AJ, Reddy VM, Shadden SC, Spilker RL, Chan FP, Taylor CA, Feinstein JA (2009) Evaluation of a novel Y-shaped extracardiac Fontan baffle using computational fluid dynamics. J Thorac Cardiovasc Surg 137(2):394–403 e2
94. Antaki JF, Ghattas O, Burgreen GW, He B (1995) Computational flow optimization of rotary blood pump components. Artif Organs 19(7):608–615

95. Quarteroni A, Rozza G (2003) Optimal control and shape optimization of aorto-coronaric bypass anastomoses. Math Model Methods Appl Sci 13(12):1801–1824
96. Pekkan K, Frakes D, De Zelicourt D, Lucas CW, Parks WJ, Yoganathan AP (2005) Coupling pediatric ventricle assist devices to the Fontan circulation: simulations with a lumped-parameter model. ASAIO J 51(5):618–628
97. de Leval MR, Dubini G, Migliavacca F, Jalali H, Camporini G, Redington A, Pietrabissa R (1996) Use of computational fluid dynamics in the design of surgical procedures: application to the study of competitive flows in cavo-pulmonary connections. J Thorac Cardiovasc Surg 111(3):502–513
98. Rydberg A, Teien DE, Krus P (1997) Computer simulation of circulation in patient with total cavo-pulmonary connection: inter-relationship of cardiac and vascular pressure, flow, resistance and capacitance. Med Biol Eng Comput 35(6):722–728
99. Clipp RB, Steele BN (2009) Comparison of Three Types of Dynamic Boundary Conditions. Proceedings of the 2009 Summer Bioengineering Conference, Lake Tahoe, CA.
100. Steele BN (2007) Using one-dimensional finite element analysis to estimate differential pressure of renal artery stenosis. Computers in Cardiology 34:391–394
101. Spilker RL, Feinstein JA, Parker DW, Reddy VM, Taylor CA (2007) Morphometry-based impedance boundary conditions for patient-specific modeling of blood flow in pulmonary arteries. Ann Biomed Eng 35(4):546–559
102. Clipp RB, Steele BN (2009) Impedance boundary conditions for the pulmonary vasculature including the effects of geometry, compliance and respiration. IEEE Trans Biomed Eng 56:862–870
103. Orlando W, Hertzberg J, Shandas R, DeGroff C (2002) Reverse flow in compliant vessels and its implications for the Fontan procedure: numerical studies. Biomed Sci Instrum 38:321–326
104. Vigmond EJ, Clements C, McQueen DM, Peskin CS (2008) Effect of bundle branch block on cardiac output: a whole heart simulation study. Prog Biophys Mol Biol 97(2–3):520–542
105. Marsden AL, Vignon-Clementel IE, Chan FP, Feinstein JA, Taylor CA (2007) Effects of exercise and respiration on hemodynamic efficiency in CFD simulations of the total cavopulmonary connection. Ann Biomed Eng 35(2):250–263
106. Figueroa AC, Baek S, Taylor CA, Humphrey JD (September 2009) A computational framework for fluid–solid-growth modeling in cardiovascular simulations. Comput Meth Appl Mech nd Eng, 198(45–46):3583–3602

Chapter 10
Computational Analyses of Airway Flow and Lung Tissue Dynamics

David W. Kaczka, Ashley A. Colletti, Merryn H. Tawhai, and Brett A. Simon

Abstract The function of the mammalian respiratory system is the facilitation the transfer of gas exchange between the organism's environment and its internal transport medium, the blood. Evolutionary processes have optimized the anatomic structure of the lung as a tree-like branching network of airways terminating in thin-walled elastic ducts and alveoli, where this gas exchange occurs. Both dissipative and elastic properties of the respiratory system contribute to its unique static and dynamic mechanical behavior. In this chapter, we will explore the various structural and functional relationships of the respiratory system, and review several computational and modeling analyses that provide insight into the pathophysiology of common respiratory diseases. Particular emphasis is placed on studies that utilize imaging to help understand and/or validate the distributed regional nature of lung function.

10.1 Introduction

The function of the mammalian respiratory system is to facilitate the transfer of gases between the organism's environment and its internal transport medium, the blood. For biological efficiency, this gas exchange must occur over a tremendously large surface area during each breath. Evolutionary processes have optimized the anatomic structure of the lung as a reciprocating pump with a single intake/exhaust conduit, the trachea. The trachea leads to a tree-like branching network of airways terminating in thin-walled elastic ducts and alveoli, where most gas exchange occurs. The many branching airway segments provide viscous pathways for airflow, and the rapid, exponential expansion in effective airway cross-sectional area results in reduced gas velocities and changing flow regimes. The dissipative and elastic

D.W. Kaczka (✉)
Harvard Medical School, Beth Israel Deaconess Medical Center, 330 Brookline Avenue, Boston, MA 02215, USA
e-mail: dkaczka@bidmc.harvard.edu

K.B. Chandran et al. (eds.), *Image-Based Computational Modeling of the Human Circulatory and Pulmonary Systems*, DOI 10.1007/978-1-4419-7350-4_10,

properties of the respiratory system contribute to its unique static and dynamic mechanical behavior. In this chapter, we will explore the various structural and functional relationships of the respiratory system and review several computational and modeling analyses that have been used to provide insight into the pathophysiology of common respiratory diseases. Particular emphasis is placed on studies that utilize imaging to help understand and/or validate the distributed regional nature of lung function.

10.2 Basic Anatomy and Physiology

For the purposes of gas transport, the lung is divided anatomically into the conductive and respiratory zones [1, 2]. Beginning with the trachea, the conducting airways form a series of branching conduits, including the main, lobar, and segmental bronchi, down to the terminal bronchioles. The conducting airways lead inspired air to the respiratory zone, consisting of respiratory bronchioles that contain a few budding alveoli, and finally to the alveoli-lined acini (Fig. 10.1). Alveoli are the terminal gas-exchanging units of the lung. The human lung contains about 300 million alveoli, which amounts to a surface area between 50 and 100 m^2. An acinus refers to the anatomical unit of the portion of lung distal to a terminal bronchiole.

During inspiration, a pressure gradient between the airway opening and the alveoli results in the convective movement of air from the atmosphere to the terminal

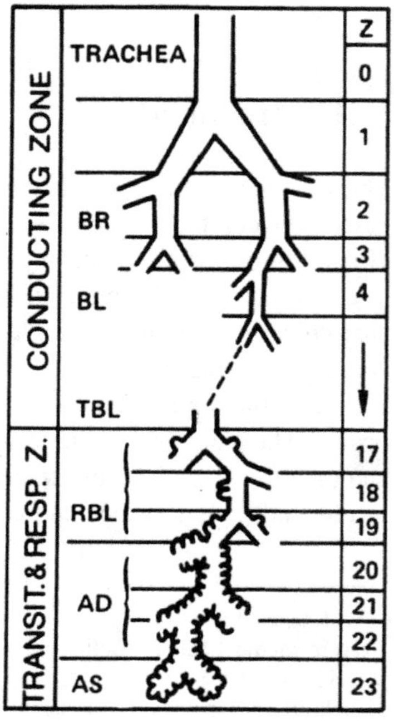

Fig. 10.1 Schematic of human bronchial tree according to Weibel [3]. Conducting airway zone consists of the first 16 generations, while transitional/respiratory zone consists of generations 17–23. BR, bronchus; BL, bronchiole; TBL, terminal bronchiole; RBL, respiratory bronchiole; AD, alveolar duct; AS, alveolar sac (modified from [2], with permission)

units of gas exchange. For spontaneous breathing, this gradient is generated by the contraction of the diaphragm and intercostal muscles, causing the chest wall to expand and decrease the intrapleural pressure between itself and the lung. Under conditions of artificial respiration, a mechanical ventilator provides positive pressure at the airway opening relative to the alveoli. In either situation, the lung will expand and air will move convectively through the bronchial tree. While convection dominates the travel of inspired gas from the trachea throughout the conductive zone, upon reaching the respiratory zone, its velocity becomes negligible and diffusion becomes the primary means of gas transport [1, 2]. This decrease in velocity is due to the rapid increase in total airway cross-sectional area over the short distance from the trachea to the respiratory zone. At the alveoli, inspired oxygen will diffuse down its partial pressure gradient into the bloodstream. Simultaneously, carbon dioxide will follow its diffusion gradient from the pulmonary capillaries into the alveoli to be expired. During exhalation, relaxation of the diaphragm and intercostal muscles allows the elastic lung to return passively to its pre-inspiratory volume. Functional residual capacity (FRC) refers to the volume at which the inward elastic recoil of the lung is exactly balanced by the outward elastic force of the chest wall.

10.3 Respiratory Mechanics

Despite its tremendous anatomic and structural complexities, the functional behavior of the respiratory system follows remarkably simple mechanical relationships under most types of ventilation, requiring the development of pressure to overcome the resistive (R), elastic (E), and (under certain conditions) inertial (I) forces opposing the convective movement of gas (or flow) into the lungs. As a first approximation, this relationship between pressure (P) and flow ($\dot{V}$) can be well-described mathematically using the equation of motion:

$$P = R\dot{V} + EV + I\ddot{V} \tag{10.1}$$

where V denotes volume and $\ddot{V}$ denotes volume acceleration. These resistive, elastic, and inertial properties may refer to the mechanics of the total respiratory system, the lung alone, or the chest wall, depending on whether the pressure being characterized is the airway pressure relative to atmosphere, airway pressure relative to pleural pressure, or pleural pressure alone, respectively [4, 5].

The resistive properties of the respiratory system arise from energy losses associated with gas flowing through a complex, bifurcating network of airway segments that are typically assumed to have a circular cross section [6–8], and the frictional losses associated with the expansion and contraction of the lung parenchyma and chest wall [9–11]. Airway resistance is defined as the pressure difference between the alveoli and airway opening divided by flow rate [2]. Theoretically, the resistance for the entire airway tree can be obtained from the sum total of all individual segment resistances arranged in the appropriate serial and parallel fashion. However, each segment resistance (R_{seg}) depends on the assumed paradigm for gas flow. For simplicity, an assumption of fully developed, laminar flow through each segment is often made, allowing for a mathematical description of resistance based on Poiseuille's law [12]:

$$R_{seg} = \frac{128\mu l}{\pi d^4} \tag{10.2}$$

where μ denotes the fluid viscosity, d is the segment diameter, and l its length. Equation (10.2) provides an approximate relationship between bronchial segment dimensions (i.e., length and diameter) and airway resistance. This can be useful in understanding how reductions in airway diameter, as occur during decreases in lung volume or contraction of bronchial smooth muscle, increase the resistance of the lung [13]. Alternatively, if the flow is non-steady, or oscillatory, a more accurate description of segment resistance can be obtained as [14, 15]:

$$R_{seg} = \mathrm{Re}\left\{\left(\frac{j4\omega\rho l}{\pi d^2}\right)\left[1 - \frac{2J_1\left(\alpha_\omega j^{3/2}\right)}{\alpha_\omega j^{3/2} J_0\left(\alpha_\omega j^{3/2}\right)}\right]^{-1}\right\} \tag{10.3}$$

where Re{} denotes the real part of the argument enclosed, ω is the oscillation frequency in radians/s, j is the unit imaginary number, ρ is the fluid density, J_0 and J_1 are the complex Bessel functions of orders 0 and 1, respectively, and α_ω is the so-called Womersley parameter [16, 17] characterizing the ratio of inertial forces to viscous forces:

$$\alpha_\omega = \frac{d}{2}\sqrt{\frac{\omega\rho}{\mu}} \tag{10.4}$$

Thus, for a given airway segment resistance, Equation (10.3) will yield a higher value for resistance compared to Equation (10.2) due to additional energy losses associated with velocity profile distortion [18]. However, during any point in the respiratory cycle and at any locus within the airway tree, the flow paradigm may be characterized as laminar or turbulent, fully developed or transitional, quasi-steady or non-steady [8]. These additional pressure drops will cause the magnitude of airway resistance to be underestimated by Equations (10.2) and (10.3). Both theory and experiment confirm that the greatest dissipative pressure loss occurs across medium-sized bronchi [7], the major source of resistance experienced by gas as it traverses the bronchial tree (Fig. 10.2).

The total resistance of the respiratory system is not solely due to airway caliber. Tissue resistance is an additional dissipative property of the parenchyma and the chest wall [9, 10, 19]. Lung tissue resistance may arise from frictional losses of connective tissue fibers moving against one another [20], the biophysical properties of surfactant [21], or energy dissipation associated with cross-bridge cycling of contractile elements [22, 23]. Since the viscous losses associated with the stretching of tissue are proportional to flow, they take a form similar to airway resistance and thus

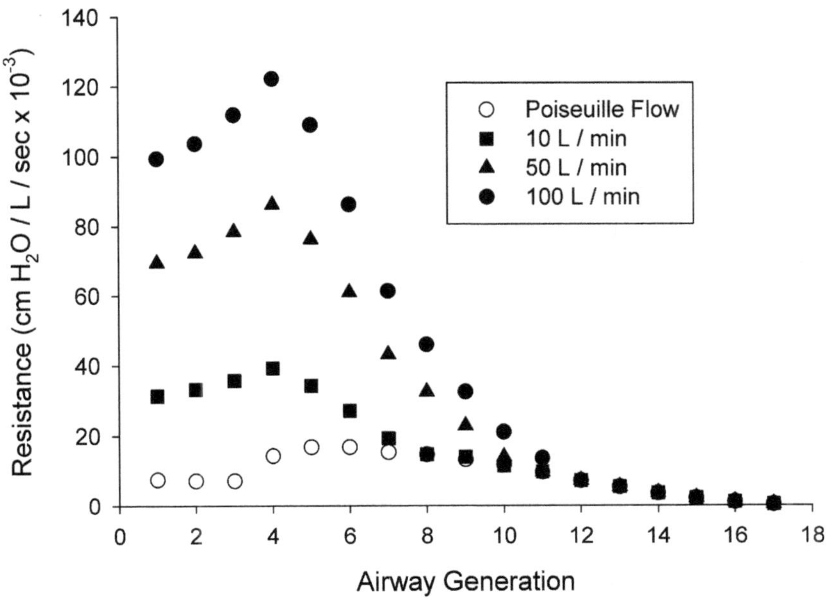

Fig. 10.2 Viscous flow resistance as a function of airway generation predicted for a human lung during steady flows of 100, 50 and 10 l/min, as well as an ideal Poiseuillean flow regime (i.e., Equation (10.2)) (modified from [7], with permission)

are frequently distinguished as a separate component of the total lung resistance. Due to viscoelasticity, tissue resistance demonstrates a strong negative dependence on breathing frequency, asymptotically approaching zero as frequency increases [24]. Studies using alveolar capsules [24–28] and forced oscillations [11, 29–31] have demonstrated that parenchymal tissue resistance may comprise 40–60% of total lung resistance at breathing frequencies (Fig. 10.3).

The elastic properties of the respiratory system describe the tendency of the parenchyma and chest wall to return to their normal resting configurations when they are deformed (or expanded) by some external force, giving rise to what is referred to as elastic recoil pressure [32]. Elastance is defined as the change in distending pressure per unit change in lung volume and may also be quantified for the total respiratory system, the lungs, or the chest wall. It can be measured under static conditions, in which the resulting transrespiratory or transpulmonary at a fixed volume [33], or during the dynamic process of ventilation, which requires the separation of these pressures into their respective resistive and elastic components [34]. Due to viscoelasticity, dynamic elastance will in general be higher than static elastance, depending on breathing frequency [35, 36]. In addition, both static and dynamic elastances will vary with lung volume, becoming stiffer as volume increases. This gives rise to the nonlinear pressure–volume curve shown in Fig. 10.4. In clinical situations, elastance is usually expressed as its reciprocal, compliance (i.e., change in volume per unit change in distending pressure).

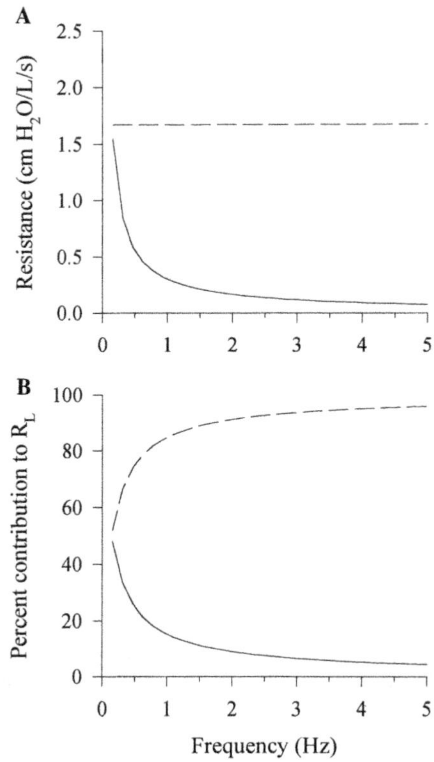

Fig. 10.3 (**a**) Schematic of human lung tissue resistance (*solid line*) and airway resistance (*dashed line*) as a function of frequency; (**b**) percent contribution of lung tissue resistance (*solid line*) and airway resistance (*dashed line*) to total lung resistance as a function of frequency (modified from [11], with permission)

Finally in certain situations, such as high-frequency oscillation or jet ventilation, an additional pressure drop is encumbered due to the acceleration of the respiratory tissues and gas contained in the central airways [38, 39]. This mechanical property is referred to as inertance and results in an additional component of impedance to flow in the lung.

In the healthy lung, ventilation distributions vary in normal gravity, with the lower regions receiving better ventilation per unit volume compared to the upper regions [2]. The cause for these regional differences in ventilation is that the lung tissue deforms under a gravitational load (i.e., the so-called slinky effect [40]), which in turn causes the intrapleural pressure to be less negative at the base of the lung compared to the apex. This results in regional compliance differences that translate into regional expansion differences for the same driving pressure [14, 41]. At normal breathing rates, the distribution of ventilation is primarily determined by local compliance [42–44]. At low lung volumes, however, the distribution of ventilation is inverted, and upper regions ventilate better than lower zones. Airway closure can result, especially in patients who have experienced loss of lung recoil due to age or chronic disease. In these cases, the lowermost regions of the lung may be only intermittently ventilated. Such scenarios assume that the pleural pressure changes are

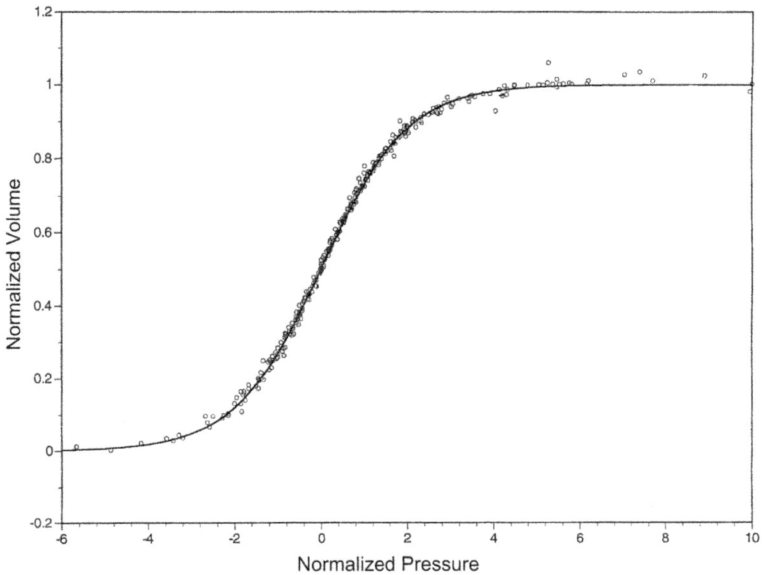

Fig. 10.4 Normalized pressure–volume curve obtained from data in healthy dogs and humans as well as during various pathologic conditions such as acute lung injury. After normalization, data points are described as a nonlinear sigmoidal function (modified from [37], with permission)

uniformly distributed around the lung. During spontaneous breathing, differences in regional pleural pressure may occur due to local variation in chest wall and/or diaphragm expansion, whether due to voluntary effort or passive properties, and this factor may further alter the ventilation distribution.

10.4 Mechanical Input Impedance

Equation (10.1) can be applied to various pressure and flow waveforms to estimate the mechanical properties of the respiratory system using multiple linear regression techniques [45–48]. However, Equation (10.1) describes a linear system with constant coefficients, implying that these fundamental mechanical properties are invariant under all conditions. In reality, R, E, and I may vary depending on breathing frequency, lung volume, volume history, and flow rate. Thus forced oscillation techniques, in which time-varying flows of different frequencies are forced into the lungs at the airway opening, have been gaining acceptance as an efficient and valid method for providing a more thorough and robust assessment of dynamic respiratory mechanics [11, 49–55]. The complex ratio of the resulting pressure to the delivered flow as a function of oscillation frequency is defined as respiratory input impedance (Z). Mathematically, Z can be represented as a vector in the complex plane, with both real (resistive) and imaginary (reactive) components. As such, it may be expressed in polar coordinates at each frequency, with a magnitude $|Z|$ to account for the amplitude of the pressure relative to flow, as well as a phase angle $\angle Z$ to indicate any time shifts between the pressure and flow oscillations. For frequencies less than 10 Hz, Z is often expressed as the mechanical properties associated with energy dissipation and energy storage during dynamic oscillation, such as respiratory resistance (R) and elastance (E):

$$R(\omega) = \mathrm{Re}\{Z(\omega)\}\,; \quad E(\omega) = -\omega\,\mathrm{Im}\{Z(\omega)\} \tag{10.5}$$

where ω is the frequency of oscillation in radians per second, and Re{ } and Im{ } denote the real and imaginary components, respectively, of the arguments enclosed. When these quantities are measured over these low frequencies, the spectral features of Z can be a sensitive indicator of serial and parallel mechanical heterogeneities [51, 53, 56–59]. Furthermore, Z can provide insight into the locus of airway obstruction [29, 50] and is useful in partitioning the mechanical properties of the airways from lung tissues [11, 24].

Deviations from normal Z behavior have been used not only to infer the presence of lung pathology but also to indicate the degree of serial or parallel time constant heterogeneity among the resistive and elastic mechanical properties of individual lung units [58, 59]. Both experimental and computer modeling studies have demonstrated how distinct patterns of heterogeneity can be uniquely identified from the frequency-dependent features of Z [14, 41, 51, 52, 55, 56, 60]. For example, parallel time constant heterogeneity will result in enhanced frequency dependence in the magnitude of impedance below 2 Hz [51, 56, 57]. Random closure or flooding of airways leads to derecruitment of lung units and less tissue in communication with the airway opening [41, 61], resulting in substantial elevations in R and E at specific respiratory rates, exclusive of any alterations in intrinsic tissue resistance or elastance.

10.4.1 Inverse Modeling of Respiratory Mechanics

A method often used to provide additional interpretations of the physiologic alterations contributing to Z is inverse model analysis [14, 41, 49, 62]. Here, mechanical analogues of the respiratory system are proposed that may adequately characterize its dynamic oscillatory behavior. These models consist of a limited number of lumped physical elements (typically 4–6), which should ideally correspond to precise physiologic quantities such as airway resistance or tissue elastance [46, 49]. These parameters are then estimated by fitting the model to impedance data using various linear or nonlinear techniques [14, 46, 63, 64], yielding functional insight into global respiratory mechanics. In recent studies, the model most frequently used for characterizing the impedance of the lungs and total respiratory system is the so-called constant-phase model [11, 20, 49, 54, 65]. This model consists of a single lumped airway compartment, with parameters for a Newtonian airway resistance

(R) and inertance (I), in series with a viscoelastic tissue compartment with parameters for dynamic hysteresivity (η) and elastance (H). It has a predicted impedance ($\hat{Z}$) as a function of frequency (ω) given as

$$\hat{Z}(\omega) = R + j\omega I + \frac{(\eta - j)H}{\omega^{\alpha}} \tag{10.6}$$

where $\alpha = (2/\pi)\tan^{-1}(1/\eta)$. One drawback of this model, however, is that it assumes the impedance of the respiratory system can be sufficiently described by serial compartments representing homogenous airways and tissues (Fig. 10.5a). Such an assumption may not always be valid, especially under pathologic conditions [51, 66].

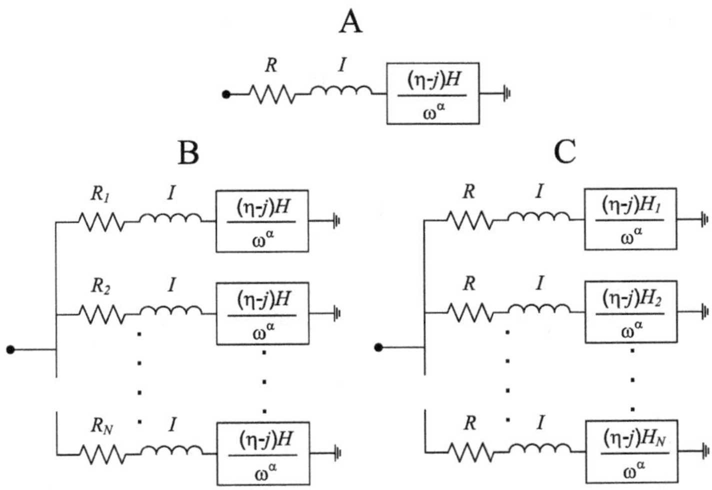

Fig. 10.5 Various inverse models used to describe respiratory input impedance. (**a**) Model with homogeneous airways and tissues; (**b**) model with distributed airways; (**c**) model with distributed tissue elastances (modified from [13, 41], with permission)

More recently, distributed inverse models have been developed and applied to impedance data obtained in pathologic states to assess heterogeneous regional mechanics associated with bronchoconstriction [13, 67, 68], emphysema [69, 70], and acute lung injury [41, 61]. While such distributed models are not meant to provide specific anatomic information on regional mechanics, they nonetheless have the ability to quantify mechanical heterogeneity and characterize dynamic regional lung function with a minimal number of physical parameters [14, 49]. These topologies generally assume that the mechanics of the respiratory system can be modeled with an arbitrary number of parallel branches (Fig. 10.5b, c). The impedance of each individual branch can be characterized by Equation (10.6), but the values of either R or H are assumed to vary from branch to branch according to some predefined continuous probability density function $P(x)$, depending on whether one seeks to quantify airway heterogeneity or tissue heterogeneity, respectively. For example, if we seek to characterize dynamic lung mechanics during heterogeneous bronchoconstriction [13, 67, 68], we may assume distributed airways for which the stochastic variations in each branch R are governed by $P(R)$ with lower and upper bounds $R_{\min}$ and $R_{\max}$, respectively (Fig. 10.5b). The predicted impedance of this model is given as

$$\hat{Z}(\omega) = \left[\int_{R_{\min}}^{R_{\max}} \frac{P(R)}{R + j\omega I + \frac{(\eta - j)H}{\omega^{\alpha}}} \mathrm{d}R \right]^{-1} \tag{10.7}$$

In contrast to the model described by Equation (10.6), with four independent parameters, this model requires the estimation of five parameters ($R_{\min}$, $R_{\max}$, I, η, and H). Likewise, for pathologies associated with variations in tissue elastance [14, 41, 69, 70], we may assume a similar topology whose predicted impedance depends on a distributed elastance parameter H (Fig. 10.5c):

$$\hat{Z}(\omega) = \left[\int_{H_{\min}}^{H_{\max}} \frac{P(H)}{R + j\omega I + \frac{(\eta - j)H}{\omega^{\alpha}}} \mathrm{d}H \right]^{-1} \tag{10.8}$$

A challenge in interpreting and validating these distributed models is the lack of reliable measurements of actual regional airway and tissue mechanical properties for comparison with model predications, since they do not reflect individual anatomy or provide information about alterations in lung structure in chronic disease or acute physiologic perturbations. As described in the next section, the use of forward modeling and novel imaging techniques is now providing quantitative, individualized subject characterizations that, combined with the types of measurements described above, allow validation and refinement of these inverse distributed models, which may have the potential to direct therapy [41, 50, 71].

10.5 Forward Morphometric Models of the Respiratory System

In contrast to inverse modeling, forward modeling of the lung involves *predicting* mechanical behavior in response to alterations in various functional *and* structural relationships. Typically, these forward models are rigorously grounded in anatomy and thus involve large computational structures with many degrees of freedom (on the order of 10^5–10^6). These include individual airway segment lengths and diameters, as well as acinar rheological properties, all of which can be altered to simulate various physiological conditions or pathological derangements.

Weibel and Gomez [3, 72] were the first to use a comprehensive morphologic approach to develop such a model of the human lung. Their lung was divided into the conductive zone, composed of larger airways and blood vessels, and the respiratory zone, containing respiratory bronchioles, alveolar ducts, alveoli, and the capillary network. They estimated that the human lung contains approximately 300 million alveoli, ventilated by an airway tree composed of an average 23 dichotomous branching generations. Weibel's symmetric model has been used extensively for computational studies of lung function. But since each airway in a given generation possesses identical dimensions, these simulations need only to consider 23 airway generations, with 16 for the conducting portion of the tree. While Weibel's model does not account for realistic asymmetry in airway branching or dimensions, it does provide a very convenient description of an average pathway in the human airway tree.

A later morphometric model that accounted for branching and dimensional asymmetry was the model of Horsfield and coworkers [73–75]. These investigators also assumed a dichotomous branching but classified airways by Horsfield order, where sequential ordering starts at the order 1 terminal airways and converges to the highest order trachea. In their canine model [75], all acinar tissue elements, totaling 150,077, were assigned to order 1, with the trachea assigned to order 47 (Fig. 10.6). While symmetrical airway models assume that the parent branch of order n bifurcates into two daughter branches of order n–1, with asymmetric branching the two daughter branches may be of two different orders. This difference between the numerical order of the daughter branches is referred to as the recursion index (Δ) and is used to characterize the degree of asymmetry within the tree [75]. Table 10.1 lists recursion indices, segment lengths, and segment diameters for each airway order in the dog lung model of Horsfield et al. [75]. Since Horsfield and coworkers counted and ordered all of the airway segments based on lengths and diameters, they assigned each airway segment a specific order, with the diameters, lengths, and recursion indices within each order being identical. This structural feature allows for powerful and efficient recursive calculations of mechanical properties, such as serial and parallel addition of the longitudinal impedances of each individual airway

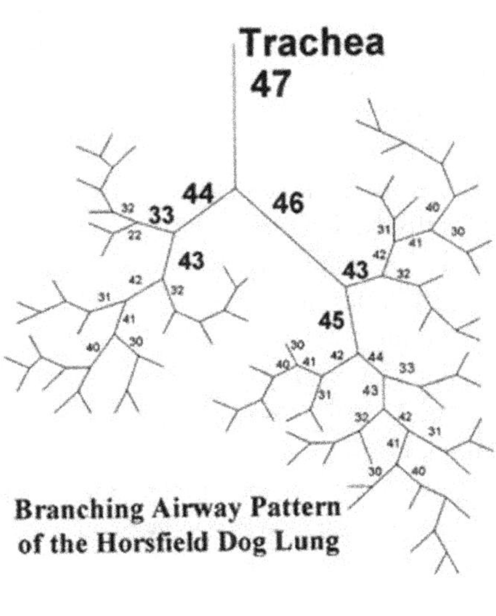

Fig. 10.6 Asymmetric dichotomous branching pattern implemented in Horsfield's model of the airways of the dog lung (modified from [51, 78, 80], with permission)

Table 10.1 Horsfield's classification of each of the 47 generations of the airways of the dog lung with a unique set of delta, diameter, and length. Values are the same for any airway segment of a given generation. Order 47, trachea; order 5, terminal bronchioles; orders 2–4, respiratory bronchioles; order 1, acinus. Delta is the difference in order between the two daughter branches (modified from [75])

Order	Delta	Diameter (cm)	Length (cm)	Order	Delta	Diameter (cm)	Length (cm)
47	2	2.10	20.00	23	4	0.274	0.328
46	2	2.10	0.75	22	4	0.255	0.311
45	2	2.03	1.80	21	4	0.238	0.294
44	10	1.15	1.01	20	4	0.223	0.279
43	10	1.08	0.961	19	4	0.208	0.264
42	10	1.01	0.911	18	4	0.194	0.251
41	10	0.939	0.863	17	4	0.181	0.237
40	10	0.876	0.818	16	4	0.169	0.225
39	10	0.818	0.775	15	4	0.158	0.213
38	10	0.764	0.734	14	4	0.148	0.202
37	10	0.714	0.696	13	4	0.139	0.191
36	10	0.666	0.660	12	4	0.124	0.181
35	10	0.622	0.625	11	4	0.110	0.172
34	10	0.581	0.592	10	4	0.0983	0.163
33	10	0.543	0.561	9	4	0.0876	0.154
32	10	0.507	0.532	8	4	0.0781	0.146
31	10	0.473	0.504	7	4	0.0697	0.139
30	10	0.442	0.478	6	0	0.0621	0.131
29	10	0.413	0.453	5	0	0.0554	0.125
28	10	0.385	0.429	4	0	0.0494	0.118
27	10	0.360	0.407	3	0	0.0440	0.112
26	10	0.336	0.385	2	0	0.0393	0.106
25	4	0.314	0.365	1			
24	4	0.293	0.346				

segment in a tree. Horsfield's structures gained popularity when numerous investigators developed computational models based on realistic morphometry, facilitating their use for the calculation of physiological quantities such as input impedance and flow distributions under various simulated pathologies [14, 51, 55, 56, 60, 76–79].

The Weibel symmetric and Horsfield asymmetric models are convenient and concise abstractions of the airway tree, but neither model is spatially distributed nor do they account for regional variability in path length, branching asymmetry, and branch dimensions. Both models are summaries of the average branching structure of mammalian lungs. Spatially distributed airway tree models have been developed by Kitaoka et al. [81] and Tawhai et al. [82, 83]. The Kitaoka model is based on Murray's law [84–86]: the relationship between parent and child diameters at a bifurcation is fixed, with branch lengths based on length-to-diameter ratio. A set of rules is used to control the rotation angle for the planes of branching and minimize the number of airways that penetrate outside of the bounding "host" lung.

Fig. 10.7 Anatomically-based and spatially-distributed model of a human subject lung and airways. Airways generated using the Tawhai et al. method [82] are colored *grey* in the right lung lobe, with the right upper lobe highlighted in *light grey*. A volumetric mesh of the left lung is overlaid over the left lung airways (colored *white*) to show the relationship between model lung and airway

The Kitaoka model was created within an idealized description of the geometry of the lung. It has not yet been used to derive models of airway geometry using anatomically accurate lung boundaries from individual subjects. The Tawhai model is developed using a volume-filling algorithm to generate a tree structure into any shaped host volume. The host volume may be an idealized shape or an accurate anatomical description of the lung lobes (in the form of a finite element mesh) derived from medical imaging (Fig. 10.7). Instead of determining the tree structure on the basis of flow distribution as in the Kitaoka model, the algorithm of the Tawhai model creates successive bifurcations such that airways or blood vessels are oriented toward the center of mass of the functional tissue units that the tree will supply. The diameters of airways or vessels are prescribed in a later step, typically using a combination of measurements from imaging data and an assumed proportionality between branch order and diameter. The advantage of the Tawhai method is that it can be used to create subject-specific models of the airway tree that can then be used to study inter-subject variability, or to match structure to function within an individual.

10.5.1 Computational Modeling Example: Airway Thermodynamics in Symmetric and Anatomical Models

A fundamental "rule" when using forward computational modeling for predicting physical processes is to adopt the simplest form suitable for the study in question. With the airways of the lung, we could choose from symmetric [72], regular asymmetry [73–75], or anatomically based [81, 83]. The appropriateness of the model can be evaluated by asking, "What additional insight will be gained by using a model that is more complex?" In this example, we consider airway thermo-fluid dynamics, simplified to a one-dimensional system of equations [87]. If air enters the bronchial tree at less than body temperature or water saturation, then the airways must supply heat and moisture to attain equilibrium. Such a situation arises during exercise in a cold dry environment or during artificial positive pressure ventilation. A system of coupled equations can be formulated to calculate the transfer of heat and moisture from and to the airway walls (i.e., recovery during expiration). These airway models contain computational domains through which the system of equations can be directly solved, making it straightforward to evaluate the influence of airway geometry on model predictions. Tawhai and Hunter [87] compared predictions of lumen temperature in symmetric and anatomical models to measurements obtained from McFadden et al. [88]. McFadden et al. measured the luminal air temperature in spontaneously breathing volunteers at several locations down a pathway to a sub-segmental bronchus of the right lower lobe, yielding an airway "thermal map." Measurements from this study are reproduced in Fig. 10.8 for ventilation with room

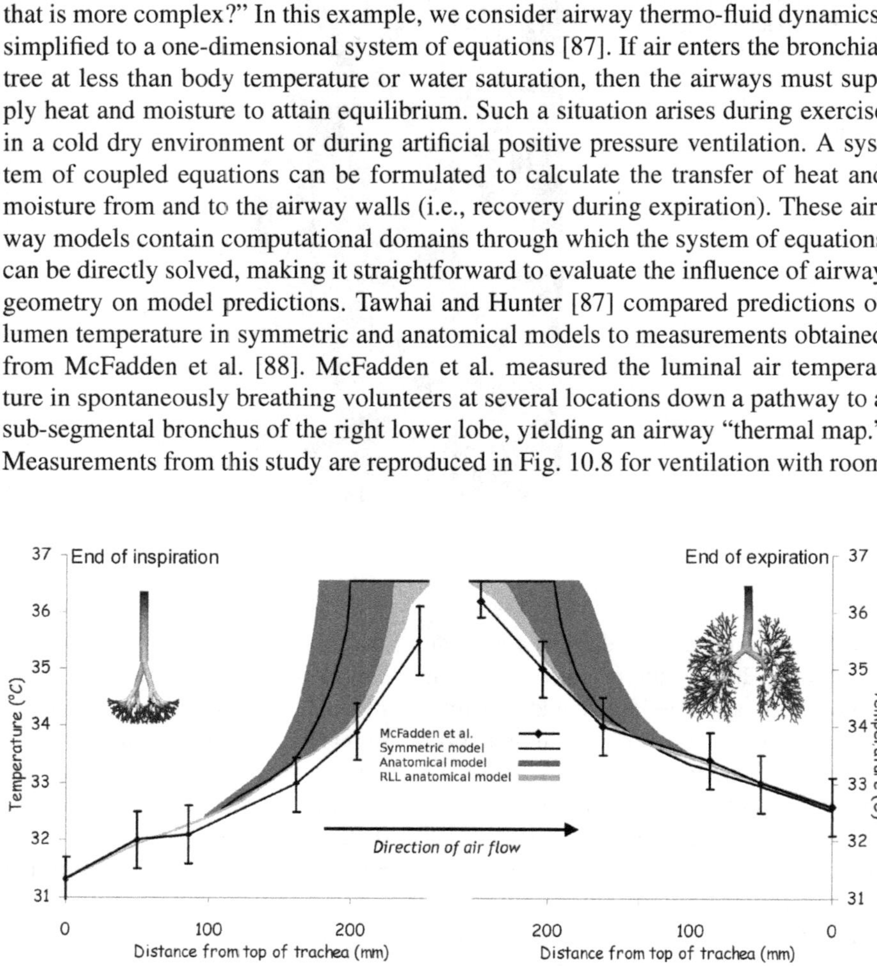

Fig. 10.8 Airway lumen temperatures predicted by a coupled water and heat transfer computational model, compared with physical measurements for ventilation with room temperature air at 15 litres/minute. Experimental measurements from McFadden et al. [88] (mean ± standard error) are compared with temperatures in a single pathway (symmetric) model, and the range of temperatures in an anatomically structured model (the whole model in *dark grey*, and right lower lobe highlighted in *light grey*). Temperature fields at end-inspiration in the symmetric and anatomical models are overlaid: symmetric model on the *left* and anatomical on the *right*.

air at 15 l/min. Prediction of the thermal gradient in the symmetric model (in red in Fig. 10.8) compared well with the measured temperatures in the trachea and main bronchi, but the rate of inspired air warming was more rapid compared to the actual measurements. Repeating the calculation in the anatomical model with a gravitational distribution of ventilation gave a wide range of temperature gradients that varied depending on the proportion of ventilation delivered through that path (in dark gray in Fig. 10.8). When the pathway to the right lower lobe in the model was isolated, the predicted temperature was in far closer agreement to the measurements. This is consistent with the lower lobes receiving proportionately more of the inspired air, since higher ventilation equates to greater wall cooling. Such results could not have been achieved using either a symmetric or a regular asymmetric model; both airway tree asymmetry and a realistic ventilation distribution—determined by a spatial relationship between tissue and airway tree—were necessary. It is important to note that the symmetric model functions as an excellent predictor for the mean behavior of the airway tree, since its temperature gradient lies midway through that of the anatomical model. The symmetric model yielded accurate predictions in the trachea. Thus in simulations for which the trachea is of primary interest, the more simplified model would be a better choice with respect to computational burden.

10.6 Application of Morphometric Models to Computational Studies of Lung Mechanics

Fredberg and colleagues [76, 77] were among the first to simulate mechanical data using computational models such as those described in the previous section. Fredberg and Hoenig [76] developed a method for computing the input impedance of complex asymmetrically branching duct networks, simulating and analyzing the dynamic mechanical response of healthy adult human lungs at high frequencies. They determined that airway walls are a principal modulator of system damping and wave dispersion. In order to achieve computational efficiency, their self-consistent structure was defined by a tree characterized by the same input impedance for every node of a given generation. In a subsequent study, Fredberg and Moore [77] simulated a distributed system response in a Horsfield-type network to characterize variations in pressure that occur with frequency, path, and position.

Several recent studies have aimed to characterize functional heterogeneity in the lung. Based on Horsfield's airway structure, Thorpe and Bates [55] developed a forward computational model of acute heterogeneous bronchoconstriction by stochastically varying airway responsiveness during a simulated bronchoprovocation time course. They concluded that heterogeneous airway constriction alone can cause significant increases in lung impedance. Lutchen and Gillis [56] created a similar morphometric model of inhomogeneous constriction in which airway diameters were changed according to designated statistical distribution. In this model, constant dynamic tissue hysteresis was assumed by fixing acinar parenchymal hysteresivity and elastance values. In their computational approach, they constructed an asymmetric airway tree from the bottom up using a stack-based procedure that did not store data from all pathways. In a later study, these investigators also demonstrated that the different patterns of heterogeneous airway smooth muscle shortening resulted in randomly distributed airway closure, which dramatically increased resistance and elastance at breathing frequencies and contributed to the hyperresponsiveness characteristic of an asthmatic response [60]. They also presented simulations of ventilation distribution during airway constriction in the Horsfield human lung model [74, 78]. Flow distributions to the acini were predicted during various degrees of heterogeneous bronchoconstriction, illustrating how uneven flow distribution could potentially damage tissue due to overventilation. More recently, Kaczka et al. [14] used a similar forward model to simulate mechanical input impedance under various degrees of tissue heterogeneity typical of acute lung injury and emphysema. They concluded that the simple inverse models of lung mechanics described above (Fig. 10.5), when applied to such impedance spectra, provide useful and accurate functional information on the mechanical status of diseased lungs.

Tgavalekos et al. [89] linked organ-level imaging with airway mechanics using the Tawhai airway model generated from a single human subject [83]. Ventilation defects in six subjects were obtained using PET imaging, which were then mapped onto the airway model. This yielded a relationship between structure and function under various constriction patterns that could mimic the observed ventilation defect. The model-predicted impedances were then evaluated to determine how closely they matched the actual impedance measurements from the subjects.

10.7 Imaging Methodology

The use of medical imaging technology has provided tremendous insight into the functional and structural relationships of the respiratory system. Of the various imaging modalities available, X-ray computed tomography (CT) has become the most widely used among researchers. Besides its routine use for clinical diagnoses, CT has provided tremendous insight into the alterations in the distribution of airway diameters during bronchoconstriction [13, 90, 91], anatomic patterns of aeration in emphysema or lung injury [92, 93], and quantification of the distribution of ventilation and perfusion in health and disease [94]. Recent advances in CT image analysis of local tissue deformation and/or grayscale density allow for the estimation of regional mechanical properties [42, 43, 95–97]. Positron emission tomography (PET) uses radioactive tracers in the inspired gas or injected into the blood to quantify regional ventilation, perfusion, lung volumes, and even inflammation [98–100]. Magnetic resonance imaging (MRI), particularly with the addition of hyperpolarized helium and xenon as inhaled contrast agents [101–103], can also provide complementary information to CT and PET regarding ventilation, perfusion, regional oxygenation concentrations, local diffusion properties, and chest wall motion [104] without exposing the subject to ionizing radiation. Each of these imaging modalities has various advantages or disadvantages with respect to temporal and spatial resolution, the volume of the lung that can be simultaneously imaged, potential toxicities of radiation and/or contrast agents.

All functional lung imaging modalities involve several key aspects that have important implications for the physiologic data obtained. First, image acquisition may be static, obtained during a constant volume breath-hold, or dynamic, in which the image is acquired during continuous steady-state breathing or an active inhalation or exhalation. Depending on the extent of the lung imaged and the speed and volumetric acquisition capability of the scanner, this may mean that the image was obtained over several seconds (static) or with a gated (dynamic) procedure over several breaths. Implications of the acquisition protocol for analysis of dynamic respiratory physiology must then be considered. For example, the distribution of a tracer gas during a single vital capacity inhalation may be different than during steady, tidal breathing, particularly if there exist substantial time constant heterogeneities within the lung. Second, the temporal and spatial resolution of the images may be important. Depending on the question under investigation, higher spatial resolution data (usually with higher noise) may not be necessary to provide important insights into physiologic processes. Finally, different imaging modalities may provide complementary information. For example, combined PET-CT scanners are now commercially available and can potentially utilize the high-spatial resolution information on airway structure and lung deformation from CT in conjunction with the lower resolution PET functional information on ventilation and perfusion distributions and metabolic or inflammatory processes. In the following section, we will review some recent applications of lung imaging relevant to the development and validation of computational models of lung airway and tissue dynamics.

10.8 Image-Based Computational Models

10.8.1 Insights into Bronchoconstriction: Airways and Interdependence

Various imaging modalities have been used to visualize airways in vivo and to infer structural and functional relationships in health and disease across a wide variety of species [13, 82, 90, 105, 106]. High-resolution CT (HRCT) has perhaps provided the most information about the distribution of airway diameters during spontaneous and pharmacologically induced bronchoconstriction [13, 90, 91, 105]. Moreover, the effects of lung inflation as well as dose and delivery route of various bronchoconstricting agonists on airway anatomy can be studied with relative efficiency. Even the distribution of a small sampling of airway diameters obtained from HRCT demonstrates significant correlations with measures of functional airway heterogeneity obtained from the forced oscillation and inverse modeling approaches

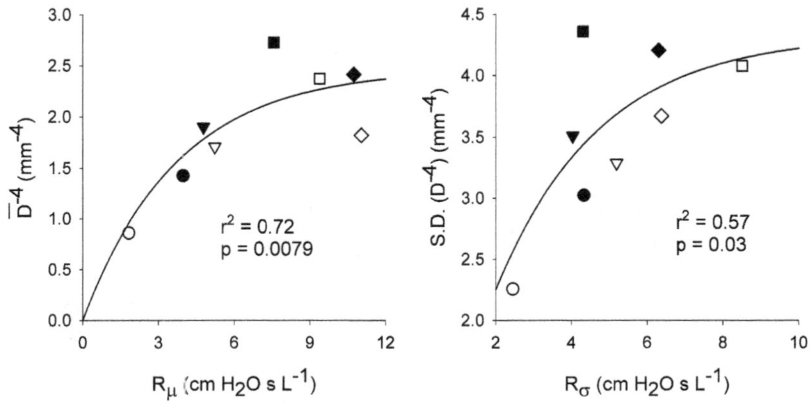

Fig. 10.9 Correlations between the means and standard deviations of a sampling (20–30 in number) of airway segment diameters raised to the inverse fourth power measured with HRCT with means and standard deviations of airway resistance parameters derived using forced oscillations and the distributed airway model (Equation (10.7) and Fig. 10.5b) for a representative dog. Data were obtained at baseline (*circles*) and during intravenous methacholine infusion rates of 17 (*inverted triangles*), 67 (*squares*), and 200 μg/min (*diamonds*) before (*white*) and 5 min following DI (*black*). *Solid line* represents rising exponential regression equation $y = a(1-e^{-bx})$. $\bar{D}^{-4}$, mean of HRCT airway diameters to the inverse fourth power; SD (D^{-4}), standard deviation of HRCT airway diameters to the inverse fourth power; R_{μ}, average airway resistance parameter from model of Fig. 10.5b; R_{σ}, standard deviation of airway resistance from model of Fig. 10.5b (modified from [13], with permission)

described above (Fig. 10.9). Nonetheless, routine clinical use of HRCT to assess airway pathology is limited due to cost, radiation exposure, and small timescale experiments.

The combination of much lower spatial resolution PET imaging with computational models has provided considerable insight into some of the more interesting aspects of the locus and heterogeneity of airway constriction and the critical role of mechanical interdependence. The Venegas group has pioneered the use of PET imaging to non-invasively determine ventilation and perfusion distributions in both experimental animals and human subjects from imaging the intravenous infusion of saline with dissolved radiolabeled ^{13}N-nitrogen (^{13}NN) [107]. The poorly soluble ^{13}NN comes out of solution into the air-filled alveoli during its first pass through the lung, providing a measure of regional perfusion, and then the labeled gas washes out of the lung during continuous breathing, the exponential washout rate of which quantifies regional ventilation. Residual tracer gas left in the lung after several minutes of breathing indicates gas trapping or greatly reduced ventilation, a phenomenon seen during bronchoconstriction when the airways leading to a given hypoventilated region constrict or even completely close (gas trapping). In early studies, these investigators noted that the bimodal ventilation–perfusion distributed traditionally noted using the multiple inert gas elimination technique (MIGET) could be explained by their observation of large, patchy regions of hypoventilation in the PET images. Further, by examining the multicompartment nature of the gas washouts within these regions, they determined that there was significant intraregional ventilation heterogeneity on a scale smaller than the 2.2 cm^3 spatial resolution of their measurement [108]. To address the long-term controversy as to whether this effect could be a result of constriction of only the large airways serving these hypoventilated regions [109, 110], they combined their imaging results with measurements of low-frequency lung input impedance (resistance and elastance from 0.15 to 8 Hz) and a distributed computational model of airway mechanics [89]. By mapping the imaged ventilation defects to a three-dimensional anatomic lung model, they could identify the largest airways that, if constricted, could account for the extent of the hypoventilated regions seen with PET. Next, they imposed distributions of airway constriction with differing degrees of heterogeneity on the remaining tree to best match the measured changes in the subject's impedance. They concluded that constriction of large airways alone could not explain both the distribution of hypoventilation and the degree of mechanical dysfunction observed, and that either heterogeneous small airway constriction alone or combined large and small airway constriction was required. Importantly, it was the combination of computational modeling with low-spatial resolution functional imaging, rather than high-resolution direct measurement of airway diameters, that enabled this conclusion.

A second observation made by these investigators from their ventilation images was that pre-existing ventilation heterogeneity appeared to predict the severity of the functional response to an inhaled bronchoconstrictor and, further, that the degree of initial regional hyperventilation was correlated with the degree of regional hypoventilation after bronchoconstriction [111]. In all these observations, clusters of hypoventilation were adjacent to lung regions of normal or increased ventilation. In addition, a deep inspiration after bronchoconstriction would restore ventilation to localized regions within the hypoventilated defects rather than open the entire region downstream of a large obstructed airway [112]. Rather than treat the airways as all independent, Venegas and coworkers constructed an idealized 2D computational model in which airway diameters were determined not only by the constriction of the smooth muscle in the wall but also by the tethering forces of the surrounding parenchyma, which serve to resist constriction as the region inflates [112, 113]. In a series of simulations, they showed that the clusters of hypoventilation seen in their images could be explained as a consequence of self-organizing patchiness deriving from local feedback between interdependent lung regions and the bistable nature of airway constriction [114]. As smooth muscle tone is uniformly increased throughout the airway tree, as soon as a small degree of ventilatory heterogeneity or a perturbation is introduced, the region of reduced local tidal volume experiences reduced tethering forces on the enclosed airways, reduced diameters, increased resistance, and so further reduced tidal volume, and so on. On the other hand, as some regions undergo a reduction in tidal volume, other regions must receive a relative increase in local tidal volume, stretching the enclosed airways, reducing their resistance, and consequently leading to further increases in local tidal volume. As would be expected, this effect is enhanced as total tidal volume is reduced. An interesting corollary of this model is that delivering inhaled bronchodilators to a patchy, constricted lung could result in preferential delivery to the less constricted regions and actually exacerbate the process of catastrophic closure in the already constricted regions [112], just as increased delivery of bronchoconstrictors to previously hyperventilated regions resulted in increased severity of subsequent regional hypoventilation [111]. Once again, the synergistic combination of regional image-derived functional data with computational modeling resulted in insights into the underlying mechanical pathophysiologic processes and generated testable hypotheses to continue to move the science forward.

Finally, another method of studying airway function using minimal imaging is computational fluid dynamics (CFD) simulations. CFD analyses of airway flow have been performed for a number of years but only recently have they been applied to image-based airway geometries [115, 116]. Image-based CFD analysis uses static imaging to define the geometry of the computational domain and then predicts pressure and flow throughout this domain to allow functional insight into what would otherwise be only structural information. For example, Lin et al. [116] evaluated the significance of the larynx in inducing a turbulent laryngeal jet and turbulent flow structures that can propagate through several generations of airway. Such analyses are important for understanding particulate transport in the airways and how this can vary between individuals or due to subtle changes in the configuration of oral structures such as the lips and tongue.

10.8.2 Regional Tissue Mechanics

Image-based attempts to quantify the heterogeneity of lung tissue mechanics have relied primarily on processing high-resolution computed tomographic (CT) scans [117, 118] or tracking the relative three-dimensional motions of implanted parenchymal beads with biplane fluoroscopy [119, 120, 121]. Several CT studies have demonstrated that distribution of regional aeration based on grayscale density is significantly widened following lung injury [117, 118, 122, 123, 124]. Changes in grayscale density can also be used to infer estimates of regional mechanical properties such as specific tissue compliance sC [42, 96, 117, 122]:

$$sC = \frac{F_2 - F_1}{F_1(1 - F_2)(P_2 - P_1)} \tag{10.9}$$

where F_1 and F_2 denote the fractional air content (based on grayscale) of the lung region of interest (ROI) at the inflation pressures P_1 and P_2, respectively. Such density-based estimates of regional compliance demonstrate good correlation with estimates of regional ventilation during xenon washin/washout [42]. However, these estimates may be biased, since regional aeration may not always correspond to regional volumes or tissue strains, especially under pathological conditions such as acute lung injury [122, 125].

Alternatively, image registration techniques, which rely on aligning three-dimensional scans taken at different points in time or inflation pressures, use anatomic features and boundaries to effectively map specific ROIs between image pairs, thus describing the deformation of the image with high spatial resolution. This deformation information can be used to compare co-localized features in the registered images as well as to quantify regional expansion, contraction, or parenchymal tissue strains. Several different registration methods exist for determining regional mechanics [43, 95, 96, 126–131], providing a direct assessment of the nonuniformity of expansion or compression at the scale of an individual voxel from the deformation field associated with CT image registration at different inflation pressures [97, 132]. Images may be obtained under static (i.e., breath-hold) conditions or using gated (end-expiratory and end-inspiratory) imaging during continuous ventilation. Here, three-dimensional warping functions map individual lung volume elements at a specified inflation pressure to corresponding volume elements at a higher inflation pressure [43, 127, 128]. This expansion or contraction of a particular ROI can be further quantified using various mathematical tools to quantify regional strains and deformations [127, 128]. For example, the determinant of the Jacobian matrix $|J|$ for a given deformation defines the differential expansion or contraction of the ROI from one state to the other [43, 95]. Since the Jacobian determinant is a scalar quantity, it can be interpreted as a ratio between the two volume states:

$$|J_i| = \frac{V_i + \Delta V_i}{V_i} \tag{10.10}$$

where V_i is the original volume of element i and ΔV_i is its change in volume to the higher inflation pressure. Thus, if the Jacobian determinant is greater than 1, the specified region has expanded, while if it is less than 1, the region has contracted.

The Jacobian can be used to quantify more traditional measures of regional mechanics. Rearranging Equation (10.10) to solve for the change in the elemental volume i between two pressures, we obtain

$$\Delta V_i = V_i(|J_i| - 1) \tag{10.11}$$

Dividing this volume change by the assumed pressure change gives us the actual local compliance for the element C_i:

$$C_i = \frac{\Delta V_i}{\Delta P} = \frac{V_i}{\Delta P}(|J_i| - 1) \tag{10.12}$$

A key assumption of Equation (10.12) is that the distending pressures (as inferred from airway opening pressures) are transmitted equally to all alveoli. This may not be the case if scans are obtained during gated (i.e., dynamic) conditions and significant mechanical heterogeneity exists, as during lung injury or other parenchymal disease processes [41, 95]. An additional limitation arises if alveolar flooding or consolidation obscures the anatomic features necessary for registration [97].

Static lung tissue mechanics can also be studied using image-based finite deformation theory. Early models of tissue mechanics analyzed infinitesimal deformations of the lung about some uniform state of inflation [133, 134], assuming linear elasticity and small strain. In reality, the lung parenchyma undergoes large deformations during breathing, with both static and dynamic properties being highly nonlinear. Finite deformation theory can be used to compute large tissue strains [135–137] assuming constitutive laws for parenchymal stress–strain a priori. To date, few studies have used finite deformation theory to study static mechanics of the parenchyma in image-based models of the lung [138, 139]. Computational evaluation of such stress distributions may provide insight into mechanisms that contribute to normal functional heterogeneity as opposed to more pathologically related structural heterogeneity.

10.9 Conclusions

In this chapter, we have reviewed various structural and functional relationships in the mammalian respiratory system and how computational modeling applied to such relationships elucidates both normal and abnormal airway and lung tissue physiology. Such modeling may combine global assessments of lung mechanical function, such as forced oscillations and inverse modeling, with anatomic and image-based structural analyses of regional airway and tissue mechanics. Combining these unique approaches allows for accurate assessments of the distribution of airway and tissue mechanics in both health and disease. While these techniques are currently impractical for routine diagnostic use, they may ultimately yield sufficient information to validate more clinically practical approaches for ventilator management and therapeutic intervention.

Acknowledgments Supported in part by National Heart, Lung, and Blood Institute Grants K08 HL-089227. The authors wish to thank Dr. Jason H.T. Bates for his many helpful suggestions.

References

1. Lumb AB (2005) Nunn's applied respiratory physiology. Elsevier/Butterworth Heinemann, Philadelphia, PA, p 501
2. West JB (1990) Respiratory physiology – the essentials. Williams & Wilkins, Baltimore,MD, p 185
3. Weibel ER (1963) Morphometry of the human lung. Springer, Berlin, p 151
4. Barnas GM, Campbell DN, Mackenzie CF, Mendham JE, Fahy BG, Runcie CJ, Mendham GE (1992) Lung, chest wall, and total respiratory resistances and elastances in the normal range of breathing. Am Rev Respir Dis 145:110–113
5. Warner DO (2000) So you want to be a pulmonary mechanic? A clinical guide. J Clin Monit Comput 16:417–423
6. Pedley TJ, Schroter RC, Sudlow MF (1970a) Energy losses and pressure drop in models of human airways. Respir Physiol 9:371–386
7. Pedley TJ, Schroter RC, Sudlow MF (1970b) The prediction of pressure drop and variation of resistance within the human bronchial airways. Respir Physiol 9:387–405
8. Schroter RC, Sudlow MF (1969) Flow patterns in models of the human bronchial airways. Respir Physiol 7:341–355
9. Fredberg JJ, Bunk D, Ingenito E, Shore SA (1993) Tissue resistance and contractile state of lung parenchyma. J Appl Physiol 74:1387–1397
10. Fredberg JJ, Stamenovic D (1989) On the imperfect elasticity of lung tissue. J Appl Physiol 67:2048–2419

11. Kaczka DW, Ingenito EP, Suki B, Lutchen KR (1997) Partitioning airway and lung tissue resistances in humans: effects of bronchoconstriction. J Appl Phys 82:1531–1541
12. Sutera SP, Skalak R (1993) The history of Poiseuille's law. Ann Rev Fluid Mech 25: 1–19
13. Kaczka DW, Brown RH, Mitzner W (2009) Assessment of heterogeneous airway constriction in dogs: a structure–function analysis. J Appl Physiol 106:520–530
14. Kaczka DW, Massa CB, Simon BA (2007) Reliability of estimating stochastic lung tissue heterogeneity from pulmonary impedance spectra: a forward-inverse modeling study. Ann Biomed Eng 35:1722–1738
15. Thurston GB (1952) Periodic fluid flow through circular tubes. J Acoust Soc Am 24: 653–656
16. Womersley JR (1955a) Method for the calculation of velocity, rate of flow and viscous drag in arteries when the pressure gradient is known. J Physiol 127:553–563
17. Womersley JR (1955b) Oscillatory motion of a viscous liquid in a thin-walled elastic tube: I. The linear approximation for long waves. Philos Mag 46:199–221
18. Finucane KE, Dawson SV, Phelan PD, Mead J (1975) Resistance of intrathoracic airways of healthy subjects during periodic flow. J Appl Physiol 38:517–530
19. Barnas GM, Stamenovic D, Fredberg JJ (1991) Proportionality between chest wall resistance and elastance. J Appl Physiol 70:511–515
20. Suki B, Barabasi, A-L, Lutchen KR (1994) Lung tissue viscoelasticity: a mathematical framework and its molecular basis. J Appl Physiol 76:2749–2759
21. Mead J, Whittenberger JL, Radford EP Jr (1957) Surface tension as a factor in pulmonary hysteresis. J Appl Physiol 10:191–196
22. Fredberg JJ, Jones KA, Nathan M, Raboudi S, Prakash YS, Shore SA, Butler JP, Sieck GC (1996) Friction in airway smooth muscle: mechanism, latch, and implications in asthma. J Appl Physiol 81:2703–2712
23. Kapanci Y, Assimacopoulos A, Irle C, Zwahlen A, Gabbiani G (1974) Contractile interstitial cells in pulmonary alveolar septa: a possible regulator of ventilation/perfusion ratio? J Cell Biol 60:375–392
24. Petak F, Hantos Z, Adamicza A, Daroczy B (1993) Partitioning of pulmonary impedance: modeling vs. alveolar capsule approach. J Appl Physiol 75:513–521
25. Barnas GM, Sprung J, Kahn R, Delaney PA, Agarwal M (1995) Lung tissue and airway impedance during pulmonary edema in the normal range of breathing. J Appl Physiol 78:1889–1897
26. Ludwig MS, Dreshaj I, Solway J, Munoz A, Ingram RH Jr (1987) Partitioning of pulmonary resistance during constriction in the dog: effects of volume history. J Appl Physiol 62: 807–815
27. Ludwig MS, Robatto FM, Simard S, Stamenovic D, Fredberg JJ (1992) Lung tissue resistance during contractile stimulation: structural damping decomposition. J Appl Physiol 72:1332–1337
28. Ludwig MS, Shore SA, Fredberg JJ, Drazen JM (1988) Differential response of tissue viscance and collateral resistance to histamine and leukotriene C4. J Appl Physiol 65:1424–1429
29. Kaczka DW, Ingenito EP, Israel E, Lutchen KR (1999) Airway and lung tissue mechanics in asthma: effects of albuterol. Am J Respir Crit Care Med 159:169–178
30. Petak F, Babik B, Asztalos T, Hall GL, Deak ZI, Sly PD, Hantos Z (2003) Airway and tissue mechanics in anesthetized paralyzed children. Pediatr Pulmonol 35:169–176
31. Tepper R, Sato J, Suki B, Martin JG, Bates JHT (1992) Low-frequency pulmonary impedance in rabbits and its response to inhaled methacholine. J Appl Physiol 73: 290–295
32. Hoppin FG Jr, Strothert JC Jr, Greaves IA, Lai, Y-L, Hildebrandt J (1986) Lung recoil: elastic and rheological properties. In: Macklem PT, Mead J (eds) Handbook of physiology. American Physiological Society, Bethesda, MD, pp 195–215

33. Agostoni E, Hyatt RE (1986) Static behavior of the respiratory system. In: Macklem PT, Mead J (eds) Handbook of physiology. American Physiological Society, Bethesda, MD, pp 113–130
34. Mead J (1961) Mechanical properties of lungs. Physiol Rev 41:281–330
35. Hildebrandt J (1969) Dynamic properties of air-filled excised cat lung determined by liquid plethysmograph. J Appl Physiol 27:246–250
36. Hildebrandt J (1970) Pressure–volume data of the cat determined by a plastoelastic, linear viscoelastic model. J Appl Physiol 28:365–372
37. Venegas JG, Harris RS, Simon BA (1998) A comprehensive equation for pulmonary pressure–volume curve. J Appl Physiol 84:389–395
38. Hager DN, Fuld M, Kaczka DW, Fessler HE, Brower RG, Simon BA (2006) Four methods of measuring tidal volume during high-frequency oscillatory ventilation. Crit Care Med 34:751–757
39. Mead J (1956) Measurement of inertia of the lungs at increased ambient pressure. J Appl Physiol 9:208–212
40. Hopkins SR, Henderson AC, Levin DL, Yamada K, Arai T, Buxton RB, Prisk GK (2007) Vertical gradients in regional lung density and perfusion in the supine human lung: the Slinky effect. J Appl Physiol 103:240–248
41. Kaczka DW, Hager DN, Hawley ML, Simon BA (2005) Quantifying mechanical heterogeneity in canine acute lung injury: impact of mean airway pressure. Anesthesiology 103:306–317
42. Fuld MK, Easely RB, Saba OI, Chon D, Reinhardt JM, Hoffman EA, Simon BA (2008) CT-measured regional specific volume change reflects regional ventilation in supine sheep. J Appl Physiol 104:1177–1184
43. Reinhardt JM, Ding K, Cao K, Christensen GE, Hoffman EA, Bodas SV (2008) Registration-based estimates of local lung tissue expansion compared to xenon CT measures of specific ventilation. Med Image Anal 12:752–763
44. Rodarte JR, Rehder K. Dynamics of respiration. In: Macklem PT, Mead J (eds) Handbook of physiology. American Physiological Society, Bethesda, MD, pp 131–144
45. Amygdalou A, Psarakis C, Vassiliou P, Dalavanga YA, Mandragos C, Constantopoulos SH, Behrakis PK, Vassiliou MP (2002) Evaluation of the end-expiratory pressure by multiple linear regression and Fourier analysis in humans. Respir Med 96:499–505
46. Kaczka DW, Barnas GM, Suki B, Lutchen KR (1995) Assessment of time-domain analyses for estimation of low-frequency respiratory mechanical properties and impedance spectra. Ann Biomed Eng 23:135–151
47. Peslin R, de Silva JF, Chabot F, Duvivier C (1992) Respiratory mechanics studied by multiple linear regression in unsedated ventilated patients. Eur Respir J 5:871–878
48. Ruiz-Ferron F, Rucabado Aguilar L, Ruiz Navarro S, Ramirez Puerta R, Parra Alonso S, Munoz Munoz JL (2001) Results of respiratory mechanics analysis in the critically ill depend on the method employed. Intensive Care Med 27:1487–1495
49. Bates JHT, Lutchen KR (2005) The interface between measurement and modeling of peripheral lung mechanics. Respir Physiol Neurobiol 148:153–164
50. Kaczka DW, Ingenito EP, Body SC, Duffy SE, Mentzer SJ, DeCamp MM, Lutchen KR (2001) Inspiratory lung impedance in COPD: effects of PEEP and immediate impact of lung volume reduction surgery. J Appl Physiol 90:1833–1841
51. Lutchen KR, Greenstein JL, Suki B (1996) How inhomogeneities and airway walls affect frequency dependence and separation of airway and tissue properties. J Appl Physiol 80:1696–1707
52. Lutchen KR, Hantos Z, Petak F, Adamicza A, Suki B (1996) Airway inhomogeneities contribute to apparent lung tissue mechanics during constriction. J Appl Physiol 80: 1841–1849
53. Lutchen KR, Jensen A, Atileh H, Kaczka DW, Israel E, Suki B, Ingenito EP (2001) Airway constriction pattern is a central component of asthma severity: the role of deep inspirations. Am J Respir Crit Care Med 164:207–215

54. Lutchen KR, Suki B (1996) Understanding pulmonary mechanics using the forced oscillations technique. In: Khoo MCK (ed) Bioengineering approaches to pulmonary physiology and medicine. Plenum Press, New York, NY, pp 227–253
55. Thorpe CW, Bates JHT (1997) Effect of stochastic heterogeneity on lung impedance during acute bronchoconstriction: a model analysis. J Appl Physiol 82:1616–1625
56. Lutchen KR, Gillis H (1997) Relationship between heterogeneous changes in airway morphometry and lung resistance and elastance. J Appl Physiol 83:1192–1201
57. Lutchen KR, Hantos Z, Jackson AC (1988) Importance of low-frequency impedance data for reliably quantifying parallel inhomogeneities of respiratory mechanics. IEEE Trans Biomed Eng 35:472–481
58. Mead J (1969) Contribution of compliance of airways to frequency-dependent behavior of lungs. J Appl Physiol 26:670–673
59. Otis AB, McKerrow CB, Bartlett RA, Mead J, McIlroy MB, Selverstone NJ, Radford EP Jr (1956) Mechanical factors in the distribution of pulmonary ventilation. J Appl Physiol 8:427–443
60. Gillis HL, Lutchen KR (1999a) Airway remodeling in asthma amplifies heterogeneities in smooth muscle shortening causing hyperresponsiveness. J Appl Physiol 86:2001–2012
61. Bellardine C, Ingenito EP, Hoffman A, Lopez F, Sanborn W, Suki B, Lutchen KR (2005) Heterogeneous airway versus tissue mechanics and their relation to gas exchange function during mechanical ventilation. Ann Biomed Eng 33:626–641
62. Bates JHT (1991) Lung mechanics – the inverse problem. Aust Phys Eng Sci Med 14: 197–203
63. Daroczy B, Hantos Z (1982) An improved forced oscillatory estimation of respiratory impedance. Int J Biomed Comput 13:221–235
64. Haase M, Schmalisch G, Meffert B (2007) Convergence properties of a new technique for estimating parameters in a nonlinear viscoelastic lung model in newborns. Comput Biol Med 37:1750–1758
65. Hantos Z, Daroczy B, Suki B, Nagy S, Fredberg JJ (1992) Input impedance and peripheral inhomogeneity of dog lungs. J Appl Physiol 72:168–178
66. Bates JHT, Allen G (2006) The estimation of lung mechanics parameters in the presence of pathology: a theoretical analysis. Ann Biomed Eng 34:384–392
67. Suki B, Yuan H, Zhang Q, Lutchen KR (1997) Partitioning of lung tissue response and inhomogeneous airway constriction at the airway opening. J Appl Physiol 82:1349–1359
68. Yuan H, Suki B, Lutchen KR (1998) Sensitivity analysis for evaluating nonlinear models of lung mechanics. Ann Biomed Eng 26:230–241
69. Ito S, Ingenito EP, Arold SP, Parameswaran H, Tgavalekos NT, Lutchen KR, Suki B (2004) Tissue heterogeneity in the mouse lung: effects of elastase treatment. J Appl Physiol 204–212
70. Sakai H, Ingenito EP, Mora R, Abbay S, Cavalcante FSA, Lutchen KR, Suki B (2001) Hysteresivity of the lung and tissue strip in the normal rat: effects of heterogeneities. J Appl Physiol 91:737–747
71. Kaczka DW, Ingenito EP, Lutchen KR (1999) Technique to determine inspiratory impedance during mechanical ventilation: implications for flow-limited patients. Ann Biomed Eng 27:340–355
72. Weibel ER, Gomez DM (1962) Architecture of the human lung. Science 137:577–585
73. Horsfield K (1986) Morphometry of airways. In: Fishman AP, Macklem PT, Mead J, Geiger SR (eds) Handbook of physiology – section 3: the respiratory system. The American Physiology Society, Bethesda, MD, pp 75–88
74. Horsfield K, Dart G, Olson DE, Filley GF (1971) Cumming G Models of the human bronchial tree. J Appl Physiol 31:207–217
75. Horsfield K, Kemp W, Phillips S (1982) An asymmetrical model of the airways of the dog lung. J Appl Physiol 52:21–26
76. Fredberg JJ, Hoenig A (1978) Mechanical response of the lungs at high frequencies. J Biomech Eng 100:57–66

77. Fredberg JJ, Moore JA (1978) The distributed response of complex branching duct networks. J Acoust Soc Am 63:954–961
78. Gillis HL, Lutchen KR (1999b) How heterogeneous bronchoconstriction affects ventilation distribution in human lungs: a morphometric model. Ann Biomed Eng 27:14–22
79. Jackson AC, Tabrizi M, Kotlikoff MI, Voss JR (1984) Airway pressures in an asymmetrically branched airway model of the dog respiratory system. J Appl Physiol 57:1222–1230
80. Sidell RS, Fredberg JJ (1978) Noninvasive interface of airway network geometry from broadband lung reflection data. J Biomech Eng 100:131–138
81. Kitaoka H, Takaki R, Suki B (1999) A three-dimensional model of the human airway tree. J Appl Physiol 87:2207–2217
82. Tawhai MH, Hunter P, Tschirren J, Reinhardt JM, McLennan G, Hoffman EA (2004) CT-based geometry analysis and finite element models of the human and ovine bronchial tree. J Appl Physiol 97:2310–2321
83. Tawhai MH, Pullan AJ, Hunter PJ (2000) Generation of an anatomically based three-dimensional model of the conducting airways. Ann Biomed Eng 28:793–802
84. Murray CD (1926) The physiological principle of minimum work: I. The vascular system and the cost of blood volume. Proc Natl Acad Sci 12:207–214
85. Murray CD (1927) A relationship between circumference and weight in trees and its bearing on branching angles. J Gen Physiol 10:725–729
86. Sherman TF (1981) On connecting large vessels to small: the meaning of Murray's law. J Gen Physiol 78:431–453
87. Tawhai MH, Hunter PJ (2004) Modeling water vapor and heat transfer in the normal and the intubated airway. Ann Biomed Eng 32:609–622
88. McFadden ER Jr, Pichurko BM, Bowman HF, Ingenito E, Burns S, Dowling N, Solway J (1985) Thermal mapping of the airways in humans. J Appl Physiol 58:564–570
89. Tgavalekos NT, Tawhai M, Harris RS, Musch G, Vidal-Melo M, Venegas JG, Lutchen KR (2005) Identifying airways responsible for heterogeneous ventilation and mechanical dysfunction in asthma: an image functional modeling approach. J Appl Physiol 99:2388–2397
90. Brown R, Mitzner W (2003a) Functional imaging of airway narrowing. Respir Physiol Neurobiol 137:327–337
91. Brown RH, Mitzner W (2003b) Understanding airway pathophysiology with computed tomography. J Appl Physiol 95:854–862
92. Gattinoni L, Chiumello D, Biondetti P, Carlesso E (2005) CT Ventilation imaging: technical background and impact in acute lung injury and ARDS management. In: Lipson DA, van Beek EJR (eds) Functional lung imaging. CRC Press, Boca Raton, FL, pp 33–61
93. Stavngaard T, Mortensen J, Dirksen A (2005) Emphysema/alpha-1 antitrypsin deficiency. In: Lipson DA, van Beek EJR (eds) Functional lung imaging. CRC Press, Boca Raton, FL, pp 453–478
94. Hoffman EA, Tajik JK, Kugelmass SD (1995) Matching pulmonary structure and perfusion via combined dynamic multislice CT and thin-slice high resolution CT. Comput Med Imaging Graph 19:101–112
95. Christensen GE, Song JH, Lu W, El Naqa I, Low DA (2007) Tracking lung tissue motion and expansion/compression with inverse consistent image registration and spirometry. Med Phys 34:2155–2163
96. Guerrero T, Castillo R, Sanders K, Price R, Komaki R, Cody D (2006) Novel method to calculate pulmonary compliance images in rodents from computed tomography acquired at constant pressures. Phys Med Biol 51:1101–1112
97. Kaczka DW, Kumar D, Christensen GE, Massa CB, Simon BA (2006) Assessment of regional mechanics in acute lung injury using 3D image registration. Proc Am Thorac Soc 3:A788 (abstract)

98. Musch G, Venegas JG (2005) Positron emission tomography imaging of regional pulmonary perfusion and ventilation. Proc Am Thorac Soc 2:522–527
99. Musch G, Venegas JG (2006) Positron emission tomography imaging of regional lung function. Minerva Anesthesiol 72:363–367
100. Musch G, Venegas JG, Bellani G, Winkler T, Schroeder T, Petersen B, Harris RS, Vidal Melo MF (2007) Regional gas exchange and cellular metabolic activity in ventilator-induced lung injury. Anesthesiology 106:723–735
101. Hoffman EA, Clough AV, Christensen GE, Lin, C-L, McLennan G, Reinhardt JM, Simon BA, Sonka M, Tawhai MH, van Beek EJR, Wang G (2004) The comprehensive imaging-based analysis of the lung: a forum for team science. Acad Radiol 11:1370–1380
102. Millis GH, Wild JM, Eberle B, van Beek EJR (2003) Functional magnetic resonance imaging of the lung. Br J Anaesth 91:16–30
103. Tzeng Y-S, Hoffman E, Cook-Granroth J, Gereige J, Mansour J, Washko G, Cho M, Stepp E, Lutchen K, Albert M (2008) Investigation of hyperpolarized ^{3}He magnetic resonance imaging utility in examining human airway diameter behavior in asthma through comparison with high-resolution computed tomography. Acad Radiol 15:799–808
104. van Beek EJR, Hoffman EA (2008) Functional imaging: CT and MRI. Clin Chest Med 29:195–216
105. Brown RH, Kaczka DW, Fallano K, Chen S, Mitzner W (2008) Temporal variability in responses of individual canine airways to methacholine. J Appl Physiol 104:1381–1386
106. Mitzner W, Brown R, Lee W (2001) In vivo measurement of lung volumes in mice. Physiol Genomics 4:215–221
107. Vidal Melo MF, Layfield D, Harris RS, O'Neill K, Musch G, Richter T, Winkler T, Fischman AJ, Venegas JG (2003) Quantification of regional ventilation–perfusion ratios with PET. J Nucl Med 44:1982–1991
108. Vidal Melo MF, Harris RS, Layfield JD, Venegas JG (2005) Topographic basis of bimodal ventilation–perfusion distributions during bronchoconstriction in sheep. Am J Respir Crit Care Med 171:714–721
109. Klarreich E (2003) Take a deep breath. Nature 424:873–874
110. Samee S, Altes T, Powers P, de Lange EE, Knight-Scott J, Rakes G, Mugler JP 3rd, Ciambotti JM, Alford BA, Brookeman JR, Platts-Mills TA (2003) Imaging the lungs in asthmatic patients by using hyperpolarized helium-3 magnetic resonance: assessment of response to methacholine and exercise challenge. J Allergy Clin Immunol 111:1205–1211
111. Venegas JG, Schroeder T, Harris S, Winkler RT, Vidal Melo MF (2005) The distribution of ventilation during bronchoconstriction is patchy and bimodal: a PET imaging study. Respir Physiol Neurobiol 148:57–64
112. Venegas JG, Winkler T, Musch G, Vidal Melo MF, Layfield D, Tgavalekos N, Fischman AJ, Callahan RJ, Bellani G, Harris RS (2005) Self-organized patchiness in asthma as a prelude to catastrophic shifts. Nature 434:777–782
113. Winkler T, Venegas JG (2007) Complex airway behavior and paradoxical responses to bronchoprovocation. J Appl Physiol 103:655–663
114. Anafi R, Wilson TA (2001) Airway stability and heterogeneity in the constricted lung. J Appl Physiol 91:1185–1192
115. De Backer JW, Vos WG, Gorlé CD, Germonpré P, Partoens B, Wuyts FL, Parizel PM, De Backer W (2008) Flow analyses in the lower airways: patient-specific model and boundary conditions. Med Eng Phys 30:872–879
116. Lin, C-L, Tawhai MH, McLennan G, Hoffman EA (2007) Characteristics of the turbulent laryngeal jet and its effect on airflow in the human intra-thoracic airways. Respir Physiol Neurobiol 157:295–309
117. Simon BA (2000) Non-invasive imaging of regional lung function using X-ray computed tomography. J Clin Monit Comput 16:433–442
118. Simon BA (2005) Regional ventilation and lung mechanics using X-ray CT. Acad Radiol 12:1414–1422

119. Hubmayr RD, Walters BJ, Chevalier PA, Rodarte JR, Olson LE (1983) Topographical distribution of regional volume in anesthetized dogs. J Appl Physiol 54:1048–1056
120. Martynowicz MA, Minor T, Walters BJ, Hubmayr RD (1999) Regional expansion of oleic acid-injured lungs. Am J Respir Crit Care Med 160:250–258
121. Martynowicz MA, Walters BJ, Hubmayr RD (2001) Mechanisms of recruitment in oleic acid-injured lungs. J Appl Physiol 90:1744–1753
122. Bellardine-Black CL, Hoffman AM, Tsai L, Ingenito EP, Suki B, Kaczka DW, Simon BA, Lutchen KR (2007) Relationship between dynamic respiratory mechanics and disease heterogeneity in sheep lavage injury. Crit Care Med 35:870–878
123. Gattinoni L, Pelosi P, Crotti S, Valenza F (1995) Effects of positive end-expiratory pressure on regional distribution of tidal volume and recruitment in adult respiratory distress syndrome. Am J Respir Crit Care Med 151:1807–1814
124. Gattinoni L, Vagginelli F, Chiumello D, Taccone P, Carlesso E (2003) Physiologic rationale for ventilator setting in acute lung injury/acute respiratory distress syndrome patients. Crit Care Med 31:S300–S304
125. Hubmayr RD (2002) Perspective on lung injury and recruitment: a skeptical look at the opening and collapse story. Am J Respir Crit Care Med 165:1647–1653
126. Chen C-T (2003) Radiologic image registration: old skills and new tools. Acad Radiol 10:239–241
127. Christensen GE, Johnson HJ (2001) Consistent image registration. IEEE Trans Med Imaging 20:568–582
128. Dougherty L, Asmuth JC, Gefter WB (2003) Alignment of CT lung volumes with an optical flow method. Acad Radiol 10:249–254
129. Guerrero T, Zhang G, Huang TC, Lin KP (2004) Intrathoracic tumour motion estimation from CT imaging using the 3D optical flow method. Phys Med Biol 49:4147–4161
130. Johnson HJ, Christensen GE (2002) Consistent landmark and intensity-based image registration. IEEE Trans Med Imaging 21:450–461
131. Pan Y, Kumar D, Hoffman EA, Christensen GE, McLennan G, Song JH, Ross A, Simon BA, Reinhardt JM (2005) Estimation of regional lung expansion via 3D image registration. Proc SPIE 5764:453–464
132. Simon BA, Christensen GE, Low DA, Reinhardt JM (2005) Computed tomography studies of lung mechanics. Proc Am Thorac Soc 2:517–521
133. Ganesan S, Rouch KE, Lai-Fook SJ (1995) A finite element analysis of the effects of the abdomen on regional lung expansion. Respir Physiol 99:341–353
134. Matthews FL, West JB (1972) Finite element displacement analysis of a lung. J Biomech 5:591–600
135. Kowalczyk P, Klieber M (1994) Modeling and numerical-analysis of stresses and strains in the human lung including tissue-gas interaction. Eur J Mech Solids 13:367–393
136. Lee GC, Tseng NT, Yuan YM (1983) Finite element modeling of lungs including interlobar fissures and the heart cavity. J Biomech 16:679–690
137. Tani J, Nakamura M, Sasaki H, Okubo T, Takishima T, Hildebrandt J (1982) Mechanical behavior of lung parenchyma as a compressible continuum: a theoretical analysis. Tohoku J Exp Med 137:125–136
138. Tawhai MH, Nash MP, Hoffman EA (2006) An imaging-based computational approach to model ventilation distribution and soft-tissue deformation in the ovine lung. Acad Radiol 13:113–120
139. Tawhai MH, Nash MP, Tschirren J, Hoffman EA, Hunter PJ (2005) An image-based computational model of ovine lung mechanics and ventilation distribution. In: Amini AA, Manduca A (eds) SPIE medical imaging. SPIE, San Diego, CA, pp 84–91

Chapter 11
Native Human and Bioprosthetic Heart Valve Dynamics

Hyunggun Kim, Jia Lu, and K.B. Chandran

Abstract Native human heart valves undergo complex deformation during a cardiac cycle and the tissue leaflets are subjected to regions of stress concentrations particularly during the opening and closing phases. Diseases of the heart valves include stenosis and valvular incompetence and the valves in the left heart (aortic and mitral valves) subjected to higher pressure loads are more prone to these diseases. A correlation has been established between regions of high stress concentration on the leaflets and regions of calcification and tissue failure. Computational simulations play a significant role in the determination of stress distribution on the leaflets during a cardiac cycle. In this chapter, the development of state-of-the-art structural analysis of the biological leaflet valves as well as fluid–structure interaction algorithms for the analysis of biological tissue valve dynamics are described. The potential application of the computational analyses on improving the design of biological heart valve prostheses is discussed. The need for further advancements in multiscale simulation for increasing our understanding of the effect of mechanical stresses on the leaflet microstructure is also pointed out.

11.1 Human Heart Valves

The human circulatory system provides blood supply without interruption to the various organs in the body. Flowing blood transports nutrients, hormones, and gases for the metabolic needs of the cells, removes the waste products for excretion, and regulates the body temperature. The energy required for the maintenance of the blood flow is supplied by the contracting of the heart. The human heart consists of two pumps in series that maintain blood flow through the pulmonary and systemic

K.B. Chandran (✉)
Department of Biomedical Engineering, College of Engineering, 1138 Seamans Center, The University of Iowa, Iowa City, IA 52242, USA
e-mail: chandran@engineering.uiowa.edu

K.B. Chandran et al. (eds.), *Image-Based Computational Modeling of the Human Circulatory and Pulmonary Systems*, DOI 10.1007/978-1-4419-7350-4_11,

circulation. The right heart, the low-pressure pump, circulates the oxygen-depleted venous blood through the lungs where carbon dioxide is removed and the blood is replenished with oxygen from the air we breathe. The oxygenated blood returns to the left heart, the high-pressure pump that circulates blood through the systemic circulation, providing oxygen and nutrients to the cells through various tissues and organs in the body. Four heart valves ensure that the blood flows only in the forward direction by opening at the appropriate time during a cardiac cycle and closing efficiently to prevent backflow or regurgitation of the blood (Fig. 11.1).

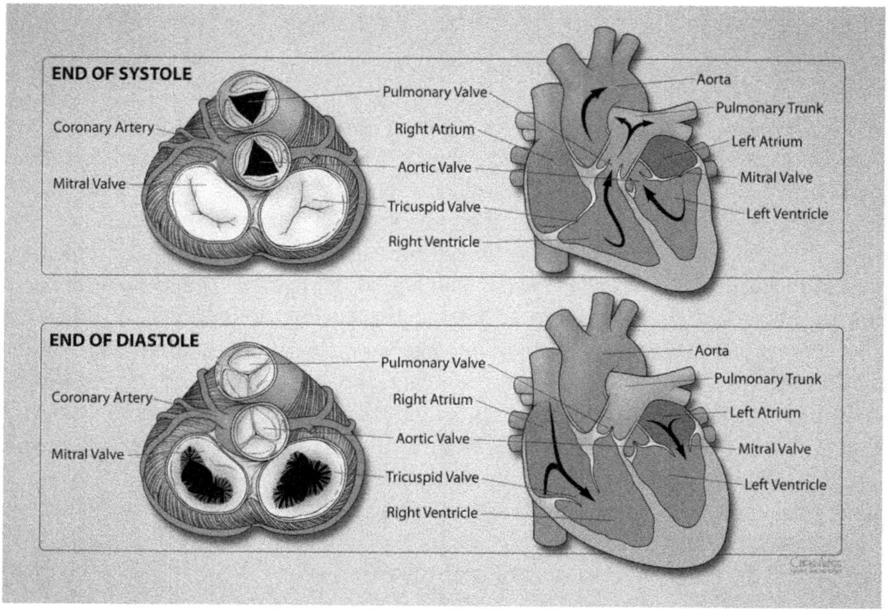

Fig. 11.1 Opening and closing of the heart valves at the end of diastole and at the end of systole. The aortic and pulmonary valves are fully closed and the mitral and tricuspid valves are fully open at the end of diastole. On the contrary, the mitral and tricuspid valves are fully closed and the aortic and pulmonary valves are fully open at the end of systole. (Courtesy of H. J. Safi, MD and illustrated by C. Akers, University of Texas Health Science Center at Houston)

The venous blood flows continuously from the inferior and superior vena cavae to the right atrium. During the ventricular diastole, the tricuspid valve between the right atrium and the right ventricle opens, allowing the blood to fill the right ventricle. During the ventricular contraction, the tricuspid valve closes and the pulmonic valve opens, directing the blood through the pulmonary artery into the lungs. The oxygenated blood from the lungs is transported by the pulmonary veins and continuously fills the left atrium. During diastole, the mitral valve (bicuspid valve) between the left atrium and the left ventricle opens, allowing the blood to fill the left ventricle. During systole, the mitral valve closes and the aortic valve opens with the blood flowing through the aorta to various parts of the body in the systemic circulation. The aortic and pulmonary valves are anatomically similar and so are the tricuspid and mitral valves. The valves open and close more than 100,000 times per day and more than three billion times over the average human life span. The valve leaflets undergo complex deformation during the opening and closing phases of the cardiac cycle and are subjected to in-plane and bending stresses during each cardiac cycle. The valves on the right side are subjected to a pressure load of about 30 mmHg when the valves are fully closed, whereas the aortic and mitral valve leaflets bear significantly higher pressures. Diseases of the valves are more often encountered in the aortic and mitral valves and hence the relationship between the dynamics of valvular function and valvular diseases are of more interest regarding the valves of the left heart. In this chapter, we will consider the dynamics of the aortic and mitral valves, the relationship between mechanical stresses and valvular diseases, as well as the problems associated with biological prostheses that are commonly used as replacements.

11.2 Aortic Valve

The aortic valve complex consists of the aortic root, valve leaflets, and aortic sinuses acting as a functional unit. The aortic root is a fibrous annular ring that separates the aorta from the left ventricle at the end of the ventricular outflow tract. The valve is composed of three semicircular (semilunar) leaflets attached to the fibrous root. The cusps meet at the three commissures that are equally spaced along the circumference at the supraaortic ridge (Fig. 11.2).

This ridge is the thickening of the aorta at which the cusps insert and there is no continuity of tissue from one cusp to the other. Each of the leaflets is lined with valvular endothelial cells that are morphologically different from the vascular endothelial cells. The leaflets consist of three layers: ventricularis, spongiosa, and fibrosa [1–3]. The layer on the aortic side is the fibrosa and is the major fibrous layer within the belly of the leaflet. This layer is composed predominantly of a dense network of collagen and elastin fibers and is the major stress-bearing layer, as this surface of the leaflet is subjected to the diastolic aortic pressure during the part of the cardiac cycle when the valve is fully closed. The ventricularis, on the ventricular aspect of the leaflet, also consists of a network of collagen and elastin fibers. The spongiosa contains a high concentration of proteoglycans and this layer apparently does not bear the mechanical load on the leaflet. However, numerous interconnecting fibrous structures couple the fibrosa and ventricularis layers. Interstitial cells permeate all three layers with larger concentration in the spongiosa. Collagen fibers can withstand high tensile forces, but have lower torsional and flexural stiffness [3]. In the leaflets, the collagen fibers are predominantly in the circumferential direction [4]. Furthermore, the collagen fibers are crimped at lower transvalvular pressures and the fibers become taut with increase in the same. The elastin fibers are oriented primarily along the radial direction. Based on the structural characteristics of the leaflets due to elastin and collagen fiber orientations, the leaflets are stiffer along the circumferential direction compared to the radial direction, resulting in

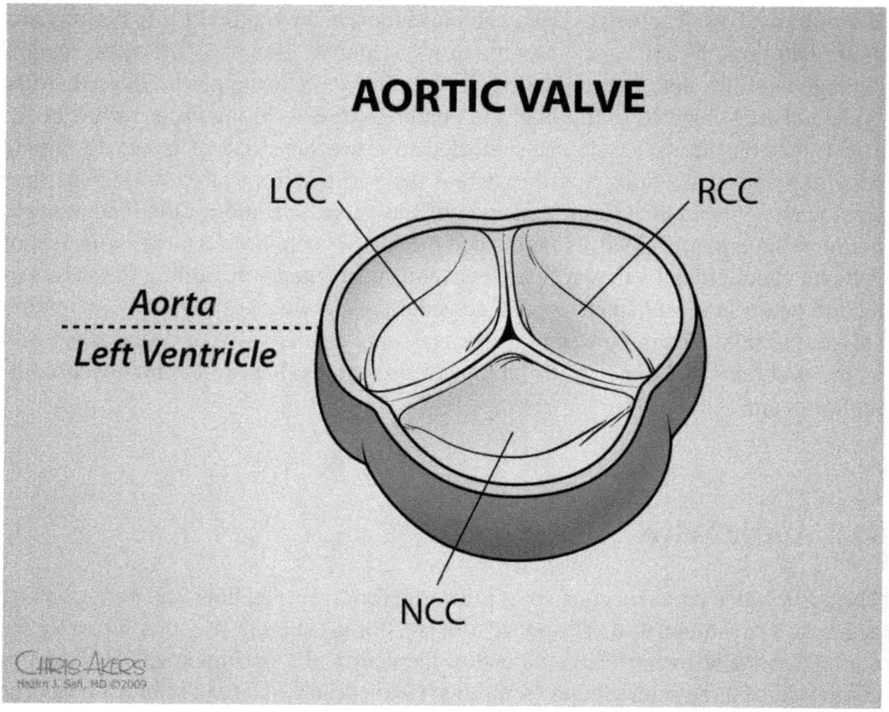

Fig. 11.2 The aortic sinuses and valve in the closed position. The noncoronary cusp (NCC) is in front. The left and right coronary cusps (LCC and RCC) are positioned as marked. The aorta is above the closed valve in this orientation and the left ventricle is below the *dashed line*. (Courtesy of H. J. Safi, MD and illustrated by C. Akers, University of Texas Health Science Center at Houston)

anisotropic material properties. In addition, with increase in transvalvular pressure, the leaflets become stiffer due to the straightened collagen fibers taking more tensile load. Biaxial [5–8] and flexural tests [9, 10] on the leaflets have been reported to determine the in-plane and flexural behavior of the leaflets for constitutive modeling. The sinus of Valsalva is attached to the fibrous annular ring on the aortic side. The sinus consists of three bulges at the root aligned with the belly or the central part of the leaflet. The ostia for the coronary arteries feeding blood supply to the muscles of the heart arise from the left and right sinuses, and the third is the noncoronary sinus. In the fully open position, the valve leaflets extend to the upper ridges of the sinuses. The length of the leaflets is greater than the radius of the annulus and a small overlap of tissue from each leaflet protrudes and forms a coaptation surface within the aorta when the valve is closed. Such an overlap ensures that the valve is sealed in the closed position, preventing regurgitation of blood back into the left ventricle during diastole [1, 11]. When the valve is closed, the total length of the free edges of the three leaflets is approximately six times the radius of the annulus [4] and the three-leaflet geometry results in a circular opening without significant changes in leaflet length.

The aortic valve opens when the left ventricular pressure exceeds the aortic pressure during early systole and the leaflets move to the fully open position in about 20–30 ms. Blood accelerates rapidly through the valve and peak blood velocity is reached during the first third of systole, after which the flow begins to decelerate. Since the heart rotates and translates during the contraction–relaxation cycle, the base of the aorta translates along the axis of the aorta and changes in size during a cardiac cycle. The base perimeter is the largest at end diastole and decreases to a minimum at end systole with the perimeter variation of about 22% during a cardiac cycle [12]. The velocity profile at the aortic root has been shown to be relatively flat with a slight skewing toward the septal wall. The peak forward flow velocity in the aorta is about 1.35 m/s. Even though during the peak acceleration the Reynolds number for flow in the aorta exceeds the critical value for flow through a cylindrical tube, velocity measurements at the aortic root have shown that disturbed flow is present during the deceleration phase but fully developed turbulence is not anticipated with a functional healthy aortic valve [13]. During systole, vortices develop in all three sinuses behind the leaflets. It has been suggested that the vortices and the relatively higher pressure of blood in the sinuses compared to the central core flow region of the valve help move the leaflets toward closure even when the forward flow persists during the latter half of systole [14]. In addition, the adverse pressure difference within the aorta also causes the low inertial fluid in the boundary layer to reverse direction and push the belly of the leaflet to move away from the aortic wall. The net result is an efficient closure of the leaflets at the end of systole, resulting in very little regurgitation (<2 mL/beat) of blood back into the left ventricle. The leaflets in the fully closed position are subjected to a pressure of about 80 mmHg on the aortic side, while the ventricular pressure is about 0 mmHg.

The pulmonic valve is anatomically similar to the aortic valve and slightly larger. The leaflets of the pulmonic valve are subjected to smaller pressure loads compared to those of the aortic valve and the velocity magnitudes of blood flow past the pulmonic valve are also smaller compared to those past the aortic valve.

11.3 Mitral Valve

The mitral valve consists of two main leaflets (cusps): the posterior and anterior leaflets [15] (Fig. 11.3).

The two leaflets of the mitral valve are continuous tissue with two regularly placed indentations called the commissures. The combined surface area of both leaflets is approximately twice that of the valve orifice and hence the leaflets coaptate during valve closure [11]. The leaflets are inserted into the dense connective tissue of the saddle-shaped annulus. Chordae tendinae emanating from the two papillary muscles in the left ventricle are attached at several locations of the free edge of the leaflets. The tethering of the chordae tendinae during a cardiac cycle plays an important role in the dynamics of the mitral valve leaflets and the chordae also prevent the leaflets from prolapsing into the left atrium during ventricular ejection.

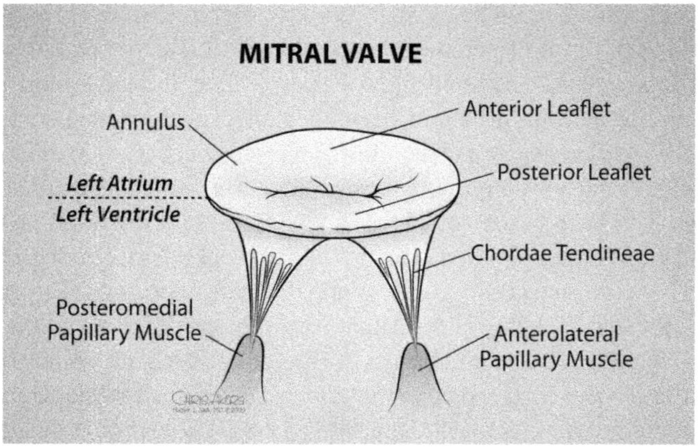

Fig. 11.3 Schematic of the mitral valve showing the valve leaflets, papillary muscles, and chordae tendinae. (Courtesy of H. J. Safi, MD and illustrated by C. Akers, University of Texas Health Science Center at Houston)

The posterior leaflet encircles two-thirds of the annulus and is quadrangular shaped, while the anterior leaflet is roughly triangular in shape with a round apical contour [16]. The anterior leaflet is thin and translucent at the base with a thick opaque area bordering the free edge where the chordae tendinae insert into the ventricular surface of the leaflet. The free edge of the posterior leaflet has one or two indentations resulting in a scalloped appearance and also has a rough zone where the chordae insert into the leaflet-free edges. The rough zones of both the leaflets appose each other during valve closure. Unlike the aortic valve leaflets, substantial numbers of cardiac muscle fibers have been demonstrated in the human mitral valve [16]. These appear to be similar to the atrial muscles and hence are believed to be activated in conjunction with the atrial wall and may contribute toward the valve function.

At the beginning of diastole, the ventricular relaxation and the reduction of ventricular pressure below that of the left atrium result in the opening of the mitral valve. In addition, the shortening of the papillary muscles during and after the isovolumic relaxation enables the separation of the leaflets for ventricular filling [16]. Ventricular filling results in the generation of two vortices behind each of the leaflets [17]. Due to the shearing forces on the leaflets by the vortices, the leaflets move toward closure even before the ventricular contraction. With atrial systole and additional blood moving from the atrium into the ventricle, the leaflets once again move apart. Ventricular contraction and increase in pressure on the ventricular surface of the leaflets move the valve to closure. The leaflets come into contact at their corresponding rough zones and are stabilized by the action of the chordae tendinae. The remaining leaflet surface billows into the left atrium due to the high transvalvular pressure [16].

The tricuspid valve between the right atrium and the right ventricle is functionally similar to the mitral valve even though the valve has three dominant cusps and the pressures across the leaflets are smaller in magnitude compared to those in the left-side chambers.

11.4 Diseases of the Heart Valves

Problems associated with heart valves can be due to either valvular diseases or congenital malformation. Valvular stenosis, incompetence, or a combination of both are generally the diseases associated with heart valves and are more common for the aortic and mitral valves under the high-pressure environment as well as for the tricuspid valve in the right heart. Stenosis due to calcification of the leaflets results in stiffer leaflets with a higher transvalvular pressure gradient being required to open the valve. In addition, the orifice area of the fully open area also reduces significantly. Aortic sclerosis due to aging can result in stenotic valves in some patients, and rheumatic heart disease also results in stenosis, particularly with the mitral valve [4, 17]. Commissural fusion can also result in valvular stenosis that can be surgically repaired. In the case of the mitral valve, thickened and fused chordae may also result in considerable obstruction to flow [18]. Incomplete closure of the leaflets, decrease in leaflet area due to rheumatic disease, or leaflet perforations due to bacterial endocarditis can result in regurgitation. Structural alterations such as loss of commissural support or aortic root dilatation can result in incompetent aortic valve. Mitral annulus dilatation, chordal rupture, or papillary muscle abnormality can also result in mitral regurgitation.

Approximately 1–2% of humans are born with bicuspid aortic valves with an approximate 2:1 male predominance, while unicuspid or quadric-cuspid aortic valves are rare. Among the patients with bicuspid valves, it is estimated that 30–50% will require surgical intervention some time in their lives [19]. Surgery in patients with bicuspid valve is due to the presence of aortic stenosis, aortic insufficiency, endocarditis, ascending aortic aneurysm, and aortic dissection. Aortic valve replacement is a common treatment for patients with aortic stenosis, and bicuspid aortic valve is the major cause of aortic stenosis. Most of the patients under 50 years of age undergoing aortic valve replacement to treat valvular stenosis have bicuspid aortic valves, and even under 70 years of age, the majority of patients with aortic stenosis have bicuspid aortic valves [20, 21]. Nearly one-half of the patients requiring aortic valve replacement to treat aortic insufficiency or resection of the ascending aortic aneurysm have bicuspid aortic valves. Thus, the presence of a bicuspid aortic valve is a strong risk factor for valvular disease or ascending aortic pathology.

Valvular reconstructive surgery or repair, rather than valvular replacement, results in lower risk of mortality and recurrence [22]. Mitral valve regurgitation due to the prolapse of the posterior leaflet is generally treated with valvular repair. In the case of regurgitation due to annular dilatation, implantation of a prosthetic ring to improve leaflet coaptation is a common treatment. Mitral endocarditis with valvular or chordal lesions is also surgically repaired rather than repaired with valvular replacement. The predominant cause for aortic valve regurgitation is also dilatation of the aortic root and prolapse of the leaflets.

11.5 Biological Valve Prostheses

In the case of severe valvular diseases and where surgical repair is not an option, replacement of the diseased valve with valve prosthesis has been a viable treatment modality for more than five decades. With the development of the technique for bypassing the blood from the lungs and the heart through cardio-pulmonary bypass and the availability of cold potassium cardioplegia to arrest the heart to perform open-heart surgery in the 1950s, valve replacement surgery has become a common treatment modality. Transplantation of freshly explanted human heart valves from donors who died of noncardiovascular disease is the most ideal replacement for diseased valves. However, even transplanted valves lack cellular regeneration capability and hence are vulnerable to deterioration over long-term use. Furthermore, options to use homograft valves are limited due to the lack of availability of donors and hence this option is not generally practical.

Currently available valve prostheses can be broadly classified into mechanical and biological prostheses. The mechanical valve prostheses currently available for implantation, as well as the advantages and problems associated with those implants, are addressed in Chapter 12, which discusses simulation of mechanical heart valve dynamics. In this chapter, our discussion is focused on the simulations of the dynamics of the native heart valves, as well as on the biological valve prostheses. Hence, we will briefly consider the development of biological prostheses below [1, 11, 23].

Biological valves made of autologous fascia lata tissue (a membrane that encases the thigh muscles) were initially attempted. This was not successful due to the inadequate strength of the tissue for long-term cyclic loading, in addition to the fact that the tissue was prone to deterioration. Attempts to make valves from human duramater tissue suffered from the lack of availability for commercial manufacture in sufficient quantities. Biological valve prostheses from animal tissues have been successfully used as replacement and include excised bovine pericardial tissue made in the form of trileaflet valves or excised and treated porcine aortic valves. Early attempts on the use of xenograft valves employed tissue treatment with organic mercurial salts or formaldehyde to overcome problems of rejection of foreign tissue by the body. Formaldehyde is used by histologists to fix and preserve excised tissue and results in shrinkage and stiffening of the specimens. Formaldehyde-treated valves suffered from durability and 60% of the implants failed within 2 years after implantation. Thus, glutaraldehyde became the preferred medium for preservation of biological valve prostheses in maintaining the structural integrity of the tissue. Glutaraldehyde also reduces the antigenicity of foreign tissue implants and the valves can be implanted without significant immunological reaction.

Currently available biological valve prostheses are either preserved porcine aortic valves or trileaflet valves made of preserved bovine pericardial tissue. Porcine aortic valves are harvested from pigs with the aortic root and shipped to the valve manufacturers in chilled saline solution. The valves are cleaned, trimmed, fitted, and sewn to supporting stents. These supporting stents, in the form of three upright wide posts with a base ring, were earlier made of metal, and subsequently made of flexible material such as polypropylene in a rigid base. The stent is covered in a fabric and a sewing flange is attached to the base for suturing the valve in place during implantation. The stented valves are preserved in glutaraldehyde in concentrations ranging from 0.2 to 0.625% in low-pressure (<4 mmHg) fixation. A supra-annular valve design allows the valve to be implanted on top of the aortic annulus while aligning the internal diameter of the valve to the patient's annulus, thus making it possible to implant a larger valve for a given annular size. The tissue valves are also treated with antimineralization solutions such as sodium dodecyl sulfate (SDS) to reduce the tendency for leaflet calcification. Simulations have shown that regions of stress concentration are present at the leaflet–stent junction that may be prone to failure. In order to avoid such regions of stress concentration, more recently, stentless bioprostheses have been introduced. The absence of stents results in less obstruction to flow across the valve. Larger size valves can also be implanted for a given aortic orifice in order to improve the hemodynamics. At present, stentless valves are implanted only in the aortic position since implanting of stentless valves in the mitral position is more complex. Pericardial valves are made by forming three leaflets from the bovine pericardial tissue in a geometrical configuration similar to the human aortic valve. In these valves, harvested bovine pericardial tissue is debrided of fatty deposits and trimmed to remove nonusable tissue before being fixed in glutaraldehyde. After fixation, leaflets are cut from selected areas and sewn to cloth-covered stents appropriately in order to obtain coapting and fully sealing cusps. Stentless pericardial valves are also being introduced into the market. Pictures of typical pericardial and porcine bioprosthetic valves are shown in Fig. 11.4.

In the case of implanted mechanical valve prostheses, thrombus deposition on the valve structures and subsequent embolic complications are significant; thus, patients with implanted mechanical valves require a long-term anticoagulant therapy. The biological prostheses are generally not prone to thrombus deposition and have the advantage of the patients not requiring anticoagulants on a long-term basis. However, in spite of the improved design and tissue preservation techniques, the biological prostheses are prone to leaflet calcification and tissue failure. Hence, tissue valves suffer from durability problems and only last for an average of 10–12 years, after which these valves need to be replaced [2, 11]. As preserved tissue leaflets, these valves also undergo complex deformation during a cardiac cycle, resulting in regions of stress concentration that have been linked to the failure of the valves.

11.6 Experimental Studies on Valve Dynamics

During a cardiac cycle, the heart valve leaflets undergo complex deformation and regions of stress concentration are present, especially during the opening and closing phases of the valve function. From a mechanical point of view, it can be anticipated that regions of high stress concentration are prone to structural failure and microstructural alterations resulting in calcification and leaflet tearing. Alterations in the geometry of the valve complex, such as enlargement of the valve annulus, can

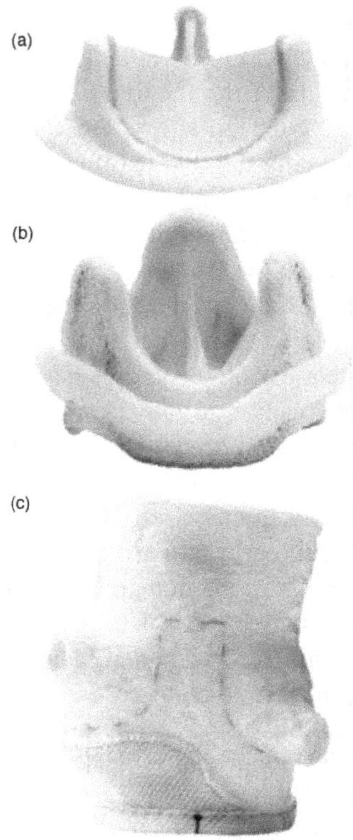

Fig. 11.4 Bioprosthetic aortic valves: (**a**) pericardial, (**b**) stented porcine, and (**c**) stentless porcine valves. (From Edwards Lifesciences Co. website: http://www.edwards.com with permission)

also result in abnormal valve function and valvular diseases. In addition, alterations in the deformation characteristics of the leaflets can result in abnormal flow dynamics in the vicinity of the valve structures inducing additional pathologies. Hence, it is of interest to understand the complex dynamics of the valve structures and the flow past the valves in order to understand the physiology and pathophysiology of the valve function. Such an understanding and comparison of the dynamics of native valves with biological valve prostheses can also be very useful in identifying the mechanical bases for loss of durability of the prostheses. The information gained from such an analysis can also be used to advantage in improved designs of the prostheses to increase the durability of the implants. Numerous experimental studies, both in vivo and in vitro, have been employed to analyze the dynamics of native heart valves as well as prosthetic valves.

The rapid motion of the leaflets during the opening and closing of an aortic valve has been studied by aortic root cine-angiography in which radiopaque dye is injected into the aortic root and the motion of the dye is recorded using x-ray. Such studies with the human aortic valve have shown that the normal aortic valve opens to either a circular or triangular orifice [24], while other studies have shown that the valve opens to a circular orifice particularly in early systole [25]. Van Steenhoven et al.

[26] have shown that the aortic valve initially opens to a circular orifice and changes to a triangular orifice at the end of systole. A common method employed in vivo in animal models to study the motion of the leaflets and other valvular apparatus is to place small radiopaque markers on the leaflets; the movement of the markers is then observed under x-ray [4, 27]. Thubrikar et al. [25, 28] have employed the radiopaque marker motion tracking method to study the motion of the leaflets, commissures, and the base of the valve in detail. The complex geometric variations in the mitral valve such as annular, leaflet, and papillary muscle displacements have been measured using radiopaque marker placement in normal valve function and in the presence of mitral regurgitation [27, 29]. Markers of sufficient mass are required to be visible in x-ray imaging and hence the leaflet motion will be affected by the inertial effect in this method. In addition, only the markers are visible in the images, and hence the complex motion of the leaflets cannot be studied with sufficient accuracy using this technique. Sonomicrometry array localization technique has also been effectively employed to quantitatively track the three-dimensional motion of the mitral valve complex [30]. Implanted piezoelectric crystals provide the ability to measure distances between the crystals in an aqueous medium. In addition, various imaging modalities such as high-speed videography [31], two-dimensional [32, 33] and three-dimensional [34–39] echocardiographic imaging techniques have been employed to assess the mitral valve structure and function in vivo. Saito et al. [31] have observed the motion of the mitral valve in isolated and working swine hearts through high-speed videographic images obtained with an endoscope. Their studies revealed integrated movement of the mitral leaflets, chords, and annulus, and the effect of preload on the mitral valve function. Echocardiographic images are generally employed to assess the shape and function of the annulus, the motion of the leaflets, and the alterations in function in the presence of mitral regurgitation, cardiomyopathy, and annuloplasty ring. The in vivo assessment of valvular function, such as those described above, has been exploited to study the dynamics of valvular function as well as alterations in function in the presence of diseases such as valvular regurgitation. In vivo studies have also been employed in combination with numerical simulation to evaluate related parameters such as annulus shape and stresses developed on the leaflets [39].

Even though in vivo studies provide gross alterations in function in the presence of diseases, such methods do not provide details of the stress and strain distribution on the leaflets and other structures in the valvular complex. Several in vitro studies have also been attempted to gain additional insights into the mechanics of the leaflets with physiological loading. Adamczyk and Vesely [40] measured the strain patterns on isolated porcine aortic valve cusps and roots with pressure applied on the valve. A pattern of dots was marked on the ventricular side of the cusp and a pair of cameras was employed to measure the structural distortion and to compute the strain distribution. This study demonstrated that the noncoronary cusp strain pattern was different from the right and left coronary cusps. A noncontacting laser-light projection method was employed by Iyengar et al. [41] to measure the motion of a bioprosthetic valve leaflet under dynamic loading. This study showed that leaflets exhibited significant flexural deformation and the shapes of the leaflets during the opening and closing phases were different. However, since this technique uses a projected grid on the leaflets that is not material points, these measurements will not yield information on the strain pattern on the leaflet. In vitro analyses of the mitral valve complex, including the chordae and papillary muscles, have been analyzed extensively with the valves mounted in an in vitro physiological flow loop system [42–44]. These studies have demonstrated the effect of papillary muscle position on the dynamic strain on the mitral valve leaflets, the effect of the saddle-shaped annulus on the mitral valve function and chordal force distribution, as well as the dynamic strain behavior on the valve leaflets under physiological loading. However, measurements of strain distribution on the leaflets require suturing of markers on the leaflets that can potentially alter the motion of the leaflets.

As is evident from the review of the dynamics of the native and biological prosthetic heart valves, one of the significant problems with the valvular function is leaflet calcification and structural failure. In order to understand the relationship between the dynamics of the leaflets and the resulting microstructural alterations such as calcification and failure, detailed information on the dynamic stress and strain distribution on the leaflets is necessary. Even though in vivo and in vitro experiments can provide us with some information on the stress and strain distribution, the data are limited due to physical constraints in the measurements, as well as to the fact that these experiments are often time consuming and expensive. Numerical simulations, once validated, can be used to advantage on detailed measurements of the deformations and strains on the valvular structures under dynamic loading. These simulations can be used to understand the alterations in the valve dynamics under diseases, and to assess the effect of implants such as annuloplasty rings on the mitral valve dynamics. The simulations will also prove to be valuable in improving the designs of biological prostheses and developing tissue-engineered heart valves. We describe the details, to date, of computational simulations on heart valve dynamics.

11.7 Three-Dimensional Geometrical Reconstruction of the Aortic and Mitral Valves

Noninvasive three-dimensional (3D) cardiac imaging modalities can be generally categorized into ultrasound imaging (echocardiography), computed tomography (CT), and magnetic resonance imaging (MRI). In this section, advantages and limitations of each imaging modality are briefly described.

Currently, the most widely used noninvasive imaging modality to examine heart valve function for clinical purposes is echocardiography. Echocardiography not only offers patients a cost-effective valve examination but also provides physicians both morphologic and functional information on the heart valves. With good spatial and temporal resolutions, echocardiography can accurately capture detailed morphologic change of the fast-moving valve structure over the cardiac cycle. Another advantage of echocardiography is that functional information of the blood flow such as flow direction and velocity around the valves can be identified using color Doppler ultrasound. Primary disadvantages of echocardiography include a restricted window view in some patients, low signal-to-noise ratio compared to CT and MRI, acoustic reflection, ultrasound attenuation, and relatively high interobserver variability.

CT and MRI can provide less noisy images compared to echocardiography, with a high signal-to-noise ratio, as well as comprehensive cross-sectional images of the patient's entire body regardless of tissue types. Primary disadvantages of CT and MRI in heart valve imaging are the difficulty in capturing the moving valve leaflets and relatively high cost. In addition, CT imaging modality has critical disadvantages such as radiation exposure and potential risk of contrast administration compared to the other imaging modalities. Both CT and MRI examinations are generally limited by patient size. Most MRI and CT scanners can accommodate patients up to 350 and 450 lb, respectively, while ultrasound imaging has less difficulty with overweight patients. Ultrasound imaging also has an advantage for patients with severe claustrophobia who may not undergo CT or MRI imaging. Because of these limitations, CT and MRI have been less popular for heart valve assessment as a primary diagnostic modality. However, with dramatically rapid improvement in technology, these two imaging modalities are increasingly playing an important role in helping physicians better understand the morphology and function of the heart valves, in addition to echocardiographic imaging data.

All of these imaging modalities can provide 3D images using 3D rendering techniques (Fig. 11.5).

In general, 3D data are created by registering a series of 2D slice images acquired by any of these imaging modalities. Following a specific registration pattern for each imaging modality, voxel elements are generated, and this process is referred to as a volume rendering. Volume rendering can be performed as either real-time image processing or image postprocessing. Real-time 3D echocardiography (RT3DE) utilizes a matrix phased-array transducer directly creating pyramidal volume data in real time; thus, image postprocessing is not additionally required for 3D visualization.

3D geometric data of the heart valves from these imaging modalities can be utilized for structural and/or fluid analyses in computational studies. Validation studies on computational analyses of the heart valve function can be conducted using information obtained from these noninvasive imaging systems.

11.7.1 3D Echocardiography

Echocardiography has been the primary standard clinical imaging modality for examining cardiac function and valvular diseases. Transesophageal echocardiography (TEE) can provide relatively better image acquisition than transthoracic echocardiography (TTE) due to the closer distance between the ultrasound probe and the heart. However, TEE may give some patients discomfort since the probe

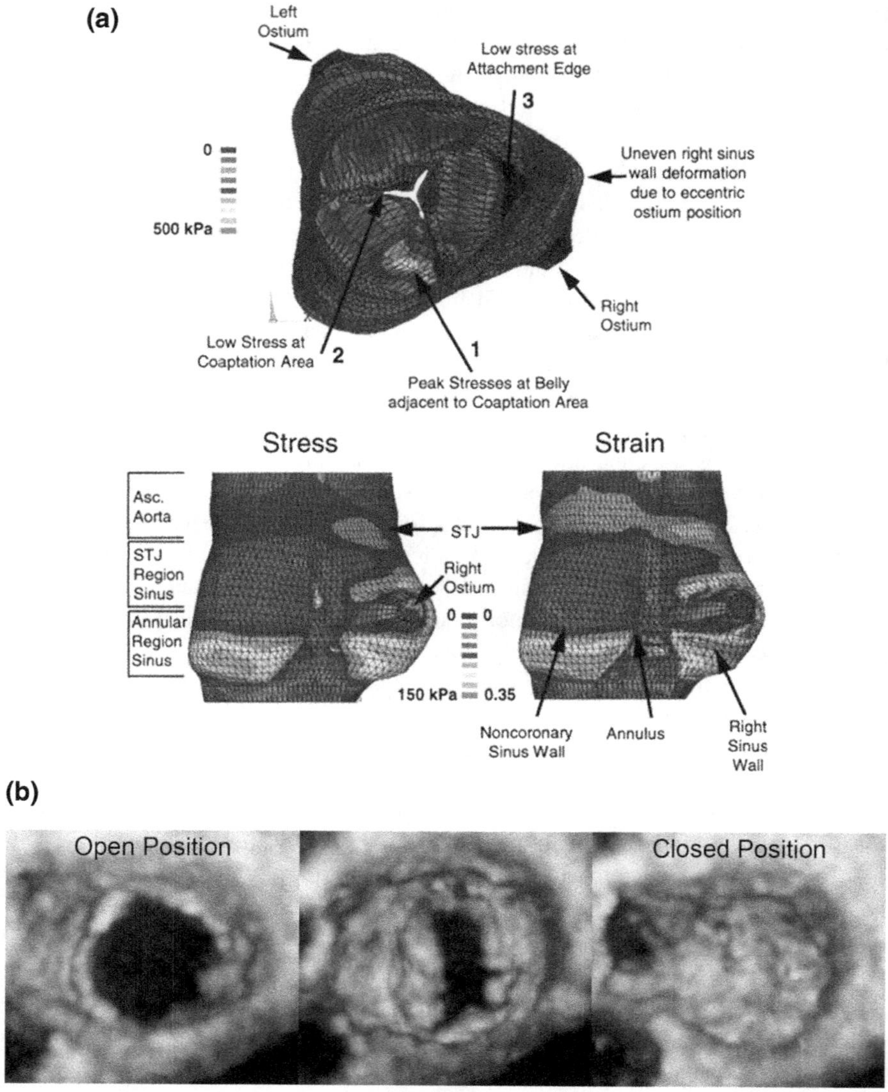

Fig. 11.5 Examples of 3D geometrical reconstruction and simulation: (**a**) FE model of the aortic valve and root created from MRI data (from Grande et al. [72] with permission from Springer, Netherlands) and (**b**) volumetric 3D images of the mitral valve obtained from transesophageal echocardiography

has to be inserted into the esophagus. With technical advances during the last two decades, echocardiography can provide effective 3D images of the heart and valves, helping physicians improve the diagnosis of valvular diseases. Current RT3DE systems utilize more than 6,000 imaging elements in the matrix array, allowing excellent 3D images of the heart valves in real time without volume rendering [45,

46]. Because RT3DE creates intrinsic 3D volume data, 2D images are derived from the 3D volume data by utilizing 2D cutting planes.

Although 2D echocardiography can provide effective cross-sectional views of the heart to examine valvular diseases, 3D assessment is more suitable for evaluation of valve function. In particular, 3D echocardiography has significantly benefited mitral valve evaluation as the complex mitral valve structure includes the asymmetric leaflets, chordae tendinae, papillary muscles, and saddle-shaped annulus. 3D echocardiography has been utilized to characterize mitral annular shape [34, 35], diagnose mitral valve prolapse and leaflet perforation [47, 48], and measure mitral valve area [49–51]. 3D echocardiography is now widely being used for cardiac surgeons to evaluate mitral valve geometry and focus the surgical procedures prior to mitral valve repair.

Ryan et al. [37, 38, 52] have developed a methodology for quantitative characterization of human mitral valve geometry using RT3DE and demonstrated that their image-processing and graphic-rendering techniques can provide a complete description of 3D mitral valve geometry in human subjects. Chaput et al. have also performed a similar study [53]. Vehrey et al. [54] have developed a rapid transforming technique of the mitral valve geometry data from 3D echocardiography to a finite element (FE) analysis model.

11.7.2 3D Computed Tomography

The most popular application area of cardiac CT imaging is coronary CT angiography. CT technology has been significantly advanced over the last two decades, and new state-of-the-art multislice CT systems such as 16 or 64 multidetector (MDCT) systems can provide both anatomical and functional cardiac information. To date, typical MDCT can provide an in-plane spatial resolution of approximately 0.5 mm with slice thickness from 0.4 to 0.6 mm [55].

With the retrospective electrocardiogram (ECG) gating technique, a series of 2D CT images are obtained over the cardiac cycle, enabling ECG-gated 3D reconstruction with a relatively low temporal resolution after image postprocessing. Gantry rotation speed has been improved from 500 to 330 ms/rotation, resulting in improved temporal resolution of 150–200 ms. This temporal resolution can provide 3D data at 5–10% intervals over the cardiac cycle with the newest MDCT systems [56].

Morgan-Hughes et al. [57] have recently demonstrated that aortic valve calcification can be quantitatively evaluated using the MDCT system. They have measured and visualized 3D aortic valve calcification volume and demonstrated that this technique is reproducible. Boughner et al. [58] have reported excellent 3D images of an explanted porcine aortic bioprosthetic valve using their custom-designed microtomography system, which can provide a spatial resolution with a voxel size of 0.003 mm^3. This technique, however, cannot be translated to the clinical arena due to the size of the microtomography system.

11.7.3 3D Magnetic Resonance Imaging

MRI has been offering useful clinical tools to assess various cardiac functions. Cardiac MRI can also be incorporated with the ECG gating technique to capture moving heart images and large deformation of the heart valves. Most currently available MRI in the market can provide a spatial resolution of 1–2 mm and temporal resolution of 20–50 ms. MRI can, therefore, provide better images of leaflet motion, though less quality images of valve morphology, compared to CT imaging. Phase-contrast MRI can provide quantitative information on blood flow around the valves. Vogel-Claussen et al. [59] have reported a detailed review comparing MRI and 64 MDCT systems on 3D heart valve assessment.

Dowsey et al. [60] have developed a novel technique to overcome the difficulty in capturing large movements of the heart valves and demonstrated 3D aortic valve image acquisition over the cardiac cycle. Kaji et al. [61] have also performed a clinical 3D MRI study on the measurement of the mitral annulus and papillary muscle geometry.

11.8 Computational Simulations of the Native Valves

11.8.1 Aortic Valve

The total annual mortality due to valvular heart diseases was approximately 20,000 patients per year in 2008 [62]. This mortality is predominantly related to aortic valve diseases, with the remainder related to mitral valve diseases, endocarditis, and unspecified valvular diseases. Aortic stenosis is generally caused by the degenerative calcific changes of a trileaflet valve, the secondary calcification of a congenital abnormal valve or rheumatic valve disease, while aortic regurgitation is caused commonly by abnormalities in the valve leaflets and aortic root [63]. It has been a traditionally accepted belief that calcification in the aortic valve is distinct from the pathogenesis of atherosclerosis, but recently several studies have been reported suggesting shared mechanistic features between calcific aortic stenosis and atherosclerosis [64]. Even though the pathologic mechanism of calcification in the aortic valve is still under active discussion, many researchers have been studying it to better determine the relationship between structural deterioration and excessive calcification of the aortic valve leading to valvular stenosis. FE analysis is an effective method for structural evaluation of the aortic valves. This is due in part to our understanding that localized concentration of mechanical stress and large flexural deformation are closely related to the tissue degeneration and calcification in the aortic valve [28, 65–67].

The normal aortic valve and root structure is composed of the three leaflets, three sinuses, and aortic base. Most FE simulation studies of the native aortic valve [68–71] have utilized an FE model of the aortic valve structure using a computer-aided design (CAD) software based on typical geometric parameters of the natural aortic valve reported by Thubrikar [4]. One-sixth of the valve structure is generally simulated with the CAD-designed aortic valve model due to the symmetrical structural design. On the other hand, Grande et al. [72–75] have imaged an entire ex vivo homograft specimen of the aortic valve and root under MRI scanner, created an FE model by converting the image data into structural data, and performed a series of studies on the aortic valve simulation. They have demonstrated the effect of asymmetry of the aortic valve anatomy [72], the loss of coaptation, and increased stiffness by aging [73], and progressive increases of leaflet stress and strain with aortic root dilatation [74, 75] using a linear elastic material model for the final closing phase of the cardiac cycle. Others [68, 69, 71] have emphasized the importance of dynamic FE analysis rather than static analysis. These simulation studies presented the relationship between the orifice area and material stiffness over time or under different pressure loads. However, all of these studies have not provided appropriate experimental data to validate their FE simulations.

There are several effective experimental methods to generate data for validation of computational aortic valve dynamics simulations. Haj-Ali et al. [76] have demonstrated an experimental flow system allowing qualitative and quantitative measurements of deformation and transvalvular pressures of the aortic valve under physiological pulsatile flow conditions. Robicsek et al. [77] have imaged morphologic changes of the human bicuspid aortic valves in a heart simulator such as orifice shape, leaflet deformation, and root function using high-speed video cameras and intravascular ultrasound imaging system. Sacks et al. [43, 78] have demonstrated a novel methodology to experimentally measure the strain distribution on the valve leaflets under physiological conditions. The advent of more advanced experimental systems is expected to effectively provide comprehensive characteristics of the valve dynamics resulting in allowing more accurate validation data for computational analyses in the near future.

The role of collagen fiber architecture in the aortic valve leaflets has also been studied using FE analysis. Driessen et al. [79–81] have performed a series of FE analyses to investigate the interaction between collagen fiber remodeling and mechanical loading conditions in the aortic valve leaflets. They have demonstrated a strong resemblance of the computer-predicted fiber directions to those in the native aortic valve. More recently, fluid–structure interaction (FSI) analysis has been utilized for aortic valve simulations and the details of aortic valve studies with FSI models are described later in this chapter.

11.8.2 Mitral Valve

Mitral valve repair and replacement are the major indications for cardiac valve surgery. The mitral valve apparatus has a complex anatomical structure consisting of two asymmetric leaflets, a saddle-shaped annulus, chordae tendinae, and papillary muscles. Abnormal mitral valve morphology due to pathologic tissue alterations, improper valve repair, or replacement may lead to abnormally high stress concentration on the mitral valve leaflets resulting in tissue damage and valve failure. It is important to assess physiologic characteristics of the mitral valve for accurate diagnosis and appropriate treatment of mitral valve diseases, such as mitral regurgitation and mitral prolapse [34]. The most common treatment for early mitral valve pathology is surgical repair, with valve replacement performed if the mitral valve is severely diseased [82, 83]. A thorough understanding of the dynamics of mitral valve function and structural characteristics of the mitral valve apparatus is imperative for accurate diagnosis and focused treatment to mitral valve pathology.

Mitral valve repair in the early days of cardiac surgery was limited to a small percentage of mitral valve cases, but is now employed by surgeons in more than 90% of cases [84, 85]. In the early days, a repair involving plication of the commissures between the anterior and posterior leaflets was popular for those cases with central directed jets of mitral insufficiency [86]. More involved techniques with application of an annuloplasty ring to decrease annular diameter and improve coaptation of anterior and posterior leaflets increased the rate of repair in patients undergoing surgery for mitral valve insufficiency [87]. Recent surgical advancements including replacement of the chordae tendinae and chordal shortening, in addition to more sophisticated annuloplasty ring implants, have resulted in an increasing repair rate of more than 90% [88, 89]. The unsolved problem in mitral valve repair surgery is predicting which repair will be optimal for each patient. Much of the difficulty lies in not precisely understanding the mitral valve physiology, which predisposes it toward dysfunction and insufficiency. The majority of cases have complex pathophysiologic involvement combining multiple pathologies including mitral valve annular enlargement, chordal lengthening, chordal rupture, calcification of the mitral valve structures, lack of leaflet coaptation. Conventional imaging techniques alone cannot accurately determine which pathology is exactly present and which repair will produce the least stress and tension on the leaflets.

Computational simulation can evaluate mitral valve function to help better understand structural characteristics as well as individual roles and intercorrelation between functional components of the mitral valve. With the state-of-the-art noninvasive imaging systems, computer-predicted structural information on the mitral valve apparatus and fluid dynamic information around the mitral valve and left atrium from computational analyses may allow improved diagnosis and therapeutic approaches to mitral valve repair. In the long term, computational analysis combined with cardiac imaging systems may allow identification of structural mitral valve pathology in vivo in real time, determination of hemodynamic characteristics of blood flow around the mitral valve over the cardiac cycle, improvement of prosthetic valve design, development of novel mitral valve repair strategies, and ultimately provide a comprehensive noninvasive diagnostic modality.

Although FE analysis has been the most popular computational method to perform structural evaluation of the heart valves, FE analysis has been limited to the aortic and bioprosthetic valves due to the complicated anatomical structure and asymmetrical leaflets of the mitral valve. Only a few groups have published 3D FE simulations of the mitral valve function. Kunzelman et al. [90–93] have reported on a series of FE studies on the mitral valve. In these early studies, they created a linear elastic, anisotropic, dynamic FE model of a porcine mitral valve by manually tracing photographs of the excised mitral valve apparatus, and performed FE simulations with model alterations such as multiple types of valve modeling with or without annulus or papillary muscle, increased annular circumference, increased or decreased collagen concentration, increased leaflet thickness, and increased annular stiffness. They noted that the effects of annular and papillary muscle contraction [90], increased tissue thickness with decreased stiffness [91], decreased annular circumference [92], and annuloplasty with a flexible ring [93] would be beneficial to reduce stress concentration. Salgo et al. [39] created a mathematical phantom model of the mitral valve leaflets with a range of extent of the annular height to commissural width ratio, performed FE analyses with a linear elastic, orthotropic, static model, and correlated the results with experimentally measured annular data of humans, sheep, and baboons. Lim et al. [94] performed linear, isotropic, quasi-static FE simulation of a mathematically designed mitral valve phantom model with dynamic boundary conditions to study the effect of asymmetry.

All the FE mitral valve simulations reported above have been performed with linear materials model. The intrinsically complicated structure of the mitral valve apparatus may have led to the difficulty in nonlinear material modeling of the mitral valve. Nonlinear material properties of the porcine mitral valve leaflets [95–98] and chordae tendinae [99] have been experimentally investigated, and several FE analyses have been performed with these material properties. In these studies, a biaxial mechanical testing system was utilized for leaflet characterization and a uniaxial test was conducted for chordae tendinae. With the experimentally determined nonlinear anisotropic material properties, a more realistic FE model of the mitral valve leaflets can be achieved. The details on nonlinear hyperelastic material modeling of biological tissues are described in Chapter 4.

FE simulations of the mitral valve function can be applicable to more specific clinical cases. Unlike the aortic valve pathology, the annular shape and geometric distribution of the chordae tendinae play an important role with respect to structural abnormalities in the mitral valve functions [39, 100, 101]. Kunzelman et al. [93] have compared flexible versus rigid ring annuloplasty for annular dilatation using their FE model previously described, and demonstrated the flexible ring reduced coaptation and stresses closer to normal than the rigid ring. Votta et al. [101, 102] have utilized the same FE mitral valve model and demonstrated that a dogbone-shaped ring prosthesis provides better performance in the correction of functional mitral regurgitation than standard D-shaped ring prostheses.

Another active clinical application with computational mitral valve simulation is the edge-to-edge surgical repair technique to correct mitral valve prolapse. Mitral valve prolapse is one of the frequent mitral valvular diseases induced by abnormal elongation of chordae tendinae or other complicated multiple pathologies resulting in mitral regurgitation. The edge-to-edge technique restores valvular competence by anchoring the free edge of the prolapsing segment to the corresponding free edge of the opposing leaflet [103]. Several FE simulations have been performed to evaluate structural effects of the edge-to-edge repairs on the mitral valve with different suture sizes [104, 105] and suture positions [104, 106].

11.9 Biological Valve Prostheses

11.9.1 Quasi-Static and Dynamic FE Analyses

Bioprosthetic valves are currently used for more than a half of valve replacement surgeries [107]. Data reported in similar populations suggest that pericardial and porcine bioprostheses presently have equivalent structural durability [108]. However, it has been widely accepted that structural degeneration is frequently accompanied by excessive calcification on bioprosthetic valve leaflets causing stenosis and cuspal tears even though the detailed mechanism of calcification is still not clearly understood [109–111]. Moreover, stress concentration within the bioprosthetic valve leaflets may cause valve failure through noncalcific structural degeneration [67, 109, 112]. Cuspal commissures and basal attachments are the common locations of the deposition of calcified substances in bioprosthetic valves [109, 113, 114]. Calcification can also occur near the areas with large leaflet flexion where both tensile and compressive in-plane stresses are induced [109, 111]. It was also reported that large deflection induced substantial bending stresses in the bioprosthetic valve leaflets during the cardiac cycle, contributing significantly to the total resultant stresses [41]. Therefore, it is important to understand where and why the regions of extreme stress concentration occur during the entire cardiac cycle via computational simulations of bioprosthetic valve models. Better durability and performance of bioprosthetic valves can be anticipated if we can evaluate detailed structural characteristics with dynamic simulations and improve prosthetic valve designs and chemical treatment to reduce the resultant local stress concentration.

Material models employed in the FE simulations of bioprosthetic valve function in previous studies can be generally categorized into four groups: linear isotropic [115–117], nonlinear isotropic [69, 118–120], linear anisotropic [72–75], and nonlinear anisotropic [121–125]. Early studies used relatively simpler linear elastic or viscoelastic material models, but bioprosthetic tissues have significantly nonlinear characteristics and involve large deformations under physiological pressure loads. Patterson et al. [120] have demonstrated that a nonlinear model is clearly more responsive to the time-varying pressure loads and shows more complex deformation, thus producing significantly higher maximum principal stresses than a linear model. Burriesci et al. [126] have studied the influence of the orthotropy of chemically modified pericardium using a hyperelastic FE model and found that even a relatively small degree of orthotropy causes a significant effect on stress distributions within the leaflets as well as deformed shape. Huang et al. [127] have demonstrated that bioprosthetic valve leaflets go through large deflections and bending strains during the cardiac cycle and the bending components make a significant contribution to the resultant in-plane (membrane) stresses. Shell element modeling has been used since then for bioprosthetic valve simulation rather than membrane element. Accordingly, recent FE studies on bioprosthetic valves have successfully employed all the realistic material characteristics described above such as nonlinearity, anisotropy, and bending effect.

There have been only a few attempts to perform dynamic simulations of the bioprosthetic valve function using FE analysis. Most of the FE studies of bioprosthetic valves have performed static analyses at the closed position or quasi-static analyses. FE simulation has an excellent capability to execute dynamic analysis with advantages of capturing the consequent information in microseconds. An advanced dynamic FE analysis accurately simulating the bioprosthetic valve function during the opening and closing phases over the entire cardiac cycle can determine the levels and locations of abnormally large in-plane and bending stresses in the moving bioprosthetic valve leaflets and correlate the same with leaflet calcification and tissue degeneration. Dynamic FE analysis, therefore, may provide important and distinctive spatial and temporal information of bioprosthetic valve function, which may not be obtained from static FE analysis.

Sun et al. [125] have performed biaxial mechanical tests on pericardial tissues, implemented the material behavior into a Fung-elastic material model, and simulated pericardial bioprosthetic valve leaflet deformations under quasi-static transvalvular pressure loads. They have also measured strain distribution on the leaflets experimentally using the 3D marker coordinate reconstruction technique, and utilized the determined strain data to validate quasi-static FE analysis of the bioprosthetic valve deformation. Kim et al. [121–123] have further improved this FE model, developing a novel, fully experiment-based, shell element model for bioprosthetic valves, and demonstrating a series of dynamic FE studies on pericardial bioprosthetic heart valve simulation (Fig. 11.6).

Unlike the other previous FE heart valve models, they have implemented an experimentally determined bending moment–curvature relationship directly into the bending components of the FE shell model combined with the same in-plane

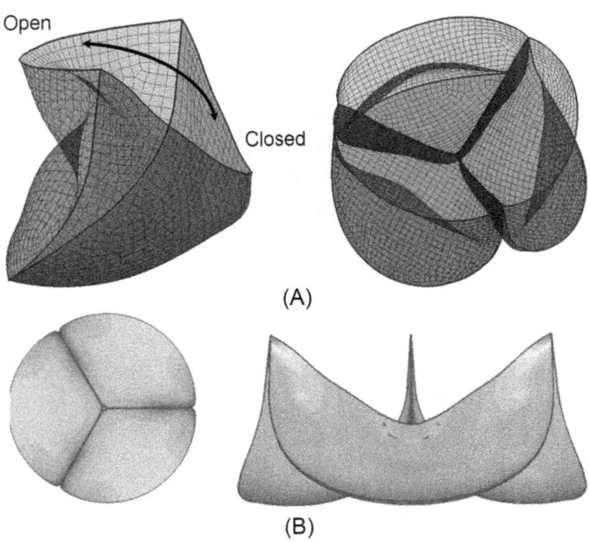

Fig. 11.6 FE simulation of a pericardial bioprosthetic valve: (**a**) leaflet displacements at the fully open and fully closed positions and (**b**) von Mises stress distribution within the leaflets at the fully closed position. (From Kim et al. [123] with permission from Springer, Netherlands)

material characteristics utilized in the study by Sun et al. [125], based on the classical shell theory.

One of the typical limitations in most FE studies on bioprosthetic valve function is the assumption of uniform pressure loads on the leaflet and surrounding structure. In order to accomplish a more realistic dynamic simulation of the fluid-induced bioprosthetic valve operation, a comprehensive FSI model is necessary.

11.9.2 Fluid–Structure Interaction Analysis

In the static or dynamic FE structural analysis of the valvular complex, particularly on the deformation and stress (or strain) distribution on the leaflets, a uniform pressure load is generally applied as the load representing the hydrostatic pressure exerted by the blood on the valve structures at the appropriate surfaces. Even though uniform pressure loads represent a reasonable approximation, such an analysis ignores the distribution of the normal and shear stresses acting on the valvular structures induced by the blood flow dynamics in the vicinity of the structures and its effects on the valve dynamics. In addition, alterations in hemodynamics due to valvular malformations such as the bicuspid aortic valve, variations in material properties of the leaflets induced by calcification, or the presence of raphes cannot be analyzed by FE structural analysis alone. FSI analysis provides a more realistic analysis of native and bioprosthetic valve function taking into consideration the effects of fluid motion and stresses on the structures, and the deformation of the valvular structures and vice versa. FSI analysis requires the simultaneous solution of the governing equations of motion for the fluid and the solid. In order to enforce the kinematic compatibility and the equilibrium at the fluid–structure interface, the solutions for the fluid and the solid must be coupled. The details of the FSI analysis, including various strategies for the coupling, are included in Chapter 5.

Peskin and colleagues pioneered the immersed boundary method for the FSI analysis of the left ventricle and mitral valve motion [128–131]. In this method, the forces exerted by the solid interface on the fluid are incorporated as a source term in the momentum equations.

Fluid–structure interaction analysis requires the analysis of fluid domain using an Eulerian reference frame, while the solid structural analysis is performed using a Lagrangian formulation, and these two formulations are incompatible [132]. The numerical difficulties associated with this method are described in Chapter 4, and the simulations employing this method have been restricted to nonphysiological flow regimes. In order to overcome this difficulty, an arbitrary Lagrangian–Eulerian (ALE) method has been employed in which the fluid mesh is continuously adapted without modifying the mesh topology. The ALE method has been successfully employed in arterial flow dynamics. However, in heart valve dynamics in which thin leaflets undergo large deformations, mesh adaptation is difficult without loss of mesh quality or changing the mesh topology. Alternatively, remeshing can be performed continuously, but may result in artificial diffusivity, and the computations are extremely difficult for 3D analysis. In order to overcome these difficulties, De Hart and colleagues [132–134] have employed a fictitious domain method to implement the interaction between the leaflets and the fluid. Although the fictitious domain method has been employed for the 2D and 3D analyses of biological heart valves, physiological Reynolds number in the flow domain has not been achieved in these simulations due to the numerical instabilities with higher Reynolds number flows [135]. In addition, the accuracy of the fictitious domain method in the treatment of the interface is lower and hence its usefulness in the computation of the detailed stress distribution on the leaflets, a quantity of interest with biological valve prostheses, is limited [136]. Makhijani et al. [137] have presented a strongly coupled 3D FSI analysis for a pericardial aortic valve function. The fluid solutions were obtained using a finite-volume flow solver while FE code was employed for the structural components. An implicit influence coefficient technique was employed for the strong coupling at the interface. The results from the dynamic analysis are compared with in vitro experimental results on the valve orifice configurations for validation. Velocity vectors and leaflet stress contours during the opening and closing phases are presented, even though detailed validation of the results is lacking in this work. Kunzelman et al. [138] have reported on an FSI analysis of the mitral valve dynamics in which anisotropic nonlinear material property was specified for the leaflets while blood was modeled as compressible fluid employing an artificial compressibility approach. However, the bulk modulus for blood was modified by several orders of magnitudes to render the magnitude to be similar to that of air for computational efficiency. Carmody et al. [139] have presented a two-stage approach to the FSI analysis of aortic valve dynamics. In the first stage, the analysis considers a left ventricular model and the inlet velocity profiles for the aortic valve are computed using the moving boundary specification of the left ventricular boundaries from experimental data and the ventricular chamber geometry. These velocity profiles are specified as inlet boundary conditions for the FSI analysis in which the 3D geometry of an idealized aortic valve is employed. A commercial FE package is employed in the FSI analysis that includes contact element analysis for simulation of leaflet coaptation. They have suggested that such an analysis is valuable in parametric studies for valve design. Nicosia et al. [140] have presented an FSI simulation of the aortic valve utilizing a MRI-based model previously developed by Grande et al. [72]. Weinberg et al. [141] have presented multiscale simulations of the aortic valve dynamics at the cell, tissue, and organ length scales. They have performed an FSI simulation of the aortic cusps and root over the cardiac cycle at the organ level, an FE analysis of the cusp comprising three distinct layers at the tissue level, and a cell-level simulation to determine cellular deformations of individual valvular interstitial cells within the cusps. Even though the FSI analyses described above have been employed to analyze the complex aortic and mitral valve dynamics, the numerical challenges encountered have generally necessitated the solutions to be limited to nonphysiological flow regimes and inadequate grid resolutions [135].

Vigmostad [135] has reported on an FSI algorithm development that overcomes the difficulties encountered in the FSI analysis in order to analyze the dynamics of biological tissue valves in detail at physiological Reynolds numbers with physiologically realistic material property specification for the leaflets. A fixed Cartesian grid flow solver is employed in order to avoid the remeshing that becomes necessary with complex motion of the leaflets. The solid interfaces in the computational domain are identified by a sharp-interface level-set technique and by updating the level-set field; the solid boundaries can move freely through the computational domain without mesh distortion. The current method affords the advantage of handling both thin and volumetric embedded objects in a unified fashion (see Chapter 5 for the details of the computational scheme). Experimentally determined nonlinear material property for the tissue leaflet is specified and computations of the deformation and stresses of the valvular structures are achieved through FE analysis with a strong coupling between the solid and the fluid at the interface. In order to adequately resolve the flow dynamics in regions such as the leaflet boundaries, a local mesh refinement algorithm is incorporated into the flow solver that automatically refines or coarsens the mesh based on the computed velocity and vorticity gradients. The local mesh refinement algorithm has been previously exploited in the analysis of flow dynamics in the regions of interest such as the clearance region between the leaflet edge and the valve housing during the closing phase [142] and the hinge region of bileaflet valves [143]. Fig. 11.7 demonstrates a two-dimensional FSI simulation of an aortic valve leaflet during the opening phase with physiologically realistic ventricular pressures.

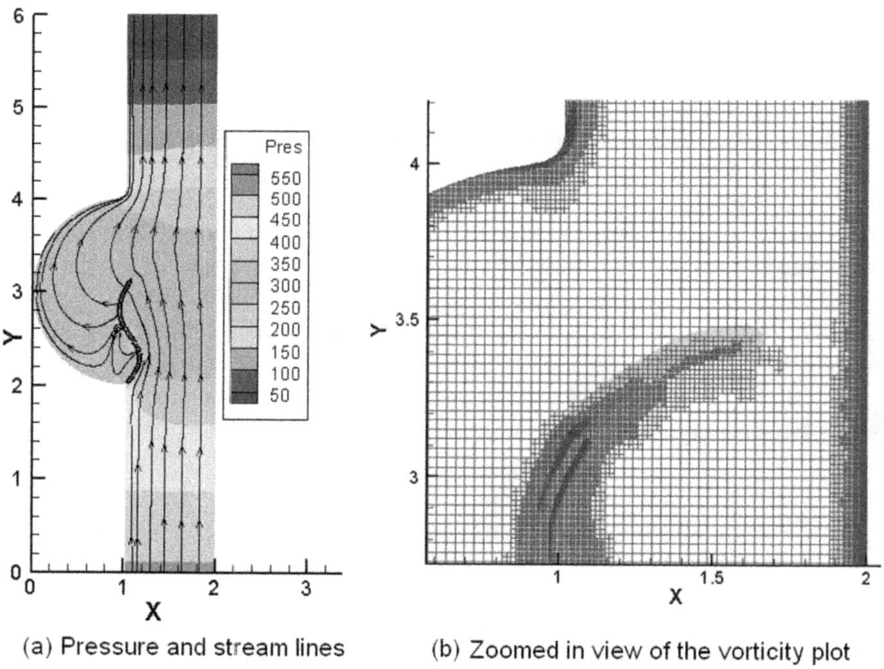

Fig. 11.7 Two-dimensional FSI simulation of aortic valve at the end of the valve opening phase: (**a**) pressure contours with superposed flow stream lines and (**b**) zoomed-in view of the vorticity profiles near the surface of the leaflets that are accurately resolved through a local mesh refinement algorithm

Experimentally determined material property specification for a pericardial bioprosthetic valve leaflet has been employed in the analysis. The leaflet deformation during the opening phase demonstrates significantly different results from De Hart et al [133], who employed a much stiffer leaflet material in their analysis. Fig. 11.7(a) represents the pressure contour with the flow streamlines superposed at the instant with the valve in the fully open position. The mean Reynolds number based on the orifice opening is around 2,000, demonstrating that the algorithm can be used to simulate physiologically realistic flow rates. Fig. 11.7(b) depicts a zoomed-in view of the vorticity plot very near the leaflets computed accurately with the local mesh refinement algorithm. Due to the high mesh density near the moving boundaries, extending this analysis to 3D requires parallelization techniques employing several processors for heavy computation.

11.10 Need for Multiscale Simulations

Advanced simulations employing physiologically realistic material properties of the valvular structures and strongly coupled FSI algorithms are becoming a reality. We now have the ability to perform complex 3D simulations of tissue valve function and compute the in-plane and bending stresses in the leaflets. A strong correlation is demonstrated between regions of relatively high mechanical stresses on the leaflets and tissue calcification and structural disruption observed with native as well as bioprosthetic heart valves. However, we still lack the knowledge of the precise mechanism through which the microstructures of the leaflets are affected by the induced mechanical stresses. The mechanical properties of the various layers of the aortic valve leaflets are now being experimentally determined and the effect of mechanical stresses on the tissue response at the cellular level is also being studied. Development of multiscale analysis strategies will aid in improving our understanding of the effect of mechanical stresses on the alterations of the biological tissues at the microstructural level. One possible approach to the problem is to develop an accurate FSI model for the analysis of valvular function at the organ level. Regions of interest with relatively large deformation and high mechanical stresses can be identified with such an analysis. A microscale analysis is then performed in the "zoomed-in" regions of interest to determine the detailed dynamics of the tissue microstructure. Strategies and appropriate algorithms are in need of development for concurrent analysis at the organ and microscale levels and appropriate transfer of information among the multiscale analyses. Development of such multiscale simulations will be important to understand the precise pathophysiologic mechanisms of valvular diseases and structural failure. Such efforts can improve our understanding on the etiology of valvular diseases and aid in developing better strategies for managing and treating these diseases. These efforts will also help to improve biological prosthetic valve design toward increased durability as well as development of tissue-engineered valve substitutes.

11.11 Summary

In this chapter, we have reviewed the current status of computational simulations of native human and bioprosthetic heart valve dynamics. There have been substantial advances in computational tools for heart valve simulation with extensive improvement of both computer hardware and simulation software over the last three decades. The ultimate goals of computational simulation of heart valve function are to accurately and comprehensively determine the characteristics of structural information and fluid dynamics during the heart valve function, allowing better understanding of native heart valve pathophysiology and better prosthetic valve design, and to evaluate the abnormality and severity of valvular diseases in vivo, allowing the most beneficial patient-specific surgical treatment. Despite the significant progress of computational techniques to date, no computational study has yet been reported to create the geometric models of aortic or mitral valve in vivo using a noninvasive imaging modality and simultaneously to perform computational analysis on the model with appropriate hemodynamic data from in vivo measurements. Although achieving such a comprehensive computational simulation technique is still a long-term goal at present, many researchers are currently working hard to develop more advanced diagnostic methodologies and there is no doubt that this will be available in the near future.

References

1. Chandran KB, Rittgers SE, Yoganathan AP (2007) Biofluid mechanics: the human circulation. CRC/Taylor & Francis, Boca Raton, FL
2. Sacks MS, Yoganathan AP (2007) Heart valve function: a biomechanical perspective. Philos Trans R Soc Lond B Biol Sci 362(1484):1369–1391
3. Yoganathan AP, Lemmon JD, Ellis JT (2000) Heart valve dynamics. In: Bronzino JD (ed) The biomedical engineering handbook. CRC Press and IEEE Press, Boca Raton, FL, pp 29.21–29.15
4. Thubrikar M (1990) The aortic valve. CRC Press, Boca Raton, FL
5. Billiar KL, Sacks MS (2000) Biaxial mechanical properties of the natural and glutaraldehyde treated aortic valve cusp – part I: experimental results. J Biomech Eng 122(1):23–30
6. Sacks MS (1999) A method for planar biaxial mechanical testing that includes in-plane shear. J Biomech Eng 121(5):551–555
7. Sacks MS, Sun W (2003) Multiaxial mechanical behavior of biological materials. Annu Rev Biomed Eng 5:251–284
8. Sun W, Sacks MS, Sellaro TL, Slaughter WS, Scott MJ (2003) Biaxial mechanical response of bioprosthetic heart valve biomaterials to high in-plane shear. J Biomech Eng 125(3): 372–380
9. Engelmayr GC, Hildebrand DK, Sutherland FW, Mayer JE, Sacks MS (2003) A novel bioreactor for the dynamic flexural stimulation of tissue engineered heart valve biomaterials. Biomaterials 24(14):2523–2532
10. Mirnajafi A, Raymer J, Scott MJ, Sacks MS (2005) The effects of collagen fiber orientation on the flexural properties of pericardial heterograft biomaterials. J. Biomech Eng 26: 795–804
11. Chandran KB (2006) Heart valve prostheses. In: Webster JG (ed) Encyclopedia of medical devices and instrumentation. Wiley, New York, NY, pp 407–426

12. Thubrikar M, Piepgrass WC, Shaner TW, Nolan SP (1981) The design of the normal aortic valve. Am J Physiol 241(6):H795–H801
13. Seed WA, Wood NB (1971) Velocity patterns in the aorta. Cardiovasc Res 5(3): 319–330
14. Bellhouse BJ, Bellhouse FH (1968) Mechanism of closure of the aortic valve. Nature 217(123):86–87
15. Ho SY (2002) Anatomy of the mitral valve. Heart 88(Suppl. iv):iv5–iv10
16. Barlow JB, Antunes MJ (1987) Functional anatomy of the mitral valve, perspectives on the mitral valve. F. A. Davis Company, Philadelphia, PA, pp 1–14
17. Reul H, Talukder N, Muller EW (1981) Fluid mechanics of the natural mitral valve. J Biomech 14(5):361–372
18. Barlow JB, Lakier JB, Pocock WA (1987) Mitral stenosis, perspectives on the mitral valve. F.A. Davis Company, Philadelphia, PA
19. Lewin MB, Otto CM (2005) The bicuspid aortic valve: adverse outcomes from infancy to old age. Circulation 111:832–834
20. Roberts WC, Ko JM (2005) Frequency by decades of unicuspid, bicuspid, and tricuspid aortic valves in adults having isolated aortic valve replacement for aortic stenosis with or without aortic insufficiency. Circulation 111:920–925
21. Roberts WC, Ko JM, Moore TR, Jones III WH (2006) Causes of pure aortic regurgitation in patients having isolated aortic valve replacement at a single US Tertiary Hospital (1993–2005). Circulation 114:422–429
22. Aazami M, Schafers HJ (2003) Advances in heart valve surgery. J Interv Cardiol 16(6): 535–541
23. Yoganathan AP (2000) Cardiac valve prostheses. In: Bronzino JD (ed) Biomedical engineering handbook. CRC Press and IEEE Press, Boca Raton, FL, pp 127.121–127.123
24. Stein PD (1971) Roentgenographic method for measurement of the cross-sectional area of the aortic valve. Am Heart J 81(5):622–634
25. Thubrikar M, Carabello BA, Aouad J, Nolan SP (1982) Interpretation of aortic root angiography in dogs and in humans. Cardiovasc Res 16(1):16–21
26. Van Steenhoven AA, Verlaan CW, Veenstra PC, Reneman RS (1981) In vivo cinematographic analysis of behavior of the aortic valve. Am J Physiol 240(2):H286–H292
27. Green GR, Dagum P, Glasson JR, Daughters GT, Bolger AF, Foppiano LE, Berry GJ, Ingels NB Jr, Miller DC (1999) Mitral annular dilatation and papillary muscle dislocation without mitral regurgitation in sheep. Circulation, 100(19 Suppl):II95–II102
28. Thubrikar MJ, Deck JD, Aouad J, Nolan SP (1983) Role of mechanical stress in calcification of aortic bioprosthetic valves. J Thorac Cardiovasc Surg 86(1):115–125
29. Tibayan FA, Rodriguez F, Zasio MK, Bailey L, Liang D, Daughters GT, Langer F, Ingels NB Jr, Miller DC (2003) Geometric distortions of the mitral valvular-ventricular complex in chronic ischemic mitral regurgitation. Circulation 108(Suppl 1):II116–II121
30. Gorman JH 3rd, Gupta KB, Streicher JT, Gorman RC, Jackson BM, Ratcliffe MB, Bogen DK, Edmunds LH Jr (1996) Dynamic three-dimensional imaging of the mitral valve and left ventricle by rapid sonomicrometry array localization. J Thorac Cardiovasc Surg 112(3): 712–726
31. Saito S, Araki Y, Usui A, Akita T, Oshima H, Yokote J, Ueda Y (2006) Mitral valve motion assessed by high-speed video camera in isolated swine heart. Eur J Cardiothorac Surg 30(4):584–591
32. Otsuji Y, Handschumacher MD, Kisanuki A, Tei C, Levine RA (1998) Functional mitral regurgitation. Cardiologia 43(10):1011–1016
33. Patel AR, Mochizuki Y, Yao J, Pandian NG (2000) Mitral regurgitation: comprehensive assessment by echocardiography. Echocardiography 17(3):275–283
34. Flachskampf FA, Chandra S, Gaddipatti A, Levine RA, Weyman AE, Ameling W, Hanrath P, Thomas JD (2000) Analysis of shape and motion of the mitral annulus in subjects with and without cardiomyopathy by echocardiographic 3-dimensional reconstruction. J Am Soc Echocardiogr 13:277–287

35. Kaplan SR, Bashein G, Sheehan FH, Legget ME, Munt B, Li XN, Sivarajan M, Bolson EL, Zeppa M, Arch MZ, Martin RW (2000) Three-dimensional echocardiographic assessment of annular shape changes in the normal and regurgitant mitral valve. Am Heart J 139(3): 378–387
36. Otsuji Y, Handschumacher MD, Schwammenthal E, Jiang L, Song JK, Guerrero JL, Vlahakes GJ, Levine RA (1997) Insights from three-dimensional echocardiography into the mechanism of functional mitral regurgitation: direct in vivo demonstration of altered leaflet tethering geometry. Circulation 96(6):1999–2008
37. Ryan LP, Jackson BM, Eperjesi TJ, Plappert TJ, St John-Sutton M, Gorman RC, Gorman JH 3rd (2008) A methodology for assessing human mitral leaflet curvature using real-time 3-dimensional echocardiography. J Thorac Cardiovasc Surg 136(3):726–734
38. Ryan LP, Jackson BM, Hamamoto H, Eperjesi TJ, Plappert TJ, St John-Sutton M, Gorman RC, Gorman JH 3rd (2008) The influence of annuloplasty ring geometry on mitral leaflet curvature. Ann Thorac Surg 86(3):749–760; discussion 749–760
39. Salgo IS, Gorman JH 3rd, Gorman RC, Jackson BM, Bowen FW, Plappert T, St John Sutton MG, Edmunds LH Jr (2002) Effect of annular shape on leaflet curvature in reducing mitral leaflet stress. Circulation 106(6):711–717
40. Adamczyk MM, Vesely I (2002) Characteristics of compressive strains in porcine aortic valves cusps. J Heart Valve Dis 11(1):75–83
41. Iyengar AKS, Sugimoto H, Smith DB, Sacks MS (2001) Dynamic in vitro quantification of bioprosthetic heart valve leaflet motion using structured light projection. Ann Biomed Eng 29(11):963–973
42. He Z, Ritchie J, Grashow JS, Sacks MS, Yoganathan AP (2005) In vitro dynamic strain behavior of the mitral valve posterior leaflet. J Biomech Eng 127(3):504–511
43. He Z, Sacks MS, Baijens L, Wanant S, Shah P, Yoganathan AP (2003) Effects of papillary muscle position on in-vitro dynamic strain on the porcine mitral valve. J Heart Valve Dis 12(4):488–494
44. Jimenez JH, Soerensen DD, He Z, He S, Yoganathan AP (2003) Effects of a saddle shaped annulus on mitral valve function and chordal force distribution: an in vitro study. Ann Biomed Eng 31(10):1171–1181
45. Correale M, Ieva R, Di Biase M (2008) Real-time three-dimensional echocardiography: an update. Eur J Intern Med 19(4):241–248
46. Hung J, Lang R, Flachskampf F, Shernan SK, McCulloch ML, Adams DB, Thomas J, Vannan M, Ryan T (2007) 3D echocardiography: a review of the current status and future directions. J Am Soc Echocardiogr 20(3):213–233
47. Ahmed S, Nanda NC, Miller AP, Nekkanti R, Yousif AM, Pacifico AD, Kirklin JK, McGiffin DC (2003) Usefulness of transesophageal three-dimensional echocardiography in the identification of individual segment/scallop prolapse of the mitral valve. Echocardiography 20(2):203–209
48. Schwalm SA, Sugeng L, Raman J, Jeevanandum V, Lang RM (2004) Assessment of mitral valve leaflet perforation as a result of infective endocarditis by 3-dimensional real-time echocardiography. J Am Soc Echocardiogr 17(8):919–922
49. Binder TM, Rosenhek R, Porenta G, Maurer G, Baumgartner H (2000) Improved assessment of mitral valve stenosis by volumetric real-time three-dimensional echocardiography. J Am Coll Cardiol 36(4):1355–1361
50. Xie MX, Wang XF, Cheng TO, Wang J, Lu Q (2005) Comparison of accuracy of mitral valve area in mitral stenosis by real-time, three-dimensional echocardiography versus two-dimensional echocardiography versus Doppler pressure half-time. Am J Cardiol 95(12):1496–1499
51. Zamorano J, Cordeiro P, Sugeng L, Perez de Isla L, Weinert L, Macaya C, Rodriguez E, Lang RM (2004) Real-time three-dimensional echocardiography for rheumatic mitral valve stenosis evaluation: an accurate and novel approach. J Am Coll Cardiol 43(11): 2091–2096

52. Ryan LP, Jackson BM, Enomoto Y, Parish L, Plappert TJ, St John-Sutton MG, Gorman RC, Gorman JH 3rd (2007) Description of regional mitral annular nonplanarity in healthy human subjects: a novel methodology. J Thorac Cardiovasc Surg 134(3):644–648
53. Chaput M, Handschumacher MD, Tournoux F, Hua L, Guerrero JL, Vlahakes GJ, Levine RA (2008) Mitral leaflet adaptation to ventricular remodeling: occurrence and adequacy in patients with functional mitral regurgitation. Circulation 118(8):845–852
54. Verhey JF, Nathan NS, Rienhoff O, Kikinis R, Rakebrandt F, D'Ambra MN (2006) Finite-element-method (FEM) model generation of time-resolved 3D echocardiographic geometry data for mitral-valve volumetry. Biomed Eng Online 5:17
55. Flamm SD (2007) Cross-sectional imaging studies: what can we learn and what do we need to know? Semin Vasc Surg 20(2):108–114
56. Woodard PK, Bhalla S, Javidan-Nejad C, Gutierrez FR (2006) Non-coronary cardiac CT imaging. Semin Ultrasound CT MR 27(1):56–75
57. Morgan-Hughes GJ, Owens PE, Roobottom CA, Marshall AJ (2003) Three dimensional volume quantification of aortic valve calcification using multislice computed tomography. Heart 89(10):1191–1194
58. Boughner DR, Thornton M, Dunmore-Buyze J, Holdsworth DW (2000) The radiographic quantitation of aortic valve calcification: implications for assessing bioprosthetic valve calcification in vitro. Physiol Meas 21(3):409–416
59. Vogel-Claussen J, Pannu H, Spevak PJ, Fishman EK, Bluemke DA (2006) Cardiac valve assessment with MR imaging and 64-section multi-detector row CT. Radiographics 26(6):1769–1784
60. Dowsey AW, Keegan J, Lerotic M, Thom S, Firmin D, Yang GZ (2007) Motion-compensated MR valve imaging with COMB tag tracking and super-resolution enhancement. Med Image Anal 11(5):478–491
61. Kaji S, Nasu M, Yamamuro A, Tanabe K, Nagai K, Tani T, Tamita K, Shiratori K, Kinoshita M, Senda M, Okada Y, Morioka S (2005) Annular geometry in patients with chronic ischemic mitral regurgitation: three-dimensional magnetic resonance imaging study. Circulation 112(9 Suppl):I409–I414
62. Rosamond W, Flegal K, Furie K, Go A, Greenlund K, Haase N, Hailpern SM, Ho M, Howard V, Kissela B, Kittner S, Lloyd-Jones D, McDermott M, Meigs J, Moy C, Nichol G, O'Donnell C, Roger V, Sorlie P, Steinberger J, Thom T, Wilson M, Hong Y (2008) Heart disease and stroke statistics–2008 update: a report from the American Heart Association Statistics Committee and Stroke Statistics Subcommittee. Circulation 117(4):e25–e146
63. Otto CM (2004) Valvular heart disease. Saunders, Philadelphia, PA
64. Schoen FJ (2008) Evolving concepts of cardiac valve dynamics: the continuum of development, functional structure, pathobiology, and tissue engineering. Circulation 118(18): 1864–1880
65. Aupart MR, Babuty DG, Guesnier L, Meurisse YA, Sirinelli AL, Marchand MA (1996) Double valve replacement with the Carpentier-Edwards pericardial valve: 10-year results. J Heart Valve Dis 5(3):312–316
66. Sacks MS (2001) The biomechanical effects of fatigue on the porcine bioprosthetic heart valve. J Long Term Eff Med Implants 11(3–4):231–247
67. Sacks MS, Schoen FJ (2002) Collagen fiber disruption occurs independent of calcification in clinically explanted bioprosthetic heart valves. J Biomed Mater Res 62(3): 359–371
68. Gnyaneshwar R, Kumar RK, Balakrishnan KR (2002) Dynamic analysis of the aortic valve using a finite element model. Ann Thorac Surg 73(4):1122–1129
69. Howard IC, Patterson EA, Yoxall A (2003) On the opening mechanism of the aortic valve: some observations from simulations. J Med Eng Technol 27(6):259–266
70. Ranga A, Mongrain R, Mendes Galaz R, Biadillah Y, Cartier R (2004) Large-displacement 3D structural analysis of an aortic valve model with nonlinear material properties. J Med Eng Technol 28(3):95–103; discussion 104

71. Sripathi VC, Kumar RK, Balakrishnan KR (2004) Further insights into normal aortic valve function: role of a compliant aortic root on leaflet opening and valve orifice area. Ann Thorac Surg 77(3):844–851
72. Grande KJ, Cochran RP, Reinhall PG, Kunzelman KS (1998) Stress variations in the human aortic root and valve: the role of anatomic asymmetry. Ann Biomed Eng 26(4): 534–545
73. Grande KJ, Cochran RP, Reinhall PG, Kunzelman KS (1999) Mechanisms of aortic valve incompetence in aging: a finite element model. J Heart Valve Dis 8(2):149–156
74. Grande KJ, Cochran RP, Reinhall PG, Kunzelman KS (2000) Mechanisms of aortic valve incompetence: finite element modeling of aortic root dilatation. Ann Thorac Surg 69(6):1851–1857
75. Grande-Allen KJ, Cochran RP, Reinhall PG, Kunzelman KS (2001) Mechanisms of aortic valve incompetence: finite-element modeling of Marfan syndrome. J Thorac Cardiovasc Surg 122(5):946–954
76. Haj-Ali R, Dasi LP, Kim HS, Choi J, Leo HW, Yoganathan AP (2008) Structural simulations of prosthetic tri-leaflet aortic heart valves. J Biomech 41(7):1510–1519
77. Robicsek F, Thubrikar MJ, Cook JW, Fowler B (2004) The congenitally bicuspid aortic valve: how does it function? Why does it fail? Ann Thorac Surg 77(1):177–185
78. Sacks MS, He Z, Baijens L, Wanant S, Shah P, Sugimoto H, Yoganathan AP (2002) Surface strains in the anterior leaflet of the functioning mitral valve. Ann Biomed Eng 30(10): 1281–1290
79. Driessen NJ, Boerboom RA, Huyghe JM, Bouten CV, Baaijens FP (2003) Computational analyses of mechanically induced collagen fiber remodeling in the aortic heart valve. J Biomech Eng 125(4):549–557
80. Driessen NJ, Bouten CV, Baaijens FP (2005) Improved prediction of the collagen fiber architecture in the aortic heart valve. J Biomech Eng 127(2):329–336
81. Driessen NJ, Bouten CV, Baaijens FP (2005) A structural constitutive model for collagenous cardiovascular tissues incorporating the angular fiber distribution. J Biomech Eng 127(3):494–503
82. Acar C, Farge A, Ramsheyi A, Chachques JC, Mihaileanu S, Gouezo R, Gerota J, Carpentier AF (1994) Mitral valve replacement using a cryopreserved mitral homograft. Ann Thorac Surg 57(3):746–748
83. Gillinov AM, Cosgrove DM, Blackstone EH, Diaz R, Arnold JH, Lytle BW, Smedira NG, Sabik JF, McCarthy PM, Loop FD (1998) Durability of mitral valve repair for degenerative disease. J Thorac Cardiovasc Surg 116(5):734–743
84. David TE, Ivanov J, Armstrong S, Rakowski H (2003) Late outcomes of mitral valve repair for floppy valves: implications for asymptomatic patients. J Thorac Cardiovasc Surg 125(5):1143–1152
85. Hayek E, Gring CN, Griffin BP (2005) Mitral valve prolapse. Lancet 365(9458): 507–518
86. Kay GL, Aoki A, Zubiate P, Prejean CA Jr, Ruggio JM, Kay JH (1994) Probability of valve repair for pure mitral regurgitation. J Thorac Cardiovasc Surg 108(5):871–879
87. Braunberger E, Deloche A, Berrebi A, Abdallah F, Celestin JA, Meimoun P, Chatellier G, Chauvaud S, Fabiani JN, Carpentier A (2001) Very long-term results (more than 20 years) of valve repair with Carpentier's techniques in nonrheumatic mitral valve insufficiency. Circulation 104(12 Suppl 1):I8–I11
88. Lawrie GM (1998) Mitral valve repair vs replacement. Current recommendations and long-term results. Cardiol Clin 16(3):437–448
89. Lawrie GM (2006) Mitral valve: toward complete repairability. Surg Technol Int 15:189–197
90. Kunzelman KS, Cochran RP, Chuong C, Ring WS, Verrier ED, Eberhart RD (1993) Finite element analysis of the mitral valve. J Heart Valve Dis 2(3):326–340
91. Kunzelman KS, Quick DW, Cochran RP (1998) Altered collagen concentration in mitral valve leaflets: biochemical and finite element analysis. Ann Thorac Surg 66(6 Suppl): S198–S205

92. Kunzelman KS, Reimink MS, Cochran RP (1997) Annular dilatation increases stress in the mitral valve and delays coaptation: a finite element computer model. Cardiovasc Surg 5(4):427–434
93. Kunzelman KS, Reimink MS, Cochran RP (1998) Flexible versus rigid ring annuloplasty for mitral valve annular dilatation: a finite element model. J Heart Valve Dis 7(1): 108–116
94. Lim KH, Yeo JH, Duran CM (2005) Three-dimensional asymmetrical modeling of the mitral valve: a finite element study with dynamic boundaries. J Heart Valve Dis 14(3):386–392
95. Grashow JS, Sacks MS, Liao J, Yoganathan AP (2006) Planar biaxial creep and stress relaxation of the mitral valve anterior leaflet. Ann Biomed Eng 34(10):1509–1518
96. Grashow JS, Yoganathan AP, Sacks MS (2006) Biaxial stress-stretch behavior of the mitral valve anterior leaflet at physiologic strain rates. Ann Biomed Eng 34(2):315–325
97. May-Newman K, Yin FC (1995) Biaxial mechanical behavior of excised porcine mitral valve leaflets. Am J Physiol 269(4 Pt 2):H1319–H1327
98. May-Newman K, Yin FC (1998) A constitutive law for mitral valve tissue. J Biomech Eng 120(1):38–47
99. Ritchie J, Jimenez J, He Z, Sacks MS, Yoganathan AP (2006) The material properties of the native porcine mitral valve chordae tendineae: an in vitro investigation. J Biomech 39(6):1129–1135
100. He S, Weston MW, Lemmon J, Jensen M, Levine RA, Yoganathan AP (2000) Geometric distribution of chordae tendineae: an important anatomic feature in mitral valve function. J Heart Valve Dis 9(4):495–501; discussion 502–493
101. Maisano F, Redaelli A, Soncini M, Votta E, Arcobasso L, Alfieri O (2005) An annular prosthesis for the treatment of functional mitral regurgitation: finite element model analysis of a dog bone-shaped ring prosthesis. Ann Thorac Surg 79(4):1268–1275
102. Votta E, Maisano F, Bolling SF, Alfieri O, Montevecchi FM, Redaelli A (2007) The Geoform disease-specific annuloplasty system: a finite element study. Ann Thorac Surg 84(1): 92–101
103. Alfieri O, Maisano F, De Bonis M, Stefano PL, Torracca L, Oppizzi M, La Canna G (2001) The double-orifice technique in mitral valve repair: a simple solution for complex problems. J Thorac Cardiovasc Surg 122(4):674–681
104. Dal Pan F, Donzella G, Fucci C, Schreiber M (2005) Structural effects of an innovative surgical technique to repair heart valve defects. J Biomech 38(12):2460–2471
105. Votta E, Maisano F, Soncini M, Redaelli A, Montevecchi FM, Alfieri O (2002) 3-D computational analysis of the stress distribution on the leaflets after edge-to-edge repair of mitral regurgitation. J Heart Valve Dis 11(6):810–822
106. Avanzini A (2008) A computational procedure for prediction of structural effects of edge-to-edge repair on mitral valve. J Biomech Eng 130(3):031015
107. El Oakley R, Kleine P, Bach DS (2008) Choice of prosthetic heart valve in today's practice. Circulation 117(2):253–256
108. Bach DS (2003) Choice of prosthetic heart valves: update for the next generation. J Am Coll Cardiol 42(10):1717–1719
109. Schoen FJ, Levy RJ (2005) Calcification of tissue heart valve substitutes: progress toward understanding and prevention. Ann Thorac Surg 79(3):1072–1080
110. Stock UA, Vacanti JP, Mayer JE, Wahlers T (2002) Tissue engineering of heart valves – current aspects. Thorac Cardiovasc Surg 50(3):184–193
111. Vyavahare NR, Hirsch D, Lerner E, Baskin JZ, Zand R, Schoen FJ, Levy RJ (1998) Prevention of calcification of glutaraldehyde-crosslinked porcine aortic cusps by ethanol preincubation: mechanistic studies of protein structure and water-biomaterial relationships. J Biomed Mater Res 40(4):577–585
112. Vesely I, Barber JE, Ratliff NB (2001) Tissue damage and calcification may be independent mechanisms of bioprosthetic heart valve failure. J Heart Valve Dis 10(4):471–477
113. Schoen FJ (1998) Pathologic findings in explanted clinical bioprosthetic valves fabricated from photooxidized bovine pericardium. J Heart Valve Dis 7(2):174–179

114. Schoen FJ, Levy RJ (1994) Pathology of substitute heart valves: new concepts and developments. J Card Surg 9(2 Suppl):222–227
115. Chandran KB, Kim SH, Han G (1991) Stress distribution on the cusps of a polyurethane trileaflet heart valve prosthesis in the closed position. J Biomech 24(6):385–395
116. Hamid MS, Sabbah HN, Stein PD (1986) Influence of stent height upon stresses on the cusps of closed bioprosthetic valves. J Biomech 19(9):759–769
117. Rousseau EP, van Steenhoven AA, Janssen JD (1988) A mechanical analysis of the closed Hancock heart valve prosthesis. J Biomech 21(7):545–562
118. Black MM, Howard IC, Huang X, Patterson EA (1991) A three-dimensional analysis of a bioprosthetic heart valve. J Biomech 24(9):793–801
119. Krucinski S, Vesely I, Dokainish MA, Campbell G (1993) Numerical simulation of leaflet flexure in bioprosthetic valves mounted on rigid and expansile stents. J Biomech 26(8): 929–943
120. Patterson EA, Howard IC, Thornton MA (1996) A comparative study of linear and nonlinear simulations of the leaflets in a bioprosthetic heart valve during the cardiac cycle. J Med Eng Technol 20(3):95–108
121. Kim H, Chandran KB, Sacks MS, Lu J (2007) An experimentally derived stress resultant shell model for heart valve dynamic simulations. Ann Biomed Eng 35(1):30–44
122. Kim H, Lu J, Sacks MS, Chandran KB (2006) Dynamic simulation pericardial bioprosthetic heart valve function. J Biomech Eng 128(5):717–724
123. Kim H, Lu J, Sacks MS, Chandran KB (2008) Dynamic simulation of bioprosthetic heart valves using a stress resultant shell model. Ann Biomed Eng 36(2):262–275
124. Li J, Luo XY, Kuang ZB (2001) A nonlinear anisotropic model for porcine aortic heart valves. J Biomech 34(10):1279–1289
125. Sun W, Abad A, Sacks MS (2005) Simulated bioprosthetic heart valve deformation under quasi-static loading. J Biomech Eng-T Asme 127(6):905–914
126. Burriesci G, Howard IC, Patterson EA (1999) Influence of anisotropy on the mechanical behaviour of bioprosthetic heart valves. J Med Eng Technol 23(6):203–215
127. Huang X, Black MM, Howard IC, Patterson EA (1990) A two-dimensional finite element analysis of a bioprosthetic heart valve. J Biomech 23(8):753–762
128. Peskin CS (1982) The fluid-dynamics of heart-valves – experimental, theoretical, and computational methods. Annu Rev Fluid Mech 14:235–259
129. Peskin CS (2002) The immersed boundary method. Acta Numerica 11:479–517
130. Peskin CS, Mcqueen DM (1989) A 3-dimensional computational method for blood-flow in the heart. 1. Immersed elastic fibers in a viscous incompressible fluid. J Comput Phys 81(2):372–405
131. Peskin, CS, Printz BF (1993) Improved volume conservation in the computation of flows with immersed elastic boundaries. J Comput Phys 105:33–46
132. De Hart J, Peters GW, Schreurs PJ, Baaijens FP (2000) A two-dimensional fluid-structure interaction model of the aortic valve. J Biomech 33(9):1079–1088
133. De Hart J, Baaijens FP, Peters GW, Schreurs PJ (2003) A computational fluid-structure interaction analysis of a fiber-reinforced stentless aortic valve. J Biomech 36(5):699–712
134. De Hart J, Peters GW, Schreurs PJ, Baaijens FP (2003) A three-dimensional computational analysis of fluid-structure interaction in the aortic valve. J Biomech 36(1):103–112
135. Vigmostad S (2007) A sharp interface fluid-structure interaction for bioprosthetic heart valves. Ph.D. Dissertation, The University of Iowa, Iowa City, IA
136. Dumont K, Stijnen JM, Vierendeels J, van de Vosse FN, Verdonck PR (2004) Validation of a fluid-structure interaction model of a heart valve using the dynamic mesh method in fluent. Comput Methods Biomech Biomed Eng 7(3):139–146
137. Makhijani VB, Yang HQ, Dionne PJ, Thubrikar MJ (1997) Three-dimensional coupled fluid-structure simulation of pericardial bioprosthetic aortic valve function. Asaio J 43(5):M387–M392
138. Kunzelman KS, Einstein DR, Cochran RP (2007) Fluid-structure interaction models of the mitral valve: function in normal and pathological states. Philos Trans R Soc Lond B Biol Sci 362(1484):1393–1406

139. Carmody CJ, Burriesci G, Howard IC, Patterson EA (2006) An approach to the simulation of fluid-structure interaction in the aortic valve. J Biomech 39(1):158–169
140. Nicosia MA, Cochran RP, Einstein DR, Rutland CJ, Kunzelman KS (2003) A coupled fluid-structure finite element model of the aortic valve and root. J Heart Valve Dis 12(6):781–789
141. Weinberg EJ, Kaazempur Mofrad MR (2007) Transient, three-dimensional, multiscale simulations of the human aortic valve. Cardiovasc Eng (Dordrecht, Netherlands) 7(4):140–155
142. Krishnan S, Udaykumar HS, Marshall JS, Chandran KB (2006) Two-dimensional dynamic simulation of platelet activation during mechanical heart valve closure. Ann Biomed Eng 34(10):1519–1534
143. Govindarajan V, Udaykumar HS, Chandran KB (2009) Two-dimensional simulation of flow and platelet dynamics in the hinge region of a mechanical heart valve. J Biomech Eng 131:031002–1

Chapter 12
Mechanical Valve Fluid Dynamics and Thrombus Initiation

Tom Claessens, Joris Degroote, Jan Vierendeels, Peter Van Ransbeeck, Patrick Segers, and Pascal Verdonck

Abstract Heart valve, and subsequently cardiac function, may be seriously compromised as a result of stenosis or regurgitation. If necessary, the native heart valve is surgically replaced with an artificial substitute, which is in about 40% of the cases a bileaflet mechanical heart valve (BMHV). While generally showing excellent hemodynamic performance in the short term, current BMHVs are not free of clinical complications, which are induced by thrombus formation and hemolysis. Computational fluid dynamic (CFD) modeling is now considered a powerful and extremely useful tool to investigate blood flow in existing BMHVs and to reduce the costs associated with the development of new prototypes. A prerequisite for performing realistic heart valve simulations is the implementation of a fluid–structure interaction (FSI) algorithm that accounts for the mechanical interaction between the valve leaflets and the ambient blood. Provided the numerical resolution is sufficiently high, three-dimensional CFD-FSI models are able to compute the complex flow structures that exist in the vicinity of a BMHV. In order to get information about the valve's potential for platelet activation and blood hemolysis, these CFD models must be accompanied by appropriate mathematical models that describe the relation between fluid dynamic variables and the damage to blood corpuscles. This chapter provides an overview on the state of the art in computational flow simulations in BMHV and discusses various approaches taken to integrate blood damage accumulation models into flow simulations.

P. Verdonck (✉)
bioMMeda Research Group, Faculty of Engineering, Ghent University, B-9000 Ghent, Belgium
e-mail: pascal.verdonck@ugent.be

K.B. Chandran et al. (eds.), *Image-Based Computational Modeling of the Human Circulatory and Pulmonary Systems*, DOI 10.1007/978-1-4419-7350-4_12,

12.1 Background

12.1.1 Heart Valve Disease

The terminology "heart valve disease" refers to various pathological conditions that prevent one or more of the valves in the heart from opening and closing properly. The diseased valve may be incompetent, which means that it is not able to fully close and that it allows some blood to leak backward across the valve (regurgitation) rather than to flow forward. On the other hand, the valve may not be able to open properly, due to stiffened, thickened, or fused leaflets (stenosis). In those cases, the blood is obstructed to enter or leave the heart. Therefore, both valvular stenosis and regurgitation put a serious burden on the cardiac chambers.

If advanced valvular heart disease is left untreated, it may lead to hypertrophy, heart failure, stroke, or even sudden death due to sudden cardiac arrest. When the function of a heart valve is severely compromised, heart surgery is traditionally performed to repair or replace the valve with an artificial heart valve. Such implants are well-engineered devices that mimic the function of a natural valve.

12.1.2 Artificial Heart Valves

The first implant of an artificial heart valve to replace the aortic valve in its anatomic position took place in 1960. Since then, more than 50 different designs of artificial heart valves have been introduced. Despite years of research and many design improvements by valve manufacturers, the ideal prosthetic valve still does not exist. The main problems and clinical complications associated with the implantation of artificial heart valves include [1–4] the following:

- Thrombosis and thromboembolisms
- Anticoagulant-related hemorrhage
- Tissue overgrowth
- Endocarditis and other infections
- Paravalvular leaks due to healing defects
- Valve failure due to material fatigue or chemical change

At present, more than 290,000 artificial heart valves are implanted throughout the world each year [5] and the demand is increasing rapidly at 10–12% per year. Two types of artificial heart valves are equally widely used in clinical applications: *mechanical* devices made from synthetic materials and *bioprosthetic* valves made from animal or human tissue. A comprehensive overview of the basic geometrical characteristics and hemodynamic performance of these valves can be found in [2]. The bioprosthetic valves have been dealt with in the previous chapter in detail and thus will only be briefly touched upon hereafter.

12.1.2.1 Mechanical Heart Valves

Mechanical heart valves (MHVs) are meticulously designed structures, entirely man-made from biocompatible and durable synthetic materials, such as ceramics or polymers. The various designs of MHVs can be categorized according to the occluder type: a ball or a disk in cage, a tilting disk, or two semicircular hinged leaflets (Fig. 12.1a–c.). The ball-in-cage design (e.g., Starr–Edwards) uses a ball enclosed within a metal cage. During forward blood flow, the rubber ball is pushed forward to open the orifice, while during the closing phase the ball moves retrogradely to close the valve and prevent backflow. The cage assures that the ball is retained in position. The hemodynamic performance of these artificial valves is rather poor, due to their high-profile heights and significant turbulent flow characteristics distal to the valve occluder. The caged-disk valves (e.g., Kay Shiley and Beall) offer less resistance to flow than do the ball-in-cage valves, but are only used in mitral position.

A drastic improvement in valve hemodynamics was obtained by the introduction of the tilting disk and the bileaflet MHVs. With the tilting disk valves, the disk (e.g., Bjork-Shiley, Lillehei-Kaster, Medtronic Hall) completely occludes the valve orifice in the closed position, whereas it tilts to an angle depending on the design of the disc-retaining struts during forward flow. Today, the most popular mechanical heart valve design in use by far is the bileaflet mechanical heart valve (BMHV), due to its superior hemodynamic characteristics. It has two semicircular hinged (often pyrolytic carbon) occluders, called leaflets, which divide the area available for forward flow

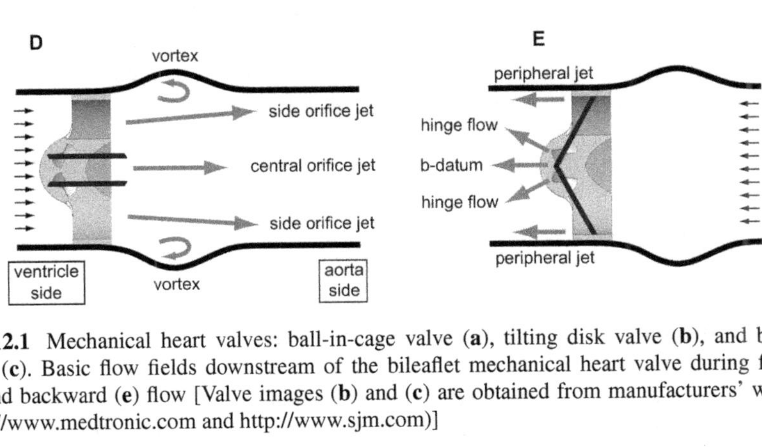

Fig. 12.1 Mechanical heart valves: ball-in-cage valve (**a**), tilting disk valve (**b**), and bileaflet valve (**c**). Basic flow fields downstream of the bileaflet mechanical heart valve during forward (**d**) and backward (**e**) flow [Valve images (**b**) and (**c**) are obtained from manufacturers' websites (http://www.medtronic.com and http://www.sjm.com)]

into three regions: two lateral orifices and a central orifice. When the leaflets are closed, blood flow regurgitates through the small gaps between both leaflets (called the b-datum gap) and between the leaflets and the housing itself. However, the main part of regurgitation occurs in the gap at the hinge region. The most common BMHVs are the SJM valve (Saint Jude Medical), Carbomedics valve (Sorin Group, Austin TX, USA), ATS Open Pivot valve (Minneapolis, MN, USA), the Parallel valve (Medtronic) and the ON-X valve (Medical Carbon Research Institute, Austin, TX, USA). Approximately 80% of the mechanical valves implanted today are BMHVs [6]. The geometry and basic flow structures of a typical BMHV design are depicted in Fig. 12.1d, e.

Mechanical heart valves are generally preferred because of their high mechanical durability. On the other hand, despite many investigations, current MHV designs are still subject to thrombus deposition and ensuing complications as a result of emboli. As such, the use of mechanical valves remains associated with two major disadvantages, being the need for long-term anticoagulation therapy and the accompanying problems of bleeding.

12.1.2.2 Bioprosthetic Heart Valves

Biological prosthetic heart valves (or bioprostheses) are in part made from porcine tissue or calf pericardium after proper modification using physico-chemical treatment and are usually mounted on flexible valve stents. The major advantage of bioprosthetic valves is their geometric similarity to human native valves, yielding overall relatively smooth and undisturbed blood flow. These valves offer less resistance to blood flow and patients do not require anticoagulation therapy ("blood thinners"). However, mainly due to structural changes such as calcification and leaflet wear, many bioprostheses suffer from valve failure. This explains their relatively short longevity of 15 years.

Recently a hybrid kind of valve has been proposed, which combines the advantages of both of the above-mentioned artificial heart valve prostheses: long-term durability without the necessity of permanent anticoagulation [7]. This so-called bio-mechanical valve is a completely artificial polymeric valve structure with flexibility comparable to a biological valve. It is believed to lead to a less disturbed flow pattern compared to MHVs and thus be less thromboresistant.

12.1.3 Design and Performance Issues

Hemodynamic performance, structural durability, and biocompatibility arguably constitute the most critical aspects in MHV design and determine the clinical success of these devices.

All of the currently implanted MHVs provide satisfactory results in terms of their *short-term* hemodynamic performance, which means that they have the following:

- *Low transvalvular pressure gradient*: the pressure drop over the valve (ΔP) should be as small as possible in order to minimize the load on the ventricle during ejection and as such to reduce the oxygen consumption of the ventricle.
- *High effective orifice area*: the effective orifice area (EOA) is a standard parameter for the clinical assessment of the severity of the valve stenosis, but it is also commonly used to compare the efficiency of various artificial heart valves. EOA is based on the Gorlin equation, which is based on the principle of energy conservation [8]:

$$\text{EOA} = \frac{Q}{51.5 \cdot \sqrt{\Delta P}} \tag{12.1}$$

  with Q being the mean forward flow in milliliter per s and ΔP expressed in mmHg. A large EOA signifies that the valve uses its internal orifice very efficiently while only minimally obstructing the flow of blood through the valve [2]. Note, however, that EOA might vary in patients with the same prosthesis, as reported in [9].
- *Small regurgitation volume*: regurgitation results from reverse flow created during valve closure (closing volume) and from backward leakage during diastole (leakage volume) after the valves are completely closed. A small regurgitation volume is obviously preferred, as it maximizes the net cardiac output through the valve (Fig. 12.2).

The *long-term* hemodynamic performance is more closely related to the risk for clinical complications such as hemolysis and platelet activation and ensuing thrombus formation, which in turn are the result of mechanical damage to red blood cells

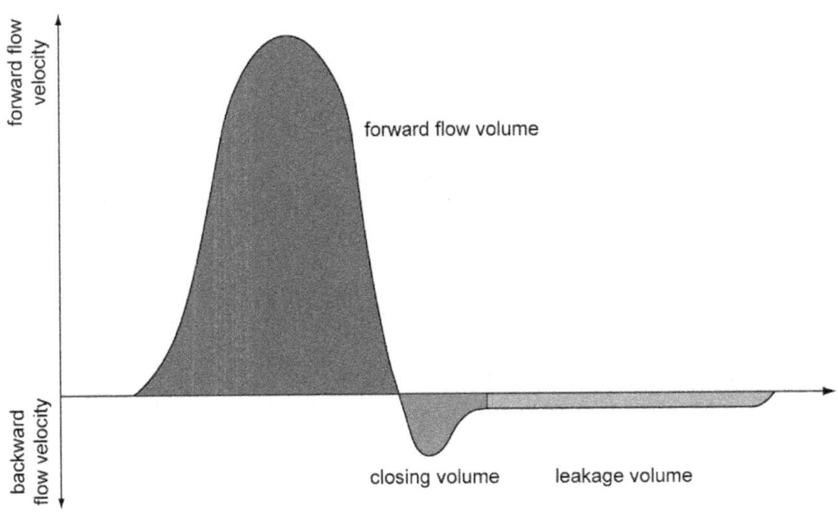

Fig. 12.2 Flow rate as a function of time during the cardiac cycle. Regurgitation volume is composed of closing volume (during the valve closure phase) and the leakage volume (when the valve is in its closed position)

and platelets and activation of the coagulation cascade. The exact mechanisms that lead to these undesirable complications are yet to be fully understood, but it is well agreed that damage to blood corpuscles is associated with the highly disturbed flow field in the vicinity of the MHV with regions of high shear stress [2, 10, 11]. Valve designers therefore need to attempt to [1, 12, 13]

- minimize high velocity gradients and ensuing shear stress
- minimize turbulence
- avoid zones of flow stagnation and separation

The structural durability (i.e., the resistance to mechanical wear) represents another critical factor for the long-term performance of the valve. It is a function of its material properties as well as its geometrical design. The latter, in conjunction with the local flow dynamics, is what ultimately determines the load distribution on the valve leaflets.

Lastly, all materials used need to be biocompatible, so they should not degrade in the physiological environment and neither absorb blood constituents nor release foreign substances to the blood and activate the immune system [14].

The remainder of this chapter is organized as follows. First, we focus on computational fluid dynamics as a tool to investigate heart valve hemodynamics and highlight the complexities that are frequently involved, in particular the implementation of fluid–structure interaction. We then continue with an overview of the state of the art on current (mainly) three-dimensional fluid–structure computational models of bileaflet mechanical heart valves in the aortic position. Where available, the computational results are accompanied by validation results obtained with experimental flow measurement techniques. In the subsequent paragraphs, we briefly talk about the typical clinical complications with heart valves and how they are related with hemodynamics. After providing an overview of the existing mathematical models for platelet activation and damage to red blood cells, we show how these models are integrated into existing CFD simulations. In the last section, some guidelines for future work on computational modeling of mechanical heart valves are provided.

12.1.4 Computational Fluid Dynamics

Fluid dynamics analysis of (artificial) heart valves is a challenging research field in biomedical engineering, even for the most advanced computational and experimental techniques. In the last 20 years, the use of computational fluid dynamics (CFD) has gained widespread acceptance as a virtual BMHV performance assessment and prototype development tool. Modern CFD analyses allow design changes to be precisely investigated in a cost-effective, safe, and non-invasive way before actual prototype development and experimental (in vitro or in animo) assessment, provided that the underlying models faithfully represent the physical phenomena.

Complexities that are frequently encountered in CFD analyses of BMHV include the following:

- Complex geometry of the valve, in particular the valve leaflets and the hinges
- Dynamic interaction between the blood and the valve leaflets due to pulsatile blood flow (i.e., fluid–structure interaction)
- Highly unsteady borderline turbulent and transitional flow inducing complex three-dimensional flow phenomena, such as flow separation and reversal, stagnation, and vortex shedding
- Non-Newtonian properties of blood which are critical when studying blood flow in narrow flow zones, such as the hinge region, and in stagnation zones
- Considerable difference in length scales from fluid motion (e.g., aortic orifice vs. hinge region)
- Difficulties in defining appropriate inlet and outlet boundary conditions

Current numerical algorithms are powerful enough to resolve the bulk flow dynamics in the vicinity of the BMHV and to capture some of the small-scale flow peculiarities, provided that grid resolution is fine enough. However, none of the current CFD models are able to account for all of the above-mentioned aspects simultaneously, which is, in part, due to constraints in computational resources. Many investigators therefore tend to study all aspects separately, thereby neglecting their mutual interaction. Yet, computational models that address even a few of these issues have provided valuable insights into the process of thrombus formation and blood damage directly associated with the failure of prosthetic devices currently in use [15]. Nevertheless, it should be acknowledged that sound engineering judgment is ultimately required to evaluate the results of the simplified CFD analyses.

12.1.5 Experimental Fluid Dynamics

It is highly recommended for CFD algorithms to undergo comprehensive validation with experimental data. Validation may expose shortcomings of the numerical models and identify areas for model refinement [16]. It requires the use of high-resolution and preferably non-intrusive flow measuring devices and appropriate post-processing techniques. Flow visualization and/or velocity measurement inside physical models is a very effective method of research and testing. However, it is generally very costly and time consuming in part due to the calibration of the experimental setup, especially if several tests must be conducted [17].

Laser Doppler anemometry (LDA) and particle image velocimetry (PIV) are two optical, non-intrusive flow velocity measurement techniques. They measure the velocity of a fluid element indirectly by means of the measurement of the velocity of tracer particles within the flow, which have been added to the flow before the experiment starts. They are considered highly accurate, both in terms of spatial and temporal resolution, and have been shown to be able to capture many of the fine scale flow characteristics associated with BMHVs [18–30]. While the temporal resolution of LDA is higher than that of PIV, owing to the detection of the passage of single particles in the region of interest, LDA is more time consuming, since only one point in space can be investigated at a time. As opposed to LDA, PIV allows the ability to provide full-field information in an extended region of the flow [29], which makes PIV the technique of choice for analyzing BMHV fluid mechanics. The spatial resolution is usually a bit poorer compared to LDA, since a certain number of particles must be imaged at the same time to obtain a statistically significant measurement (Fig. 12.3a) [17].

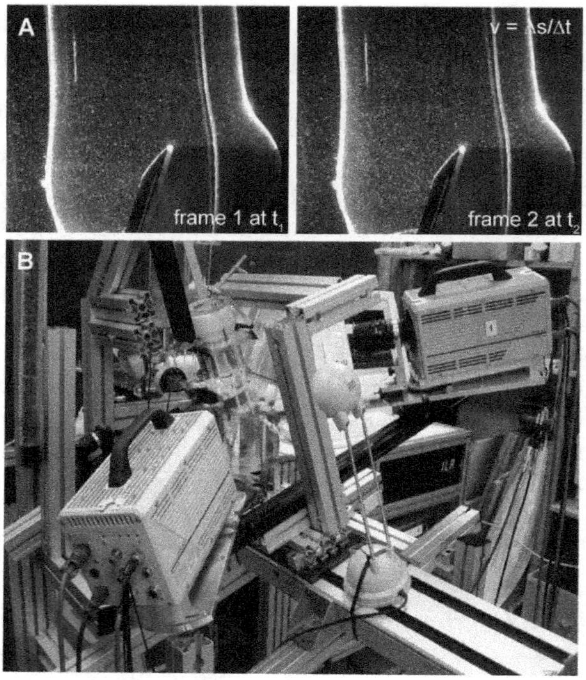

Fig. 12.3 (**a**) Basic principles underlying particle image velocimetry. The fluid is seeded with tracer particles that are assumed to follow the fluid flow. Two consecutive laser pulses illuminate the measurement plane and the images are captured with a high-resolution camera. The displacement (Δs) divided by the time difference (Δt) between both frames yields velocity (v) of the particles. (**b**) Example of a stereoscopic PIV setup that allows extraction of the out-of-plane velocity component

PIV is known to have a high cost (in particular because of the synchronizer, the cameras, and the data-processing unit), and the models need to be compatible with PIV in order to ensure optical access to the region of interest. In regions in the in vitro setup where illumination by the laser light is suboptimal, flow measurements cannot be easily performed. Stereographic PIV (Fig. 12.3b) is now becoming more common in heart valve studies [31]. It uses two cameras with different viewing angles to extract the out-of-plane velocity component, however, with inherently less accuracy than the in-plane components [32].

In some experimental studies, the focus is on valve leaflet kinematics (i.e., the opening angles, the velocity, and the acceleration of the leaflets) rather than on measuring flow velocity. In those studies, high-speed cameras are commonly used to track the position of the leaflets [33].

12.2 Fluid–Structure Interaction

12.2.1 The Need for Fluid–Structure Interaction

In many previous 2D and 3D MHV investigations, leaflets were kept in a fixed (fully open or closed) position under stationary flow conditions, or their motion was prescribed as a function of time [34–46]. Even though these studies have undeniably provided valuable information about flow dynamics, shear stress levels, and leakage flow, it should be clear that such simulations are unable to reflect physical reality, as they neglect the inherent mechanical interplay between the leaflets and the surrounding blood. Large rotations (and to a lesser extent translations and deformations) of the valve leaflets are a direct consequence of the load applied on them by the ambient blood [47]. Nowadays, modeling fluid–structure interaction (FSI) is an important issue in computational fluid dynamics studies on heart valves. In real life, FSI is a well-known phenomenon which can be found in several areas of engineering, including amongst others marine engineering, aeronautical engineering, construction engineering, and biomedical engineering. With respect to BMHV mechanics, implementation of FSI algorithms in CFD allows researchers to

- analyze fluid dynamics in a more realistic way compared to non-FSI simulations;
- provide information about valve leaflet kinematics during the opening and the closing stages;
- determine the stress inside the heart valve leaflets;
- study the closing and rebound stages of the leaflets and the ensuing problems of cavitation, which in turn are conducive to pitting of erosion of the leaflets.

The governing equation for the motion of the leaflets models a single degree of freedom system and is given by the equation

$$I \cdot \frac{\mathrm{d}^2\theta}{\mathrm{d}t^2} + f \cdot \frac{\mathrm{d}\theta}{\mathrm{d}t} = M_\mathrm{p} + M_\tau + M_\mathrm{w} \tag{12.2}$$

where I represents the angular moment of inertia of the leaflet and θ the rotating angle; f is the friction coefficient for the hinges. M_p and M_τ are the moments of the leaflet about the pivotal axis generated by the pressure and the wall shear stress, respectively. M_w represents the momentum resulting from gravitational force and buoyancy. Note that the effect of buoyancy should be taken into account only if gravitation is not accounted (for?) while calculating the pressure field. $\mathrm{d}^2\theta/\mathrm{d}t^2$ and $\mathrm{d}\theta/\mathrm{d}t$ are the angular acceleration and the velocity, respectively. In many cases, hinge friction, as well as the effect of shear stress and gravity, is neglected, and the equation to solve then reduces to

$$I \cdot \frac{\mathrm{d}^2\theta}{\mathrm{d}t^2} = M_\mathrm{p} \tag{12.3}$$

In literature, a considerable number of different numerical methods with various degrees of accuracy have been devised to tackle the complex FSI problem. They can be broadly categorized into *fixed grid methods* and *moving grid methods*. The fixed grid methods include the immersed boundary method, the fictitious domain method, and many derivatives, while the (body-conforming) moving grid methods are better known as the arbitrary Lagrangian–Eulerian (or ALE)-based methods.

Some of the key issues in solving FSI problems are outlined below. However, for more in-depth information about various FSI approaches, the reader is referred to Chapter 5, in particular to Sections 5.3 and 5.15.

12.2.1.1 Monolithic vs. Partitioned Methods and Loose vs. Strong Coupling

Fluid–structure interaction is a multiphysics problem and there are different ways to treat the coupled interaction problem: using a *monolithic* [48] (or direct) or a *partitioned* (or iterative) approach [49]. In the monolithic approach the flow and structural equations are combined and solved simultaneously with a single dedicated code, whereas the fluid and structural equations are solved with separate codes and coupled through the exchange of boundary conditions in the partitioned approach. Hence, monolithic codes take into account the interaction between the fluid and the solid automatically, while a coupling algorithm is required for partitioned simulations. The latter approach allows the re-use of well established existing software in the respective fields and, as such, permits the use of an optimal solution strategy to be selected independently for both the fluid and structural equations [47]. In a monolithic code, the discretized flow equations and structural equations result in a system of coupled equations which are generally nonlinear. The interaction between fluid and structure is considered by solving for all variables at once, often with Newton–Raphson iterations [50–54].

In addition, in partitioned techniques the coupling can be modeled in a *loose* and *strong* fashion. The major difference between the loose and the strong coupling is that they integrate the governing equations of the structure explicitly and implicitly in time. In the strong coupling approaches, equilibrium of fluid and solid stress and velocity on the interface is enforced in every time step. So, within each time step a number of sub-iteration steps are carried out in a subroutine to update the solution of the fluid dynamic field and to obtain a stable solution at every physical time step (i.e., difference between the external momentum and the angular acceleration converges to zero). Iterations between the flow solver and the structural solver within the time step of the Gauss–Seidel or Jacobi type converge slowly if at all, especially if the fluid is incompressible, but the convergence of Gauss-Seidel iterations can be accelerated by means of Aitken relaxation [55]. In the loose coupling, an internal FSI subroutine is absent, and equilibrium of the fluid and solid stress and velocity on the fluid–structure interface is not imposed. The latter results in restrictions on the time step for stability reasons; however, it considerably reduces computational costs [56]. It was shown in [57] that loose coupling methods cannot be used for heart valve simulations. The density ratio of the fluid and the solid and geometrical factors (such as the gap size between the valve and the housing) are the factors that determine the stability of the coupling iterations and not the time step. With under-relaxation, stability can be obtained and the motion of the valve (position and velocity) can be predicted well, but the acceleration of the valves and the pressure field suffer from unphysical wiggles as a function of time. It was also shown that when different time integration schemes are used in the CFD and the CSD solver, these unphysical wiggles can occur.

The following overview focuses on, but is not limited to, three-dimensional flow simulations of BMHV incorporating FSI algorithms. In the majority of these studies, resolving the relatively large-scale flow dynamics was considered the most important aspect. Analysis of small-scale flow dynamics by means of multiscale modeling is briefly touched upon in the last section.

Direct quantitative comparison of the computed dynamic flow variables between the various simulations is very intricate and can arguably be considered irrelevant, given the wide disparity in imposed (pressure or flow velocity) boundary conditions, flow regimes (Reynolds numbers), valve type, and geometry of inflow and outflow configurations.

12.2.1.2 Moving Grid Methods

The Eulerian reference frame is commonly used to describe the motion of the fluid phase, although it is incompatible with the Lagrangian formulation, which is more appropriate for the structural phase. To tackle this incompatibility, Donea et al. [58] proposed the arbitrary Lagrangian–Eulerian (ALE) method, which combines both Lagrangian and Eulerian formulations. In ALE the fluid mesh is continuously updated according to the movement of the Lagrangian grid without a change in the topology [59]. In the case of mechanical heart valves, however, large rotations of the leaflet typically occur, and the ambient fluid grid is subsequently subject to significant deformation, potentially resulting in degradation of the grid quality (ill-shaped elements or negative volumes) when the original topology is to be maintained. In such cases, the proposed ALE method is inadequate, and remeshing of the fluid domain must be performed to preserve the quality of the mesh. The resulting change in topology during remeshing necessitates interpolation techniques to transfer state variables from the old mesh to the newly generated mesh. Not only does this interpolation induce artificial diffusivity and loss of accuracy, but also it is time consuming to perform, in particular for 3D FSI problems [47, 60]. The advantage of this approach is that it allows controlling mesh quality and density next to the leaflet surface, and thus an accurate prediction of the flow field and shear stress is possible.

In the past 5 years, the ALE method in combination with remeshing has been successfully applied in a number of 3D FSI studies of BMHVs (Table 12.1). Redaelli et al. [33] used a weakly coupled FSI algorithm to study in particular the valve kinematics of a 27-mm St. Jude Medical Hemodynamic Plus mechanical valve. The motion of the leaflets was validated experimentally using an ultrafast cinematographic technique. The reported small delay in computed leaflet motion in this study could not be ascribed to the weak coupling according to the authors. Dumont et al. [61] used a strongly coupled, partitioned FSI model. They investigated clinically

Table 12.1 List of state-of-the-art three-dimensional computational fluid dynamics studies on bileaflet mechanical heart valves implementing fluid–structure interaction

Reference	Method	Coupling	Valve type	Reynolds	Phase	Validation
Moving grids methods						
[33]	ALE	Loose	27-mm SJM	2,400	Opening and closing	Leaflet kinematics
[61]	ALE	Strong	22-mm ATS	± 4,125	Opening and closing	2D PIV (3D PIV)
[62]	ALE	Strong	22-mm ATS, 21-mm SJM	± 4,125	Opening, closing, and leakage	2D PIV (3D PIV)
[15]	ALE	Loose and strong	27-mm SJM	± 2,040	Opening and closing	Leaflet kinematics
[64]	ALE	Strong	27-mm SJM	6,000	Opening and closing	Leaflet kinematics
Fixed grid methods						
[75]	IOM	Loose	29-mm SJM	1,893	Opening	No
[77]	CURVIB	Loose and strong	Generic	6,000	Opening and closing	2D PIV
[79]	IMM	Strong	29-mm SJM	4,000	Opening and closing	No

useful valve performance indices, such as the transvalvular pressure gradients and shear stress, as well as the valve kinematics of the 22-mm AP ATS Open Pivot ATS bileaflet valve mounted in a straight (aortic) and expanding (mitral) conduit. In combination with a simplified platelet activation model, the validated FSI model was later employed to investigate and compare the thrombogenic potential of two different BMHVs: a 22-mm Open Pivot (OP) ATS and a 21-mm St. Jude Medical Regent valve [62] (Fig. 12.4a, b). Herein, they reported that the ATS valve may offer a lower thrombogenic potential owing to its different hinge mechanism design. Nobili et al. [15] compared the strong and weak FSI algorithms for the simulation of the opening phase of a 27-mm St. Jude Hemodynamic Plus mechanical heart valve. The weak coupling method appeared accurate enough to determine the angular velocity of the leaflets, but it was noted that the strong coupling model provides a more consistent physical description of the interaction when looking at small-scale features like stress close to geometrical singularities. Later, Nobili et al. [63] used the strongly coupled algorithm to analyze the leaflet kinematics of the 27-mm St. Jude Hemodynamic Plus as well as the fluid dynamics using direct numerical simulations under physiological flow conditions.

A considerable drawback in the above-mentioned ALE-based simulations constitutes the fact that the valve is not allowed to close completely, since the fluid elements between the wall and the leaflets will be compressed. The problem is frequently circumvented by reducing the dimensions of the leaflets [62, 64]. While such simplifications are not expected to dramatically influence fluid dynamics during forward flow, there is a general consensus among the investigators that leakage

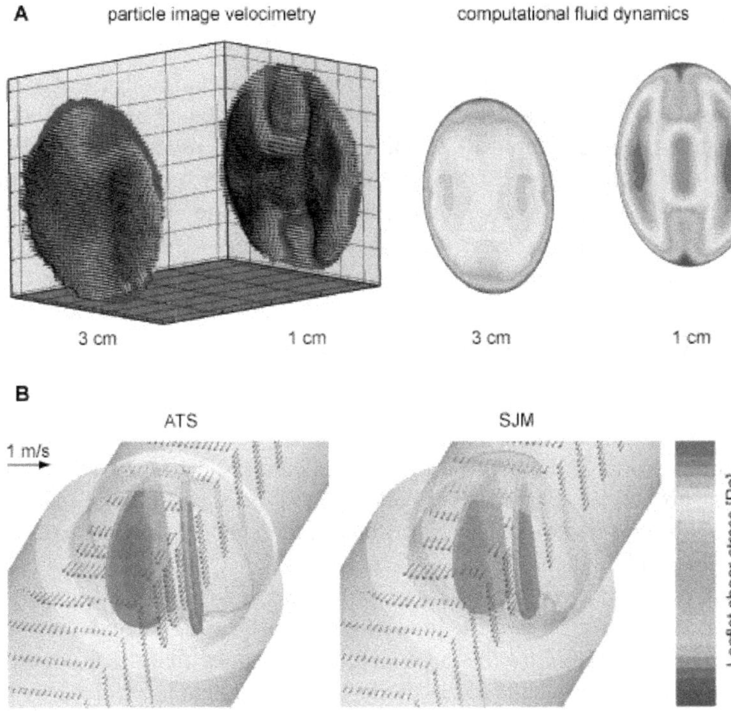

Fig. 12.4 (**a**): Flow velocity magnitude at 1 and 3 cm distal from a bileaflet mechanical valve computed with a three-dimensional fluid–structure algorithm and validated with stereoscopic particle image velocimetry. (**b**): Comparison of the calculated flow field during forward flow acceleration between a 22-mm Open Pivot (OP) ATS and a 21-mm St. Jude Medical Regent valve. Shear stress levels are indicated on scale bar

flow will be (highly) overestimated. This problem does not necessarily apply to the immersed boundary techniques nor to the fictitious domain methods, which are described below.

12.2.1.3 Fixed Grid Methods

Fixed grid methods have the advantage that they avoid mesh movement, thus preserving the original mesh quality, but they generally prove to be less accurate at the interface between the fluid and the leaflets compared to the moving grid methods. However, as the mesh is independent of the position of the valve, the valve surface is not represented in the mesh and an accurate prediction of the flow field next to the leaflets is difficult, leading to low accuracy of wall shear stresses. Well-known so-called *diffuse* fixed grid methods are the *immersed boundary method* and the *fictitious domain method.* The limitations in accuracy are largely overcome by the introduction of the *sharp interface* fixed grid methods, like the curvilinear immersed boundary method (CURVIB), the immersed membrane method, and the immersed interface method. The reader is referred to specialized literature on these algorithms to appreciate their (dis)advantages [35, 65–67].

In the past, fixed grid methods have been frequently used to study flexible (native) heart valves. Only more recently, these methods are also being used to model fluid dynamics in stiff valves (Table 12.1). De Hart et al. [68] used a method based on the fictitious domain method (FDM) [69], which is closely related to the immersed boundary method [70]. The coupling of the fluid and the structure was done by putting velocity constraints using Lagrange multipliers. They first studied flow through the native aortic valve in two dimensions. Later, they extended their code to be able to perform flow and structural analyses in 3D valves [60] and to investigate the effect of collagen fibers on the native valve mechanics and hemodynamics [71]. Stijnen et al. [72] successfully used the FDM for evaluating the motion of a stiff prosthetic heart valve leaflet in two dimensions. Van Loon et al. also used FDM for studying flexible valves but extended it with an algorithm which adapts the fluid mesh to the position of the solid in order to calculate shear stress with enhanced accuracy at both sides of the valve leaflet [73]. Later, they also implemented a Lagrange multiplier-based contact algorithm which enables the interaction of the solid with the fluid domain [74].

Tai et al. [75] employed the *immersed object method* (IOM) [76] to investigate the opening phase of a 29-mm St. Jude Medical aortic BMHV, under a moderate Reynolds number of 1893. In this approach, it is assumed that the fluid within an object is frozen and moves like a solid body, a condition that is enforced by adding source terms to the momentum equations for the Eulerian nodes lying inside the object. Borazjani et al. [77] extended the sharp interface CURVIB method developed in [35] and successfully studied the opening and closing phases of a generic, yet realistic, model of a BMHV, under physiological flow conditions (Re 6,000) and with very high numerical resolution. Their results were validated with 2D PIV experiments described in another study [78]. They were the first to report the existence of asymmetries in the computed leaflet kinematics as a result of asymmetries in the flow due to natural flow instabilities (Fig. 12.5). In Xia et al. [79] an *immersed membrane method* [65] was applied to calculate the opening and closing phases of a 29-mm St. Jude BMHV under pulsatile inflow conditions. In order to concentrate on the blood–leaflet interaction in this work, a moderate Reynolds number of only 1,200 was chosen to limit the flow condition to unsteady laminar flow, even though the sharp interface method allows for higher Reynolds numbers. The focus of their study was on leaflet dynamics rather than on fluid mechanics phenomena. It was shown that the moment generated by the fluid pressure is the main part of the total moment, while the moment due to the shear stresses is almost negligible.

Table 12.1 summarizes current state of the art on three-dimensional FSI models of bileaflet mechanical heart valves.

ALE, Arbitrary Lagrangian–Eulerian; IOM, immersed object method; IMM, immersed membrane method; CURVIB, curvilinear immersed boundary; PIV, particle image velocimetry. *Note*: The ALE approach presented in [61, 62] has been validated qualitatively with 3D PIV (unpublished, see Fig. 12.4a).

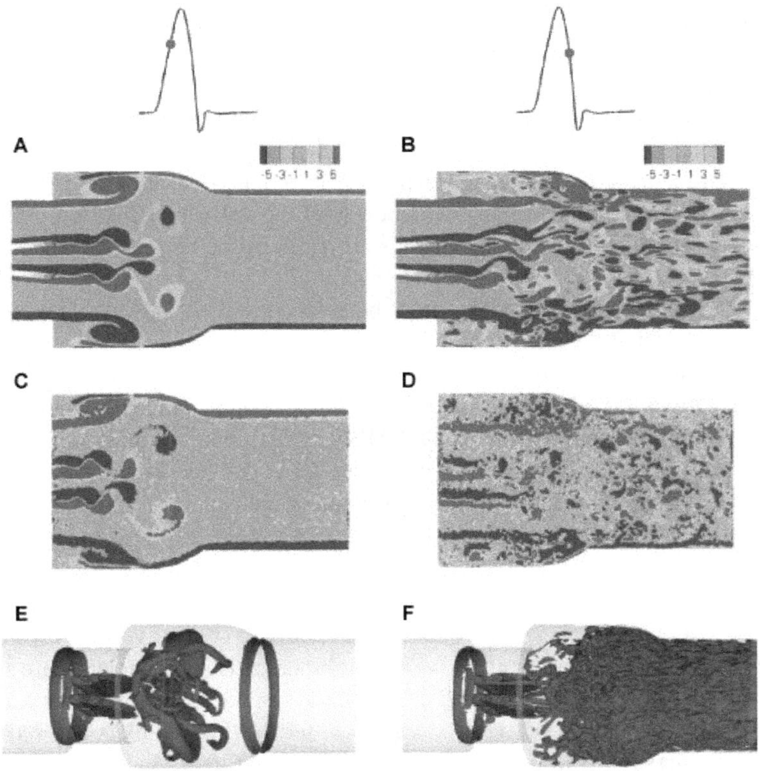

Fig. 12.5 Out-of-plane vorticity contours in a generic bileaflet mechanical heart valve computed with CFD (**a** and **b**) and validated with two-dimensional PIV (**c** and **d**) at two time instants during opening of the leaflets. Corresponding vorticity isosurfaces are shown in (**e**) and (**f**). Taken with permission from Borazjani et al. [77]

12.3 Modeling Damage to Blood Cells

12.3.1 Thrombus Formation and Hemolysis

In recent years, it has been recognized that the clinical "long–term" success of BMHVs is largely determined by their susceptibility to thromboembolic events and, to a lesser extent, hemolysis. In BMHVs, thrombus formation presents a major concern and is mostly observed in the hinge region and the valve housing [80–82]. The exact mechanisms behind hemolysis and the cascade of thrombus formation as a result of platelet activation are not yet fully understood. However, it is well agreed that both are the consequence of a three-way interaction between blood biochemistry, the blood contacting foreign material (such as the leaflets), and the very non-physiological and unfavorable blood flow in the close vicinity of the BMHV. BMHV implantation induces regions with high-velocity gradients and shear stress, especially in the leakage jet flow after the valve has closed. These elevated shear regions potentially are sites for platelet activation and (sub)lethal blood damage. On the other hand, platelet activation may as well be the result of an imbalance of the homeostatic system due to chronic blood damage [16, 83]. When activated platelets enter regions of flow stagnation and recirculation with long residence times adjacent to the valves, deposition and aggregation of damaged blood elements is promoted, leading to thrombus formation and subsequent emboli [84]. Direct mechanical interaction with the valve housing and other support structures such as the hinges may also be responsible for the damage on blood cells [16].

Various mathematical models have been proposed to describe the relation between (sub)lethal damage to blood cells (i.e., platelets and red blood cells) and mechanical shear stress. It is anticipated that these models, when integrated into CFD simulations, could furnish valuable information about damage accumulation as a result of multiple journeys through a BMHV. It should be acknowledged, though, that the predictive capacity of such combined CFD damage accumulation models greatly relies on the resolution of the CFD simulation.

12.3.2 Modeling Blood Damage

In vitro viscometric experiments in which red blood cells and platelets were subject to very well-controlled shear stress conditions revealed that there exists a certain shear stress under which no damage can be observed. For platelets and red blood cells, these threshold values equal 10 N/m^2 (Pa) and 150–400 N/m^2, respectively. Once this threshold has been reached, the degree of damage becomes a function of shear stress and exposure time to shear stress [85, 86]. The commonly used power law model [87, 88] expresses the damage on red blood cells and on platelets as a function of exposure time and the magnitude of the shear stress, and forms the basis for several numerical models of blood damage:

$$\mathrm{HI}(\%) = \frac{\Delta \mathrm{Hb}}{\mathrm{Hb}}(\%) = 3.62 \cdot 10^{-5} \cdot t^{0.785} \cdot \tau^{2.416} \tag{12.4}$$

$$\mathrm{PLI}(\%) = \frac{\Delta \mathrm{Pl}}{\mathrm{Pl}}(\%) = 3.66 \cdot 10^{-6} \cdot t^{0.77} \cdot \tau^{3.075} \tag{12.5}$$

The hemolysis index (HI) equals the percentage of plasma-free hemoglobin concentration in the blood with respect to the total hemoglobin concentration after shear loading. The results of the damaging effects on platelets are given by the platelet lysis index (PLI), quantified as the relative change in total platelet concentration. In these equations, t represents time duration (in seconds) of blood exposure to shear stress (uniform laminar or turbulent shear stress, in pascals). Shear stress (viscous and Reynolds) acting on the blood is represented by τ [89].

It has been shown that hemolysis and platelet activation are, to some extent, similar in response, even though the process of platelet activation is far more complex than red blood cell damage. From a mathematical point of view, both can be treated in a similar way [85]. In general Equations (12.4) and (12.5) can therefore be expressed as

$$\text{Lysis index (\%)} = A \cdot \tau^{\alpha} \cdot t^{\beta} \tag{12.6}$$

The exact values of α and β and A in particular, originally proposed by Giersiepen et al. [88], are widely used. However, they remain a matter of discussion, as the power law has been shown to overestimate the degree of blood damage [90, 91].

Equations (12.4) and (12.5) are derived from experiments with blood cell exposure to a constant shear stress for a fixed time interval. Consequently, they are clearly not appropriate for conditions where blood cells are subject to time-varying stress magnitude and history, such as during their transit through the BMHV. Damage to blood cells is not an "all or none" process [92], and sublethal damage is likely to occur in case of a relatively low shear stress and/or a short exposure time. The cumulative (total) damage to platelets or red blood cells, often referred to as a damage index (DI), can therefore be quantified numerically by summating all sublethal (differential) contributions to blood cell damage [93]. This strategy for determining damage accumulation yields the following equations, commonly found in literature [30, 94–96]:

$$\text{DI} = \sum_{i=1}^{N} C \cdot \Delta t_i^{a} \cdot \tau_i^{b} \tag{12.7}$$

$$\text{DI} = \sum_{i=1}^{N} a \cdot C \cdot t_i^{a-1} \cdot \tau_i^{b} \cdot \Delta t_i \tag{12.8}$$

with τ_i being the shear stress value acting during the ith observation interval Δt_i and N the number of time steps from t_0 to time t [93]. The appropriateness of the above power law-based formulations (Equations (12.7) and (12.8)) for predicting red blood cell and platelet lysis under time-varying shear stress has been seriously questioned by Goubergrits [89] and Grigioni et al. [97], as neither of these equations incorporates the effect of loading history. Grigioni et al. [93] therefore proposed a physically more sound model, based on the principle of dosimetry, comparable to the model of Goubergrits et al. [89]:

$$\text{DI} = \sum_{i=1}^{N} C \cdot a \cdot \left[\sum_{j=1}^{i} \tau(t_j)^{\frac{b}{a}} \cdot \Delta t_j + D(t_0) \right]^{a-1} \cdot \tau(t_i)^{\frac{b}{a}} \cdot \Delta t_i \tag{12.9}$$

with Δt_j being the time interval in which the shear stress is assumed constant. $D(t_0)$ equals the mechanical dose at the starting observation time; $\tau(t_j)$ is the shear stress value acting during the jth observation time interval Δt_j, and t_j is the time taken by the blood particle to reach the jth observation point. While appearing correct from a theoretical point of view, the need for validation of the proposed model was clearly underscored by the authors. However, experimental validation is cumbersome, as it is virtually impossible to subject blood to simultaneously controlled shear and exposure time. A number of difficulties inherent to experimental setup have been mentioned elsewhere [98].

An alternative blood damage accumulation model has been suggested by Yeleswarapu et al. [98]. Their model also takes into account initial damage, stress loading history, and dependence on loading rate. It is described by the following equation:

$$D(t_i) = D(t_{i-1}) + \left(\frac{\sigma(t_i)}{\sigma_0}\right)^{\tau} \frac{\Delta t}{[1 - D(t_{i-1})]^k} \tag{12.10}$$

with $D(t_0)$ being the damage accumulated during previous passages through the valve, σ a scalar stress value derived from the stress tensor, and k and r two empirically calibrated parameters.

Recently, a critically important finding with respect to shear stress used in blood cell damage models was reported by Ge et al. [10]. Using two-dimensional PIV in combination with fully 3D (non-FSI) CFD simulations, they were able to estimate viscous and Reynolds shear stress separately in the wake region of a St. Jude Medical 23-mm Regent BMHV. They found that viscous stress was borderline to the experimentally determined shear stress threshold for platelet damage, but well below the shear stress to damage for red blood cells. This finding is completely consistent with thrombus formation being more common in BMHV than hemolysis. The authors argued that instantaneous viscous stress, which is derived from the spatial derivative of the velocity field, is the dominant mechanical load imparted by the blood on the blood cells. While in prevailing literature, Reynolds stress has been frequently considered a key quantity in assessing blood damage, Ge et al. state that it is merely a statistical quantity that should not be seen as a mechanical shear force on the cell membrane. The size of red blood cells and platelets is around 10 and 2 μm, respectively, and thus smaller than the Kolmogorov scale (i.e., the scale of the smallest turbulent eddies possible in a given flow), which has been estimated to be between 20 and 70 μm [99, 100] in BMHV. As a result, the dynamic pressure caused by turbulence effects cannot significantly contribute to cell damage.

12.3.3 Implementation of Blood Damage Models in CFD

A single universally applicable damage accumulation model which accounts for the interacting complex biochemical and fluid dynamical processes involved in blood damage is elusive. Platelet activation in particular depends on many factors other than purely mechanical and which potentially accelerate the activation process. Nevertheless, even the above-mentioned blood damage and platelet activation models, in which so far only mechanical stimuli are considered, have been shown capable of

- identifying regions that are prone to thrombus formation
- assessing the degree of hemolysis as a result of successive passages through the valve

The use of the Lagrangian viewpoint, in which blood elements are followed along their trajectories, has been dominant for determining cumulative effects. Numerical tracking of the particle's position is the only way to examine the load and exposure time simultaneously, which is virtually impossible to accomplish using experimental techniques. In addition, the numerical techniques permit computation of the cell residence time, which can be used as a surrogate measure of thrombus formation potential [6].

The past 5 years have seen an increasing number of CFD-FSI studies in combination with the Lagrangian tracking approach. Dumont et al. [62] computed stress accumulation in two valve designs (22-mm Open pivot ATS and a 21-mm SJM Regent valve). A highly simplified approach was taken, in which accumulated damage is derived from discrete integration (i.e., summation) of shear stress along a fluid particle's path line [101, 102] for as long as it remains within the computational domain:

$$\sum \tau \times \Delta t \tag{12.11}$$

Despite the fact that a fully implicit FSI scheme was implemented in their CFD model, platelet activation was only computed assuming fixed leaflet positions. During forward flow (valve leaflets fully open), they found that platelet activation was lower than the 3.5 Pa threshold, according to Hellums criterion. During the regurgitation flow phase (valve leaflets completely closed), on the other hand, it was observed that 0.81% of the platelets in the SJM valve experience a stress accumulation higher than 3.5 Pa, compared with 0.63% for the ATS valve. In these simulations, shear stress during regurgitation was likely underestimated, because the leaflets were rescaled to 98% of their original size to facilitate meshing. However, for comparison purposes, such a simplification was considered justified. Alemu et al. [103] performed fully 3D FSI simulations in a St. Jude Medical bileaflet MPHV in aortic position, with a simplified leaflet hinge region to facilitate meshing. Blood was assumed a two-phase, non-Newtonian fluid with spherical solid particles, representing platelets. Platelet damage was calculated with the cumulative damage model proposed by Yeleswarapu et al. [98] (Equation (12.10)). The principle of "perfect memory" was employed, which means that for each passage through the valve, the cumulative damage at the end of that trajectory was applied as the new initial damage value for the subsequent passage. The potential advantageous effects of a damage recovery mechanism [104] between subsequent passages were not accounted for (Fig. 12.6). Later, this model was improved by the same authors to include reversed flow dynamics in diastole [105].

Grid resolution is a critically important parameter in CFD models if all physically important length scales need to be resolved. This is especially true if macroscale (such as bulk flow through the orifice) and microscale (such as flow through the

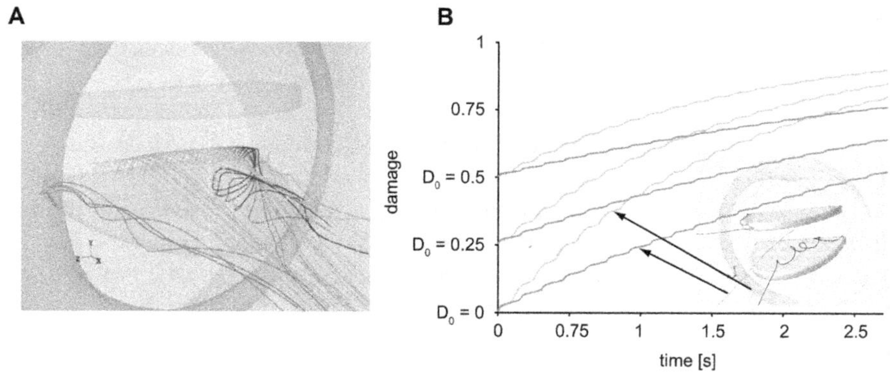

Fig. 12.6 (**a**) Smoothed path lines depicting the complex flow dynamics and helical vortices developing in the flow field above the lower leaflet. (**b**) Damage accumulation during repeated passages past the valve. Taken with permission from Alemu et al. [103]

leaflet gaps) must be solved simultaneously with a sufficiently high resolution to compute shear stress and ensuing platelet activation. At the same time, multiscale simulations would be convenient to improve hinge mechanisms for better washout to reduce the problem of thrombus formation [106]. Krishnan et al. [83] used a 2D FSI model with a Cartesian grid meshing technique to depict the detailed flow dynamics through the small gaps between the leaflets and the housing and its effect on platelet activation during valve closure and rebound. The large disparity in cell size between the hinge region and the valve orifice was accounted for with a local mesh refinement algorithm. Relying on an easy-to-implement damage accumulation model [101], it was shown that platelet activation and subsequent deposition is more likely to occur in the gap between the leaflets and the valve housing than in the gap between the leaflets. One limitation of their 2D study was that it was impossible to integrate the rather complex hinge geometry into the flow simulations.

12.4 Concluding Remarks

The ideal mechanical valve combining superior hemodynamic performance with high durability, but without the risk for thromboembolisms or the need for life-long anticoagulation, does not exist yet [107]. Precise knowledge of the complete three-dimensional (small and large scale) flow field in the vicinity of a BMHV is key to understanding and predicting the risk for platelet activation and thrombus formation and hemolysis. Implementation of highly accurate fluid structure algorithms in CFD, which account for the mechanical interaction between the leaflets and the surrounding blood, is one of the main prerequisites to accomplish this.

In this chapter, the state of the art in computational flow modeling of BMHV was highlighted. Using Lagrangian particle-tracking algorithms in combination with damage accumulations models, current CFD simulations have considerably broadened our understanding of BMHV flow dynamics and can be used as an initial tool to assess the long-term valve performance in a virtual environment. In spite of that, we envision that in the near future a number of advancements in the following fields need to be considered:

Multiscale modeling: Multiscale modeling is essential for capturing all relevant micro- and macroscale flow structures simultaneously, as well as the interaction between them. It can clarify how bulk flow within the valve orifice mechanically affects microscale flow within the complex hinge structures [83] with a level of detail that cannot be reached even with advanced experimental flow-measuring techniques. In addition, multiscale CFD models will also aid in computing the actual load imparted on the blood cells, as well as in investigating the mechanical interaction between red blood cells and platelets [108]. This is particularly important in regions where the dimensions of the blood cells are comparable to the size of the flow geometry [83] and where non-Newtonian effects should be taken into account in the governing equations. Furthermore, knowledge of shear and pressure forces and resulting deformation is desirable to improve our insight into blood cell damage. Such information cannot be obtained from models where blood cells are considered as rigid point particles suspended in (non-) Newtonian fluid and in which physical dimensions of the platelets are not accounted for.

Multiphysics modeling: More comprehensive models describing the complete biochemical cascade from platelet activation to thrombus formation under realistic conditions are necessary. Current damage accumulation models account only for mechanical stimuli (i.e., shear stress) on platelet activation and should only be regarded as an initial assessment of the thrombus formation process. Platelet activation models with integration of biochemical processes and cell deposition and aggregation models are still lacking. However, only after validation with meticulously performed (in vitro) experiments will the use of these models be justified.

Boundary conditions: Future CFD simulations need to consider patient-specific boundary conditions. A patient-specific geometry, both proximal and distal to the valve, can be obtained from various medical imaging modalities, such as computed tomography, ultrasound, angiography, and magnetic resonance imaging. Continuous left ventricular volume measurement with either ultrasound or MRI permits the definition of the inlet flow rate as a function of time. The spatial inlet flow profile, on the other hand, can be computed indirectly by investigating the intraventricular flow velocity field.

If all of the above-mentioned aspects get implemented efficiently in CFD codes, physiologically realistic and patient-specific BMHV flow simulations should be within the reach of high-performance parallel computing clusters of the near future.

Acknowledgments Joris Degroote gratefully acknowledges funding by a Ph.D. fellowship of the Research Foundation – Flanders (FWO). The authors also acknowledge the assistance of Mathias Vermeulen in creating Fig. 12.1d,e.

References

1. Giddens DP, Yoganathan AP, Schoen FJ (1993) Prosthetic cardiac valves. Cardiovasc Pathol 2(3):S167–S177
2. Yoganathan AP, He Z, Casey Jones S (2004) Fluid mechanics of heart valves. Ann Rev Biomed Eng 6:331–362
3. Bodnar E, Grunkemeier GL, Gabbay S (1999) Heart valve replacement: a statistical review of 35 years' results – discussion. J Heart Valve Dis 8(5):470–471
4. Butchart EG et al (2003) A new scoring system to determine thromboembolic risk after heart valve replacement. Circulation 108(10):68–74
5. Yacoub N, Takkenberg J (2005) Will heart valve tissue engineering change the world? Nat Clin Pract Cardiovasc Med 2(2):60–61
6. Dasi LP et al (2009) Fluid mechanics of artificial heart valves. Clin Exp Pharmacol Physiol 36(2):225–237
7. Daebritz SH et al (2004) Introduction of a flexible polymeric heart valve prosthesis with special design for aortic position. Eur J Cardiothorac Surg 25(6):946–952
8. Yoganathan AP et al (1984) Bileaflet, tilting disk and porcine aortic-valve substitutes – in vitro hydrodynamic characteristics. J Am Coll Cardiol 3(2):313–320
9. Pibarot P, Dumesnil JG (2000) Hemodynamic and clinical impact of prosthesis–patient mismatch in the aortic valve position and its prevention. J Am Coll Cardiol 36(4):1131–1141
10. Ge L et al (2008) Characterization of hemodynamic forces induced by mechanical heart valves: Reynolds vs. viscous stresses. Ann Biomed Eng 36(2):276–297
11. Yin W et al (2004) Flow-induced platelet activation in bileaflet and monoleaflet mechanical heart valves. Ann Biomed Eng 32(8):1058–1066
12. Grigioni M et al (1999) A discussion on the threshold limit for hemolysis related to Reynolds shear stress. J Biomech 32(10):1107–1112
13. Lu PC, Lai HC, Liu JS (2001) A reevaluation and discussion on the threshold limit for hemolysis in a turbulent shear flow. J Biomech 34(10):1361–1364
14. Bronzino J (2000) Biomedical engineering handbook, 3rd edn. CRC Press, Boca Raton, FL
15. Nobili M, Passoni G, Redalli A (2007) Two fluid–structure approaches for 3D simulation of St. Jude Medical bileaflet valve opening. J Appl Biomater Biomech 5(1):49–59
16. Yoganathan AP, Chandran KB, Sotiropoulos F (2005) Flow in prosthetic heart valves: state-of-the-art and future directions. Ann Biomed Eng 33(12):1689–1694
17. Grigioni M et al (2004) Innovative technologies for the assessment of cardiovascular medical devices: state-of-the-art techniques for artificial heart valve testing. Expert Rev Med Dev 1(1):81–93
18. Willert CE, Gharib M (1991) Digital particle image velocimetry. Exp Fluids 10(4): 181–193
19. Akutsu T, Modi VJ (1997) Unsteady fluid dynamics of several mechanical prosthetic heart valves using a two component laser Doppler anemometer system. Artif Organs 21(10): 1110–1120
20. Kaminsky R et al (2007) PIV validation of blood–heart valve leaflet interaction modelling. Int J Artif Organs 30(7):640–648
21. Kaminsky R et al (2007) Time-resolved PIV technique for high temporal resolution measurement of mechanical prosthetic aortic valve fluid dynamics. Int J Artif Organs 30(2):153–162
22. Browne P et al (2000) Experimental investigation of the steady flow downstream of the St. Jude bileaflet heart valve: a comparison between laser Doppler velocimetry and particle image velocimetry techniques. Ann Biomed Eng 28(1):39–47
23. Yoganathan AP, Woo YR, Sung HW (1986) Turbulent shear–stress measurements in the vicinity of aortic heart-valve prostheses. J Biomech 19(6):433–442
24. Chandran KB et al (1983) Laser anemometry measurements of pulsatile flow past aortic-valve prostheses. J Biomech 16(10):865–873

25. Fontaine AA et al (1996) In vitro assessment of prosthetic valve function in mitral valve replacement with chordal preservation techniques. J Heart Valve Dis 5(2):186–198
26. Manning KB et al (2003) Regurgitant flow field characteristics of the St. Jude bileaflet mechanical heart valve under physiologic pulsatile flow using particle image velocimetry. Artif Organs 27(9):840–846
27. Brucker C et al (2002) Unsteady flow through a new mechanical heart valve prosthesis analysed by digital particle image velocimetry. Meas Sci Technol 13(7):1043–1049
28. Leo HL et al (2006) Fluid dynamic assessment of three polymeric heart valves using particle image velocimetry. Ann Biomed Eng 34(6):936–952
29. Lim WL et al (1994) Particle image velocimetry in the investigation of flow past artificial-heart valves. Ann Biomed Eng 22(3):307–318
30. Lim WL et al (2001) Pulsatile flow studies of a porcine bioprosthetic aortic valve in vitro: PIV measurements and shear-induced blood damage. J Biomech 34(11):1417–1427
31. Kaminsky R et al (2007) Flow visualization through two types of aortic prosthetic heart valves using stereoscopic high-speed particle image velocimetry. Artif Organs 31(12): 869–879
32. Adrian RJ (2005) Twenty years of particle image velocimetry. Exp Fluids 39(2):159–169
33. Redaelli A et al (2004) 3-D simulation of the St. Jude Medical Bileaflet valve opening process: fluid–structure interaction study and experimental validation. J Heart Valve Dis 13(5):804–813
34. Baccani B, Domenichini F, Pedrizzetti G (2003) Model and influence of mitral valve opening during the left ventricular filling. J Biomech 36(3):355–361
35. Ge L, Sotiropoulos F (2007) A numerical method for solving the 3D unsteady incompressible Navier–Stokes equations in curvilinear domains with complex immersed boundaries. J Comput Phys 225(2):1782–1809
36. Shim EB, Chang KS (1997) Numerical analysis of three-dimensional Bjork–Shiley valvular flow in an aorta. J Biomech Eng Trans Asme 119(1):45–51
37. Huang ZJ et al (1994) Numerical-simulation of unsteady laminar-flow through a tilting disk heart-valve – prediction of vortex shedding. J Biomech 27(4):391–402
38. Yang JM, Balaras E (2006) An embedded-boundary formulation for large-eddy simulation of turbulent flows interacting with moving boundaries. J Comput Phys 215(1):12–40
39. Underwood FN, Mueller TJ (1977) Numerical study of steady axisymmetric flow through a disk-type prosthetic heart-valve in a constant diameter chamber. J Biomech Eng Trans Asme 99(2):91–97
40. Underwood FN, Mueller TJ (1979) Numerical study of the steady axisymmetric flow through a disk-type prosthetic heart-valve in an aortic-shaped chamber. J Biomech Eng Trans Asme 101(3):198–204
41. Kiris C et al (1997) Computational approach for probing the flow through artificial heart devices. J Biomech Eng Trans Asme 119(4):452–460
42. Krafczyk M et al (1998) Analysis of 3D transient blood flow passing through an artificial aortic valve by lattice-Boltzmann methods. J Biomech 31(5):453–462
43. King MJ et al (1996) A three-dimensional, time-dependent analysis of flow through a bileaflet mechanical heart valve: comparison of experimental and numerical results. J Biomech 29(5):609–618
44. King MJ, David T, Fisher J (1997) Three-dimensional study of the effect of two leaflet opening angles on the time-dependent flow through a bileaflet mechanical heart valve. Med Eng Phys 19(3):235–241
45. Ge L et al (2003) Numerical simulation of flow in mechanical heart valves: grid resolution and the assumption of flow symmetry. J Biomech Eng Trans Asme 125(5):709–718
46. Ge L et al (2005) Flow in a mechanical bileaflet heart valve at laminar and near-peak systole flow rates: CFD simulations and experiments. J Biomech Eng Trans Asme 127(5):782–797
47. Forsythe N, Mueller JD (2008) Validation of a fluid–structure interaction model for a bileaflet mechanical heart valve. Int J Comput Fluid Dyn 22(8):541–553

48. Blom FJ (1998) A monolithical fluid–structure interaction algorithm applied to the piston problem. Comput Methods Appl Mech Eng 167(3–4):369–391
49. Piperno S, Farhat C, Larrouturou B (1995) Partitioned procedures for the transient solution of coupled aeroelastic problems 1. Model problem, theory and 2-dimensional application. Comput Methods Appl Mech Eng 124(1–2):79–112
50. Bathe KJ, Zhang H (2004) Finite element developments for general fluid flows with structural interactions. Int J Numer Methods Eng 60:213–232
51. Heil M (2004) An efficient solver for the fully coupled solution of large-displacement fluid–structure interaction problems. Comput Methods Appl Mech Eng 193:1–23
52. Hron J, Turek S (2006) A monolithic FEM/multigrid solver for ALE formulation of fluid structure interaction with application in biomechanics. In: Bungartz HJ, Schäfer M (eds) Fluid–structure interaction – modelling, simulation, optimisation. pp 146–170
53. Scotti CM, Finol EA (2007) Compliant biomechanics of abdominal aortic aneurysms: a fluid–structure interaction study. Comput Struct 85:1097–1113
54. Degroote J, Bathe KJ, Vierendeels J (2009) Performance of a new partitioned procedure versus a monolithic procedure in fluid–structure interaction. Comput Struct 87(11–12): 793–801
55. Küttler U, Wall WA (2008) Fixed-point fluid–structure interaction solvers with dynamic relaxation. Comput Mech 43:61–72
56. Farhat C, van der Zee K, Geuzaine P (2006) Provably second-order time-accurate loosely-coupled solution algorithms for transient nonlinear computational aeroelasticity. Comput Methods Appl Mech Eng 195:1973–2001
57. Vierendeels J, Dumont K, Verdonck PR (2008) A partitioned strongly coupled fluid–structure interaction method to model heart valve dynamics. J Comput Appl Math 215(2):602–609
58. Donea J, Guiliani S, Halleux JP (1982) An arbitrary Lagrangian–Eulerian finite-element method for transient dynamic fluid structure interactions. Comput Methods Appl Mech Eng 33(1–3):689–723
59. Le Tallec P, Mouro J (2001) Fluid structure interaction with large structural displacements. Comput Methods Appl Mech Eng 190(24–25):3039–3067
60. De Hart J et al (2003) A three-dimensional computational analysis of fluid–structure interaction in the aortic valve. J Biomech 36(1):103–112
61. Dumont K et al (2005) Predicting ATS Open Pivot (TM) heart valve performance with computational fluid dynamics. J Heart Valve Dis 14(3):393–399
62. Dumont K et al (2007) Comparison of the hemodynamic and thrombogenic performance of two bileaflet mechanical heart valves using a CFD/FSI model. J Biomech Eng Trans Asme 129(4):558–565
63. Nobili M et al (2008) Numerical simulation of the dynamics of a bileaflet prosthetic heart valve using a fluid–structure interaction approach. J Biomech 41(11):2539–2550
64. Nobili M et al (2008) Numerical simulation of the dynamics of a bileaflet prosthetic heart valve using a fluid–structure interaction approach. J Biomech 41(11):2539–2550
65. Xia GH, Zhao Y, Yeo JH (2005) Numerical simulation of 3D fluid–structure interaction using an immersed membrane method. In: International symposium on physics of fluids, Huangshan, China
66. Gilmanov A, Sotiropoulos F (2005) A hybrid Cartesian/immersed boundary method for simulating flows with 3D, geometrically complex, moving bodies. J Comput Phys 207(2):457–492
67. Mittal R, Iaccarino G (2005) Immersed boundary methods. Ann Rev Fluid Mech 37: 239–261
68. De Hart J et al (2000) A two-dimensional fluid–structure interaction model of the aortic value. J Biomech 33(9):1079–1088
69. Glowinski R, Pan TW, Periaux J (1994) A fictitious domain method for Dirichlet problem and applications. Comput Methods Appl Mech Eng 111(3–4):283–303

70. Peskin CS (1972) Flow patterns around heart valves – numerical method. J Comput Phys 10(2):252–271
71. De Hart J et al (2003) A computational fluid–structure interaction analysis of a fiber-reinforced stentless aortic valve. J Biomech 36(5):699–712
72. Stijnen JMA et al (2004) Evaluation of a fictitious domain method for predicting dynamic response of mechanical heart valves. J Fluids Struct 19(6):835–850
73. van Loon R et al (2004) A combined fictitious domain/adaptive meshing method for fluid–structure interaction in heart valves. Int J Numer Methods Fluids 46(5):533–544
74. van Loon R, Anderson PD, van de Vosse FN (2006) A fluid–structure interaction method with solid–rigid contact for heart valve dynamics. J Comput Phys 217(2):806–823
75. Tai CH, Liew KM, Zhao Y (2007) Numerical simulation of 3D fluid–structure interaction flow using an immersed object method with overlapping grids. Comput Struct 85(11–14):749–762
76. Tai CH, Zhao Y, Liew KM (2005) Parallel computation of unsteady incompressible viscous flows around moving rigid bodies using an immersed object method with overlapping girds. J Comput Phys 207(1):151–172
77. Borazjani I, Ge L, Sotiropoulos F (2008) Curvilinear immersed boundary method for simulating fluid structure interaction with complex 3D rigid bodies. J Comput Phys 227(16):7587–7620
78. Dasi LP et al (2007) Vorticity dynamics of a bileaflet mechanical heart valve in an axisymmetric aorta. Phys Fluids 19(6):067105
79. Xia GH, Zhao Y, Yeo JH (2009) Parallel unstructured multigrid simulation of 3D unsteady flows and fluid–structure interaction in mechanical heart valve using immersed membrane method. Comput Fluids 38(1):71–79
80. Ellis JT et al (1996) Velocity measurements and flow patterns within the hinge region of a Medtronic Parallel(TM) bileaflet mechanical valve with clear housing. J Heart Valve Dis 5(6):591–599
81. Ellis JT et al (1996) An in vitro investigation of the retrograde flow fields of two bileaflet mechanical heart valves. J Heart Valve Dis 5(6):600–606
82. Ellis JT, Travis BR, Yoganathan AP (2000) An in vitro study of the hinge and near-field forward flow dynamics of the St. Jude Medical (R) Regent (TM) bileaflet mechanical heart valve. Ann Biomed Eng 28(5):524–532
83. Krishnan S et al (2006) Two-dimensional dynamic simulation of platelet activation during mechanical heart valve closure. Ann Biomed Eng 34(10):1519–1534
84. Woo YR et al (1986) Pulsatile flow visualization studies with aortic and mitral mechanical valve prostheses. Chem Eng Commun 47(1–3):23–48
85. Leverett LB et al (1972) Red blood-cell damage by shear-stress. Biophys J 12(3):257–273
86. Sutera SP (1977) Flow-induced trauma to blood-cells. Circ Res 41(1):2–8
87. Blackshear PL Jr, Dorman FD, Steinbach JH (1965) Some mechanical effects that influence hemolysis. Trans Am Soc Artif Intern Organs 11:112–117
88. Giersiepen M et al (1990) Estimation of shear stress-related blood damage in heart-valve prostheses – in vitro comparison of 25 aortic valves. Int J Artif Organs 13(5):300–306
89. Goubergrits L (2006) Numerical modeling of blood damage: current status, challenges and future prospects. Expert Rev Med Dev 3(5):527–531
90. De Wachter D, Verdonck P (2002) Numerical calculation of hemolysis levels in peripheral hemodialysis cannulas. Artif Organs 26(7):576–582
91. Schima H et al (1993) Mechanical blood traumatization by tubing and throttles in in vitro pump tests – experimental results and implications for hemolysis theory. Artif Organs 17(3):164–170
92. Goubergrits L, Affeld K (2004) Numerical estimation of blood damage in artificial organs. Artif Organs 28(5):499–507
93. Grigioni M et al (2005) A novel formulation for blood trauma prediction by a modified power-law mathematical model. Biomech Model Mechanobiol 4(4):249–260

94. Chan WK et al (2002) Numerical investigation of the effect of blade geometry on blood trauma in a centrifugal blood pump. Artif Organs 26(9):785–793
95. Song XW et al (2003) Studies of turbulence models in a computational fluid dynamics model of a blood pump. Artif Organs 27(10):935–937
96. Zimmer R et al (2000) Velocities, sheer stresses and blood damage potential of the leakage jets of the Medtronic Parallel (TM) bileaflet valve. Int J Artif Organs 23(1):41–48
97. Grigioni M et al (2004) The power-law mathematical model for blood damage prediction: analytical developments and physical inconsistencies. Artif Organs 28(5):467–475
98. Yeleswarapu KK et al (1995) A mathematical-model for shear-induced hemolysis. Artif Organs 19(7):576–582
99. Liu JS, Lu PC, Chu SH (2000) Turbulence characteristics downstream of bileaflet aortic valve prostheses. J Biomech Eng Trans Asme 122(2):118–124
100. Travis BR et al (2004) An in vivo method for measuring turbulence in mechanical prosthesis leakage jets. J Biomech Eng Trans Asme 126(1):26–35
101. Bluestein D et al (1997) Fluid mechanics of arterial stenosis: relationship to the development of mural thrombus. Ann Biomed Eng 25(2):344–356
102. Bluestein D et al (2004) Flow-induced platelet activation in mechanical heart valves. J Heart Valve Dis 13(3):501–508
103. Alemu Y, Bluestein D (2007) Flow-induced platelet activation and damage accumulation in a mechanical heart valve: numerical studies. Artif Organs 31(9):677–688
104. Sheriff J, Jesty J, Bluestein D (2008) Platelet damage accumulation and recovery due to hemodynamic shear stress: an in vitro study. Proc ASME Summer Bioeng Conf Pts A and B 2009:825–826
105. Alemu Y et al (2008) Damage accumulation model of prosthetic heart valves in forward and reverse flow phases. Proc ASME Summer Bioeng Conf Pts A and B 2009:1097–1098
106. Akutsu T et al (2008) Correlation between ventricular flow field and valve closing sound of mechanical mitral prostheses. J Artif Organs 11(2):67–74
107. El Oakley R, Kleine P, Bach DS (2008) Choice of prosthetic heart valve in today's practice. Circulation 117(2):253–256
108. AlMomani T et al (2008) Micro-scale dynamic simulation of erythrocyte–platelet interaction in blood flow. Ann Biomed Eng 36(6):905–920

Subject Index

K.B. Chandran et al. (eds.), *Image-Based Computational Modeling of the Human Circulatory and Pulmonary Systems*, DOI 10.1007/978-1-4419-7350-4,

GPSR Compliance
The European Union's (EU) General Product Safety Regulation (GPSR) is a set of rules that requires consumer products to be safe and our obligations to ensure this.

If you have any concerns about our products, you can contact us on

ProductSafety@springernature.com

In case Publisher is established outside the EU, the EU authorized representative is:

Springer Nature Customer Service Center GmbH
Europaplatz 3
69115 Heidelberg, Germany

www.ingramcontent.com/pod-product-compliance
Lightning Source LLC
Chambersburg PA
CBHW060748150926
51501CB00003B/126
* 9 7 8 1 4 4 1 9 7 3 4 9 8 *